1989

IMPORTANT FORMULAS

- **Distance** between $P_1(x_1, y_1)$ and $P_2(x_2, y_2)$:

$$d(P_1, P_2) = \sqrt{(x_2 - x_1)^2 + (y_2 - y_1)^2}$$

- **Equation of a circle** with center $C(h,k)$ and radius r:

$$(x - h)^2 + (y - k)^2 = r^2$$

- **Slope** m of the line through $P_1(x_1, y_1)$ and $P_2(x_2, y_2)$:

$$m = \frac{y_2 - y_1}{x_2 - x_1}$$

- **Point-slope form** for a line of slope m through $P_1(x_1, y_1)$:

$$y - y_1 = m(x - x_1)$$

- **Slope-intercept form** for a line of slope m and y-intercept b:

$$y = mx + b$$

- **Linear function** f: $\quad f(x) = ax + b$

- **Quadratic function** f: $\quad f(x) = ax^2 + bx + c$

- **Polynomial function** f:

$$f(x) = a_n x^n + a_{n-1} x^{n-1} + \cdots + a_1 x + a_0$$

- **Rational function** f:

$$f(x) = \frac{h(x)}{g(x)}$$

where h and g are polynomial functions

- **Pythagorean Theorem**

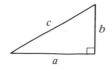

$$c^2 = a^2 + b^2$$

Formulas for area A, circumference C, volume V, and curved surface area S:

TRIANGLE

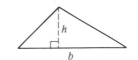

$$A = \frac{1}{2}bh$$

CIRCLE

$$A = \pi r^2$$

$$C = 2\pi r$$

RIGHT CIRCULAR CYLINDER

$$V = \pi r^2 h$$

$$S = 2\pi rh$$

PARALLELOGRAM

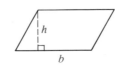

$$A = bh$$

SPHERE

$$V = \frac{4}{3}\pi r^3$$

$$S = 4\pi r^2$$

RIGHT CIRCULAR CONE

$$V = \frac{1}{3}\pi r^2 h$$

$$S = \pi r\sqrt{r^2 + h^2}$$

512
S979f
6ed.

132,948

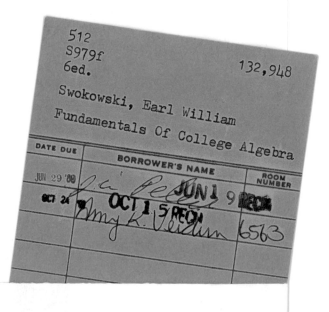

Fundamentals of
College Algebra

Fundamentals of College Algebra

Sixth Edition

Earl W. Swokowski
Marquette University

 Prindle, Weber & Schmidt ■ Boston

PWS PUBLISHERS

Prindle, Weber & Schmidt • 🐱 • Duxbury Press • ♠ • PWS Engineering • 🔺 • Breton Publishers • ⚙
20 Park Plaza • Boston, Massachusetts 02116

PWS Publishers is a division of Wadsworth, Inc.

89 88 87 86 — 10 9 8 7 6 5 4 3 2

Printed in the United States of America

ISBN 0-87150-879-6

LIBRARY OF CONGRESS CATALOGING
IN PUBLICATION DATA

   Swokowski, Earl William.
      Fundamentals of college algebra.

     Includes index.
     1. Algebra. I. Title.
  QA154.2.S96   1986   512.9   85-24375
  ISBN 0-87150-879-6

Production and design by Kathi Townes
Technical artwork by J & R Services
Text composition by Syntax International Pte. Ltd.
Text printed and bound by Halliday Lithograph
Cover photograph © Russ Kinne/Photo Researchers, Inc.
Cover printed by John P. Pow Company, Inc.

Preface

This book is a major revision of the previous five editions of *Fundamentals of College Algebra.* One of my goals was to maintain the mathematical soundness of earlier editions, while making discussions somewhat less formal by rewriting, placing more emphasis on graphs, and by adding new examples and figures. Another objective was to stress the usefulness of the subject matter through a great variety of applied problems from many different disciplines. Finally, suggestions for improvements from users of previous editions led me to change the order in which certain topics are presented. The comments that follow highlight some of the changes and features of this edition.

CHANGES FOR THE SIXTH EDITION

- Chapter 1 is streamlined. The concepts that formerly appeared in eight sections are now presented in four.
- The Binomial Theorem is discussed in Section 1.5 instead of late in the book.
- Complex numbers are defined in Section 2.4, allowing for a complete treatment of quadratic equations in Chapter 2.

- Greater emphasis is given to graphical interpretations for the domain and range of a function in Section 3.4.
- Section 3.6, *Operations on Functions,* has not appeared in previous editions.
- The discussion of inverse functions in Section 3.7 is simplified and integrated with the concept of a one-to-one function.
- The material on division of polynomials, synthetic division, and zeros of polynomials has been reorganized and moved forward in the text.
- Oblique asymptotes for graphs of rational functions are discussed in Section 4.6.
- Chapter 5 has been completely rewritten, with much more attention given to the natural exponential and logarithmic functions and their applications.
- Logarithmic tables have been deemphasized, since calculators are much more efficient and accurate. Teachers who feel that students should be instructed on the use of tables and the technique of linear interpolation will find suitable material in Appendix I.

- In Chapter 6 greater emphasis is given to finding solutions of systems of linear equations by means of the echelon form of a matrix.
- Partial fractions are introduced in Section 6.4 as an application of systems of linear equations.

FEATURES

Applications Previous editions contained applied problems, but most of them were in the fields of engineering, physics, chemistry, or biology. In this revision other subjects are also considered, such as physiology, medicine, sociology, ecology, oceanography, marine biology, business, and economics.

Examples Each section contains carefully chosen examples to help students understand and assimilate new concepts. Whenever feasible, applications are included to demonstrate the usefulness of the subject matter.

Exercises Exercise sets begin with routine drill problems and gradually progress to more difficult types. As a rule, applied problems appear near the end of the set, to allow students to gain confidence in manipulations and new ideas before attempting questions that require an analysis of practical situations.

There is a review section at the end of each chapter, consisting of a list of important topics and pertinent exercises.

Answers to the odd-numbered exercises are given at the end of the text. Instructors may obtain an answer booklet for the even-numbered exercises from the publisher.

Calculators Calculators are given much more emphasis in this edition. It is possible to work most of the exercises without the aid of a calculator; however, instructors may wish to encourage their use to shorten numerical computations. Some sections contain problems labeled *Calculator Exercises,* for which a calculator should definitely be employed.

Text Design A change in page size has made it possible to place most figures in margins, as close as possible to where they are first mentioned in the text. A new use of a second color for figures and statements of important facts should make it easier to follow discussions and remember major ideas. Graphs are usually labeled and color-coded to clarify complex figures. Many figures have been added to exercise sets to help visualize important aspects of applied problems.

Flexibility Hundreds of syllabi from schools that used previous editions attest to the flexibility of the text. Sections and chapters can be rearranged in many different ways, depending on the objectives and the length of the course.

SUPPLEMENTS

Instructors may obtain a manual containing worked-out solutions for approximately one-third of the exercises, authored by Stephen Rodi of Austin Community College. Test banks that can be used for quizzes and examinations are also available from the publisher. Students who need additional help may purchase, from their bookstore, *A Programmed Study Guide* by Roy Dobyns of Carson-Newman College. This guide is designed to assist self-study by reinforcing the mathematics presented in the lectures and the text.

ACKNOWLEDGMENTS

I wish to thank Michael Cullen of Loyola Marymount University for supplying all the new exercises dealing with applications. This large assortment of problems provides strong motivation for the mathematical concepts introduced in the text. Because of his significant input on exercise sets, Michael should be considered as a coauthor of this edition.

This revision has benefited from the comments of users of previous editions. I wish to thank the following individuals, who reviewed the manuscript and offered many helpful suggestions:

James Arnold, University of Wisconsin – Milwaukee
Thomas A. Atchison, Stephen F. Austin State University
William E. Coppage, Wright State University
Franklin Demana, The Ohio State University
Phillip Eastman, Boise State University
Carol Edwards, St. Louis Community College
Leonard E. Fuller, Kansas State University
James E. Hall, Westminster College
E. John Hornsby, Jr., University of New Orleans
Anne Hudson, Armstrong State College
E. Glenn Kindle, Front Range Community College
Robert H. Lohman, Kent State University
William H. Robinson, Ventura College
Albert R. Siegrist, University College of the University of Cincinnati
George L. Szoke, University of Akron
Michael D. Taylor, University of Central Florida

I also wish to single out the following mathematics educators, who met with me and representatives from my publisher for several days during the summer of 1984: Michael Eames of Keene State College, Carol Edwards of St. Louis Community College, John Hornsby of the University of New Orleans, and William Robinson of Ventura College. Their general comments about pedagogy and their specific recommendations about the content of mathematics courses were extremely helpful in improving the book.

I am thankful to John Spellmann of Southwest Texas State University, who checked the answers to the exercises.

I am grateful for the excellent cooperation of the staff of Prindle, Weber & Schmidt. Two people in the company deserve special mention. They are Senior Editor David Pallai and my production editor Kathi Townes. The present form of the book was greatly influenced by their efforts, and I owe them both a debt of gratitude. Moreover, their personal friendship has often been a source of comfort during the years we have worked together.

In addition to all of the persons named here, I express my sincere appreciation to the many unnamed students and teachers who have helped shape my views on mathematics education.

EARL W. SWOKOWSKI

Contents

Fundamentals of College Algebra

CHAPTER 1

Fundamental Concepts of Algebra

The material in this chapter is basic to the study of algebra. ■ We begin by discussing properties of real numbers. ■ Next we turn our attention to exponents and radicals, and how they may be used to simplify complicated algebraic expressions.

1.1 What Is Algebra?

A good foundation in algebra is essential for advanced courses in mathematics, the natural sciences, and engineering. It is also required for solving problems in business, industry, statistics, and many other fields. Indeed, algebraic methods can be applied to every situation that makes use of numerical processes.

Algebra evolved from the operations and rules of arithmetic. The study of arithmetic begins with addition, multiplication, subtraction, and division of numbers, such as

$$4 + 7, \quad (37)(681), \quad 79 - 22, \quad \text{and} \quad 40 \div 8.$$

In *algebra* we introduce symbols or letters such as a, b, c, d, x, y to denote *arbitrary* numbers and, instead of special cases, we often consider *general* statements, such as

$$a + b, \quad cd, \quad x - y, \quad \text{and} \quad x \div a.$$

This *language of algebra* serves a twofold purpose. First, it may be used as a shorthand to abbreviate and simplify long or complicated statements. Second, it is a convenient means of generalizing many specific statements. To illustrate, at an early age, children learn that

$$2 + 3 = 3 + 2, \quad 4 + 7 = 7 + 4, \quad 1 + 8 = 8 + 1,$$

and so on. In words, this property may be phrased "if two numbers are added, then the order of addition is immaterial; that is, the same result is obtained whether the second number is added to the first, or the first number is added to the second." This lengthy description can be shortened and at the same time made easier to understand by means of the algebraic statement

$$a + b = b + a,$$

where a and b denote arbitrary numbers.

Illustrations of the generality of algebra may be found in formulas used in science and industry. For example, if an airplane flies at a constant rate of 300 mph (miles per hour) for two hours, then the distance it travels is

$$(300)(2) \quad \text{or} \quad 600 \text{ miles}.$$

If the airplane's rate is 250 mph and the elapsed time is 3 hours, then the distance traveled is

$$(250)(3) \quad \text{or} \quad 750 \text{ miles}.$$

If we introduce symbols and let r denote the constant rate, t the elapsed time, and d the distance traveled, then the two situations we have described are special cases of the general algebraic formula

$$d = rt.$$

When specific numerical values for r and t are given, the distance d may be found readily by substitution in the formula. The formula may also be used to solve related problems. For example, suppose the distance between two cities is 645 miles, and we wish to find the constant rate that would enable an airplane to cover that distance in 2 hours and 30 minutes. Thus, we are given

$$d = 645 \text{ miles} \quad \text{and} \quad t = 2.5 \text{ hours},$$

and we must find r. Since $d = rt$, it follows that

$$r = \frac{d}{t}$$

and hence, for our special case,

$$r = \frac{645}{2.5} = 258 \text{ mph}.$$

Thus, if an airplane flies at a constant rate of 258 mph, then it will travel 645 miles in 2 hours and 30 minutes. In like manner, if we are given r, we can find the time t required to travel a distance d by means of the formula

$$t = \frac{d}{r}.$$

This indicates how the introduction of an algebraic formula allows us not only to solve special problems conveniently but also to enlarge the scope of our knowledge by suggesting new problems that can be considered.

We have given several elementary illustrations of the value of algebraic methods. There are an unlimited number of situations where a symbolic approach may lead to insights and solutions that would be impossible to obtain using only numerical processes. As you proceed through this text and go on either to more advanced courses in mathematics or to fields that employ mathematics, you will become even more aware of the importance and the power of algebraic techniques.

1.2 Real Numbers

Real numbers are employed in all phases of mathematics, and you are probably well acquainted with symbols that are used to represent them, such as

$$1, \quad 73, \quad -5, \quad \tfrac{49}{12}, \quad \sqrt{2}, \quad 0, \quad \sqrt[3]{-85}, \quad 0.33333\ldots, \quad \text{and} \quad 596.25.$$

The real numbers are said to be **closed** relative to operations of addition (denoted by $+$) and multiplication (denoted by $\cdot$). This means that to every pair a, b of real numbers there corresponds a unique real number $a + b$, called the **sum** of a and b, and a unique real number $a \cdot b$ (also written ab), called the **product** of a and b. If a and b denote the same real number, we write $a = b$ (read "a equals b"). An expression of this type is called an **equality.**

The special numbers 0 and 1, referred to as **zero** and **one**, respectively, have the properties $a + 0 = a$ and $a \cdot 1 = a$ for every real number a. Each real number a has a **negative,** denoted by $-a$, such that $a + (-a) = 0$, and each nonzero real number a has a **reciprocal,** $\dfrac{1}{a}$, such that $a\left(\dfrac{1}{a}\right) = 1$.

These and other important properties are included in the following list, in which a, b, and c denote arbitrary real numbers.

COMMUTATIVE PROPERTIES
$$a + b = b + a, \qquad ab = ba$$

ASSOCIATIVE PROPERTIES
$$a + (b + c) = (a + b) + c, \qquad a(bc) = (ab)c$$

IDENTITIES
$$a + 0 = a = 0 + a, \qquad a \cdot 1 = a = 1 \cdot a$$

INVERSES
$$a + (-a) = 0 = (-a) + a, \qquad a\left(\frac{1}{a}\right) = 1 = \left(\frac{1}{a}\right)a \quad \text{if } a \neq 0$$

DISTRIBUTIVE PROPERTIES
$$a(b + c) = ab + ac, \qquad (a + b)c = ac + bc$$

The real numbers 0 and 1 are sometimes referred to as the **additive identity** and **multiplicative identity,** respectively. The negative, $-a$, is also called the **additive inverse** of a and, if $a \neq 0$, $\dfrac{1}{a}$ is called the **multiplicative inverse** of a. The symbol a^{-1} may be used in place of $\dfrac{1}{a}$, as indicated by the following definition.

DEFINITION OF a^{-1}

$$a^{-1} = \frac{1}{a}; \quad a \neq 0$$

Since $a + (b + c)$ and $(a + b) + c$ are always equal, we may, without ambiguity, use the symbol $a + b + c$ to denote this real number. Similarly, the notation abc is used for either $a(bc)$ or $(ab)c$. An analogous situation exists if four or more real numbers are added or multiplied. Thus, if a, b, c, and d are real numbers, then we write $a + b + c + d$ for their sum and $abcd$ for their product, regardless of how the numbers are grouped or interchanged.

The Distributive Properties are useful for finding products of many different types of expressions. The next example provides one illustration.

EXAMPLE 1   If a, b, c, and d denote real numbers, show that

$$(a + b)(c + d) = ac + bc + ad + bd.$$

SOLUTION   Using the two Distributive Properties,

$$\begin{aligned}
(a + b)(c + d) &= (a + b)c + (a + b)d \\
&= (ac + bc) + (ad + bd) \\
&= ac + bc + ad + bd.
\end{aligned}$$ ∎

If $a = b$ and $c = d$, then $a + c = b + d$ and $ac = bd$. This is often called the **substitution principle,** since we may think of replacing a by b and c by d in the expressions $a + c$ and ac. As a special case, using the fact that $c = c$ gives us the following rules.

> If $a = b$, then $a + c = b + c$.
> If $a = b$, then $ac = bc$.

We sometimes refer to those rules by the statements "Any number c may be added to both sides of an equality" and "Both sides of an equality may be multiplied by the same number c."

The following theorem can be proved. (See Exercises 54–55.)

THEOREM

$$a \cdot 0 = 0 \text{ for every real number } a.$$
$$\text{If } ab = 0, \text{ then either } a = 0 \text{ or } b = 0.$$

The preceding theorem implies that $ab = 0$ *if and only if* either $a = 0$ or $b = 0$. The phrase "if and only if," which is used throughout mathematics, always has a twofold character. Here it means that if $ab = 0$, then $a = 0$ or $b = 0$ and, *conversely*, if $a = 0$ or $b = 0$, then $ab = 0$. Consequently, if both $a \neq 0$ and $b \neq 0$, then $ab \neq 0$; that is, *the product of two nonzero real numbers is always nonzero*.

The following properties can also be proved.

PROPERTIES OF NEGATIVES

$$-(-a) = a$$
$$(-a)b = -(ab) = a(-b)$$
$$(-a)(-b) = ab$$
$$(-1)a = -a$$

The operation of **subtraction** (denoted by $-$) is defined as follows:

**DEFINITION OF
SUBTRACTION**

$$a - b = a + (-b)$$

Division (denoted by $\div$) is defined as follows:

DEFINITION OF DIVISION

$$a \div b = a\left(\frac{1}{b}\right) = ab^{-1}; \qquad b \neq 0$$

The symbol a/b or $\dfrac{a}{b}$ is often used in place of $a \div b$, and we refer to it as the **quotient of a by b** or the **fraction a over b.** The numbers a and b are called the **numerator** and **denominator,** respectively, of the fraction. It is important to note that since 0 has no multiplicative inverse, a/b is not defined if $b = 0$; that is, *division by zero is not permissible.* Also note that if $b \neq 0$, then

$$1 \div b = \frac{1}{b} = b^{-1}.$$

The following properties of quotients may be established, where all denominators are nonzero real numbers.

PROPERTIES OF QUOTIENTS

$$\frac{a}{b} = \frac{c}{d} \quad \text{if and only if} \quad ad = bc$$

$$\frac{a}{b} = \frac{ad}{bd}, \qquad \frac{a}{-b} = \frac{-a}{b} = -\frac{a}{b}$$

$$\frac{a}{b} + \frac{c}{b} = \frac{a + c}{b}, \qquad \frac{a}{b} + \frac{c}{d} = \frac{ad + bc}{bd}$$

$$\frac{a}{b} \cdot \frac{c}{d} = \frac{ac}{bd}, \qquad \frac{a}{b} \div \frac{c}{d} = \frac{a}{b} \cdot \frac{d}{c} = \frac{ad}{bc}$$

The **positive integers,** $1, 2, 3, 4, \ldots$, may be obtained by adding the real number 1 successively to itself. The numbers $-1, -2, -3, -4, \ldots$ are called **negative integers.** The **integers** consist of all positive and negative integers together with the real number 0.

If a, b, and c are integers and $c = ab$, then a and b are called **factors,** or **divisors,** of c. For example, since

$$6 = 2 \cdot 3 = (-2)(-3) = 1 \cdot 6 = (-1)(-6),$$

we see that $1, -1, 2, -2, 3, -3, 6$, and -6 are factors of 6.

A positive integer p different from 1 is **prime** if its only positive factors are 1 and p. The first few primes are 2, 3, 5, 7, 11, 13, 17, and 19. The **Fundamental Theorem of Arithmetic** states that every positive integer different from 1 can be expressed as a product of primes in one and only one way (except for order of factors). Some examples are:

$$12 = 2 \cdot 2 \cdot 3, \qquad 126 = 2 \cdot 3 \cdot 3 \cdot 7, \qquad 540 = 2 \cdot 2 \cdot 3 \cdot 3 \cdot 3 \cdot 5.$$

A **rational number** is a real number of the form a/b, where a and b are integers and $b \neq 0$. Real numbers that are not rational are called **irrational numbers.** The ratio of the circumference of a circle to its diameter is irrational and is denoted by π. It is often approximated by the decimal 3.1416 or by the rational number $\frac{22}{7}$. We use the notation $\pi \approx 3.1416$ to indicate that π *is approximately equal to* 3.1416.

There is no rational number b such that $b^2 = 2$, where b^2 denotes $b \cdot b$. However, there is an *irrational* number, denoted by $\sqrt{2}$, such that $(\sqrt{2})^2 = 2$. In general, we have the following definition.

DEFINITION OF SQUARE ROOT

> Let a be a nonnegative real number. The **principal square root of a,** denoted by $\sqrt{a}$, is the *nonnegative* real number b such that $b^2 = a$.

We often refer to $\sqrt{a}$ simply as the *square root of a*. Some square roots are rational. For example,

$$\sqrt{25} = 5, \qquad \sqrt{\tfrac{9}{4}} = \tfrac{3}{2}, \qquad \sqrt{16} = 4.$$

Other square roots, such as $\sqrt{3}$, $\sqrt{5}$, and $\sqrt{\tfrac{7}{2}}$, are irrational.

Real numbers may be represented by decimal expressions. Decimal representations for rational numbers either are terminating or are nonterminating and repeating. For example, we can show by division that

$$\tfrac{5}{4} = 1.25 \qquad \text{and} \qquad \tfrac{177}{55} = 3.2181818\ldots,$$

where the digits 1 and 8 repeat indefinitely. Decimal representations for irrational numbers are always nonterminating and nonrepeating.

Real numbers may be represented geometrically by points on a line l in such a way that for each real number a there corresponds one and only one point on l, and conversely, to each point P on l there corresponds precisely one real number. Such an association is called a **one-to-one correspondence.** We first choose an arbitrary point O, called the **origin,** and associate with it the real number 0. Points associated with the integers are then determined by laying off successive line segments of equal length on either side of O as illustrated in Figure 1.1. The points corresponding to rational numbers, such as $\frac{23}{5}$ and $-\frac{1}{2}$, are obtained by subdividing the equal line segments. Points associated with certain irrational numbers,

FIGURE 1.1

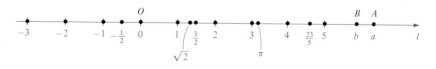

such as $\sqrt{2}$, can be found by geometric construction. (See Exercise 47.) To every irrational number there corresponds a unique point on l, and conversely, every point that is not associated with a rational number corresponds to an irrational number.

The number a that is associated with a point A on l is called the **coordinate** of A. An assignment of coordinates to points on l is called a **coordinate system** for l, and l is called a **coordinate line,** or a **real line.** A direction can be assigned to l by taking the **positive direction** along l to the right and the **negative direction** to the left. The positive direction is noted by placing an arrowhead on l as shown in Figure 1.1.

The numbers that correspond to points to the right of O in Figure 1.1 are called **positive real numbers,** whereas numbers that correspond to points to the left of O are **negative real numbers.** *The real number* 0 *is neither positive nor negative.*

If a and b are real numbers and $a - b$ is positive, we say that **a is greater than b** and we write $a > b$. An equivalent statement is that **b is less than a,** written $b < a$. The symbols $>$ or $<$ are called **inequality signs** and expressions such as $a > b$ or $b < a$ are called **inequalities.** From the manner in which we constructed the coordinate line l in Figure 1.1, we see that if A and B are points with coordinates a and b, respectively, then $a > b$ (or $b < a$) *if and only if A lies to the right of B.* The following definition is stated for reference, where a and b denote real numbers.

DEFINITIONS OF $>$ AND $<$

> $a > b$ means $a - b$ is positive.
>
> $b < a$ means $a - b$ is positive.

The expressions $a > b$ and $b < a$ have exactly the same meaning. As illustrations we may write

$$5 > 3 \quad \text{since} \quad 5 - 3 = 2 \text{ is positive;}$$
$$-6 < -2 \quad \text{since} \quad -2 - (-6) = -2 + 6 = 4 \text{ is positive;}$$
$$-\sqrt{2} < 1 \quad \text{since} \quad 1 - (-\sqrt{2}) = 1 + \sqrt{2} \text{ is positive;}$$
$$2 > 0 \quad \text{since} \quad 2 - 0 = 2 \text{ is positive;}$$
$$-5 < 0 \quad \text{since} \quad 0 - (-5) = 5 \text{ is positive.}$$

The last two illustrations are special cases of the following general properties.

$$a > 0 \quad \text{if and only if } a \text{ is positive.}$$
$$a < 0 \quad \text{if and only if } a \text{ is negative.}$$

It should be clear from our discussion that if a and b are real numbers, then *one and only one of the following statements is true.*

TRICHOTOMY LAW

$$a = b, \quad a > b, \quad \text{or} \quad a < b.$$

We refer to the **sign** of a real number as being positive or negative if the number is positive or negative, respectively. We also say that two real numbers *have the same sign* if both are positive or both are negative. The numbers have *opposite signs* if one is positive and the other is negative. The next result about the signs of products and quotients of two real numbers a and b can be proved using properties of negatives and quotients.

LAWS OF SIGNS

(i) If a and b have the same sign, then ab and a/b are positive.

(ii) If a and b have opposite signs, then ab and a/b are negative.

The converses of the Laws of Signs are also true. For example, if a quotient is negative, then the numerator and denominator have opposite signs.

The notation $a \geq b$ (which is read *a is greater than or equal to b*) means that either $a > b$ or $a = b$ (but not both). For example, $a^2 \geq 0$ for every real number a. The symbol $a \leq b$ is read *a is less than or equal to b* and means that either $a < b$ or $a = b$. The expression $a < b < c$ means that both $a < b$ and $b < c$, in which case we say that *b is* **between** *a* **and** *c.* Similarly, the expression $c > b > a$ means that both $c > b$ and $b > a$. Thus,

$$1 < 5 < \tfrac{11}{2}, \qquad -4 < \tfrac{2}{3} < \sqrt{2}, \qquad 3 > -6 > -10.$$

Other variations of the inequality notation are used. For example, $a < b \leq c$ means both $a < b$ and $b \leq c$. Similarly, $a \leq b < c$ means both $a \leq b$ and $b < c$. Finally, $a \leq b \leq c$ means both $a \leq b$ and $b \leq c$.

If a is a real number, then it is the coordinate of some point A on a coordinate line l, and the symbol $|a|$ is used to denote the number of units (or distance) between A and the origin, without regard to direction. The nonnegative number $|a|$ is called the *absolute value* of a. Referring to Figure 1.2, we see that for the point with coordinate -4, we have $|-4| = 4$. Similarly, $|4| = 4$. In general, *if a is negative we change its sign to find* $|a|$, *whereas if a is nonnegative, then* $|a| = a$. The next definition summarizes this discussion.

FIGURE 1.2

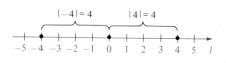

DEFINITION

If a is a real number, then the **absolute value** of a, denoted by $|a|$, is

$$|a| = \begin{cases} a & \text{if } a \geq 0. \\ -a & \text{if } a < 0. \end{cases}$$

The use of this definition is illustrated in the following example.

EXAMPLE 2   Find $|3|, |-3|, |0|, |\sqrt{2} - 2|$, and $|2 - \sqrt{2}|$.

SOLUTION   Since $3, 2 - \sqrt{2}$, and 0 are nonnegative,

$$|3| = 3, \quad |2 - \sqrt{2}| = 2 - \sqrt{2}, \quad \text{and} \quad |0| = 0.$$

Since -3 and $\sqrt{2} - 2$ are negative, we use the formula $|a| = -a$ to obtain

$$|-3| = -(-3) = 3 \quad \text{and} \quad |\sqrt{2} - 2| = -(\sqrt{2} - 2) = 2 - \sqrt{2}. \quad \blacksquare$$

Note that in Example 2, $|-3| = |3|$ and $|2 - \sqrt{2}| = |\sqrt{2} - 2|$. It can be shown in general that

$$|a| = |-a| \quad \text{for every real number } a.$$

We shall use the concept of absolute value to define the distance between any two points on a coordinate line. First note that the distance between the points with coordinates 2 and 7, shown in Figure 1.3, equals 5 units on l. This distance is the difference, $7 - 2$, obtained by subtracting the smaller coordinate from the larger. If we employ absolute values, then since $|7 - 2| = |2 - 7|$, it is unnecessary to be concerned about the order of subtraction. We shall use this as our motivation for the next definition.

FIGURE 1.3

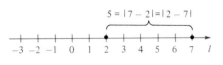

DEFINITION

Let a and b be the coordinates of two points A and B, respectively, on a coordinate line l. The **distance between A and B**, denoted by $d(A, B)$, is

$$d(A, B) = |b - a|.$$

The number $d(A, B)$ is also called the **length of the line segment AB.**

Observe that since $d(B, A) = |a - b|$ and $|b - a| = |a - b|$, then

$$d(A, B) = d(B, A).$$

Also note that the distance between the origin O and the point A is

$$d(O, A) = |a - 0| = |a|,$$

which agrees with the geometric interpretation of absolute value illustrated in Figure 1.2. The formula $d(A, B) = |b - a|$ is true regardless of the signs of a and b, as illustrated in the next example.

EXAMPLE 3 Let A, B, C, and D have coordinates -5, -3, 1, and 6, respectively, on a coordinate line l (see Figure 1.4). Find $d(A, B)$, $d(C, B)$, $d(O, A)$, and $d(C, D)$.

FIGURE 1.4

SOLUTION Using the definition of the distance between points, we obtain the distances:

$$d(A, B) = |-3 - (-5)| = |-3 + 5| = |2| = 2$$
$$d(C, B) = |-3 - 1| = |-4| = 4$$
$$d(O, A) = |-5 - 0| = |-5| = 5$$
$$d(C, D) = |6 - 1| = |5| = 5$$

These answers can be checked by referring to Figure 1.4. ■

The concept of absolute value has uses other than that of finding distances between points. Generally, it is employed whenever we are interested in the "magnitude" or "numerical value" of a real number without regard to its sign.

Exercises 1.2

1 If $x < 0$ and $y > 0$, determine the sign of the following real numbers:

(a) x^2y (b) xy^2 (c) $\dfrac{x}{y} + x$

(d) $y - x$ (e) $\dfrac{x - y}{xy}$

2 The rational numbers $\frac{22}{7}$ and $\frac{355}{113}$ are two useful approximations to π. Calculate the distances $d(\frac{22}{7}, \pi)$ and $d(\frac{355}{113}, \pi)$ to decide which approximation is closer to the actual value of π.

In Exercises 3–6 replace the symbol □ with either $<$, $>$, or $=$.

3 (a) $-7\ \square\ -4$

   (b) $3\ \square\ -1$

   (c) $1 + 3\ \square\ 6 - 2$

5 (a) $\frac{1}{3}\ \square\ 0.33$

   (b) $\frac{125}{57}\ \square\ 2.193$

   (c) $\frac{22}{7}\ \square\ \pi$

4 (a) $-3\ \square\ -5$

   (b) $-6\ \square\ 2$

   (c) $\frac{1}{4}\ \square\ 0.25$

6 (a) $\frac{1}{7}\ \square\ 0.143$

   (b) $\frac{3}{4} + \frac{2}{3}\ \square\ \frac{19}{12}$

   (c) $\sqrt{2}\ \square\ 1.4$

Express the statements in Exercises 7–18 in terms of inequalities.

7 -8 is less than -5.

8 2 is greater than 1.9.

9 0 is greater than -1.

10 $\sqrt{2}$ is less than π.

11 x is negative.

12 y is positive.

13 a is between 5 and 3.

14 b is between $\frac{1}{10}$ and $\frac{1}{3}$.

15 b is greater than or equal to 2.

16 x is less than or equal to -5.

17 c is not greater than 1.

18 d is nonnegative.

Rewrite the numbers in Exercises 19–22 without using symbols for absolute value.

19 (a) $|4 - 9|$

   (b) $|-4| - |-9|$

   (c) $|4| + |-9|$

21 (a) $3 - |-3|$

   (b) $|\pi - 4|$

   (c) $(-3)/|-3|$

20 (a) $|3 - 6|$

   (b) $|0.2 - \frac{1}{5}|$

   (c) $|-3| - |-4|$

22 (a) $|8 - 5|$

   (b) $-5 + |-7|$

   (c) $(-2)|-2|$

In Exercises 23–26 the given numbers are coordinates of three points A, B, and C (in that order) on a coordinate line l. For each problem find (a) $d(A, B)$; (b) $d(B, C)$; (c) $d(C, B)$; and (d) $d(A, C)$.

23 $-6, -2, 4$

24 $3, 7, -5$

25 $8, -4, -1$

26 $-9, 1, 10$

Rewrite the expressions in Exercises 27–30 without using symbols for absolute value.

27 $|5 - x|$ if $x > 5$

28 $|x^2 + 1|$

29 $|-4 - x^2|$

30 $|a - b|$ if $a < b$

In Exercises 31–40 replace the symbol $\square$ with either $=$ or $\neq$ for all real numbers a, b, and c, where the expressions are defined. Give reasons for your answers.

31 $\dfrac{ab + ac}{a}\ \square\ b + ac$

32 $\dfrac{ab + ac}{a}\ \square\ b + c$

33 $\dfrac{b + c}{a}\ \square\ \dfrac{b}{a} + \dfrac{c}{a}$

34 $\dfrac{a}{b + c}\ \square\ \dfrac{a}{b} + \dfrac{a}{c}$

35 $(a \div b) \div c\ \square\ a \div (b \div c)$

36 $(a - b) - c\ \square\ a - (b - c)$

37 $(ab)^{-1}\ \square\ a^{-1}b^{-1}$

38 $(a - b)^{-1}\ \square\ a^{-1} - b^{-1}$

39 $\dfrac{a - b}{b - a}\ \square\ -1$

40 $-(a + b)\ \square\ -a + b$

Prove the rules in Exercises 41–46.

41 $a = -a$ if and only if $a = 0$.

42 $a \cdot 0 = 0$ (*Hint:* Write $a \cdot 0 = a \cdot (0 + 0) = a \cdot 0 + a \cdot 0$ and then add $-(a \cdot 0)$ to both sides.)

43 If $ab = 0$, then either $a = 0$ or $b = 0$.

44 If $a + b = 0$, then $b = -a$.

45 $\left(\dfrac{a}{b}\right)^{-1} = \dfrac{b}{a}$

46 $\dfrac{a}{b} + \dfrac{c}{b} = \dfrac{a + c}{b}$ (*Hint:* Write $\dfrac{a}{b} + \dfrac{c}{b} = ab^{-1} + cb^{-1}$ and use the Distributive Properties.)

47 The point on a coordinate line corresponding to $\sqrt{2}$ may be constructed by forming a right triangle with sides of length 1 as shown in the figure. Construct the points

FIGURE FOR EXERCISE 47

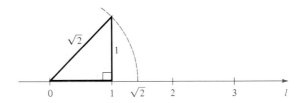

that correspond to $\sqrt{3}$ and $\sqrt{5}$, respectively. (*Hint:* Use the Pythagorean Theorem.)

48 A circle of radius 1 rolls along a coordinate line in the positive direction. (See figure.) If point P is initially at the origin, what is the coordinate of P after one complete rotation?

FIGURE FOR EXERCISE 48

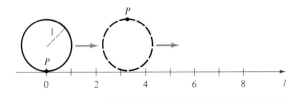

49 Geometric proofs of properties of real numbers were first given by the ancient Greeks. Compute the area of the rectangle shown in the figure in two ways to establish the Distributive Property $a(b + c) = ab + ac$ for positive real numbers a, b, and c.

FIGURE FOR EXERCISE 49

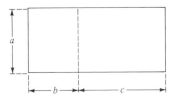

50 Rational approximations to square roots can be found using a formula discovered by the ancient Babylonians. Let x_1 be the first rough rational approximation for $\sqrt{n}$, and let

$$x_2 = \frac{1}{2}\left(x_1 + \frac{n}{x_1}\right).$$

Then x_2 will be a better approximation and we can repeat the computation with x_2 replacing x_1. Starting with $x_1 = \frac{17}{10}$, find two more rational approximations to $\sqrt{3}$.

1.3 Exponents and Radicals

The symbol a^2 denotes the real number $a \cdot a$, and a^3 is used in place of $a \cdot a \cdot a$. In general, if n is any positive integer,

$$a^n = \underbrace{a \cdot a \cdots \cdot a}_{n \text{ factors}}$$

when n factors, all equal to a, appear on the right-hand side of the equal sign. The positive integer n is called the **exponent** of a in the expression a^n, and a^n is read "a to the nth **power**" or simply "a to the n." Note that $a^1 = a$. Some numerical examples of exponents are:

$$\left(\tfrac{1}{2}\right)^5 = \tfrac{1}{2} \cdot \tfrac{1}{2} \cdot \tfrac{1}{2} \cdot \tfrac{1}{2} \cdot \tfrac{1}{2} = \tfrac{1}{32}$$

$$(-3)^3 = (-3)(-3)(-3) = -27$$

$$(\sqrt{2})^4 = \sqrt{2}\sqrt{2}\sqrt{2}\sqrt{2} = (\sqrt{2})^2(\sqrt{2})^2 = (2)(2) = 4$$

It is important to remember that if n is a positive integer, then an expression such as $3a^n$ means $3(a^n)$ and not $(3a)^n$. The real number 3 is called the **coefficient** of a^n in the expression $3a^n$. Similarly, $-3a^n$ means $(-3)a^n$, not $(-3a)^n$. For example, we observe that

$$5 \cdot 2^3 = 5 \cdot 8 = 40 \quad \text{and} \quad -5 \cdot 2^3 = -5 \cdot 8 = -40.$$

If $a \neq 0$, the extension to nonpositive exponents is made by defining

$$a^0 = 1 \quad \text{and} \quad a^{-n} = \frac{1}{a^n}.$$

The following laws are indispensable when we work with exponents.

LAWS OF EXPONENTS

If a and b are real numbers and m and n are integers, then

(i) $a^m a^n = a^{m+n};$ (ii) $(a^m)^n = a^{mn};$ (iii) $(ab)^n = a^n b^n;$

(iv) $\left(\dfrac{a}{b}\right)^n = \dfrac{a^n}{b^n}$ if $b \neq 0;$ (v) If $a \neq 0,$ then $\dfrac{a^m}{a^n} = a^{m-n} = \dfrac{1}{a^{n-m}}.$

To prove (i) for positive integers m and n, we write

$$a^m a^n = \underbrace{a \cdot a \cdot \cdots \cdot a}_{m \text{ factors}} \cdot \underbrace{a \cdot a \cdot \cdots \cdot a}_{n \text{ factors}}.$$

Since the total number of factors a on the right is $m + n$, this expression is equal to a^{m+n}.

Similarly, for (ii) we write

$$(a^m)^n = \underbrace{a^m \cdot a^m \cdot \cdots \cdot a^m}_{n \text{ factors } a^m}$$

and count the number of times a appears as a factor on the right-hand side. Since $a^m = a \cdot a \cdot \cdots \cdot a$, where a occurs as a factor m times, and since the number of such groups of m factors is n, the total number of factors is $m \cdot n$. Proofs for the cases $m \leq 0$ and $n \leq 0$ are left to the reader.

Laws (iii) and (iv) can be proved in similar fashion. Law (v) is clear if $m = n$. For the case $m > n$, the integer $m - n$ is positive, and we may write

$$\frac{a^m}{a^n} = \frac{a^n a^{m-n}}{a^n} = \frac{a^n}{a^n} \cdot a^{m-n} = 1 \cdot a^{m-n} = a^{m-n}.$$

A similar argument can be used if $n > m$.

We can, of course, use symbols for real numbers other than a and b when applying laws of exponents, as in the following illustrations.

$$x^5 x^6 = x^{5+6} = x^{11} \qquad (y^5)^7 = y^{5 \cdot 7} = y^{35}$$

$$(rs)^7 = r^7 s^7 \qquad \left(\frac{p}{q}\right)^{10} = \frac{p^{10}}{q^{10}}$$

$$\frac{c^8}{c^3} = c^{8-3} = c^5 \qquad \frac{u^3}{u^8} = u^{3-8} = u^{-5} = \frac{1}{u^5}$$

The Laws of Exponents can be extended to rules such as $(abc)^n = a^n b^n c^n$ and $a^m a^n a^p = a^{m+n+p}$. To simplify statements in all problems involving exponents, we shall assume that symbols that appear in denominators represent *nonzero* real numbers.

EXAMPLE 1 Simplify each expression:

(a) $(3x^3 y^4)(4xy^5)$ \qquad (b) $(2a^2 b^3 c)^4$

(c) $\left(\frac{2r^3}{s}\right)^2 \left(\frac{s}{r^3}\right)^3$ \qquad (d) $(u^{-2} v^3)^{-3}$

SOLUTION

(a)
$$(3x^3 y^4)(4xy^5) = (3)(4)x^3 x y^4 y^5$$
$$= 12x^4 y^9$$

(b)
$$(2a^2 b^3 c)^4 = 2^4 (a^2)^4 (b^3)^4 c^4$$
$$= 16a^8 b^{12} c^4$$

(c)
$$\left(\frac{2r^3}{s}\right)^2 \left(\frac{s}{r^3}\right)^3 = \left(\frac{2^2 r^6}{s^2}\right)\left(\frac{s^3}{r^9}\right)$$
$$= 2^2 \left(\frac{r^6}{r^9}\right)\left(\frac{s^3}{s^2}\right)$$
$$= 4\left(\frac{1}{r^3}\right)(s) = \frac{4s}{r^3}$$

(d)
$$(u^{-2} v^3)^{-3} = u^6 v^{-9} = \frac{u^6}{v^9} \qquad \blacksquare$$

The following rules are often useful for problems involving negative exponents, where $m > 0$, $n > 0$, and $xy \neq 0$:

$$\frac{x^{-m}}{y^{-n}} = \frac{y^n}{x^m} \quad \text{and} \quad \left(\frac{x}{y}\right)^{-n} = \left(\frac{y}{x}\right)^n.$$

The proofs follow readily from the definition of negative exponents and properties of quotients. Thus,

$$\frac{x^{-m}}{y^{-n}} = \frac{1/x^m}{1/y^n} = \frac{1}{x^m} \cdot \frac{y^n}{1} = \frac{y^n}{x^m} \quad \text{and} \quad \left(\frac{x}{y}\right)^{-n} = \frac{x^{-n}}{y^{-n}} = \frac{y^n}{x^n}.$$

EXAMPLE 2 Eliminate negative exponents and simplify:

(a) $\dfrac{8x^3y^{-5}}{4x^{-1}y^2}$ (b) $\left(\dfrac{u^2}{2v}\right)^{-3}$

SOLUTION

(a)
$$\frac{8x^3y^{-5}}{4x^{-1}y^2} = \frac{8x^3}{4y^2} \cdot \frac{y^{-5}}{x^{-1}} = \frac{8x^3}{4y^2} \cdot \frac{x}{y^5} = \frac{2x^4}{y^7}$$

(b)
$$\left(\frac{u^2}{2v}\right)^{-3} = \left(\frac{2v}{u^2}\right)^3 = \frac{8v^3}{u^6} \qquad\blacksquare$$

Roots of real numbers are defined by the statement

$$\sqrt[n]{a} = b \quad \text{if and only if} \quad b^n = a,$$

provided that both a and b are nonnegative real numbers and n is a positive integer, or that both a and b are negative and n is an odd positive integer. The number $\sqrt[n]{a}$ is called the **principal nth root** of a. If $n = 2$, it is customary to write $\sqrt{a}$ instead of $\sqrt[2]{a}$ and to call $\sqrt{a}$ the (principal) **square root** of a. The number $\sqrt[3]{a}$ is referred to as the **cube root** of a. Some examples of roots are:

$$\sqrt[5]{\tfrac{1}{32}} = \tfrac{1}{2} \qquad \text{since} \qquad (\tfrac{1}{2})^5 = \tfrac{1}{32}$$

$$\sqrt[3]{-8} = -2 \quad \text{since} \quad (-2)^3 = -8$$

$$\sqrt{16} = 4 \qquad \text{since} \qquad 4^2 = 16$$

$$\sqrt[4]{81} = 3 \qquad \text{since} \qquad 3^4 = 81.$$

Complex numbers (see Section 2.4) are needed to define $\sqrt[n]{a}$ if $a < 0$ and n is an *even* positive integer, since for all real numbers b, $b^n \geq 0$ whenever n is even.

It is important to observe that if $\sqrt[n]{a}$ exists, it is a *unique* real number. More generally, if $b^n = a$ for a positive integer n, then b is called *an nth root* of a. For example, both 4 and -4 are square roots of 16, since $4^2 = 16$ and also $(-4)^2 = 16$. However, the *principal* square root of 16 is 4, which we write as $\sqrt{16} = 4$.

To complete our terminology, the expression $\sqrt[n]{a}$ is called a **radical,** the number a is called the **radicand,** and n is called the **index** of the radical. The symbol $\sqrt{}$ is called a **radical sign.**

If $\sqrt{a} = b$, then $b^2 = a$; that is, $(\sqrt{a})^2 = a$. Similarly, if $\sqrt[3]{a} = b$, then $b^3 = a$, or $(\sqrt[3]{a})^3 = a$. In general, if n is any positive integer, then

$$(\sqrt[n]{a})^n = a.$$

It also follows that if $a > 0$, or if $a < 0$ and n is an odd positive integer, then

$$\sqrt[n]{a^n} = a.$$

For example,

$$\sqrt{5^2} = 5, \qquad \sqrt[3]{(-2)^3} = -2, \qquad \sqrt[4]{3^4} = 3.$$

A common algebraic error is to replace $\sqrt{x^2}$ by x for *every* x; however, this is not true if x is negative. For example,

$$\sqrt{(-3)^2} = \sqrt{9} = 3 = |-3|.$$

In general, we may write

$$\sqrt{x^2} = |x| \quad \text{for every real number } x.$$

Furthermore, if n is any even positive integer, then $\sqrt[n]{x^n} = |x|$. Note that $\sqrt{x^2} = x$ if and only if $x \geq 0$.

The following laws can be proved for positive integers m and n, where it is assumed that the indicated roots exist.

LAWS OF RADICALS

(i) $\sqrt[n]{ab} = \sqrt[n]{a}\,\sqrt[n]{b};$ (ii) $\sqrt[n]{\dfrac{a}{b}} = \dfrac{\sqrt[n]{a}}{\sqrt[n]{b}};$ (iii) $\sqrt[m]{\sqrt[n]{a}} = \sqrt[mn]{a}.$

EXAMPLE 3 Show that each statement is true.

(a) $\sqrt{50} = 5\sqrt{2}$ (b) $\sqrt[3]{-108} = -3\sqrt[3]{4}$ (c) $\sqrt[3]{\sqrt{64}} = 2$

SOLUTION

(a) $$\sqrt{50} = \sqrt{25 \cdot 2} = \sqrt{25}\sqrt{2} = 5\sqrt{2}$$

(b) $$\sqrt[3]{-108} = \sqrt[3]{(-27)(4)} = \sqrt[3]{-27}\sqrt[3]{4} = -3\sqrt[3]{4}$$

(c) Applying (iii) of the Laws of Radicals with $m = 3$ and $n = 2$,

$$\sqrt[3]{\sqrt{64}} = \sqrt[6]{64} = 2.$$

As a check, we have $\qquad \sqrt[3]{\sqrt{64}} = \sqrt[3]{8} = 2.$ ∎

If c is a real number and c^n occurs as a factor in a radical of index n, then c can be removed from the radicand provided the sign of c is taken into account. For example, if $c > 0$, or if $c < 0$ and n is odd, then

$$\sqrt[n]{c^n d} = \sqrt[n]{c^n}\,\sqrt[n]{d} = c\sqrt[n]{d}.$$

If $c < 0$ and n is *even*, then

$$\sqrt[n]{c^n d} = \sqrt[n]{c^n}\,\sqrt[n]{d} = |c|\sqrt[n]{d}.$$

In order to avoid considering positive and negative cases separately in the examples and exercises to follow, *we shall assume that letters, such as a, b, c, d, always represent positive real numbers.*

If we use the term "simplify" when referring to a radical, we mean to proceed as just explained until the radicand contains no factors with exponent greater than or equal to the index of the radical. Moreover, no fractions should appear under the final radical sign and denominators should be rationalized as illustrated in Example 5. The index should also be as low as possible.

EXAMPLE 4   Simplify the following, where all letters denote positive real numbers.

(a) $\sqrt[3]{320}$ (b) $\sqrt[3]{16x^3y^8z^4}$ (c) $\sqrt{3a^2b^3}\sqrt{6a^5b}$

SOLUTION

(a) $$\sqrt[3]{320} = \sqrt[3]{64 \cdot 5} = \sqrt[3]{4^3 \cdot 5} = \sqrt[3]{4^3}\sqrt[3]{5} = 4\sqrt[3]{5}$$

(b) $$\sqrt[3]{16x^3y^8z^4} = \sqrt[3]{(2^3x^3y^6z^3)(2y^2z)}$$
$$= \sqrt[3]{(2xy^2z)^3(2y^2z)}$$
$$= \sqrt[3]{(2xy^2z)^3}\sqrt[3]{2y^2z}$$
$$= 2xy^2z\sqrt[3]{2y^2z}$$

(c)
$$\sqrt{3a^2b^3}\sqrt{6a^5b} = \sqrt{18a^7b^4}$$
$$= \sqrt{(9a^6b^4)(2a)}$$
$$= \sqrt{(3a^3b^2)^2(2a)}$$
$$= \sqrt{(3a^3b^2)^2}\sqrt{2a}$$
$$= 3a^3b^2\sqrt{2a} \qquad\blacksquare$$

An important type of simplification involving radicals is **rationalizing a denominator.** The process involves beginning with a quotient that contains a radical in the denominator and then multiplying numerator and denominator by some expression so that the resulting denominator contains no radicals. The next example illustrates this technique.

EXAMPLE 5 Rationalize the denominator in each of the following:

(a) $\dfrac{1}{\sqrt{5}}$ (b) $\sqrt{\dfrac{2}{3}}$ (c) $\sqrt[3]{\dfrac{x}{y}}$

SOLUTION We may proceed as follows. (Supply reasons.)

(a)
$$\frac{1}{\sqrt{5}} = \frac{1}{\sqrt{5}}\cdot\frac{\sqrt{5}}{\sqrt{5}} = \frac{\sqrt{5}}{(\sqrt{5})^2} = \frac{\sqrt{5}}{5}$$

(b)
$$\sqrt{\frac{2}{3}} = \sqrt{\frac{2}{3}\cdot\frac{3}{3}} = \sqrt{\frac{6}{3^2}} = \frac{\sqrt{6}}{\sqrt{3^2}} = \frac{\sqrt{6}}{3}$$

(c)
$$\sqrt[3]{\frac{x}{y}} = \sqrt[3]{\frac{x}{y}\cdot\frac{y^2}{y^2}} = \frac{\sqrt[3]{xy^2}}{\sqrt[3]{y^3}} = \frac{\sqrt[3]{xy^2}}{y} \qquad\blacksquare$$

If a calculator is used to find decimal approximations of radicals, there is no advantage in writing $1/\sqrt{5} = \sqrt{5}/5$ or $\sqrt{2/3} = \sqrt{6}/3$ as we did in the preceding example. However, for *algebraic* simplifications, changing expressions to such forms is often desirable.

Radicals may be used to define *rational exponents.* First, if n is a positive integer and a is a real number, we define

$$a^{1/n} = \sqrt[n]{a},$$

provided that $\sqrt[n]{a}$ is a real number. We now proceed as follows.

DEFINITION OF RATIONAL EXPONENTS

If m/n is a rational number and n is a positive integer, and if a is a real number such that $\sqrt[n]{a}$ exists, then

$$a^{m/n} = (\sqrt[n]{a})^m = \sqrt[n]{a^m}.$$

We may also write $a^{m/n} = (a^{1/n})^m = (a^m)^{1/n}.$

It can be shown that the Laws of Exponents are true for rational exponents.

EXAMPLE 6   Simplify each expression.

(a) $(-27)^{2/3}(4)^{-5/2}$ (b) $(4\sqrt[3]{a})(2\sqrt{a})$

(c) $(r^2 s^6)^{1/3}$ (d) $\left(\dfrac{2x^{2/3}}{y^{1/2}}\right)^2 \left(\dfrac{3x^{-5/6}}{y^{1/3}}\right)$

SOLUTION

(a) $(-27)^{2/3}(4)^{-5/2} = (\sqrt[3]{-27})^2(\sqrt{4})^{-5} = (-3)^2(2)^{-5} = \dfrac{9}{32}$

(b) $(4\sqrt[3]{a})(2\sqrt{a}) = (4a^{1/3})(2a^{1/2}) = 8a^{1/3+1/2} = 8a^{5/6} = 8\sqrt[6]{a^5}$

(c) $(r^2 s^6)^{1/3} = (r^2)^{1/3}(s^6)^{1/3} = r^{2/3}s^2$

(d) $\left(\dfrac{2x^{2/3}}{y^{1/2}}\right)^2 \left(\dfrac{3x^{-5/6}}{y^{1/3}}\right) = \left(\dfrac{4x^{4/3}}{y}\right)\left(\dfrac{3x^{-5/6}}{y^{1/3}}\right) = \dfrac{12x^{1/2}}{y^{4/3}}$ ∎

In Chapter 5 we shall consider powers involving irrational exponents such as $3^{\sqrt{2}}$ or 5^π.

In the sciences it is often necessary to work with numbers that are very large or very small. It is customary to write these numbers in the form $a \cdot 10^n$, where a is a decimal such that $1 \le a < 10$ and n is an integer. This is referred to as the **scientific form** for real numbers. The symbol $\times$ is usually used for multiplication and we write $a \times 10^n$ instead of $a \cdot 10^n$. Scientific form enables the reader to determine quickly (without counting zeros) the relative magnitudes of large or small quantities.

The distance a ray of light travels in one year is approximately 5,900,000,000,000 miles. This number may be written in scientific form as 5.9×10^{12}. The positive exponent 12 indicates that the decimal point should be moved 12 places to the *right*. The notation works equally well for small numbers. To illustrate, the weight of an oxygen molecule is estimated to be 0.000000000000000000000053 grams, or in scientific form, 5.3×10^{-23} grams. The negative exponent indicates that the decimal point should be moved 23 places to the *left*. Some other illustrations of scientific form are:

$$513 = 5.13 \times 10^2, \qquad 20,700 = 2.07 \times 10^4, \qquad 92,000,000 = 9.2 \times 10^7$$

$$0.000648 = 6.48 \times 10^{-4}, \qquad 0.00000000043 = 4.3 \times 10^{-10}$$

Many calculators employ scientific form in their display panels: in $a \times 10^n$ the number 10 is suppressed and only the exponent is shown. For example, to find $(4,500,000)^2$ on a typical calculator, we could enter the integer 4,500,000 and press the x^2 (or squaring) key. The display panel would show

`2.025 13` or `2.025E13`

We would translate this as 2.025×10^{13}. Thus,

$$(4,500,000)^2 = 20,250,000,000,000.$$

Many calculators also use scientific form in the entry of numbers. The owner of a calculator should consult the user's manual for details.

As a final remark, applied problems often include numbers that are obtained by various types of measurements and, hence, are approximations to exact values. In such cases answers should be rounded off, since the final result of a calculation cannot be more accurate than the data that has been used. For example, if the length and width of a rectangle are measured to two-decimal-place accuracy, we cannot expect more than two-decimal-place accuracy in the calculated value of the area of the rectangle. If a number x is written in scientific form as $x = a \times 10^k$, where $1 \le a < 10$, and if a is rounded off to n decimal places, then we say that x is accurate (or has been rounded off) to $n + 1$ **significant figures.** For example, given $x = 37.2638$, we have the following:

Number of significant figures	5	4	3	2	1
Approximation to x	37.264	37.26	37.3	37	40

In Exercises 1.3, whenever an index of a radical is even (or a rational exponent m/n with n even is employed), we will assume that the letters that appear underneath the corresponding radical sign denote positive real numbers.

Exercises 1.3

Express the numbers in Exercises 1–10 in the form a/b, where a and b are integers.

1 $\left(-\frac{2}{3}\right)^4$

2 $(-3)^3$

3 $\dfrac{2^{-3}}{3^{-2}}$

4 $\dfrac{2^0 + 0^2}{2 + 0}$

5 $(-2)^3 + 3^{-2}$

6 $\left(-\frac{3}{2}\right)^4 - 2^{-4}$

7 $16^{-3/4}$

8 $9^{5/2}$

9 $(-0.008)^{2/3}$

10 $(0.008)^{-2/3}$

In Exercises 11–46 eliminate negative exponents and simplify.

11 $(\frac{1}{2}x^4)(16x^5)$

12 $(-3x^{-2})(4x^4)$

13 $\dfrac{(2x^3)(3x^2)}{(x^2)^3}$

14 $\dfrac{(2x^2)^3}{4x^4}$

15 $(\frac{1}{6}a^5)(-3a^2)(4a^7)$

16 $(-4b^3)(\frac{1}{6}b^2)(-9b^4)$

17 $\dfrac{(6x^3)^2}{(2x^2)^3}$

18 $\dfrac{(3y^3)(2y^2)^2}{(y^4)^3}$

19 $(3u^7v^3)(4u^4v^{-5})$

20 $(x^2yz^3)(-2xz^2)(x^3y^{-2})$

21 $(8x^4y^{-3})(\frac{1}{2}x^{-5}y^2)$

22 $\left(\dfrac{4a^2b}{a^3b^2}\right)\left(\dfrac{5a^2b}{2b^4}\right)$

23 $(\frac{1}{3}x^4y^{-3})^{-2}$

24 $(-2xy^2)^5\left(\dfrac{x^7}{8y^3}\right)$

25 $(3y^3)^4(4y^2)^{-3}$

26 $(-3a^2b^{-5})^3$

27 $(-2r^4s^{-3})^{-2}$

28 $(2x^2y^{-5})(6x^{-3}y)(\frac{1}{3}x^{-1}y^3)$

29 $(5x^2y^{-3})(4x^{-5}y^4)$

30 $(-2r^2s)^5(3r^{-1}s^3)^2$

31 $\left(\dfrac{3x^5y^4}{x^0y^{-3}}\right)^2$

32 $(4a^2b)^4\left(\dfrac{-a^3}{2b}\right)^2$

33 $(4a^{3/2})(2a^{1/2})$

34 $(-6x^{7/5})(2x^{8/5})$

35 $(3x^{5/6})(8x^{2/3})$

36 $(8r)^{1/3}(2r^{1/2})$

37 $(27a^6)^{-2/3}$

38 $(25z^4)^{-3/2}$

39 $(8x^{-2/3})x^{1/6}$

40 $(3x^{1/2})(-2x^{5/2})$

41 $\left(\dfrac{-8x^3}{y^{-6}}\right)^{2/3}$

42 $\left(\dfrac{-y^{3/2}}{y^{-1/3}}\right)^3$

43 $\left(\dfrac{x^6}{9y^{-4}}\right)^{-1/2}$

44 $\left(\dfrac{c^{-4}}{16d^8}\right)^{3/4}$

45 $\dfrac{(x^6y^3)^{-1/3}}{(x^4y^2)^{-1/2}}$

46 $a^{4/3}a^{-3/2}a^{1/6}$

Rewrite the expressions in Exercises 47–52 in terms of rational exponents.

47 $\sqrt[4]{x^3}$

48 $\sqrt[3]{x^5}$

49 $\sqrt[3]{(a+b)^2}$

50 $\sqrt{a+\sqrt{b}}$

51 $\sqrt{x^2+y^2}$

52 $\sqrt[3]{r^3-s^3}$

Rewrite the expressions in Exercises 53–56 in terms of radicals.

53 (a) $4x^{3/2}$ (b) $(4x)^{3/2}$

54 (a) $4 + x^{3/2}$ (b) $(4+x)^{3/2}$

55 (a) $8 - y^{1/3}$ (b) $(8-y)^{1/3}$

56 (a) $8y^{1/3}$ (b) $(8y)^{1/3}$

57 Prove that for all real numbers a and b, $|ab| = |a||b|$. (*Hint:* Write $|ab| = \sqrt{(ab)^2} = \sqrt{a^2b^2} = \sqrt{a^2}\sqrt{b^2}$ and use the fact that $\sqrt{x^2} = |x|$.)

58 Prove that $|a/b| = |a|/|b|$ if $b \neq 0$.

Simplify the expressions in Exercises 59–78.

59 $\sqrt{81}$

60 $\sqrt[3]{-125}$

61 $\sqrt[5]{-64}$

62 $\sqrt[4]{256}$

63 $\dfrac{1}{\sqrt[3]{2}}$

64 $\sqrt{\dfrac{1}{7}}$

65 $\sqrt{9x^{-4}y^6}$

66 $\sqrt{16a^8b^{-2}}$

67 $\sqrt[3]{8a^6b^{-3}}$

68 $\sqrt[4]{81r^5s^8}$

69 $\sqrt[3]{\dfrac{54a^7}{b^2}}$

70 $\sqrt[5]{\dfrac{-96x^7}{y^3}}$

71 $\sqrt{\dfrac{1}{3u^3v}}$

72 $\sqrt[3]{\dfrac{1}{4x^5y^2}}$

73 $\sqrt[4]{(3x^5y^{-2})^4}$

74 $\sqrt[6]{(2u^{-3}v^4)^6}$

75 $\sqrt[5]{\dfrac{8x^3}{y^4}}\sqrt[5]{\dfrac{4x^4}{y^2}}$

76 $\sqrt{5xy^7}\sqrt{10x^3y^3}$

77 $\sqrt[3]{3t^4v^2}\sqrt[3]{-9t^{-1}v^4}$

78 $\sqrt[3]{(2r-s)^3}$

In Exercises 79–84 replace the symbol □ with either = or ≠ and give reasons for your answers.

79 $(a^r)^2 \;\square\; a^{(r^2)}$

80 $(a^2+1)^{1/2} \;\square\; a+1$

81 $a^x b^y \;\square\; (ab)^{xy}$

82 $\sqrt{a^r} \;\square\; (\sqrt{a})^r$

83 $\sqrt[n]{\dfrac{1}{c}} \;\square\; \dfrac{1}{\sqrt[n]{c}}$

84 $a^{1/k} \;\square\; \dfrac{1}{a^k}$

85 The mass of a hydrogen atom is approximately 0.00000000000000000000000017 grams. Express this number in scientific form.

86 The mass of an electron is approximately 9.1×10^{-31} kilograms. Express this number in decimal form.

Express the numbers in Exercises 87 and 88 in scientific form.

87 (a) 427,000

(b) 0.000000098

(c) 810,000,000

88 (a) 85,200

(b) 0.0000055

(c) 24,900,000

Express the numbers in Exercises 89 and 90 in decimal form.

89 (a) 8.3×10^5

(b) 2.9×10^{-12}

(c) 5.63×10^8

90 (a) 2.3×10^7

(b) 7.01×10^{-9}

(c) 1.23×10^{10}

91 In astronomy, distances to stars are measured in light years, where 1 light year is the distance a ray of light travels in one year. If the speed of light is 186,000 miles per second, approximate 1 light year.

92 (a) It is estimated that the Milky Way galaxy contains 100 billion stars. Express this number in scientific form.

(b) The diameter d of the Milky Way galaxy is estimated as 100,000 light years. Express d in miles. (Refer to Exercise 91.)

93 The number of hydrogen atoms in a mole is Avogadro's number, 6.02×10^{23}. If one mole of the gas weighs 1.01 grams, estimate the mass of a hydrogen atom.

94 The population dynamics of many fish are characterized by extremely high fertility rates among adults and very low survival rates among the young. A mature halibut may lay as many as 2.5 million eggs, but only 0.00035% of the offspring survive to the age of 3 years. Use scientific form to calculate the number that live to age 3.

95 The longest (and undoubtedly most boring) movie ever made is a 1970 British film that runs for 48 hours. Assuming that the film speed is 24 frames per second, approximate the total number of frames in this film. Express your answer in scientific form.

96 One of the largest known prime numbers is $2^{44497} - 1$. It took one of the world's fastest computers about 60 days to verify that the number is prime. This computer is capable of performing 2×10^{11} calculations per second. Use scientific form to estimate the number of calculations needed to perform this feat.

Calculator Exercises 1.3

1 Simplify $[(x^2)^2]^2$. Use this result to explain how to calculate an eighth power using only the squaring key on a calculator. How can a sixteenth power be calculated?

2 Use a law of exponents to explain how to calculate each of the following powers using only the reciprocal and squaring keys:

(a) x^{-2}

(b) x^{32}

(c) x^{-16}

3 What happens when the reciprocal key is pressed three times in succession after entering a value? Explain using a law of exponents.

4 Simplify $(x^{1/2})^{1/2}$. Use the result to explain how to calculate fourth roots using only the square root key. How could you calculate an eighth root $\sqrt[8]{x}$?

5 One of the oldest banks in the U.S. is the Bank of

America, founded in 1812. If $200 had been deposited at that time into an account that paid simple 4% annual interest, then 170 years later the amount would have grown to $200(1.04)^{170}$ dollars. Calculate this large sum.

6 The distance d (in miles) that can be seen from the top of a tall building of height h feet can be approximated by $d = 1.2\sqrt{h}$. Approximate the distance that can be seen from the top of the Chicago Sears Tower, which is 1454 feet tall.

7 The length-weight relationship for the Pacific halibut is given by the formula $L = 0.46\sqrt[3]{W}$, where W is given in kg, and L in meters. The largest documented catch is a halibut that weighed 230 kg. Estimate its length.

8 The length-weight relationship for the sei whale is given by the formula $W = 0.0016L^{2.43}$, where W is in tons and L is in feet. A ship has sighted a sei whale that is about 25 feet long. Estimate its weight.

9 O'Carroll's formula is used to handicap weightlifters in different weight classes. If b is the body weight in kg, and w is the weight lifted, then the handicapped weight h is given by

$$h = \frac{w}{\sqrt[3]{b - 35}}.$$

A lifter weighing 75 kg has just clean-and-jerked 180 kg, while a 120 kg super-heavyweight has lifted 250 kg. Who would be judged the superior lifter in the overall competition?

10 The surface area S of the human body (in square feet) can be estimated from the height h (in inches) and weight w (in pounds) using the formula

$$S = (0.1091)w^{0.425}h^{0.725}.$$

This formula is used to estimate total body fat.

(a) Compute the body surface area of an individual 6 feet tall, weighing 175 lb.

(b) If you are 5 feet 6 inches tall, what is the effect of a 10% increase in weight on your body surface area?

1.4 # Algebraic Expressions

It is sometimes convenient to use the notation and terminology of sets. A **set** may be thought of as a collection of objects of some type. The objects are called **elements** of the set. Capital letters, such as $A, B, C, R, S, \ldots$, are often used to denote sets. Lowercase letters such as $a, b, x, y, \ldots$ usually represent elements of sets. Throughout our work $\mathbb{R}$ will denote the set of real numbers, and $\mathbb{Z}$ the set of integers. If S is a set, then $a \in S$ means that a is an element of S, whereas $a \notin S$ signifies that a is not an element of S. If every element of S is also an element of a set T, then S is called a **subset** of T. For example, $\mathbb{Z}$ is a subset of $\mathbb{R}$. Two sets S and T are said to be **equal,** written $S = T$, if S and T contain precisely the same elements. The notation $S \neq T$ means that S and T are not equal.

We frequently use symbols to represent arbitrary elements of a set. For example, we may use x to denote a real number, although no *particular* real number is specified. A letter that is used to represent any element of a set is sometimes called a **variable.** A symbol that represents a *specific* element is called a **constant.** In most of our work, letters near the end of the alphabet, such as $x, y,$ and z, will be used for variables, whereas letters

132,948

such as a, b, and c will denote constants. Throughout this text, unless otherwise specified, variables represent real numbers. The **domain of a variable** is the set of real numbers represented by the variable. To illustrate, $\sqrt{x}$ is a real number if and only if $x \geq 0$, and hence the domain of x is the set of nonnegative real numbers. Similarly, given the expression $1/(x - 2)$, we must exclude $x = 2$ to avoid division by zero, and consequently, in this case the domain is the set of all real numbers different from 2.

If the elements of a set S have a certain property, we sometimes write $S = \{x: \ \}$, where the property describing the variable x is stated in the space after the colon. For example, $\{x : x > 3\}$ represents the set of all real numbers greater than 3.

One way we may identify finite sets is to list all the elements within braces. Thus, if the set T consists of the first five positive integers, we may write $T = \{1, 2, 3, 4, 5\}$. When we describe sets in this way, the order used in listing the elements is considered irrelevant, so we could also write $T = \{1, 3, 2, 4, 5\}$, $T = \{4, 3, 2, 5, 1\}$, and so on.

If we begin with any collection of variables and real numbers, then an **algebraic expression** is the result obtained by applying additions, subtractions, multiplications, divisions, or the taking of roots. The following are examples of algebraic expressions:

$$x^3 - 2x + \frac{3^{1/9}}{\sqrt{2x}}, \qquad \frac{2xy + 3x}{y - 1}, \qquad \frac{4yz^{-2} + \left(\dfrac{-7}{x + w}\right)^5}{\sqrt[3]{y^2 + 5z}},$$

where x, y, z, and w are variables. If specific numbers are substituted for the variables in an algebraic expression, the resulting real number is called the **value** of the expression for these numbers. To illustrate, the value of the second expression above for $x = -2$ and $y = 3$ is

$$\frac{2(-2)(3) + 3(-2)}{3 - 1} = \frac{-12 - 6}{2} = -9.$$

When working with algebraic expressions, we will assume that domains are chosen so that variables do not represent numbers that make the expressions meaningless. Thus, we assume that denominators are not zero, roots always exist, etc.

Certain algebraic expressions are given special names. If x is a variable, then a **monomial** in x is an expression of the form ax^n, where the coefficient a is a real number and n is a nonnegative integer. A *polynomial in x* is a sum of monomials in x. Another way of stating this is as follows:

DEFINITION

A **polynomial in** x is an expression of the form

$$a_n x^n + a_{n-1} x^{n-1} + \cdots + a_1 x + a_0,$$

where n is a nonnegative integer and each coefficient a_k is a real number.

Each of the expressions $a_k x^k$ in the sum is called a **term** of the polynomial. If a coefficient a_k is zero, we usually delete the term $a_k x^k$. The coefficient a_n of the highest power of x is the **leading coefficient** of the polynomial and, if $a_n \neq 0$, we say that the polynomial has **degree** n. By definition, two polynomials are **equal** if and only if they have the same degree and corresponding coefficients are equal. If all the coefficients of a polynomial are zero, it is called the **zero polynomial** and is denoted by 0. It is customary not to assign a degree to the zero polynomial.

The following are some examples of polynomials:

$$3x^4 + 5x^3 + (-7)x + 4 \qquad \text{(degree 4)}$$

$$x^8 + 9x^2 + (-2)x \qquad \text{(degree 8)}$$

$$5x^2 + 1 \qquad \text{(degree 2)}$$

$$7x + 2 \qquad \text{(degree 1)}$$

$$5 \qquad \text{(degree 0)}$$

If some coefficients are negative, we often use minus signs between appropriate terms. To illustrate, instead of $3x^2 + (-5)x + (-7)$, we write $3x^2 - 5x - 7$ for this polynomial of degree 2. Polynomials in other variables may also be considered. For example, $\frac{2}{5}z^2 - 3z^7 + 8 - \sqrt{5}z^4$ is a polynomial in z of degree 7. We often arrange the terms in order of decreasing powers of the variable; thus, for this polynomial we write $-3z^7 - \sqrt{5}z^4 + \frac{2}{5}z^2 + 8$.

According to the definition of degree, if c is a nonzero real number, then c is a polynomial of degree 0. Such polynomials (together with the zero polynomial) are called **constant polynomials.**

A polynomial in x may be thought of as an algebraic expression obtained by employing only additions, subtractions, and multiplications involving x. In particular, the expressions

$$\frac{1}{x} + 3x, \qquad \frac{x-5}{x^2+2}, \qquad 3x^2 + \sqrt{x} - 2$$

are not polynomials since they involve divisions by variables or contain roots of variables.

Coefficients of polynomials may be chosen from some mathematical system other than the set of real numbers. However, in this text, unless mentioned otherwise, *the terminology "polynomial" will always refer to polynomials with real coefficients.*

Since polynomials, and the monomials that make up polynomials, represent real numbers, all of the properties in Section 1.2 can be applied. Thus, if additions, multiplications, and subtractions are carried out with polynomials, we may simplify the result by using various properties of real numbers.

EXAMPLE 1 Find the sum of $x^3 + 2x^2 - 5x + 7$ and $4x^3 - 5x^2 + 3$.

SOLUTION Rearranging terms and using properties of real numbers gives us

$$(x^3 + 2x^2 - 5x + 7) + (4x^3 - 5x^2 + 3)$$
$$= x^3 + 4x^3 + 2x^2 - 5x^2 - 5x + 7 + 3$$
$$= (1 + 4)x^3 + (2 - 5)x^2 + (-5)x + (7 + 3)$$
$$= 5x^3 - 3x^2 - 5x + 10 \qquad \blacksquare$$

Example 1 illustrates the fact that the sum of any two polynomials in x can be obtained by adding coefficients of like powers of x.

The difference of two polynomials is found by subtracting coefficients of like powers, as indicated by the next example.

EXAMPLE 2 Subtract $4x^3 - 5x^2 + 3$ from $x^3 + 2x^2 - 5x + 7$.

SOLUTION

$$(x^3 + 2x^2 - 5x + 7) - (4x^3 - 5x^2 + 3)$$
$$= x^3 + 2x^2 - 5x + 7 - 4x^3 + 5x^2 - 3$$
$$= x^3 - 4x^3 + 2x^2 + 5x^2 - 5x + 7 - 3$$
$$= (1 - 4)x^3 + (2 + 5)x^2 - 5x + (7 - 3)$$
$$= -3x^3 + 7x^2 - 5x + 4 \qquad \blacksquare$$

The intermediate steps in the previous solution were used for completeness. You may omit them after you become proficient with such manipulations. In order to multiply two polynomials, we use the Distributive

Properties together with the Laws of Exponents and combine like terms, as illustrated in the next example.

EXAMPLE 3 Find the product of $x^2 + 5x - 4$ and $2x^3 + 3x - 1$.

SOLUTION

$$
\begin{aligned}
(x^2 + 5x - 4&)(2x^3 + 3x - 1) \\
&= x^2(2x^3 + 3x - 1) + 5x(2x^3 + 3x - 1) - 4(2x^3 + 3x - 1) \\
&= 2x^5 + 3x^3 - x^2 + 10x^4 + 15x^2 - 5x - 8x^3 - 12x + 4 \\
&= 2x^5 + 10x^4 + (3 - 8)x^3 + (-1 + 15)x^2 + (-5 - 12)x + 4 \\
&= 2x^5 + 10x^4 - 5x^3 + 14x^2 - 17x + 4
\end{aligned}
$$

We may also arrange this multiplication as follows:

$$
\begin{array}{r}
2x^3 + 3x - 1 \\
x^2 + 5x - 4 \\
\hline
2x^5 \qquad + 3x^3 - \quad x^2 \\
10x^4 \qquad + 15x^2 - \quad 5x \\
- 8x^3 \qquad - 12x + 4 \\
\hline
2x^5 + 10x^4 - 5x^3 + 14x^2 - 17x + 4
\end{array}
$$

We may consider polynomials in more than one variable. For example, a polynomial in *two* variables, x and y, is a sum of terms, each of the form $ax^m y^k$ for some real number a and nonnegative integers m and k. An example is

$$3x^4y + 2x^3y^5 + 7x^2 - 4xy + 8y - 5.$$

Polynomials in three variables, x, y, z, or for that matter, in *any* number of variables may also be considered. Addition, subtraction, and multiplication are performed using properties of real numbers, as illustrated in the following example.

EXAMPLE 4 Find the product of $x^2 + xy + y^2$ and $x - y$.

SOLUTION

$$
\begin{aligned}
(x^2 + xy + y^2)(x - y) &= (x^2 + xy + y^2)x - (x^2 + xy + y^2)y \\
&= x^3 + x^2y + xy^2 - x^2y - xy^2 - y^3 \\
&= x^3 - y^3
\end{aligned}
$$

The next example illustrates division by a monomial.

EXAMPLE 5   Divide $6x^2y^3 + 4x^3y^2 - 10xy$ by $2xy$.

SOLUTION   Using properties of quotients and laws of exponents,

$$\frac{6x^2y^3 + 4x^3y^2 - 10xy}{2xy} = \frac{6x^2y^3}{2xy} + \frac{4x^3y^2}{2xy} - \frac{10xy}{2xy}$$

$$= 3xy^2 + 2x^2y - 5. \qquad \blacksquare$$

Certain products occur so frequently in algebra that they deserve special attention. We list some of these next, using letters to represent real numbers. The validity of each formula can be checked by actually carrying out the multiplications.

PRODUCT FORMULAS

> (i) $(x + y)(x - y) = x^2 - y^2$
>
> (ii) $(ax + b)(cx + d) = acx^2 + (ad + bc)x + bd$
>
> (iii) $(x + y)^2 = x^2 + 2xy + y^2$
>
> (iv) $(x - y)^2 = x^2 - 2xy + y^2$
>
> (v) $(x + y)^3 = x^3 + 3x^2y + 3xy^2 + y^3$
>
> (vi) $(x - y)^3 = x^3 - 3x^2y + 3xy^2 - y^3$

Since the symbols x and y in these formulas represent real numbers, they may be replaced by algebraic expressions, as illustrated in the next example.

EXAMPLE 6   Find the following products:

(a) $(2r^2 - \sqrt{s})(2r^2 + \sqrt{s})$ (b) $\left(\sqrt{c} + \dfrac{1}{\sqrt{c}}\right)^2$ (c) $(2a - 5b)^3$

SOLUTION
(a) Using Product Formula (i) with $x = 2r^2$ and $y = \sqrt{s}$,

$$(2r^2 - \sqrt{s})(2r^2 + \sqrt{s}) = (2r^2)^2 - (\sqrt{s})^2$$

$$= 4r^4 - s.$$

(b) Using Product Formula (iii) with $x = \sqrt{c}$ and $y = 1/\sqrt{c}$,

$$\left(\sqrt{c} + \frac{1}{\sqrt{c}}\right)^2 = (\sqrt{c})^2 + 2\sqrt{c} \cdot \frac{1}{\sqrt{c}} + \left(\frac{1}{\sqrt{c}}\right)^2$$

$$= c + 2 + \frac{1}{c}.$$

(c) Applying Product Formula (vi) with $x = 2a$ and $y = 5b$,

$$(2a - 5b)^3 = (2a)^3 - 3(2a)^2(5b) + 3(2a)(5b)^2 - (5b)^3$$

$$= 8a^3 - 60a^2b + 150ab^2 - 125b^3.$$ ∎

If a polynomial is written as a product of other polynomials, then each polynomial in the product is called a **factor** of the original polynomial. The process of expressing a polynomial as a product is called **factoring.** For example, since $x^2 - 9 = (x + 3)(x - 3)$, we see that $x + 3$ and $x - 3$ are factors of $x^2 - 9$.

Factoring plays an important role in mathematics, since it may be used to reduce the study of a complicated expression to the study of several simpler expressions. For example, properties of the polynomial $x^2 - 9$ can be determined by examining the factors $x + 3$ and $x - 3$.

We shall be interested primarily in **nontrivial factors** of polynomials; that is, factors that contain polynomials of degree greater than zero. An exception to this rule is that if the coefficients are restricted to *integers*, then we usually remove a common integral factor from each term of the polynomial as follows:

$$4x^2y + 8z^3 = 4(x^2y + 2z^3).$$

An integer $a > 1$ is prime if it cannot be written as a product of two positive integers greater than 1. In similar fashion, if S denotes a set of numbers, then a polynomial with coefficients in S is said to be **prime,** or **irreducible** over S, if it cannot be written as a product of two polynomials of positive degree with coefficients in S. A polynomial may be irreducible over one set S but not over another. For example, $x^2 - 2$ is irreducible over the rational numbers, since it cannot be expressed as a product of two polynomials of positive degree that have *rational* coefficients. If we allow the factors to have *real* coefficients, then $x^2 - 2$ is not prime, since

$$x^2 - 2 = (x + \sqrt{2})(x - \sqrt{2}).$$

Similarly, $x^2 + 1$ is irreducible over the real numbers but, as we shall see in Section 2.4, not over the complex numbers. It can be shown that every polynomial $ax + b$ of degree 1 is irreducible.

Before carrying out factorizations of polynomials, it is necessary to specify the system from which the coefficients of the factors are to be chosen. In this chapter we shall use the rule that *if a polynomial has integer coefficients, then the factors should be polynomials with integer coefficients.* For example,

$$x^2 + x - 6 = (x + 3)(x - 2)$$
$$4x^2 - 9y^2 = (2x - 3y)(2x + 3y).$$

It is usually difficult to factor polynomials of degree greater than 2. In simple cases the following formulas may be useful. Each can be verified by multiplication.

FACTORING FORMULAS

$a^2 - b^2 = (a + b)(a - b)$	(Difference of two squares)
$a^3 - b^3 = (a - b)(a^2 + ab + b^2)$	(Difference of two cubes)
$a^3 + b^3 = (a + b)(a^2 - ab + b^2)$	(Sum of two cubes)

EXAMPLE 7 Factor each of the following:

(a) $25r^2 - 49s^2$ (b) $81x^4 - y^4$ (c) $16x^4 - (y - 2z)^2$

SOLUTION
(a) Applying the Difference of Two Squares Formula with $a = 5r$ and $b = 7s$ gives us

$$25r^2 - 49s^2 = (5r)^2 - (7s)^2 = (5r + 7s)(5r - 7s).$$

(b) We make two applications of the Difference of Two Squares Formula as follows:

$$81x^4 - y^4 = (9x^2)^2 - (y^2)^2$$
$$= (9x^2 + y^2)(9x^2 - y^2)$$
$$= (9x^2 + y^2)(3x + y)(3x - y).$$

(c) If we write $16x^4 = (4x^2)^2$, then the formula for $a^2 - b^2$ with $a = 4x^2$ and $b = y - 2z$ takes on the form

$$16x^4 - (y - 2z)^2 = (4x^2)^2 - (y - 2z)^2$$
$$= [(4x^2) + (y - 2z)][(4x^2) - (y - 2z)]$$
$$= (4x^2 + y - 2z)(4x^2 - y + 2z). \qquad \blacksquare$$

EXAMPLE 8   Factor the following:

(a) $a^3 + 64b^3$ (b) $8x^6 - 27y^9$

SOLUTION

(a) Using the Sum of Two Cubes Formula, with $x = a$ and $y = 4b$,

$$a^3 + 64b^3 = a^3 + (4b)^3$$
$$= (a + 4b)[a^2 - a(4b) + (4b)^2]$$
$$= (a + 4b)(a^2 - 4ab + 16b^2).$$

(b) Using the Difference of Two Cubes Formula,

$$8x^6 - 27y^9 = (2x^2)^3 - (3y^3)^3$$
$$= (2x^2 - 3y^3)[(2x^2)^2 + (2x^2)(3y^3) + (3y^3)^2]$$
$$= (2x^2 - 3y^3)(4x^4 + 6x^2y^3 + 9y^6). \qquad \blacksquare$$

A factorization of the polynomial $px^2 + qx + r$, where p, q, and r are integers, must be of the form $(ax + b)(cx + d)$, where a, b, c, and d are integers. It follows that $ac = p$, $bd = r$, and $ad + bc = q$. Evidently, there is only a limited number of choices for a, b, c, and d that satisfy these conditions. If none of the choices work, then $px^2 + qx + r$ is prime. This method is also applicable to polynomials of the form $px^2 + qxy + ry^2$.

EXAMPLE 9   Factor the following:

(a) $6x^2 - 7x - 3$ (b) $12x^2 - 36xy + 27y^2$

(c) $4x^4y - 11x^3y^2 + 6x^2y^3$

SOLUTION

(a) If we write

$$6x^2 - 7x - 3 = (ax + b)(cx + d),$$

then $ac = 6$, $bd = -3$, and $ad + bc = -7$. Trying various possibilities, we arrive at the factorization

$$6x^2 - 7x - 3 = (2x - 3)(3x + 1).$$

(b) Since each term has 3 as a factor, we begin by writing

$$12x^2 - 36xy + 27y^2 = 3(4x^2 - 12xy + 9y^2).$$

If a factorization of $4x^2 - 12xy + 9y^2$ as a product of two first-degree polynomials exists, then it must be of the form

$$4x^2 - 12xy + 9y^2 = (ax + by)(cx + dy).$$

By trial we obtain

$$4x^2 - 12xy + 9y^2 = (2x - 3y)(2x - 3y) = (2x - 3y)^2.$$

Thus,

$$12x^2 - 36xy + 27y^2 = 3(4x^2 - 12xy + 9y^2) = 3(2x - 3y)^2.$$

(c) Since each term has $x^2 y$ as a factor, we begin by writing

$$4x^4 y - 11x^3 y^2 + 6x^2 y^3 = x^2 y(4x^2 - 11xy + 6y^2).$$

Next, by trial we obtain the following factorization:

$$4x^4 y - 11x^3 y^2 + 6x^2 y^3 = x^2 y(4x - 3y)(x - 2y)$$ ■

In some cases, if terms in a sum are grouped in suitable fashion, then a factorization can be found by means of the Distributive Properties, as illustrated in the next example.

EXAMPLE 10   Factor the following:

(a) $4ac + 2bc - 2ad - bd$ (b) $3x^3 + 2x^2 - 12x - 8$

SOLUTION

(a) We group the first two terms and the last two terms and then proceed as follows:

$$4ac + 2bc - 2ad - bd = (4ac + 2bc) - (2ad + bd)$$
$$= 2c(2a + b) - d(2a + b).$$

The right-hand side has the form

$$2ck - dk, \quad \text{where} \quad k = 2a + b.$$

Since $2ck - dk = (2c - d)k$, we obtain

$$4ac + 2bc - 2ad - bd = (2c - d)(2a + b).$$

It should be noted that if the original expression had been

$$4ac - bd + 2bc - 2ad,$$

then it would have been necessary to rearrange terms before grouping.

(b) We group the first two terms and the last two terms and then proceed as follows:

$$3x^3 + 2x^2 - 12x - 8 = (3x^3 + 2x^2) - (12x + 8)$$
$$= x^2(3x + 2) - 4(3x + 2).$$

The right-hand side has the form $x^2k - 4k$, where $k = 3x + 2$. Since $x^2k - 4k = (x^2 - 4)k$, we have

$$3x^3 + 2x^2 - 12x - 8 = (x^2 - 4)(3x + 2).$$

Finally, using the Difference of Two Squares Formula, we obtain the factorization

$$3x^3 + 2x^2 - 12x - 8 = (x + 2)(x - 2)(3x + 2). \qquad \blacksquare$$

Exercises 1.4

In Exercises 1–22 perform the indicated operations and express the result as a polynomial.

1 $(4x^3 + 2x^2 - x + 5) + (x^3 - 3x^2 - 5x + 1)$

2 $(x^4 - 3x^2 + 7x + 4) + (x^3 + 3x^2 - 4x - 3)$

3 $(5x^4 - 6x^2 + 9x) - (2x^3 + 3x^2 - 8x + 4)$

4 $(2x^2 - 11x + 13) - (3x^4 + 9x^2 - 10)$

5 $(5y^3 - 6y^2 + y - 7) - (5y^3 + 6y^2 + y + 2)$

6 $(4z^2 - 4z + 1) - (2z + 1)^2$

7 $(2a^4 - 3a^2 + 5) + a(a^3 + 3a - 4)$

8 $(3u + 1)(2u - 3) + 6u(u + 5)$

9 $(3x - 4)(2x^2 + x - 5)$

10 $(4x^3 - x^2 - 7)(3x + 2)$

11 $(r^2 + 2r + 3)(3r^2 - 2r + 4)$

12 $(s + t)(s^2 - st + t^2)$

13 $(6x^3 - 3x^2 - x + 7)(2x^2 + 4x + 5)$

14 $(2x^2 - xy + y^2)(3x - y)$

15 $(r - t)(r^2 + rt + t^2)$

16 $(x + y)(x^3 - x^2y + xy^2 - y^3)$

17 $(3x + 1)(2x^2 - x + 2)(x^2 + 4)$

18 $(3c + 1)(2c^2 + 5)(c^3 + 4)$

19 $\dfrac{8x^2y^3 - 10x^3y}{2x^2y}$

20 $\dfrac{6a^3b^3 - 9a^2b^2 + 3ab^4}{3ab^2}$

21 $\dfrac{3u^3v^4 - 2u^5v^2 + (u^2v^2)^2}{u^3v^2}$

22 $\dfrac{6x^2yz^3 - xy^2z}{xyz}$

Express the products in Exercises 23–50 as polynomials.

23 $(x - 3)(2x + 1)$ **24** $(3x + 2)(3x - 5)$

25 $(2s - 7t)(4s - 5t)$ **26** $(8n - 6p)(7n - 10p)$

27 $(5x^2 + 2y)(3x^2 - 7y)$

28 $(x + 9y^2)(3x - 4y^2)$

29 $(6t - 5v)(6t + 5v)$ **30** $(8u + 3)^2$

31 $(3r + 10s)^2$ **32** $(4v - 3w)(4v + 3w)$

33 $(4x^2 - 5y^2)^2$ **34** $(10p^2 + 7q^2)^2$

35 $(2x - x)^2$ **36** $b^6(b^3 - b^{-3})^2$

37 $(\sqrt{a} + \sqrt{b})(\sqrt{a} - \sqrt{b})$

38 $(x^{1/3} - y^{1/3})(x^{2/3} + x^{1/3}y^{1/3} + y^{2/3})$

39 $(x - 2y)^3$ **40** $(4x - y)^3$

41 $(3r + 4s)^3$ **42** $(2a + 5b)^3$

43 $(x^2 + y^2)^3$ **44** $(u^2 - 3v)^3$

45 $(a^{1/3} - b^{1/3})^3$ **46** $(a + b)^2(a - b)^2$

47 $(x + y + z)(x + y - z)$

48 $(2a - b + 3c)(2a - b - 3c)$

49 $(3x + 2y + z)^2$

50 $(x^2 + y^2 + z^2)^2$

Factor the expressions in Exercises 51–110.

51 $rs + 4st$

52 $4u^2 - 2uv$

53 $3a^2b^2 - 6a^2b$

54 $10xy + 15xy^2$

55 $9x^2y^2 + 15xy^4$

56 $-8p^4qr^2 - 4p^3q^3r^2$

57 $4x^2 + 5x - 6$

58 $21x^2 + 29x - 10$

59 $4x^2 - 4x - 15$

60 $6x^2 + x - 5$

61 $16y^2 - 14y - 15$

62 $12m^2 - 17m - 14$

63 $9z^2 - 12z + 4$

64 $4x^2 + 12x + 9$

65 $12c^2 + 50c + 48$

66 $40y^2 - 82y + 40$

67 $4r^2 - 25t^2$

68 $36a^2 - 49b^2$

69 $50x^2 + 45xy - 18y^2$

70 $45x^2 + 38xy + 8y^2$

71 $64w^4 - 36s^2$

72 $75p^2 - 48v^4$

73 $36z^2 + 60z + 25$

74 $64y^2 + 112y + 49$

75 $27x^3 - y^3$

76 $x^3 + 8y^3$

77 $27a^3 + 64b^3$

78 $8r^3 - 27s^3$

79 $8x^6 - 125$

80 $216 - y^6$

81 $6ax - 3ay + 2bx - by$

82 $5ru + 10vr + 2ut + 4vt$

83 $40zw + 8x^2w - 35z - 7x^2$

84 $18ck + 4dk + 9cj + 2dj$

85 $a^3 - a^2b + ab^2 - b^3$

86 $6w^8 + 17w^4 + 12$

87 $a^6 - b^6$

88 $x^8 - 16$

89 $x^4 + 25$

90 $a^2 + a + 1$

91 $6x^2 + 42x + 60$

92 $16x^2 + 40x - 24$

93 $4x^2 - 24x + 36$

94 $60x^2 - 85x + 30$

95 $75x^2 + 120x + 48$

96 $64x^2 - 16$

97 $36 - 9x^2$

98 $18x^2 - 50$

99 $2x^2y + xy - 2xz - z$

100 $3ac - 6bd + 3ad - 6bc$

101 $12x^2z + 8y^2z - 15x^2w - 10y^2w$

102 $4x^3 + 6x^2y - 4y^2x - 6y^3$

103 $y^6 + 7y^3 - 8$

104 $8c^6 + 19c^3 - 27$

105 $(x + y)^3 - 27$

106 $(a + b)^4 - 1$

107 $x^2 + 2x + 5$

108 $4x^3 + 4x^2 + x$

109 $x^{16} - 1$

110 $x^{16} + 1$

The ancient Greeks gave geometric proofs of the factoring formulas for the difference of two squares and the difference of two cubes. In Exercises 111 and 112 establish the formulas for special cases.

111 Compute the area of regions I and II in the figure to establish the formula for the difference of two squares for the special case $x > y$.

FIGURE FOR EXERCISE 111

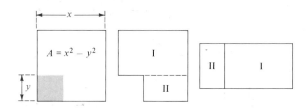

112 Compute the volumes of boxes I, II, and III in the figure to establish the formula for the difference of two cubes for the special case $x > y$.

FIGURE FOR EXERCISE 112

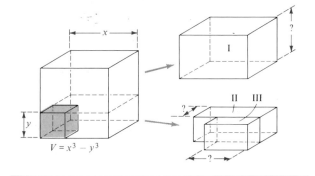

1.5 The Binomial Theorem

A sum $a + b$ is called a **binomial.** Sometimes it is necessary to consider $(a + b)^n$, where n is a large positive integer. A general formula for *expanding* $(a + b)^n$, that is, for expressing it as a sum, is given by the **Binomial Theorem.** Let us begin by considering some special cases. If we actually perform the multiplications, we obtain

$$(a + b)^2 = a^2 + 2ab + b^2$$

$$(a + b)^3 = a^3 + 3a^2b + 3ab^2 + b^3$$

$$(a + b)^4 = a^4 + 4a^3b + 6a^2b^2 + 4ab^3 + b^4$$

$$(a + b)^5 = a^5 + 5a^4b + 10a^3b^2 + 10a^2b^3 + 5ab^4 + b^5.$$

These expansions of $(a + b)^n$, for $n = 2$, 3, 4, and 5, have the following properties:

(i)   There are $n + 1$ terms, the first being a^n and the last b^n.

(ii) In going from any term to the next, the power of a decreases by 1 and the power of b increases by 1. Thus, the sum of the exponents of a and b is always n.

(iii) Each term has the form $(c)a^{n-k}b^k$, where the coefficient c is a real number and $k = 0, 1, 2, \ldots, n$.

(iv) The following formula is true for each of the first n terms of the expansion:

$$\frac{\text{(coefficient of the term)} \cdot \text{(exponent of } a)}{\text{number of the term}} = \text{coefficient of the next term}.$$

The next table illustrates property (iv) for the expansion of $(a + b)^5$.

Number of the term	Coefficient of the term	Exponent of a	Coefficient of the next term
1	1	5	$\dfrac{1 \cdot 5}{1} = 5$
2	5	4	$\dfrac{5 \cdot 4}{2} = 10$
3	10	3	$\dfrac{10 \cdot 3}{3} = 10$
4	10	2	$\dfrac{10 \cdot 2}{4} = 5$
5	5	1	$\dfrac{5 \cdot 1}{5} = 1$

Thus, as before, we get

$$(a + b)^5 = a^5 + 5a^4b + 10a^3b^2 + 10a^2b^3 + 5ab^4 + b^5.$$

Let us next consider $(a + b)^n$, for an arbitrary positive integer n. The first term is a^n, which has coefficient 1. If we assume that property (iv) is true, we obtain the successive coefficients listed in the following table.

Number of the term	Coefficient of the term	Exponent of a	Coefficient of the next term
1	1	n	$\dfrac{1 \cdot n}{1} = \dfrac{n}{1}$
2	$\dfrac{n}{1}$	$n - 1$	$\dfrac{n(n - 1)}{1 \cdot 2}$
3	$\dfrac{n(n - 1)}{1 \cdot 2}$	$n - 2$	$\dfrac{n(n - 1)(n - 2)}{1 \cdot 2 \cdot 3}$
4	$\dfrac{n(n - 1)(n - 2)}{1 \cdot 2 \cdot 3}$	$n - 3$	$\dfrac{n(n - 1)(n - 2)(n - 3)}{1 \cdot 2 \cdot 3 \cdot 4}$

The last coefficient that we calculated belongs to the fifth term. In similar fashion, the coefficient of the sixth term is

$$\frac{n(n - 1)(n - 2)(n - 3)(n - 4)}{1 \cdot 2 \cdot 3 \cdot 4 \cdot 5}.$$

It now appears likely that the coefficient of the $(k + 1)$st term in the expansion of $(a + b)^n$ is the following:

$(k + 1)$st COEFFICIENT OF $(a + b)^n$

$$\frac{n(n - 1)(n - 2)(n - 3) \cdots (n - k + 1)}{1 \cdot 2 \cdot 3 \cdot 4 \cdots k}.$$

This fraction can be written in a compact form by using **factorial notation.** If n is any positive integer, the symbol $n!$ (read "n factorial") is defined as follows:

DEFINITION OF $n!$ ($n > 0$)

$$n! = n(n - 1)(n - 2) \cdots 1$$

Thus, $n!$ is the product of the first n positive integers. Special cases are:

$$1! = 1, \qquad 2! = 2 \cdot 1, \qquad 3! = 3 \cdot 2 \cdot 1 = 6,$$

$$4! = 4 \cdot 3 \cdot 2 \cdot 1 = 24, \qquad 5! = 5 \cdot 4 \cdot 3 \cdot 2 \cdot 1 = 120.$$

To ensure that certain formulas will be true for all *nonnegative* integers, we define 0! as follows:

DEFINITION OF 0!

$$0! = 1$$

Note that for $n \geq 1$,

$$n! = n[(n-1)!] = n(n-1)!$$

For larger values of n we can write

$$n! = n(n-1)(n-2)! = n(n-1)(n-2)(n-3)!$$

and so on. The formula for the $(k+1)$st coefficient in the expansion of $(a+b)^n$ can be written as follows:

$$\frac{n(n-1)(n-2)\cdots(n-k+1)}{1 \cdot 2 \cdot 3 \cdots k} = \frac{\dfrac{n(n-1)(n-2)\cdots(n-k+1)(n-k)!}{(n-k)!}}{k!}$$

$$= \frac{\dfrac{n!}{(n-k)!}}{k!}$$

$$= \frac{n!}{k!(n-k)!}$$

These numbers are called **binomial coefficients** and are often represented by the following notation.

DEFINITION

$$\binom{n}{k} = \frac{n!}{k!(n-k)!}, \qquad k = 0, 1, 2, \ldots, n$$

The symbol $\binom{n}{k}$, which denotes the $(k + 1)$st coefficient in the expansion of $(a + b)^n$, is sometimes read "n choose k".

EXAMPLE 1

Find $\binom{5}{0}, \binom{5}{1}, \binom{5}{2}, \binom{5}{3}, \binom{5}{4}$ and $\binom{5}{5}$.

SOLUTION   These six numbers are the coefficients in the expansion of $(a + b)^5$, which we tabulated earlier in this section. By definition,

$$\binom{5}{0} = \frac{5!}{0!(5 - 0)!} = \frac{5!}{1 \cdot 5!} = 1$$

$$\binom{5}{1} = \frac{5!}{1!(5 - 1)!} = \frac{5!}{1 \cdot 4!} = \frac{5 \cdot 4!}{4!} = 5$$

$$\binom{5}{2} = \frac{5!}{2!(5 - 2)!} = \frac{5!}{2!3!} = \frac{5 \cdot 4 \cdot 3!}{2!3!} = \frac{20}{2} = 10$$

$$\binom{5}{3} = \frac{5!}{3!(5 - 3)!} = \frac{5!}{3!2!} = \frac{5 \cdot 4 \cdot 3!}{3!2!} = \frac{20}{2} = 10$$

$$\binom{5}{4} = \frac{5!}{4!(5 - 4)!} = \frac{5!}{4!1!} = \frac{5 \cdot 4!}{4!} = 5$$

$$\binom{5}{5} = \frac{5!}{5!(5 - 5)!} = \frac{1}{0!} = \frac{1}{1} = 1$$

∎

The Binomial Theorem may be stated as follows. (A proof is given in the first section of the chapter on Sequences, Series, and Probability.)

THE BINOMIAL THEOREM

$$(a + b)^n = a^n + \binom{n}{1}a^{n-1}b + \binom{n}{2}a^{n-2}b^2 + \cdots + \binom{n}{k}a^{n-k}b^k$$

$$+ \cdots + \binom{n}{n-1}ab^{n-1} + b^n$$

A statement of the Binomial Theorem without using symbols for binomial coefficients is as follows:

THE BINOMIAL THEOREM
(Alternative form)

$$(a + b)^n = a^n + na^{n-1}b + \frac{n(n-1)}{2!} a^{n-2}b^2 + \cdots$$

$$+ \frac{n(n-1)(n-2)\cdots(n-k+1)}{k!} a^{n-k}b^k$$

$$+ \cdots + nab^{n-1} + b^n$$

The following examples may be solved by using either the general formulas for the Binomial Theorem, or by repeated use of property (iv) stated at the beginning of this section.

EXAMPLE 2 Find the binomial expansion of $(2x + 3y^2)^4$.

SOLUTION Using the Binomial Theorem with $a = 2x$, $b = 3y^2$, and $n = 4$,

$$(2x + 3y^2)^4 = (2x)^4 + \binom{4}{1}(2x)^3(3y^2) + \binom{4}{2}(2x)^2(3y^2)^2$$

$$+ \binom{4}{3}(2x)(3y^2)^3 + (3y^2)^4.$$

This simplifies to

$$(2x + 3y^2)^4 = 16x^4 + 4(8x^3)(3y^2) + 6(4x^2)(9y^4) + 4(2x)(27y^6) + 81y^8$$
$$= 16x^4 + 96x^3y^2 + 216x^2y^4 + 216xy^6 + 81y^8. \quad \blacksquare$$

The next example illustrates the fact that if either a or b is negative, then the terms of the expansion are alternatively positive and negative.

EXAMPLE 3

Expand $\left(\dfrac{1}{x} - 2\sqrt{x}\right)^5$.

SOLUTION The binomial coefficients for $(a + b)^5$ were calculated in Example 1. Thus, if we let $a = 1/x$, $b = -2\sqrt{x}$, and $n = 5$ in the Binomial Theorem, then

$$\left(\frac{1}{x} - 2\sqrt{x}\right)^5 = \left(\frac{1}{x}\right)^5 + 5\left(\frac{1}{x}\right)^4(-2\sqrt{x}) + 10\left(\frac{1}{x}\right)^3(-2\sqrt{x})^2$$

$$+ 10\left(\frac{1}{x}\right)^2(-2\sqrt{x})^3 + 5\left(\frac{1}{x}\right)(-2\sqrt{x})^4 + (-2\sqrt{x})^5.$$

This simplifies to

$$\left(\frac{1}{x} - 2\sqrt{x}\right)^5 = \frac{1}{x^5} - \frac{10}{x^{7/2}} + \frac{40}{x^2} - \frac{80}{x^{1/2}} + 80x - 32x^{5/2}. \quad \blacksquare$$

Certain problems only require finding a specific term in the expansion of $(a + b)^n$. To work such problems, it is convenient to first find the exponent k that is to be assigned to b. Notice that, by the Binomial Theorem, *the exponent of b is always one less than the number of the term.* Once k is found, the exponent of a is $n - k$, and the coefficient is $\binom{n}{k}$.

EXAMPLE 4   Find the fifth term in the expansion of $(x^3 + \sqrt{y})^{13}$.

SOLUTION   Let $a = x^3$ and $b = \sqrt{y}$. The exponent of b in the fifth term is 4 and hence the exponent of a is 9. From the discussion of the preceding paragraph we obtain

$$\binom{13}{4}(x^3)^9(\sqrt{y})^4 = \frac{13!}{4!(13 - 4)!} x^{27}y^2 = \frac{13 \cdot 12 \cdot 11 \cdot 10}{4!} x^{27}y^2$$

$$= 715x^{27}y^2. \quad \blacksquare$$

Finally, there is an interesting triangular array of numbers called **Pascal's Triangle,** which can be used to obtain binomial coefficients. The numbers are arranged as follows.

$$
\begin{array}{ccccccccccccc}
 & & & & & & 1 & & & & & & \\
 & & & & & 1 & & 1 & & & & & \\
 & & & & 1 & & 2 & & 1 & & & & \\
 & & & 1 & & 3 & & 3 & & 1 & & & \\
 & & 1 & & 4 & & 6 & & 4 & & 1 & & \\
 & 1 & & 5 & & 10 & & 10 & & 5 & & 1 & \\
1 & & 6 & & 15 & & 20 & & 15 & & 6 & & 1 \\
\end{array}
$$

The numbers in the second row are the coefficients in the expansion of $(a + b)^1$; those in the third row are the coefficients determined by $(a + b)^2$; those in the fourth row are obtained from $(a + b)^3$, and so on. Each number in the array that is different from 1 can be found by adding the two numbers in the previous row that appear above and immediately to the left and right of the number.

EXAMPLE 5   Find the eighth row of Pascal's Triangle and use it to expand $(a + b)^7$.

SOLUTION   Let us rewrite the seventh row and then use the process described previously. In the following display the arrows indicate which two numbers in row seven are added to obtain the numbers in row eight.

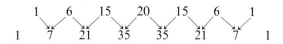

The eighth row gives us the coefficients in the expansion of $(a + b)^7$. Thus,

$$(a + b)^7 = a^7 + 7a^6b + 21a^5b^2 + 35a^4b^3$$
$$+ 35a^3b^4 + 21a^2b^5 + 7ab^6 + b^7.$$ ∎

Pascal's Triangle is useful to expand small powers of $a + b$; however, to expand large powers or to find a specific term, as in Example 4, it is better to refer to the general formula given by the Binomial Theorem.

Exercises 1.5

Find the numbers in Exercises 1–12.

1 $6!$

2 $5!$

3 $3!2!$

4 $2!0!$

5 $\dfrac{8!}{5!}$

6 $\dfrac{7!}{3!}$

7 $\dbinom{7}{3}$

8 $\dbinom{6}{2}$

9 $\dbinom{8}{5}$

10 $\dbinom{4}{4}$

11 $\dbinom{9}{4}$

12 $\dbinom{3}{2}$

Expand and simplify the expressions in Exercises 13–24.

13 $(a + b)^6$

14 $(a - b)^7$

15 $(a - b)^8$

16 $(a + b)^9$

17 $(3x - 5y)^4$

18 $(2t - s)^5$

19 $(u^2 + 4v)^5$

20 $(\tfrac{1}{2}c + d^3)^4$

21 $(r^{-2} - 2r)^6$

22 $(x^{1/2} - y^{-1/2})^6$

23 $(1 + x)^{10}$

24 $(1 - x)^{10}$

25 Find the first four terms in the binomial expansion of $(3c^{2/5} + c^{4/5})^{25}$.

26 Find the first three terms and the last three terms in the binomial expansion of $(x^3 + 5x^{-2})^{20}$.

27 Find the last two terms in the expansion of $(4b^{-1} - 3b)^{15}$.

28 Find the last three terms in the expansion of $(s - 2t^3)^{12}$.

Solve Exercises 29–40 without expanding completely.

29 Find the fifth term in the expansion of $(3a^2 + \sqrt{b})^9$.

30 Find the sixth term in the expansion of $\left(\dfrac{2}{c} + \dfrac{c^2}{3}\right)^7$.

31 Find the seventh term in the expansion of $(\tfrac{1}{2}u - 2v)^{10}$.

32 Find the fourth term in the expansion of $(2x^3 - y^2)^6$.

33 Find the middle term in the expansion of $(x^{1/3} + y^{1/3})^{12}$.

34 Find the two middle terms in the expansion of $(rs + t)^7$.

35 Find the term that does not contain x in the expansion of $\left(6x - \dfrac{1}{2x} \right)^{10}$.

36 Find the term involving x^8 in the expansion of $(y + 3x^2)^6$.

37 Find the term containing y^6 in the expansion of $(x - 2y^3)^4$.

38 Find the term containing b^9 in the expansion of $(5a + 2b^3)^4$.

39 Find the term containing c^3 in the expansion of $(\sqrt{c} + \sqrt{d})^{10}$.

40 In the expansion of $(xy - 2y^{-3})^8$ find the term that does not contain y.

41 Use the first four terms in the binomial expansion of $(1 + 0.02)^{10}$ to approximate $(1.02)^{10}$. Compare with the answer obtained using a calculator.

42 Use the first four terms in the binomial expansion of $(1 - 0.01)^4$ to approximate $(0.99)^4$. Compare with the answer obtained using a calculator.

1.6 Fractional Expressions

Quotients of algebraic expressions are called **fractional expressions.** As a special case, a quotient of two polynomials is called a **rational expression.** Some examples of rational expressions are:

$$\frac{x^2 - 5x + 1}{x^3 + 7}, \quad \frac{z^2 x^4 - 3yz}{5z}, \quad \text{and} \quad \frac{1}{4xy}.$$

Sometimes it is necessary to combine fractional expressions and then simplify the result. Since fractional expressions are quotients that contain symbols representing real numbers, the Properties of Quotients listed in Section 1.2 may be used. Of course, the letters a, b, c, and d will now be replaced by algebraic expressions. Of particular importance is the following technique:

$$\frac{ad}{bd} = \frac{a}{b} \cdot \frac{d}{d} = \frac{a}{b} \cdot 1 = \frac{a}{b}; \quad \text{that is,} \quad \frac{ad}{bd} = \frac{a}{b}.$$

This is sometimes phrased "A common factor in the numerator and denominator may be *canceled* from a quotient." To use this technique in problems involving rational expressions, we factor both the numerator and denominator into prime factors and then cancel common factors that occur in the numerator and denominator. We refer to the resulting expression as being *simplified*, or *reduced to lowest terms.*

EXAMPLE 1 Simplify:

$$\frac{3x^2 - 5x - 2}{x^2 - 4}$$

SOLUTION Factoring the numerator and denominator and canceling common factors gives us

$$\frac{3x^2 - 5x - 2}{x^2 - 4} = \frac{(3x + 1)(x - 2)}{(x + 2)(x - 2)} = \frac{3x + 1}{x + 2}.$$ ∎

In the preceding example we canceled the common factor $x - 2$; that is, we divided numerator and denominator by $x - 2$. This simplification is valid only if $x - 2 \neq 0$, that is, $x \neq 2$. However, 2 is not in the domain of x, since it leads to a zero denominator when substituted in the original expression. Hence, our manipulations are valid. We shall always assume such restrictions when simplifying rational expressions.

EXAMPLE 2 Simplify:

$$\frac{2 - x - 3x^2}{6x^2 - x - 2}$$

SOLUTION

$$\frac{2 - x - 3x^2}{6x^2 - x - 2} = \frac{(1 + x)(2 - 3x)}{(2x + 1)(3x - 2)} = \frac{-(1 + x)}{2x + 1}$$

The fact that $(2 - 3x) = -(3x - 2)$ accounts for the minus sign in the final answer, since $(2 - 3x)/(3x - 2) = -(3x - 2)/(3x - 2) = -1$. Another method is to change the form of the numerator as follows:

$$\frac{2 - x - 3x^2}{6x^2 - x - 2} = \frac{-(3x^2 + x - 2)}{6x^2 - x - 2}$$

$$= -\frac{(3x - 2)(x + 1)}{(3x - 2)(2x + 1)} = -\frac{x + 1}{2x + 1}$$ ∎

We multiply or divide rational expressions by using rules for quotients and then simplifying, as illustrated in the next example.

EXAMPLE 3 Perform the indicated operations and simplify:

(a) $\dfrac{x^2 - 6x + 9}{x^2 - 1} \cdot \dfrac{2x - 2}{x - 3}$ (b) $\dfrac{x + 2}{2x - 3} \div \dfrac{x^2 - 4}{2x^2 - 3x}$

SOLUTION

(a) $\dfrac{x^2 - 6x + 9}{x^2 - 1} \cdot \dfrac{2x - 2}{x - 3} = \dfrac{(x - 3)^2}{(x + 1)(x - 1)} \cdot \dfrac{2(x - 1)}{x - 3}$

$$= \dfrac{2(x - 3)^2(x - 1)}{(x + 1)(x - 1)(x - 3)} = \dfrac{2(x - 3)}{x + 1}$$

(b) $\dfrac{x + 2}{2x - 3} \div \dfrac{x^2 - 4}{2x^2 - 3x} = \dfrac{x + 2}{2x - 3} \cdot \dfrac{2x^2 - 3x}{x^2 - 4}$

$$= \dfrac{(x + 2)x(2x - 3)}{(2x - 3)(x + 2)(x - 2)} = \dfrac{x}{x - 2} \qquad \blacksquare$$

When adding or subtracting two rational expressions, we usually find a common denominator and use the following properties of real numbers:

$$\frac{a}{d} + \frac{c}{d} = \frac{a + c}{d}; \qquad \frac{a}{d} - \frac{c}{d} = \frac{a - c}{d}.$$

If the denominators are not the same, a common denominator may be introduced by multiplying numerator and denominator of each of the fractions by suitable polynomials. It is usually desirable to use the **least common denominator (lcd)** of the two fractions. To find the lcd we obtain the prime factorization for each denominator and then form the product of the different prime factors by using the *greatest* exponent that appears with each prime factor. Let us begin with a numerical example of this technique.

EXAMPLE 4 Express as a rational number in lowest terms:

$$\tfrac{7}{24} + \tfrac{5}{18}$$

SOLUTION The prime factorizations of the denominators of the fractions are $24 = (2^3)(3)$ and $18 = (2)(3^2)$. To find the lcd we form the product of the different prime factors by using the highest exponent associated with each factor. This gives us $(2^3)(3^2)$, or 72. We now change each fraction to an equal fraction with denominator 72 and add, as follows:

$$\tfrac{7}{24} + \tfrac{5}{18} = \tfrac{7}{24} \cdot \tfrac{3}{3} + \tfrac{5}{18} \cdot \tfrac{4}{4}$$

$$= \tfrac{21}{72} + \tfrac{20}{72} = \tfrac{41}{72}.$$

If we use the Addition Property of Quotients, stated on page 7, we obtain

$$\frac{7}{24} + \frac{5}{18} = \frac{(7)(18) + (24)(5)}{(24)(18)}$$

$$= \frac{126 + 120}{432} = \frac{246}{432} = \frac{41}{72}.$$ ∎

The method for finding the lcd for rational expressions is analogous to the process illustrated in the solution of Example 4. The only difference is that we use factorizations of polynomials instead of integers.

EXAMPLE 5 Change to a rational expression in lowest terms:

$$\frac{6}{x(3x - 2)} + \frac{5}{3x - 2} - \frac{2}{x^2}$$

SOLUTION The denominators are already in factored form. Evidently, the lcd is $x^2(3x - 2)$. In order to obtain three fractions having that denominator, we multiply numerator and denominator of the first fraction by x, of the second by x^2, and of the third by $3x - 2$. This gives us

$$\frac{6}{x(3x - 2)} + \frac{5}{3x - 2} - \frac{2}{x^2} = \frac{6x}{x^2(3x - 2)} + \frac{5x^2}{x^2(3x - 2)} - \frac{2(3x - 2)}{x^2(3x - 2)}$$

$$= \frac{6x + 5x^2 - (6x - 4)}{x^2(3x - 2)} = \frac{5x^2 + 4}{x^2(3x - 2)}.$$ ∎

EXAMPLE 6 Change to a rational expression in lowest terms:

$$\frac{2x + 5}{x^2 + 6x + 9} + \frac{x}{x^2 - 9} + \frac{1}{x - 3}$$

SOLUTION We begin by factoring denominators:

$$\frac{2x + 5}{x^2 + 6x + 9} + \frac{x}{x^2 - 9} + \frac{1}{x - 3} = \frac{2x + 5}{(x + 3)^2} + \frac{x}{(x + 3)(x - 3)} + \frac{1}{x - 3}$$

Since the lcd is $(x + 3)^2(x - 3)$, we multiply numerator and denominator of the first fraction by $x - 3$, of the second by $x + 3$, and of the third by

$(x + 3)^2$ and then add, as follows:

$$\frac{(2x + 5)(x - 3)}{(x + 3)^2(x - 3)} + \frac{x(x + 3)}{(x + 3)^2(x - 3)} + \frac{(x + 3)^2}{(x + 3)^2(x - 3)}$$

$$= \frac{(2x^2 - x - 15) + (x^2 + 3x) + (x^2 + 6x + 9)}{(x + 3)^2(x - 3)}$$

$$= \frac{4x^2 + 8x - 6}{(x + 3)^2(x - 3)} = \frac{2(2x^2 + 4x - 3)}{(x + 3)^2(x - 3)} \qquad \blacksquare$$

It is sometimes necessary to simplify quotients in which the numerator and denominator are not polynomials, as illustrated in the next example.

EXAMPLE 7   Simplify:

$$\frac{1 - \dfrac{2}{x + 1}}{x - \dfrac{1}{x}}.$$

SOLUTION

$$\frac{1 - \dfrac{2}{x + 1}}{x - \dfrac{1}{x}} = \frac{\dfrac{(x + 1) - 2}{x + 1}}{\dfrac{x^2 - 1}{x}} = \frac{\left(\dfrac{x - 1}{x + 1}\right)}{\left(\dfrac{x^2 - 1}{x}\right)}$$

$$= \frac{x - 1}{x + 1} \cdot \frac{x}{x^2 - 1} = \frac{(x - 1)x}{(x + 1)(x + 1)(x - 1)}$$

$$= \frac{x}{(x + 1)^2} \qquad \blacksquare$$

The denominators of certain fractional expressions contain sums or differences that involve radicals. Such denominators can be rationalized, as shown in the next example.

EXAMPLE 8   Rationalize the denominator of the fraction

$$\frac{1}{\sqrt{x} + \sqrt{y}}.$$

SOLUTION   Multiplying numerator and denominator by $\sqrt{x} - \sqrt{y}$, we obtain

$$\frac{1}{\sqrt{x} + \sqrt{y}} = \frac{1}{\sqrt{x} + \sqrt{y}} \cdot \frac{\sqrt{x} - \sqrt{y}}{\sqrt{x} - \sqrt{y}}$$

$$= \frac{\sqrt{x} - \sqrt{y}}{(\sqrt{x})^2 + \sqrt{y}\sqrt{x} - \sqrt{x}\sqrt{y} - (\sqrt{y})^2}$$

$$= \frac{\sqrt{x} - \sqrt{y}}{x - y}$$ ∎

Some problems in calculus require rationalizing a *numerator,* as in the following example.

EXAMPLE 9   Rationalize the numerator of the fraction

$$\frac{\sqrt{x + h} - \sqrt{x}}{h}, \quad \text{where } x > 0 \text{ and } h > 0.$$

SOLUTION   We multiply numerator and denominator by the expression $\sqrt{x + h} + \sqrt{x}$ and proceed as follows:

$$\frac{\sqrt{x + h} - \sqrt{x}}{h} = \frac{\sqrt{x + h} - \sqrt{x}}{h} \cdot \frac{\sqrt{x + h} + \sqrt{x}}{\sqrt{x + h} + \sqrt{x}}$$

$$= \frac{(\sqrt{x + h})^2 - \sqrt{x}\sqrt{x + h} + \sqrt{x + h}\sqrt{x} - (\sqrt{x})^2}{h(\sqrt{x + h} + \sqrt{x})}$$

$$= \frac{(x + h) - x}{h(\sqrt{x + h} + \sqrt{x})}$$

$$= \frac{h}{h(\sqrt{x + h} + \sqrt{x})}$$

$$= \frac{1}{\sqrt{x + h} + \sqrt{x}}$$ ∎

For certain problems in calculus it is necessary to simplify expressions of the types given in the next two examples. (Also see Exercises 35–52.)

EXAMPLE 10 Simplify the fraction

$$\frac{\dfrac{1}{(x+h)^2} - \dfrac{1}{x^2}}{h}.$$

SOLUTION

$$\frac{\dfrac{1}{(x+h)^2} - \dfrac{1}{x^2}}{h} = \frac{\dfrac{x^2 - (x+h)^2}{(x+h)^2 x^2}}{h}$$

$$= \frac{x^2 - (x^2 + 2xh + h^2)}{(x+h)^2 x^2 h}$$

$$= \frac{x^2 - x^2 - 2xh - h^2}{(x+h)^2 x^2 h}$$

$$= \frac{-h(2x + h)}{(x+h)^2 x^2 h}$$

$$= -\frac{2x + h}{(x+h)^2 x^2}$$

$\blacksquare$

EXAMPLE 11 Simplify:

$$\frac{3x^2(2x+5)^{1/2} - x^3(\tfrac{1}{2})(2x+5)^{-1/2}(2)}{[(2x+5)^{1/2}]^2}$$

SOLUTION

$$\frac{3x^2(2x+5)^{1/2} - x^3(\tfrac{1}{2})(2x+5)^{-1/2}(2)}{[(2x+5)^{1/2}]^2} = \frac{3x^2(2x+5)^{1/2} - \dfrac{x^3}{(2x+5)^{1/2}}}{2x+5}$$

$$= \frac{\dfrac{3x^2(2x+5) - x^3}{(2x+5)^{1/2}}}{2x+5}$$

$$= \frac{6x^3 + 15x^2 - x^3}{(2x+5)^{1/2}(2x+5)}$$

$$= \frac{5x^3 + 15x^2}{(2x+5)^{3/2}} = \frac{5x^2(x+3)}{(2x+5)^{3/2}}$$

An alternative solution is to eliminate the negative power $(2x + 5)^{-1/2}$ in the given expression first, by multiplying numerator and denominator by $(2x + 5)^{1/2}$ as follows:

$$\frac{3x^2(2x + 5)^{1/2} - x^3(\frac{1}{2})(2x + 5)^{-1/2}(2)}{[(2x + 5)^{1/2}]^2} \cdot \frac{(2x + 5)^{1/2}}{(2x + 5)^{1/2}}$$

$$= \frac{3x^2(2x + 5) - x^3(\frac{1}{2})(2)}{(2x + 5)(2x + 5)^{1/2}}$$

$$= \frac{6x^3 + 15x^2 - x^3}{(2x + 5)^{3/2}}$$

The remainder of the simplification is the same. ∎

Exercises 1.6

Simplify the expressions in Exercises 1–52.

1 $\dfrac{6x^2 + 7x - 10}{6x^2 + 13x - 15}$

2 $\dfrac{10x^2 + 29x - 21}{5x^2 - 23x + 12}$

3 $\dfrac{12y^2 + 3y}{20y^2 + 9y + 1}$

4 $\dfrac{4z^2 + 12z + 9}{2z^2 + 3z}$

5 $\dfrac{6 - 7a - 5a^2}{10a^2 - a - 3}$

6 $\dfrac{6y - 5y^2}{25y^2 - 36}$

7 $\dfrac{4x^3 - 9x}{10x^4 + 11x^3 - 6x^2}$

8 $\dfrac{16x^4 + 8x^3 + x^2}{4x^3 + 25x^2 + 6x}$

9 $\dfrac{6r}{3r - 1} - \dfrac{4r}{2r + 5}$

10 $\dfrac{3s}{s^2 + 1} - \dfrac{6}{2s - 1}$

11 $\dfrac{2x + 1}{2x - 1} - \dfrac{x - 1}{x + 1}$

12 $\dfrac{3u + 2}{u - 4} + \dfrac{4u + 1}{5u + 2}$

13 $\dfrac{9t - 6}{8t^3 - 27} \cdot \dfrac{4t^2 - 9}{12t^2 + 10t - 12}$

14 $\dfrac{a^2 + 4a + 3}{3a^2 + a - 2} \cdot \dfrac{3a^2 - 2a}{2a^2 + 13a + 21}$

15 $\dfrac{5a^2 + 12a + 4}{a^4 - 16} \div \dfrac{25a^2 + 20a + 4}{a^2 - 2a}$

16 $\dfrac{x^3 - 8}{x^2 - 4} \div \dfrac{x}{x^3 + 8}$

17 $\dfrac{2}{3x + 1} - \dfrac{9}{(3x + 1)^2}$

18 $\dfrac{4}{(5x - 2)^2} + \dfrac{x}{5x - 2}$

19 $\dfrac{1}{c} - \dfrac{c + 2}{c^2} + \dfrac{3}{c^3}$

20 $\dfrac{6}{3t} + \dfrac{t + 5}{t^3} + \dfrac{1 - 2t^2}{t^4}$

21 $\dfrac{5}{x - 1} + \dfrac{8}{(x - 1)^2} - \dfrac{3}{(x - 1)^3}$

22 $\dfrac{8}{x} + \dfrac{3}{2x - 4} + \dfrac{7x}{x^2 - 4}$

23 $\dfrac{2}{x} + \dfrac{7}{x^2} + \dfrac{5}{2x - 3} + \dfrac{1}{(2x - 3)^2}$

24 $\dfrac{4}{x} + \dfrac{3x^2 + 5}{x^3} - \dfrac{6}{2x + 1}$

25 $\dfrac{p^4 + 3p^3 - 8p - 24}{p^3 - 2p^2 - 9p + 18}$

26 $\dfrac{2ac + bc - 6ad - 3bd}{6ac + 2ad + 3bc + bd}$

27 $\dfrac{5}{7x - 3} - \dfrac{2}{2x + 1} + \dfrac{4x}{14x^2 + x - 3}$

28 $2 + \dfrac{3}{x} + \dfrac{7x}{3x + 10}$

29 $\dfrac{\dfrac{a}{b} - \dfrac{b}{a}}{\dfrac{1}{a} + \dfrac{1}{b}}$

30 $\dfrac{\dfrac{1}{x + 1} - 5}{\dfrac{1}{x} - x}$

31 $\dfrac{\dfrac{x}{y^2} - \dfrac{y}{x^2}}{\dfrac{1}{y^2} - \dfrac{1}{x^2}}$

32 $\dfrac{\dfrac{r}{s} + \dfrac{s}{r}}{\dfrac{r^2}{s^2} - \dfrac{s^2}{r^2}}$

33 $\dfrac{\dfrac{5}{x + 1} + \dfrac{2x}{x + 3}}{\dfrac{x}{x + 1} + \dfrac{7}{x + 3}}$

34 $\dfrac{\dfrac{3}{w} - \dfrac{6}{2w + 1}}{\dfrac{5}{w} + \dfrac{8}{2w + 1}}$

35 $\dfrac{(x + h)^2 - 7(x + h) - (x^2 - 7x)}{h}$

36 $\dfrac{(x + h)^3 + 4(x + h) - (x^3 + 4x)}{h}$

37 $\dfrac{\dfrac{1}{(x + h)^3} - \dfrac{1}{x^3}}{h}$

38 $\dfrac{\dfrac{1}{x + h} - \dfrac{1}{x}}{h}$

39 $\dfrac{\dfrac{1}{2x + 2h + 3} - \dfrac{1}{2x + 3}}{h}$

40 $\dfrac{\dfrac{7}{5x + 5h - 2} - \dfrac{7}{5x - 2}}{h}$

41 $(2x^2 - 3x + 1)(4)(3x + 2)^3(3) + (3x + 2)^4(4x - 3)$

42 $(6x - 5)^3(2)(x^2 + 4)(2x) + (x^2 + 4)^2(3)(6x - 5)^2(6)$

43 $(x^2 - 4)^{1/2}(3)(2x + 1)^2(2) + (2x + 1)^3(\tfrac{1}{2})(x^2 - 4)^{-1/2}(2x)$

44 $(3x + 2)^{1/3}(2)(4x - 5)(4) + (4x - 5)^2(\tfrac{1}{3})(3x + 2)^{-2/3}(3)$

45 $(3x + 1)^6(\tfrac{1}{2})(2x - 5)^{-1/2}(2) + (2x - 5)^{1/2}(6)(3x + 1)^5(3)$

46 $(x^2 + 9)^4(-\tfrac{1}{3})(x + 6)^{-4/3} + (x + 6)^{-1/3}(4)(x^2 + 9)^3(2x)$

47 $\dfrac{(6x + 1)^3(27x^2 + 2) - (9x^3 + 2x)(3)(6x + 1)^2(6)}{(6x + 1)^6}$

48 $\dfrac{(x^2 - 1)^4(2x) - x^2(4)(x^2 - 1)^3(2x)}{(x^2 - 1)^8}$

49 $\dfrac{(x^2 + 4)^{1/3}(3) - (3x)(\tfrac{1}{3})(x^2 + 4)^{-2/3}(2x)}{[(x^2 + 4)^{1/3}]^2}$

50 $\dfrac{(1 - x^2)^{1/2}(2x) - x^2(\tfrac{1}{2})(1 - x^2)^{-1/2}(-2x)}{[(1 - x^2)^{1/2}]^2}$

51 $\dfrac{(4x^2 + 9)^{1/2}(2) - (2x + 3)(\tfrac{1}{2})(4x^2 + 9)^{-1/2}(8x)}{[(4x^2 + 9)^{1/2}]^2}$

52 $\dfrac{(3x + 2)^{1/2}(\tfrac{1}{3})(2x + 3)^{-2/3}(2) - (2x + 3)^{1/3}(\tfrac{1}{2})(3x + 2)^{-1/2}(3)}{[(3x + 2)^{1/2}]^2}$

Rationalize the denominators in Exercises 53–58.

53 $\dfrac{\sqrt{a}}{\sqrt{a} - \sqrt{b}}$

54 $\dfrac{1}{\sqrt{c} + d}$

55 $\dfrac{\sqrt{t} - 4}{\sqrt{t} + 4}$

56 $\dfrac{4}{3 - \sqrt{w}}$

57 $\dfrac{1}{\sqrt[3]{a} - \sqrt[3]{b}}$ (*Hint*: Multiply numerator and denominator by $\sqrt[3]{a^2} + \sqrt[3]{a}\,\sqrt[3]{b} + \sqrt[3]{b^2}$.)

58 $\dfrac{1}{\sqrt[3]{x} + \sqrt[3]{y}}$

Rationalize the numerators in Exercises 59–64.

59 $\dfrac{\sqrt{a} - \sqrt{b}}{c}$

60 $\dfrac{b - \sqrt{c}}{d}$

61 $\dfrac{\sqrt{2(x + h) + 1} - \sqrt{2x + 1}}{h}$

62 $\dfrac{\sqrt{x} - \sqrt{x + h}}{h\sqrt{x}\sqrt{x + h}}$

63 $\dfrac{\sqrt{1 - x - h} - \sqrt{1 - x}}{h}$

64 $\dfrac{\sqrt[3]{x + h} - \sqrt[3]{x}}{h}$ (*Hint*: Compare Exercise 57.)

1.7 Review

Define or discuss each of the following.

1. Commutative Properties of real numbers
2. Associative Properties
3. Distributive Properties
4. Integers
5. Prime number
6. Rational and irrational numbers
7. Coordinate line
8. A number a is greater than a number b.
9. A number a is less than a number b.
10. Absolute value of a real number
11. The distance between points on a coordinate line
12. Exponent
13. Laws of Exponents
14. Principal nth root
15. Radical notation
16. Rationalizing a denominator
17. Rational exponents
18. Scientific form for real numbers
19. Variable
20. Domain of a variable
21. Algebraic expression
22. Polynomial
23. Degree of a polynomial
24. Prime polynomial
25. Irreducible polynomial
26. Factorial notation
27. The Binomial Theorem
28. Binomial coefficients
29. Pascal's Triangle
30. Rational expression

Exercises 1.7

1. Express each of the following as a rational number with least positive numerator.

   (a) $\left(\frac{2}{3}\right)\left(-\frac{5}{8}\right)$ (b) $\frac{3}{4} + \frac{6}{5}$

   (c) $\frac{5}{8} - \frac{6}{7}$ (d) $\frac{3}{4} \div \frac{6}{5}$

2. Replace $\square$ with either $<$, $>$, or $=$.

   (a) $-0.1 \;\square\; -0.01$ (b) $\sqrt{9} \;\square\; -3$

   (c) $\frac{1}{6} \;\square\; 0.166$

3. Express in terms of inequalities:

   (a) x is negative.

   (b) a is between $\frac{1}{2}$ and $\frac{1}{3}$.

   (c) The absolute value of x is not greater than 4.

4. Rewrite without using the absolute value symbol:

   (a) $|-7|$ (b) $\dfrac{|-5|}{-5}$ (c) $|3^{-1} - 2^{-1}|$

5. If points A, B, and C on a coordinate line have coordinates -8, 4, and -3, respectively, find the following:

   (a) $d(A, C)$ (b) $d(C, A)$ (c) $d(B, C)$

6. Determine whether the following are identities.

   (a) $(x + y)^2 = x^2 + y^2$

   (b) $\dfrac{1}{\sqrt{x + y}} = \dfrac{1}{\sqrt{x}} + \dfrac{1}{\sqrt{y}}$

   (c) $\dfrac{1}{\sqrt{c} - \sqrt{d}} = \dfrac{\sqrt{c} + \sqrt{d}}{c - d}$

Simplify the expressions in Exercises 7–28.

7 $(3a^2b)^2(2ab^3)$

8 $\dfrac{6r^3y^2}{2r^5y}$

9 $\dfrac{(3x^2y^{-3})^{-2}}{x^{-5}y}$

10 $\left(\dfrac{a^{2/3}b^{3/2}}{a^2b}\right)^6$

11 $(-2p^2q)^3(p/4q^2)^2$

12 $c^{-4/3}c^{3/2}c^{1/6}$

13 $\left(\dfrac{xy^{-1}}{\sqrt{z}}\right)^4 \div \left(\dfrac{x^{1/3}y^2}{z}\right)^3$

14 $\left(\dfrac{-64x^3}{z^6y^9}\right)^{2/3}$

15 $((a^{2/3}b^{-2})^3)^{-1}$

16 $\dfrac{(3u^2v^5w^{-4})^3}{(2uv^{-3}w^2)^4}$

17 $\dfrac{r^{-1}+s^{-1}}{(rs)^{-1}}$

18 $(u+v)^3(u+v)^{-2}$

19 $\sqrt[3]{(x^4y^{-1})^6}$

20 $\sqrt[3]{8x^5y^3z^4}$

21 $\dfrac{1}{\sqrt[3]{4}}$

22 $\sqrt{\dfrac{a^2b^3}{c}}$

23 $\sqrt[3]{4x^2y}\,\sqrt[3]{2x^5y^2}$

24 $\sqrt[4]{(-4a^3b^2c)^2}$

25 $\dfrac{1}{\sqrt{t}}\left(\dfrac{1}{\sqrt{t}}-1\right)$

26 $\sqrt{\sqrt[3]{(c^3d^6)^4}}$

27 $\dfrac{\sqrt{12x^4y}}{\sqrt{3x^2y^5}}$

28 $\sqrt[3]{(a+2b)^3}$

In Exercises 29 and 30 rationalize the denominator.

29 $\dfrac{1-\sqrt{x}}{1+\sqrt{x}}$

30 $\dfrac{1}{\sqrt{a}+\sqrt{a-2}}$

31 Express in scientific form:
   (a) 93,700,000,000   (b) 0.00000402

32 Express as a decimal:
   (a) 6.8×10^7 (b) 7.3×10^{-4}

33 The body of an average person contains 5.5 liters of blood and about 5 million red blood cells per cubic milliliter of blood. Remembering that 1 liter $= 10^6$ mm^3, estimate the number of red blood cells in an average person's circulatory system. Express the answer in scientific form.

34 A healthy heart beats 70 to 90 times per minute. If an individual lives to age 80, estimate the number of heart-beats in his or her lifetime. Express the answer in scientific form.

In Exercises 35–42 perform the indicated operations.

35 $(3x^3 - 4x^2 + x - 7) + (x^4 - 2x^3 + 3x^2 + 5)$

36 $(4z^4 - 3z^2 + 1) - z(z^3 + 4z^2 - 4)$

37 $(x + 4)(x + 3) - (2x - 1)(x - 5)$

38 $(4x - 5)(2x^2 + 3x - 7)$

39 $(3y^3 - 2y^2 + y + 4)(y^2 - 3)$

40 $(3x + 2)(x - 5)(5x + 4)$

41 $(a - b)(a^3 + a^2b + ab^2 + b^3)$

42 $\dfrac{9p^4q^3 - 6p^2q^4 + 5p^3q^2}{3p^2q^2}$

Find the products in Exercises 43–50.

43 $(3a - 5b)(2a + 7b)$

44 $(4r^2 - 3s)^2$

45 $(13a^2 + 4b)(13a^2 - 4b)$

46 $(a^3 - a^{-3})^2$

47 $(2a + b)^3$

48 $(c^2 - d^2)^3$

49 $(3x + 2y)^2(3x - 2y)^2$

50 $(a + b + c + d)^2$

Factor the expressions in Exercises 51–62.

51 $60xw + 70w$

52 $2r^4s^3 - 8r^2s^5$

53 $28x^2 + 4x - 9$

54 $16a^4 + 24a^2b^2 + 9b^4$

55 $2wy + 3yx - 8wz - 12zx$

56 $2c^3 - 12c^2 + 3c - 18$

57 $8x^3 + 64y^3$

58 $u^3v^4 - u^6v$

59 $p^8 - q^8$

60 $x^4 - 8x^3 + 16x^2$

61 $w^6 + 1$

62 $3x + 6$

63 Expand and simplify $(x^2 - 3y)^6$.

64 Find the first four terms in the binomial expansion of $(a^{2/5} + 2a^{-3/5})^{20}$.

65 Find the sixth term in the expansion of $(b^3 - \frac{1}{2}c^2)^9$.

66 In the expansion of $(2c^3 + 5c^{-2})^{10}$ find the term that does not contain c.

Simplify the expressions in Exercises 67–76.

67 $\dfrac{6x^2 - 7x - 5}{4x^2 + 4x + 1}$

68 $\dfrac{r^3 - t^3}{r^2 - t^2}$

69 $\dfrac{6x^2 - 5x - 6}{x^2 - 4} \div \dfrac{2x^2 - 3x}{x + 2}$

70 $\dfrac{2}{4x - 5} - \dfrac{5}{10x + 1}$

71 $\dfrac{7}{x + 2} + \dfrac{3x}{(x + 2)^2} - \dfrac{5}{x}$

72 $\dfrac{x + x^{-2}}{1 + x^{-2}}$

73 $\dfrac{1}{x} - \dfrac{2}{x^2 + x} - \dfrac{3}{x + 3}$ *lcm*

$x(x+1)(x+3)$

74 $(a^{-1} + b^{-1})^{-1}$

75 $(x^2 + 1)^{3/2}(4)(x + 5)^3 + (x + 5)^4(\tfrac{3}{2})(x^2 + 1)^{1/2}(2x)$

76 $\dfrac{(4 - x^2)(\tfrac{1}{3})(6x + 1)^{-2/3}(6) - (6x + 1)^{1/3}(-2x)}{(4 - x^2)^2}$

CHAPTER 2

Equations and Inequalities

For hundreds of years one of the main concerns in algebra has been the study of equations. ■ More recently inequalities have achieved the same degree of importance. ■ Both of these topics are used extensively in applications of mathematics. ■ In this chapter we shall discuss several methods for solving equations and inequalities.

2.1 Linear Equations

If x is a variable, the expressions

$$x + 3 = 0, \quad x^2 - 5 = 4x, \quad \text{and} \quad (x^2 - 9)\sqrt[3]{x + 1} = 0$$

are called **equations** in x. A number a is called a **solution,** or **root,** of an equation if a true statement is obtained when a is substituted for x. We also say that a **satisfies** the equation. For example, 5 is a solution of the equation $x^2 - 5 = 4x$ since substitution gives us

$$(5)^2 - 5 = 4(5), \quad \text{or} \quad 20 = 20,$$

which is a true statement. To **solve** an equation means to find all the solutions.

If every number in the domain of the variable is a solution of an equation, the equation is called an **identity.** An example of an identity is

$$\frac{1}{x^2 - 4} = \frac{1}{(x + 2)(x - 2)},$$

since this equation is true for every number in the domain of x. An equation is called a **conditional equation** if there are numbers that are *not* solutions.

The solutions of an equation depend on the type of numbers that are allowed. For example, if only rational numbers are available, then the equation $x^2 = 2$ has no solutions, since there is no rational number whose square is 2. However, if we allow *real* numbers, then the solutions are $-\sqrt{2}$ and $\sqrt{2}$. Similarly, the equation $x^2 = -1$ has no real solutions; however, we shall see later that this equation has solutions if *complex* numbers are used.

Two equations are said to be **equivalent** if they have exactly the same solutions. For example, the equations

$$x = 3, \quad x - 1 = 2, \quad 5x = 15, \quad \text{and} \quad 2x + 1 = 7$$

are all equivalent.

One method of solving an equation is to replace it with a list of equivalent equations, each in some sense simpler than the preceding one, and to end the list with an equation for which the solutions are obvious. This is accomplished by using properties of real numbers. For example, we may add the same expression to both sides of an equation without changing the solutions. Similarly, we may subtract the same expression from both sides of an equation. We can also multiply or divide both sides of an equation by an expression that represents a nonzero real number. The following example illustrates these remarks.

EXAMPLE 1 Solve the equation $6x - 7 = 2x + 5$.

SOLUTION The equations in the following list are equivalent.

$$6x - 7 = 2x + 5$$
$$(6x - 7) + 7 = (2x + 5) + 7$$
$$6x = 2x + 12$$
$$6x - 2x = (2x + 12) - 2x$$
$$4x = 12$$
$$\tfrac{1}{4}(4x) = \tfrac{1}{4}(12)$$
$$x = 3$$

Since the last equation has exactly one solution, 3, it follows that 3 is the only solution of the original equation, $6x - 7 = 2x + 5$. ■

In the process of finding solutions of equations, we may introduce errors because of incorrect manipulations or mistakes in arithmetic. Thus, it is advisable to check answers by substitution in the original equation. To apply a check in Example 1, we substitute 3 for x in the equation $6x - 7 = 2x + 5$, obtaining

$$6(3) - 7 = 2(3) + 5$$
$$18 - 7 = 6 + 5$$
$$11 = 11,$$

which is a true statement.

We sometimes say that an equation such as that given in Example 1 *has the solution x = 3*. As another illustration, the equation $x^2 = 4$ is said to have solutions $x = 2$ and $x = -2$.

If we multiply both sides of an equation by an expression that equals zero for some value of x, then the equation obtained may not be equivalent to the previous one. The following example illustrates a situation in which this may happen.

EXAMPLE 2   Solve the equation

$$\frac{3x}{x-2} = 1 + \frac{6}{x-2}.$$

SOLUTION   Multiplying both sides by $x - 2$ and simplifying leads to

$$\left(\frac{3x}{x-2}\right)(x-2) = (1)(x-2) + \left(\frac{6}{x-2}\right)(x-2)$$

$$3x = (x-2) + 6$$

$$3x = x + 4$$

$$2x = 4$$

$$x = 2.$$

Let us check to see whether 2 is a solution of the original equation. Substituting 2 for x, we obtain

$$\frac{3(2)}{2-2} = 1 + \frac{6}{2-2}, \quad \text{or} \quad \frac{6}{0} = 1 + \frac{6}{0}.$$

Since division by 0 is not permissible, 2 is not a solution. Actually, the given equation has no solutions, for we have shown that if the equation is true for some value of x, then that value must be 2. However, as we have seen, 2 is not a solution.   ∎

The preceding example indicates that *it is essential to check answers that are obtained after multiplying both sides of an equation by an expression that contains variables.*

In this section we shall concentrate on equations that can be written in the form

$$ax + b = 0$$

for some real numbers a and b, with $a \neq 0$. An equation of this type is called a **linear equation** in x. To solve the equation we first subtract b from both sides and then multiply by $1/a$ as follows:

$$ax + b = 0$$

$$ax = -b$$

$$x = \frac{-b}{a}$$

We have shown that the linear equation $ax + b = 0$ has precisely one solution, $x = -b/a$.

In Section 2.3 we shall consider equations having more than one solution.

EXAMPLE 3   Solve $(8x - 2)(3x + 4) = (4x + 3)(6x - 1)$.

SOLUTION   The following equations are equivalent:

$$(8x - 2)(3x + 4) = (4x + 3)(6x - 1)$$

$$24x^2 + 26x - 8 = 24x^2 + 14x - 3$$

$$26x - 8 = 14x - 3$$

$$26x - 14x = -3 + 8$$

$$12x = 5$$

$$x = \frac{5}{12}$$

Hence, the solution of the given equation is $\frac{5}{12}$.   ∎

EXAMPLE 4   Solve

$$\frac{3}{2x - 4} - \frac{5}{x + 3} = \frac{2}{x - 2}.$$

SOLUTION   If we rewrite the equation as

$$\frac{3}{2(x - 2)} - \frac{5}{x + 3} = \frac{2}{x - 2},$$

we see that the lcd of the three fractions is $2(x - 2)(x + 3)$. Multiplying both sides by this lcd gives us

$$\frac{3}{2(x - 2)} 2(x - 2)(x + 3) - \frac{5}{x + 3} 2(x - 2)(x + 3)$$

$$= \frac{2}{x - 2} 2(x - 2)(x + 3),$$

which reduces to

$$3(x + 3) - 10(x - 2) = 4(x + 3).$$

Simplifying, we obtain

$$3x + 9 - 10x + 20 = 4x + 12$$

$$3x - 10x - 4x = 12 - 9 - 20$$

$$-11x = -17$$

$$x = \tfrac{17}{11}.$$

Since we multiplied by an expression involving x, we must check this result in the given equation. Substituting for x, we have

$$\frac{3}{2(\frac{17}{11}) - 4} - \frac{5}{\frac{17}{11} + 3} = \frac{2}{\frac{17}{11} - 2}.$$

It can be shown that this reduces to $-\tfrac{22}{5} = -\tfrac{22}{5}$. Thus, the given equation has the solution $\tfrac{17}{11}$. ∎

Sometimes it is difficult to determine whether an equation is conditional or is an identity. An identity will often be indicated when, after applying properties of real numbers, an equation of the form $p = p$ is obtained, for some expression p. To illustrate, if we multiply both sides of the equation

$$\frac{x}{x^2 - 4} = \frac{x}{(x + 2)(x - 2)}$$

by $x^2 - 4$, we obtain $x = x$. This alerts us to the fact that we may have an identity on our hands; however, it does not prove anything. A standard method for verifying that an equation is an identity is to show, using properties of real numbers, that the expression that appears on one side of the equation can be transformed into the expression that appears on the other side. That is easy to do in the above illustration, since we know that $x^2 - 4 = (x + 2)(x - 2)$. To show that an equation is not an identity, we need only find one real number in the domain of the variable that fails to satisfy the equation.

Exercises 2.1

Solve the equations in Exercises 1–38.

1 $4x + 7 = 0$

2 $3x + 16 = 0$

3 $\sqrt{2}x - 5 = 0$

4 $\sqrt{3}x - 2 = 0$

5 $5x + 3 = 7x - 2$

6 $8x - 5 = 6x + 4$

7 $3(7y - 2) = 2(4y + 1)$

8 $2(9z + 2) - 5(z - 8) = 0$

9 $\frac{2}{3}t + 4 = 2 - \frac{5}{2}t$

10 $\frac{3}{4}u - 1 = 2 + \frac{1}{5}u$

11 $0.2(4 - 3x) + 0.3x = 2.6$

12 $0.7x - 1.2 = 0.3(2x - 1)$

13 $\dfrac{12 - 7w}{6} = \dfrac{2w + 1}{9}$

14 $\dfrac{3r + 2}{8} = 1 - \dfrac{r}{12}$

15 $\dfrac{18 - 5p}{3p + 2} = \dfrac{7}{3}$

16 $\dfrac{6}{5v - 2} = \dfrac{9}{7v + 3}$

17 $4 - \dfrac{3}{x} = 6 - \dfrac{5}{x}$

18 $\dfrac{5}{q} + \dfrac{2}{q} - \dfrac{10}{q} = 12$

19 $(6x - 5)^2 = (4x + 3)(9x - 2)$

20 $(x - 7)^2 - 4 = (x + 1)^2$

21 $(4x - 3)(2x + 3) - 8x(x - 4) = 0$

22 $(2x + 3)(3x - 2) = 6x^2 + 1$

23 $\dfrac{6s + 7}{4s - 1} = \dfrac{3s + 8}{2s - 4}$

24 $\dfrac{8t + 5}{10t - 7} = \dfrac{4t - 3}{5t + 7}$

25 $\dfrac{1}{3} + \dfrac{2}{6x + 3} = \dfrac{3}{2x + 1}$

26 $\dfrac{-3}{4x - 2} + \dfrac{2}{2x - 1} = \dfrac{7}{2}$

27 $\dfrac{5}{4a - 2} - \dfrac{1}{6a - 3} = \dfrac{4}{5}$

28 $\dfrac{8}{3b + 6} - \dfrac{1}{2b + 4} = \dfrac{3}{4}$

29 $\dfrac{1}{2x - 1} = \dfrac{3}{4x - 2}$

30 $\dfrac{6}{2x + 11} + 5 = 5$

31 $2 - \dfrac{5}{3x - 7} = 2$

32 $\dfrac{4}{5x + 2} - \dfrac{7}{15x + 6} = 0$

33 $\dfrac{7}{y^2 - 4} - \dfrac{4}{y + 2} = \dfrac{5}{y - 2}$

34 $\dfrac{4}{2u - 3} + \dfrac{10}{4u^2 - 9} = \dfrac{1}{2u + 3}$

35 $(x + 3)^3 - (3x - 1)^2 = x^3 + 4$

36 $(x - 1)^3 = (x + 1)^3 - 6x^2$

37 $\dfrac{9x}{3x - 1} = 2 + \dfrac{3}{3x - 1}$

38 $\dfrac{2x}{2x + 3} + \dfrac{6}{4x + 6} = 5$

Prove that the equations in Exercises 39–44 are identities.

39 $(3x + 1)^2 - 9x^2 = 6x + 1$

40 $(4x - 5)(3x + 2) + 7x = 12x^2 - 10$

41 $\dfrac{x^2 - 1}{x - 1} = x + 1$

42 $\dfrac{3x - 2}{x} = 3 - \dfrac{2}{x}$

43 $\dfrac{5x^2 + 2}{x} = \dfrac{2}{x} + 5x$

44 $\dfrac{9x^2 - 16}{3x + 4} = 3x - 4$

45 For what value of c is -3 a solution of the equation $3x + 1 - 5c = 2c + x - 10$?

46 For what value of b is 8 a solution of the equation $4x + 3b = 7$?

47 Determine values for a and b such that $\frac{5}{3}$ is a solution of the equation $ax + b = 0$. Are these the only possible values for a and b? Explain.

48 In each of the following, determine whether the two given equations are equivalent:

(a) $x^2 = 4, \; x = 2$

(b) $x = \sqrt{4}, \; x = 2$

(c) $2x = 4, \; x = 2$

49 (a) Find an equation in the following list that is not equivalent to the preceding equation.

$$x^2 - x - 2 = x^2 - 4$$
$$(x + 1)(x - 2) = (x + 2)(x - 2)$$
$$x + 1 = x + 2$$
$$1 = 2$$

(b) Find the solutions of the first equation in (a).

50 Find an equation in the following list that is not equivalent to the preceding equation.

$$x + 3 = 0$$
$$5x - 4x = -3$$
$$5x + 6 = 4x + 3$$
$$x^2 + 5x + 6 = x^2 + 4x + 3$$
$$(x + 3)(x + 1) = (x + 2)(x + 3)$$
$$x + 1 = x + 2$$
$$0 = 1$$

2.2 Applications

Formulas involving several variables occur in many applications of mathematics. Often it is necessary to solve for a specific variable in terms of the remaining variables that appear in the formula, as the next three examples illustrate.

EXAMPLE 1   If a sum of money P (the principal) is invested at a simple interest rate of r percent per year, then the interest I at the end of t years is given by $I = Prt$. Solve for r in terms of the remaining variables.

SOLUTION   We begin by writing

$$Prt = I.$$

To solve for r, we multiply both sides by $1/Pt$, obtaining

$$\left(\frac{1}{Pt}\right)Prt = \left(\frac{1}{Pt}\right)I.$$

It follows that
$$r = \frac{I}{Pt}. \qquad\blacksquare$$

EXAMPLE 2   The relationship between the temperature F on the Fahrenheit scale and the temperature C on the Celsius scale is expressed as follows:

$$C = \tfrac{5}{9}(F - 32).$$

Solve for F in terms of C.

SOLUTION   We may proceed as follows:

$$C = \tfrac{5}{9}(F - 32)$$
$$\tfrac{9}{5}C = F - 32$$
$$\tfrac{9}{5}C + 32 = F$$
$$F = \tfrac{9}{5}C + 32 \qquad\blacksquare$$

FIGURE 2.1

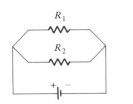

R_1

R_2

EXAMPLE 3 The formula $R = R_1R_2/(R_1 + R_2)$ is used in electrical theory to find the total resistance R when two resistances R_1 and R_2 are connected in parallel, as illustrated in Figure 2.1. Solve for R_1 in terms of R and R_2.

SOLUTION The following equations are equivalent.

$$R = \frac{R_1R_2}{R_1 + R_2}$$

$$(R_1 + R_2)R = (R_1 + R_2)\left(\frac{R_1R_2}{R_1 + R_2}\right)$$

$$R_1R + R_2R = R_1R_2$$

$$R_1R = R_1R_2 - R_2R$$

$$R_1R - R_1R_2 = -R_2R$$

$$R_1(R - R_2) = -R_2R$$

$$R_1 = \frac{-R_2R}{R - R_2}$$

An alternative form is $\quad R_1 = \dfrac{R_2R}{R_2 - R}.$ ∎

Often we can use equations to solve problems that occur in everyday life. Some problems are spoken, from one person to another. Others are stated using written words, as in textbooks. For this reason they are often called "word problems" by students and teachers of mathematics. They may also be referred to as "practical problems." We shall use the terminology "applied problem" for any problem that involves an application of mathematics to some other field.

Due to the unlimited variety of applied problems, it is difficult to state specific rules for finding solutions. However, it is possible to develop a general strategy for attacking such problems. The following guidelines may be helpful, provided the problem can be formulated in terms of an equation in one variable.

GUIDELINES: Solving Applied Problems

1 If the problem is stated in written words, read it carefully several times and think about the given facts, together with the unknown quantity that is to be found.

2 Introduce a letter to denote the unknown quantity. This is one of the most crucial steps in the solution! Phrases containing words such as "what," "find," "how much," "how far," or "when" should alert you to the unknown quantity.

3 If feasible, draw a picture and label it appropriately.

4 Make a list of known facts, together with any relationships that involve the unknown quantity. A relationship may often be described by means of an equation in which written statements, instead of letters or numbers, appear on either one or both sides of the equal sign.

5 After analyzing the list in step 4, and perhaps rereading the problem several more times, formulate an equation that describes precisely what is stated in words.

6 Solve the equation formulated in step 5.

7 Check the solutions obtained in step 6 by referring to the original statement of the problem. Carefully note whether the solution agrees with the stated conditions.

8 Don't become discouraged if you are unable to solve a given problem. It takes a great deal of effort and practice to become proficient in solving applied problems. Keep trying! ■ ■ ■

EXAMPLE 4 A student has test scores of 64 and 78. What score on a third test will give the student an average of 80?

SOLUTION We shall follow the Guidelines that precede this example. Reading the problem carefully, as suggested in step 1, we note that the unknown quantity is the score on the third test. Accordingly, as in step 2, we introduce a letter as follows:

$$x = \text{score on the third test}.$$

Drawing a picture, as mentioned in step 3, is inappropriate for this problem, so we go on to step 4 and look for relationships involving x. Since the average of the three scores is found by adding them and dividing by 3, we may write

$$\frac{64 + 78 + x}{3} = \text{average of the three scores 64, 78, and } x.$$

From the statement of the problem we see that

$$80 = \text{average desired}.$$

Thus, x must satisfy the equation

$$\frac{64 + 78 + x}{3} = 80.$$

The last equation is the one referred to in step 5 of the Guidelines. We next solve the equation (see step 6) as follows:

$$64 + 78 + x = 240$$
$$142 + x = 240$$
$$x = 240 - 142$$
$$x = 98.$$

Step 7 tells us to check the solution by referring to the original statement. If the three test scores are 64, 78, and 98, then the average is

$$\frac{64 + 78 + 98}{3} = \frac{240}{3} = 80.$$

Hence, a score of 98 on the third test will give the student an average of 80. Happily, we can ignore step 8. ∎

In the remaining examples we will not point out the explicit Guidelines that are used to arrive at solutions.

EXAMPLE 5 A store holding a clearance sale advertises that all prices have been discounted 20%. If a certain article is on sale for $28, what was its price before the sale?

SOLUTION We begin by noting that the unknown quantity is the presale price. It is convenient to arrange our work as follows, where the quantities are measured in dollars:

$$x = \text{presale price}$$
$$0.20x = \text{discount}$$
$$28 = \text{sale price}$$

The sale price is determined as follows:

$$(\text{presale price}) - (\text{discount}) = (\text{sale price}).$$

This leads to the equation

$$x - 0.20x = 28,$$

which we solve as follows:

$$0.8x = 28$$

$$x = \frac{28}{0.8} = 35.$$

Thus, the price before the sale was $35.

To check this answer we note that if a $35 article is discounted 20%, then the discount (in dollars) is $(0.20)(35) = 7$, and the selling price is $35 - 7$, or $28.   ∎

EXAMPLE 6   A man has $15,000 to invest. He plans to deposit part of it in a savings account paying 5% simple interest and the remainder in an investment fund yielding 8% simple interest. How much should he invest in each to obtain a 7% return on his money after one year?

SOLUTION   The simple interest formula $I = Prt$ was given in Example 1. In the present example, $t = 1$ and hence, the interest is given by $I = Pr$. If we let x denote the amount deposited in the savings account, then the remainder, $15,000 - x$, will be put into the investment fund. This leads to the following equalities:

$$x = \text{amount invested at } 5\%$$

$$15,000 - x = \text{amount invested at } 8\%$$

$$0.05x = \text{interest on } x \text{ dollars at } 5\%$$

$$0.08(15,000 - x) = \text{interest on } 15,000 - x \text{ dollars at } 8\%$$

$$0.07(15,000) = \text{total interest desired}.$$

Since the total interest must equal the combined interest from the two investments, we see that

$$0.05x + 0.08(15,000 - x) = 0.07(15,000)$$

$$0.05x + 1200 - 0.08x = 1050$$

$$-0.03x = -150$$

$$x = \frac{-150}{-0.03} = 5000.$$

Consequently, $5000 should be deposited in the savings account and $10,000 in the investment fund.

Checking, we see that if $5000 is placed in the savings account, then the interest obtained is $(0.05)(\$5000) = \$250$. If $10,000 is placed in the investment fund, then the interest is $(0.08)(\$10,000) = \$800$. Hence, the total interest is $1050, which is 7% of $15,000. ∎

In certain applications it is necessary to mix two substances to obtain a prescribed mixture. For such problems it is helpful to draw a picture, as illustrated in the next two examples.

EXAMPLE 7 A chemist has 10 ml of a solution that contains a 30% concentration of acid. How many ml of pure acid must be added in order to increase the concentration to 50%?

SOLUTION Since we wish to find the amount of pure acid to add, we let

$$x = \text{ml of pure acid to be added.}$$

Figure 2.2 is self-explanatory.

FIGURE 2.2

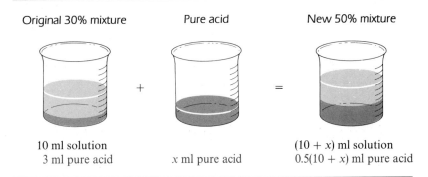

Original 30% mixture

Pure acid

New 50% mixture

+

=

10 ml solution
3 ml pure acid

x ml pure acid

$(10 + x)$ ml solution
$0.5(10 + x)$ ml pure acid

This leads to the equation

$$3 + x = 0.5(10 + x)$$

or, equivalently,

$$3 + x = 5 + 0.5x$$

$$0.5x = 2$$

$$x = \frac{2}{0.5} = 4.$$

Hence, 4 ml of the acid should be added to the original solution.

To check, we note that if 4 ml of acid are added to the given solution, then the new solution contains 14 ml, 7 of which are acid. This is the desired 50% concentration. ∎

EXAMPLE 8   A radiator contains 8 quarts of a mixture of water and antifreeze. If 40% of the mixture is antifreeze, how much of the mixture should be drained and replaced by pure antifreeze in order that the resultant mixture will contain 60% antifreeze?

SOLUTION   Let

$$x = \text{the number of quarts of mixture to be drained.}$$

Since there were 8 quarts in the original 40% mixture, we may picture the problem as shown in Figure 2.3.

FIGURE 2.3

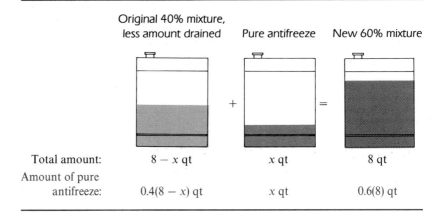

	Original 40% mixture, less amount drained	Pure antifreeze	New 60% mixture
Total amount:	$8 - x$ qt	x qt	8 qt
Amount of pure antifreeze:	$0.4(8 - x)$ qt	x qt	$0.6(8)$ qt

This gives us the equation

$$0.4(8 - x) + x = 0.6(8)$$

or, equivalently,

$$3.2 - 0.4x + x = 4.8$$

$$-0.4x + x = 4.8 - 3.2$$

$$0.6x = 1.6$$

$$x = \frac{1.6}{0.6} = \frac{16}{6} = \frac{8}{3}.$$

Thus, $\frac{8}{3}$ quarts should be drained from the original mixture.

To check, let us first note that the amount of antifreeze in the original 8-quart mixture was 0.4(8), or 3.2 quarts. In draining $\frac{8}{3}$ quarts of the original 40% mixture, we lose $0.4(\frac{8}{3})$ quarts of antifreeze, and hence, there remain $(3.2) - 0.4(\frac{8}{3})$ quarts of antifreeze. If we then add $\frac{8}{3}$ quarts of pure antifreeze, the amount of antifreeze in the final mixture is $(3.2) - 0.4(\frac{8}{3}) + \frac{8}{3}$ quarts. This reduces to 4.8, which is 60% of 8.   ∎

Many applied problems involve objects that move at a constant, or uniform, rate. If an object travels at a constant (or average) rate r, then the distance d covered in time t is given by $d = rt$. Of course, we assume that the units are properly chosen; that is, if r is in feet per second, then t is in seconds, and so on.

EXAMPLE 9   Two cities, A and B, are connected by means of a highway. A car leaves A at 1:00 P.M. and travels at a constant rate of 40 miles per hour toward B. Thirty minutes later, another car leaves A and travels toward B at a constant rate of 55 miles per hour. At what time will the second car overtake the first car?

SOLUTION   Let t denote the number of hours after 1:00 P.M. for the second car to overtake the first. Since the second car leaves A at 1:30 P.M., it has traveled $\frac{1}{2}$ hour less than the first. This leads to the following table.

	Rate (mph)	Hours traveled	Miles traveled
First car	40	t	$40t$
Second car	55	$t - \frac{1}{2}$	$55(t - \frac{1}{2})$

At time t the number of miles traveled by the two cars is equal; that is,

$$55\left(t - \frac{1}{2}\right) = 40t.$$

We now solve for t:

$$55t - \frac{55}{2} = 40t$$

$$15t = \frac{55}{2}$$

$$t = \frac{55}{30} = \frac{11}{6}.$$

Thus, $t = 1\frac{5}{6}$ hours, or equivalently, 1 hour and 50 minutes after 1:00 P.M. Consequently, the second car overtakes the first at 2:50 P.M.

To check our answer we note that at 2:50 P.M. the first car has traveled for $1\frac{5}{6}$ hours and its distance from A is $40(\frac{11}{6}) = \frac{220}{3}$ miles. At 2:50 P.M. the second car has traveled for $1\frac{1}{3}$ hours and is $55(\frac{4}{3}) = \frac{220}{3}$ miles from A. Hence, they are together at 2:50 P.M. ∎

EXAMPLE 10 Tom can do a certain job in 3 hours, and Bob can do the same job in 4 hours. If they work together, how long will it take them to do the job?

SOLUTION Let

$$t = \text{number of hours to do the job together.}$$

It is convenient to introduce the *part* of the job done in 1 hour as follows:

$$\frac{1}{3} = \text{part of job done by Tom in 1 hour}$$

$$\frac{1}{4} = \text{part of job done by Bob in 1 hour}$$

$$\frac{1}{t} = \text{part of job done working together in 1 hour.}$$

We have assumed that both Tom and Bob work at a steady pace throughout the job, and that there is no gain or loss in efficiency when they work together. It is customary to make such assumptions when a real-world situation is represented mathematically. A representation of this type is called a **mathematical model** for the situation. Since

$$\begin{pmatrix} \text{part done by} \\ \text{Tom in one hour} \end{pmatrix} + \begin{pmatrix} \text{part done by} \\ \text{Bob in one hour} \end{pmatrix} = \begin{pmatrix} \text{part done working} \\ \text{together in one hour} \end{pmatrix}$$

we obtain $\qquad \dfrac{1}{3} + \dfrac{1}{4} = \dfrac{1}{t}, \quad \text{or} \quad \dfrac{7}{12} = \dfrac{1}{t}.$

Solving for t gives us

$$7t = 12, \quad \text{or} \quad t = \frac{12}{7}.$$

Thus, by working together, Tom and Bob can do the job in 1 and $\frac{5}{7}$ hours, or approximately 1 hour 43 minutes. ∎

Exercises 2.2

The formulas in Exercises 1–24 occur in mathematics and its applications. Solve each for the indicated variable in terms of the remaining variables.

1 $I = Prt$ for P

2 $d = rt$ for t

3 $A = \frac{1}{2}bh$ for h

4 $C = 2\pi r$ for r

5 $P = 2l + 2w$ for w

6 $A = P + Prt$ for r

7 $ax + by + cz = d$ for z

8 $ax + by + c = 0$ for y

9 $R = \dfrac{E}{I}$ for I

10 $K = \dfrac{mv^2}{2g}$ for g

11 $V = \frac{1}{3}\pi r^2 h$ for h

12 $F = g\dfrac{m_1 m_2}{d^2}$ for m_1

13 $S = \dfrac{a}{1 - r}$ for r

14 $m = \dfrac{y_2 - y_1}{x_2 - x_1}$ for x_1

15 $\dfrac{1}{R} = \dfrac{1}{R_1} + \dfrac{1}{R_2} + \dfrac{1}{R_3}$ for R_2

16 $\dfrac{x}{a} + \dfrac{y}{b} = 1$ for y

17 $S = P + Prt$ for P

18 $F = \frac{9}{5}C + 32$ for C

19 $V = \frac{1}{3}\pi h^2(3r - h)$ for r

20 $s = \frac{1}{2}gt^2 + v_0 t$ for v_0

21 $S = \dfrac{a - rl}{l - r}$ for r

22 $S = a + (n - 1)d$ for n

23 $A = \frac{1}{2}(b_1 + b_2)h$ for b_1

24 $A = 2\pi r(r + h)$ for h

25 A student in an algebra course has test scores of 75, 82, 71, and 84. What score on the next test will raise the student's average to 80?

26 Going into a final exam, a student has test scores of 72, 80, 65, 78, and 60. If the final exam counts as $\frac{1}{3}$ of the final grade, what score must the student receive in order to end up with an average of 76?

27 The relationship between the temperature F on the Fahrenheit scale and the temperature C on the Celsius scale is given by $C = \frac{5}{9}(F - 32)$. Find the temperature at which the reading is the same on both scales.

28 Shown in the figure is a cross section of a two-story home for which the center height h of the second story has not yet been determined. Find h such that the second story will have the same cross-sectional area as the first story.

FIGURE FOR EXERCISE 28

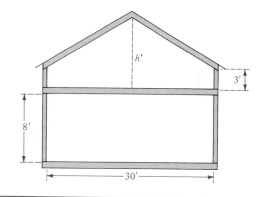

29 The prenatal growth of a fetus more than 12 weeks old can be approximated by the formula $L = 1.53t - 6.7$, where L is the length in cm and t is the age in weeks. Prenatal length can be determined by X-rays. Estimate the age of a fetus whose length is 28 cm.

30 Based on discus records from the Olympics, the winning distance can be approximated by $d = 175 + 1.75t$, where d is in feet and $t = 0$ corresponds to the year 1948.

(a) Predict the winning distance for the 1988 Olympics in Seoul, Korea.

(b) Estimate the year in which the winning toss should be about 260 feet.

31 A businesswoman wishes to invest $30,000 in two different funds that yield annual profits of 13% and $15\frac{1}{2}\%$, respectively. How much should she invest in each in order to realize a profit of $4350 after one year?

32 A college student has $3000 in two different savings accounts, paying simple interest at the rates of $5\frac{1}{2}\%$ and 6%, respectively. If the total yearly interest is $174.10, how much is deposited in each account?

33 Two boys who are 224 meters apart start walking toward each other at the same time. If they walk at rates of 1.5 and 2 meters per second, respectively, when will they meet? How far will each have walked?

34 A runner starts at the beginning of a runner's path and runs at a constant rate of 6 mph. Five minutes later a second runner begins at the same point, running at a rate of 8 mph and following the same course. How long will it take the second runner to catch up with the first?

35 After playing 100 games, a major league baseball team has a record of 0.650. The team then wins only 50% of its games for the remainder of the season. After how many additional games will the team record be 0.600?

36 A bullet is fired horizontally at a target, and the sound of its impact is heard 1.5 seconds later. If the speed of the bullet is 3300 feet per second and the speed of sound is 1100 feet per second, how far away is the target?

37 A ring that weighs 80 grams is made of silver and gold. By measuring the displacement of the ring in water, it has been determined that the ring has a volume of 5 cm^3. Gold weighs 19.3 g/cm^3 and silver weighs 10.5 g/cm^3. How many grams of gold does the ring contain?

38 A pharmacist is to prepare 15 ml of special eye drops for a glaucoma patient. The eye-drop solution must have a 2% active ingredient, but the pharmacist only has a 10% and a 1% solution in stock. How much of each type of solution should be used to fill the prescription?

39 In a certain medical test designed to measure carbohydrate tolerance, an adult drinks 7 ounces of a 30% glucose solution. When the same test is administered to a child, the glucose concentration must be decreased. How much 30% glucose solution and how much water are used to prepare 7 ounces of a 20% glucose solution?

40 Theophyline, an asthma medicine, is to be prepared from an elixir with a drug concentration of 5 mg/ml and a cherry-flavored syrup that is to be added to hide the taste of the drug. How much of each must be used to prepare a 100-ml solution with a drug concentration of 2 mg/ml?

41 British sterling silver is a copper-silver alloy that is 7.5% copper by weight. How many grams of pure copper and how many grams of British sterling silver should be used to prepare 200 grams of a copper-silver alloy that is 10% copper by weight?

42 Six hundred people attended the premiere of the latest horror film. Adult tickets cost $5, while children were admitted for $2. If box office receipts totaled $2400, how many children attended the premiere?

43 A large grain silo is to be constructed in the shape of a circular cylinder with a hemisphere attached to the top (see figure). The diameter of the silo is to be 30 ft, but the height is yet to be determined. Find the total height of the silo that will result in a capacity of 11,250π ft^3.

FIGURE FOR EXERCISE 43

44 The cross section of a drainage ditch is an isosceles trapezoid with a small base of 3 feet and a height of 1 foot (see figure). Determine a width for the larger base that would give the ditch a cross-sectional area of 5 ft^2.

FIGURE FOR EXERCISE 44

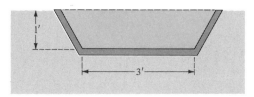

45 At 6 A.M. a snowplow, traveling at a steady rate, begins to clear a highway leading out of town. At 8 A.M. an automobile starts down the highway at a speed of 30 mph and approaches the plow 30 minutes later. How fast is the snowplow traveling?

46 An automobile 20 feet long overtakes a truck that is 40 feet long and traveling at 50 mph. At what constant speed must the automobile travel in order to pass the truck in 5 seconds?

47 It takes a boy 90 minutes to mow his father's yard, but his sister can do it in 60 minutes. How long would it take

them to mow the lawn if they worked together, using two lawnmowers?

48 Using water from one hose, a swimming pool can be filled in 8 hours. A second, larger hose used alone can fill the pool in 5 hours. How long would it take to fill the pool if both hoses were used simultaneously?

49 A salesperson purchased an automobile that was advertised as averaging 25 mi/gal in the city and 40 mi/gal on the highway. A recent sales trip that covered 1800 miles required 51 gal of gasoline. Assuming that the advertised mileage estimates were correct, how many miles were driven in the city?

50 Two boys own two-way radios that have a maximum range of 2 miles. One of the boys leaves a certain point at 1:00 P.M., walking due north at a rate of 4 mph. The second boy leaves the same point at 1:15 P.M., traveling due south at 6 mph. When will they be unable to communicate with one another?

51 A farmer plans to use 180 feet of fencing to enclose a rectangular region, using part of a straight river bank instead of fencing as one side of the rectangle (see figure). Find the area of the region if the length of the side parallel to the river bank is

(a) twice the length of an adjacent side;

(b) one-half the length of an adjacent side;

(c) the same as the length of an adjacent side.

FIGURE FOR EXERCISE 51

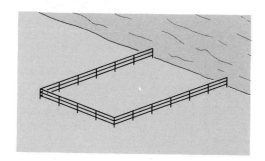

52 A consulting engineer charges $60 per hour, and his assistant receives $20 per hour. For a certain job a customer received a bill for $580. If the assistant worked 5 hours less than the engineer, how much time did each spend on the job?

53 It takes a boy 45 minutes to deliver the newspapers on his route; however, if his sister helps, it takes only 20 minutes. How long would it take his sister to deliver the newspapers by herself?

54 A water tank can be emptied by using one pump for 5 hours. A second, smaller pump can empty the tank in 8 hours. If the larger pump is started at 1:00 P.M., at what time should the smaller pump be started so that the tank will be emptied at 5:00 P.M.?

55 A boy can row a boat at a constant rate of 5 mph in still water. If he rows upstream for 15 minutes, and then rows downstream, returning to his starting point in another 12 minutes, find (a) the rate of the current; and (b) the total distance traveled.

56 A woman begins jogging at 3:00 P.M., running due north at a 6-minute-mile pace. Later she reverses direction and runs due south at a 7-minute-mile pace. If she returns to her starting point at 3:45 P.M., find the total number of miles she ran.

57 The Intelligence Quotient (IQ) is determined by multiplying the quotient of a person's mental age and chronological age by 100.

(a) Find the IQ of a 12-year-old child whose mental age is 15.

(b) Find the mental age of a person 15 years old whose IQ is 140.

58 Ohm's Law in electrical theory states that $I = E/R$, where I is the current (in amperes), E is the electromotive force (in volts), and R is the resistance (in ohms). In a certain circuit $E = 110$ and $R = 50$. If E and R are to be changed by the same numerical amount, what change in them will cause I to double?

59 The cost of installing insulation in a particular two-bedroom home is $1080. Present monthly heating costs average $60, but the insulation is expected to reduce heating costs by 10%. How many months will it take to recover the cost of the insulation?

60 A workman's basic hourly wage is $10 but he receives "time-and-a-half" for any hours worked in excess of 40 per week. If his paycheck for the week is $600, how many hours of overtime did he work?

2.3 Quadratic Equations

DEFINITION

> A **quadratic equation** is an equation of the form
>
> $$ax^2 + bx + c = 0$$
>
> where a, b, and c are real numbers with $a \neq 0$.

One technique for solving quadratic equations uses the fact that if p and q represent real numbers and $pq = 0$, then either $p = 0$ or $q = 0$. It follows that if $ax^2 + bx + c$ can be expressed as a product of two first-degree polynomials, then solutions can be found by setting each factor equal to 0, as illustrated in the next two examples.

EXAMPLE 1   Solve the equation $3x^2 = 10 - x$.

SOLUTION   First we write the equation in the form $ax^2 + bx + c = 0$, obtaining

$$3x^2 + x - 10 = 0.$$

The left side of this equation may be factored as follows:

$$(3x - 5)(x + 2) = 0$$

Setting each factor equal to 0 gives us

$$3x - 5 = 0, \qquad x + 2 = 0.$$

The solutions of these linear equations are

$$x = \tfrac{5}{3}, \qquad x = -2.$$

The fact that $\tfrac{5}{3}$ and -2 are roots of $3x^2 = 10 - x$ may be checked by substitution.   ■

The method used for solving the equation in Example 1 is called the **method of factoring.**

EXAMPLE 2 Solve $x^2 + 16 = 8x$.

SOLUTION Again we begin with an equivalent equation that has all nonzero terms on one side:

$$x^2 - 8x + 16 = 0$$

Factoring gives us

$$(x - 4)^2 = 0 \quad \text{or} \quad (x - 4)(x - 4) = 0.$$

Setting each factor equal to zero, we obtain $x - 4 = 0$ and $x - 4 = 0$. Hence, the given equation has one solution, $x = 4$. ∎

Since $x - 4$ appears as a factor twice in the previous solution, we call 4 a **double root,** or **root of multiplicity 2,** of the equation $x^2 + 16 = 8x$.

If a quadratic equation has the form $x^2 = d$, for some $d > 0$, then $x^2 - d = 0$, or equivalently,

$$(x + \sqrt{d})(x - \sqrt{d}) = 0.$$

Setting each factor equal to zero gives us the solutions $-\sqrt{d}$ and $\sqrt{d}$. We frequently use $\pm\sqrt{d}$ (read "*plus or minus $\sqrt{d}$*") as an abbreviation for these two solutions. Thus, for $d > 0$, we have proved the following:

THEOREM

$$\text{If } x^2 = d, \quad \text{then} \quad x = \pm\sqrt{d}.$$

The process of solving $x^2 = d$ as indicated in the box may be referred to by the phrase "take the square root of both sides of the equation." Note that, in so doing, we obtain a positive and a negative square root, not just the principal square root (defined in Section 1.3).

EXAMPLE 3 Solve the equation $x^2 = 5$.

SOLUTION Taking the square root of both sides gives us $x = \pm\sqrt{5}$. Thus, the solutions are $\sqrt{5}$ and $-\sqrt{5}$. ∎

In the work to follow it will be necessary to change an expression of the form $x^2 + kx$ to $(x + d)^2$, where k and d are real numbers. This procedure is called **completing the square** for $x^2 + kx$, and it may be accomplished by adding $(k/2)^2$ as follows:

$$x^2 + kx + \left(\frac{k}{2}\right)^2 = \left(x + \frac{k}{2}\right)^2.$$

In words, we *add the square of half the coefficient of* x *to* $x^2 + kx$. Let us restate this fact for reference.

COMPLETING THE SQUARE

> To complete the square for $x^2 + kx$, add $\left(\frac{k}{2}\right)^2$.

EXAMPLE 4 Complete the square for $x^2 + 5x$.

SOLUTION The square of half the coefficient of x is $(\frac{5}{2})^2$. Thus,

$$x^2 + 5x + (\tfrac{5}{2})^2 = (x + \tfrac{5}{2})^2. \qquad \blacksquare$$

Let us now consider any quadratic equation

$$ax^2 + bx + c = 0.$$

Dividing both sides by a gives us

$$x^2 + \frac{b}{a}x + \frac{c}{a} = 0,$$

or equivalently,

$$x^2 + \frac{b}{a}x = -\frac{c}{a}.$$

We next complete the square for $x^2 + (b/a)x$ by adding the square of half the coefficient of x. Of course, *to maintain equality we must add to both sides of the equation* as follows:

$$x^2 + \frac{b}{a}x + \left(\frac{b}{2a}\right)^2 = \left(\frac{b}{2a}\right)^2 - \frac{c}{a}.$$

We may now write

$$\left(x + \frac{b}{2a}\right)^2 = \frac{b^2 - 4ac}{4a^2}.$$

If $b^2 - 4ac \geq 0$, then

$$x + \frac{b}{2a} = \pm \sqrt{\frac{b^2 - 4ac}{4a^2}}$$

or

$$x = -\frac{b}{2a} \pm \sqrt{\frac{b^2 - 4ac}{4a^2}}.$$

The radical on the right-hand side of the last equation may be written

$$\pm \sqrt{\frac{b^2 - 4ac}{4a^2}} = \pm \frac{\sqrt{b^2 - 4ac}}{\sqrt{(2a)^2}} = \pm \frac{\sqrt{b^2 - 4ac}}{|2a|}.$$

Since $|2a| = 2a$ if $a > 0$, or $|2a| = -2a$ if $a < 0$, we see that in all cases

$$x = -\frac{b}{2a} \pm \frac{\sqrt{b^2 - 4ac}}{2a}.$$

We have shown that if the quadratic equation $ax^2 + bx + c = 0$ has roots, then they are given by the two numbers in the last formula. Moreover, it can be shown by direct substitution that the two numbers do satisfy the equation. This gives us the following:

QUADRATIC FORMULA

If $a \neq 0$, then the roots of the equation $ax^2 + bx + c = 0$ are given by

$$x = \frac{-b \pm \sqrt{b^2 - 4ac}}{2a}.$$

The number $b^2 - 4ac$, which appears under the radical sign in the quadratic formula, is called the **discriminant** of the quadratic equation.

The discriminant can be used to determine the nature of the roots of the equation as follows:

> (i) If $b^2 - 4ac > 0$, the equation has two real and unequal roots.
> (ii) If $b^2 - 4ac = 0$, the equation has one root of multiplicity 2.
> (iii) If $b^2 - 4ac < 0$, the equation has no real roots.

EXAMPLE 5 Solve $4x^2 + x - 3 = 0$.

SOLUTION Letting $a = 4$, $b = 1$, and $c = -3$ in the Quadratic Formula, we obtain

$$x = \frac{-1 \pm \sqrt{1 - 4(4)(-3)}}{2(4)}$$

$$= \frac{-1 \pm \sqrt{49}}{8}$$

$$= \frac{-1 \pm 7}{8}.$$

Hence, the solutions are

$$x = \frac{-1 + 7}{8} = \frac{3}{4} \quad \text{and} \quad x = \frac{-1 - 7}{8} = -1. \qquad \blacksquare$$

Example 5 can also be solved by factoring. Writing $(4x - 3)(x + 1) = 0$ and setting each factor equal to zero leads to the solutions $x = \frac{3}{4}$ and $x = -1$.

EXAMPLE 6 Solve $2x(3 - x) = 3$.

SOLUTION To use the Quadratic Formula it is necessary to write the equation in the form $ax^2 + bx + c = 0$. The following equations are equivalent:

$$2x(3 - x) = 3$$

$$6x - 2x^2 = 3$$

$$-2x^2 + 6x - 3 = 0$$

$$2x^2 - 6x + 3 = 0$$

We now let $a = 2, b = -6$, and $c = 3$ in the Quadratic Formula, obtaining

$$x = \frac{6 \pm \sqrt{(-6)^2 - 4(2)(3)}}{2(2)}$$

$$= \frac{6 \pm \sqrt{12}}{4}$$

$$= \frac{6 \pm 2\sqrt{3}}{4}$$

$$= \frac{3 \pm \sqrt{3}}{2}.$$

Hence, the solutions are $(3 + \sqrt{3})/2$ and $(3 - \sqrt{3})/2$. ∎

The following example illustrates the case of a double root.

EXAMPLE 7 Solve the equation $9x^2 - 30x + 25 = 0$.

SOLUTION Letting $a = 9, b = -30, c = 25$ in the Quadratic Formula, we obtain

$$x = \frac{30 \pm \sqrt{(-30)^2 - 4(9)(25)}}{2(9)}$$

$$= \frac{30 \pm \sqrt{900 - 900}}{18}$$

$$= \frac{30 \pm 0}{18} = \frac{5}{3}.$$

Consequently, the only solution is $\frac{5}{3}$. ∎

There are many applied problems that lead to quadratic equations. One is illustrated in the following example.

EXAMPLE 8 A box with a square base and no top is to be made from a square piece of tin by cutting out 3-inch squares from each corner and folding up the sides. If the box is to hold 48 cubic inches, what size piece of tin should be used?

FIGURE 2.4

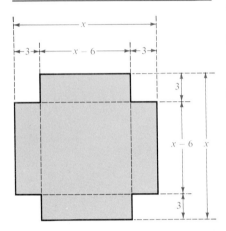

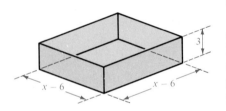

SOLUTION We begin by drawing the picture in Figure 2.4, where x denotes the length of the side of the piece of tin.

Since the area of the base is $(x - 6)^2$ and the height is 3, we obtain the following fact:

$$\text{Volume of box} = 3(x - 6)^2.$$

Since the box is to hold 48 cubic inches,

$$3(x - 6)^2 = 48.$$

We may solve for x as follows:

$$(x - 6)^2 = 16$$
$$x - 6 = \pm 4$$
$$x = 6 \pm 4.$$

Consequently, either $x = 10$ or $x = 2$.

Let us now check each of these numbers. Referring to Figure 2.4, we see that 2 is unacceptable, since no box is possible in this case. (Why?) However, if we begin with a 10-inch square of tin, cut out 3-inch corners, and fold, we obtain a box having dimensions 4 inches, 4 inches, and 3 inches. The box has the desired volume of 48 cubic inches. Thus, 10 inches is the answer to the problem. ∎

As illustrated in Example 8, even though an equation is formulated correctly, it is possible, owing to the physical nature of a given problem, to arrive at meaningless solutions. These solutions should be discarded. For example, we would not accept the answer -7 years for the age of an individual, nor $\sqrt{50}$ for the number of automobiles in a parking lot.

Exercises 2.3

Solve the equations in Exercises 1–10 by factoring.

1 $6x^2 + 11x - 10 = 0$

2 $15x^2 + x - 6 = 0$

3 $20x^2 + 7 = 33x$

4 $4x^2 + 16x + 15 = 0$

5 $4y^2 + 29y + 30 = 0$

6 $z(8z + 19) = 27$

7 $4t(t - 6) = -9$

8 $9u^2 = 30u - 25$

9 $54r^2 - 9r - 30 = 0$

10 $60s^2 - 85s - 35 = 0$

Use the Quadratic Formula to solve the equations in Exercises 11–24.

11 $2x^2 - x - 3 = 0$

12 $3x^2 = 2(x + 4)$

13 $6 = u(u + 2)$

14 $v^2 + 3v - 5 = 0$

15 $2x^2 - 4x - 5 = 0$

16 $6x - 2 = 3x^2$

17 $4y^2 = 5(4y - 5)$

18 $9t^2 + 6t + 1 = 0$

19 $\frac{5}{3}s^2 + 3s + 1 = 0$

20 $\frac{3}{2}z^2 - 4z - 1 = 0$

21 $\dfrac{x+1}{3x+2} = \dfrac{x-2}{2x-3}$

22 $\dfrac{5}{w^2} - \dfrac{10}{w} + 2 = 0$

23 $9x^2 = -4x$

24 $16x^2 - 9 = 0$

25 Given the equation $4x^2 - 4xy + 1 - y^2 = 0$, use the Quadratic Formula to solve for
 (a) x in terms of y;   (b) y in terms of x.

26 Given the equation $2x^2 - xy = 3y^2 + 1$, use the Quadratic Formula to solve for
 (a) x in terms of y;
 (b) y in terms of x.

The formulas in Exercises 27–34 occur in mathematics and its applications. Solve for the indicated variable in terms of the remaining variables (all letters denote positive numbers).

27 $V = \frac{1}{3}\pi r^2 h$ for r

28 $K = \frac{1}{2}mv^2$ for v

29 $F = g\dfrac{m_1 m_2}{d^2}$ for d

30 $V = \frac{4}{3}\pi a^2 b$ for a

31 $s = \frac{1}{2}gt^2 + v_0 t$ for t

32 $A = 2\pi r(r + h)$ for r

33 $\dfrac{x^2}{a^2} - \dfrac{y^2}{b^2} = 1$ for y

34 $s = \sqrt{(r_1 - r_2)^2 + h^2}$ for r_2

35 The diameter of a circle is 10 cm. What change in the radius will decrease the area by 16π cm^2?

36 The hypotenuse of a right triangle is 5 cm long. What are the lengths of the two legs if their sum is 6 cm?

37 A rectangular plot of ground having dimensions 26 feet by 30 feet is surrounded by a walk of uniform width. If the area of the walk is 240 ft^2, what is its width?

38 A manufacturer of tin cans wishes to construct a cylindrical can of height 20 cm and of capacity 3000 cm^3. Find the inner radius of the can.

39 An airplane flying north at 200 mph passed over a point on the ground at 2:00 P.M. Another airplane at the same altitude passed over the point at 2:30 P.M., flying east at 400 mph (see figure).

 (a) If t denotes the time (in hours) after 2:30 P.M., express the distance d between the planes in terms of t.

 (b) At what time were the airplanes 500 miles apart?

FIGURE FOR EXERCISE 39

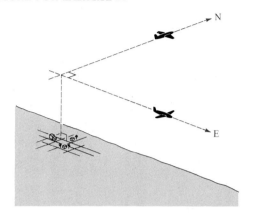

40 A box with an open top is to be constructed by cutting out 3-inch squares from a rectangular sheet of tin whose length is twice its width. What size sheet will produce a box having a volume of 60 in^3? (Compare Example 8.)

41 A piece of wire 100 inches long is cut into two pieces, and then each piece is bent into the shape of a square. If the sum of the enclosed areas is 397 in^2, find the lengths of each piece of wire.

42 A man plans to use 6 yd^3 of concrete to pour a rectangular slab for a patio. If the length of the patio is to be twice the width, and the thickness of the slab 3 inches, find the dimensions of the patio.

43 A closed cylindrical oil drum of height 4 feet is to be constructed so that the total surface area is 10π ft^2. Find the diameter of the drum.

44 When a popular brand of video disc player is priced at $300 per unit, a stereo store sells 15 units per week. Each time the price is reduced by $10, however, the sales increase by 2 per week. What selling price will result in weekly revenues of $7000?

45 Two boys with two-way radios leave the same point at 9:00 A.M., one walking due south at 4 mph and the other due west at 3 mph. How long can they communicate with one another if each radio has a maximum range of 2 miles?

46 A farmer plans to enclose a rectangular region, using part of his barn for one side and fencing for the other three sides. If he wants the side parallel to the barn to be twice the length of an adjacent side, and the area of the region to be 128 ft^2, how many feet of fencing should he purchase?

47 The speed of the current in a stream is 5 mph. It takes a girl 30 minutes longer to paddle a canoe 1.2 miles upstream than the same distance downstream. What is her rate in still water?

48 Water from two hoses can fill a swimming pool in 4 hours. If each hose is used separately, it takes 6 hours longer for the smaller hose to fill the pool than the larger hose. How long would it take to fill the pool if only the smaller hose were used?

49 A 24 × 36 inch sheet of paper is to be used for a poster, with the smaller side at the bottom. The margins at the sides and top are to have the same width, and the bottom margin is to be twice as wide as the other margins. Find the width of the margins if the printed area is to be 661.5 in^2.

50 A company sells running shoes to dealers at a rate of $20 per pair if less than 50 pairs are ordered. If 50 or more pairs are ordered (up to 600), the price per pair is reduced at a rate of 2 cents times the number ordered. How many pairs can a dealer purchase for $4200?

51 A boy throws a baseball straight upward with an initial speed of 64 feet per second. The number of feet s above the ground after t seconds is given by $s = -16t^2 + 64t$.

(a) When will the baseball be 48 feet above the ground?

(b) When will it hit the ground?

(c) What is its maximum height?

52 If a stone is thrown straight upward from a height of s_0 feet above the ground with an initial speed of v_0 feet per second, then its height s above the ground at time t is given by $s = -16t^2 + v_0 t + s_0$. Solve for t in terms of s, v_0, and s_0. When will the stone hit the ground? Express your answer in terms of v_0 and s_0.

53 The distance that a car travels between the time the driver makes the decision to hit the brakes and the time the car actually stops is called the *braking distance*. If a car is traveling v mph, the braking distance d (in feet) is approximated by $d = v + (v^2/20)$.

(a) Find the braking distance when v is 55 mph.

(b) A large cow stands peacefully in the middle of the road. If you begin to react when you are 120 feet from the cow, how fast can you be going and still avoid hitting the cow?

54 A boy drops a rock off a cliff and into the ocean below. If he hears the splash 4 seconds later, how high is the cliff? (*Hint:* Use the fact that the rock travels $16t^2$ feet after t seconds, and that the speed of sound is 1100 feet per second.)

55 The temperature T (in °C) at which water boils is related to the elevation h (in meters above sea level) by the formula

$$h = 1000(100 - T) + 580(100 - T)^2.$$

(a) At what elevation does water boil at a temperature of 98 °C?

(b) The elevation of Mt. Everest is 29,000 feet (or 8840 meters). At what temperature will water boil at the top of this mountain? (*Hint:* Let $x = 100 - T$ and use the Quadratic Formula.)

56 A square vegetable garden is to be enclosed with a fence. If the fence costs $1 per foot, and if it costs 50 cents per ft^2 to prepare the garden, determine the size of the garden that can be set up for $120.

57 In a round-robin softball tournament, each pair of teams meets once. Four games can be played each day, and the tournament organizers have rented the field for one week. How many teams should be invited? (*Hint:* The number of pairings of n teams is $n(n - 1)/2$.)

58 A *diagonal* of a polygon is a line segment joining any two nonadjacent vertices. It can be shown that the number of diagonals in a polygon with n sides is $n(n - 3)/2$. How many sides must a polygon have if the number of diagonals is 35? Is there a polygon with 100 diagonals?

59 If r_1 and r_2 are the two roots of $ax^2 + bx + c = 0$, show that $r_1 + r_2 = -b/a$ and $r_1 r_2 = c/a$.

60 Prove that the roots of $ax^2 + bx + c = 0$ are numerically equal but opposite in sign to the roots of the equation $ax^2 - bx + c = 0$.

Calculator Exercises 2.3

Approximate the solutions of the following equations to the nearest thousandth.

1 $47x^2 + 183x + 29 = 0$

2 $912x^2 - 3174x - 713 = 0$

3 $2.13x^2 + 6.04x - 5.71 = 0$

4 $6143x^2 + 5417x - 1439 = 0$

2.4 Complex Numbers

One of the reasons that real numbers are important is that we need them to find solutions of equations. To illustrate, the nonnegative integers 0, 1, 2, 3, ... do not contain solutions of equations such as $x + 5 = 0$. To remedy this defect, we need the system of integers, which also contains negatives $-1, -2, -3, \ldots$. In this expanded number system we find the solution -5 for the equation $x + 5 = 0$.

Similarly, to solve the equation $3x + 5 = 0$, we must enlarge the set of integers to the rational numbers, thereby obtaining the solution $x = -\frac{5}{3}$.

The set of rational numbers is still not large enough to solve every equation, for example, $x^2 = 5$ has no rational solutions. Thus, we must again expand our number system to include irrational numbers, such as $\sqrt{5}$ and $-\sqrt{5}$. This leads us to the real number system $\mathbb{R}$, which contains all rational and irrational numbers.

There is one more defect to overcome. Since squares of real numbers are never negative, $\mathbb{R}$ does not contain solutions of equations of the form $x^2 = -5$. To remedy this situation, we shall invent a larger number system $\mathbb{C}$, called the **complex number system,** which contains $\mathbb{R}$ and also contains numbers whose squares are negative.

We begin by introducing the **imaginary unit,** denoted by the letter i, which has the following property:

$$i = \sqrt{-1}, \quad \text{or} \quad i^2 = -1.$$

Obviously, i is not a real number. It is a new mathematical entity that will enable us to obtain the number system $\mathbb{C}$, which contains solutions of every algebraic equation.

Since i, together with $\mathbb{R}$, is to be contained in $\mathbb{C}$, it is necessary to consider products of the form bi, where b is a real number, and also expressions of the form $a + bi$, where both a and b are real numbers. This is the motivation for the next definition.

DEFINITION

> A **complex number** is an expression of the form $a + bi$, where a and b are real numbers and $i^2 = -1$.

Equality, addition, and multiplication of two complex numbers $a + bi$ and $c + di$ are defined as follows:

$$a + bi = c + di \quad \text{if and only if} \quad a = c \quad \text{and} \quad b = d.$$

$$(a + bi) + (c + di) = (a + c) + (b + d)i$$

$$(a + bi)(c + di) = (ac - bd) + (ad + bc)i$$

It is unnecessary to memorize the preceding definitions of addition and multiplication. Indeed, when working with complex numbers, *we may treat all symbols as though they represented real numbers, with exactly one exception: wherever the symbol i^2 appears, it may be replaced by -1*. Note that if we use this technique for multiplication, we obtain

$$\begin{aligned}
(a + bi)(c + di) &= (a + bi)c + (a + bi)(di) \\
&= ac + (bi)c + a(di) + (bi)(di) \\
&= ac + (bc)i + (ad)i + (bd)(i^2) \\
&= ac + (bc)i + (ad)i + (bd)(-1) \\
&= ac + (bd)(-1) + (ad)i + (bc)i \\
&= (ac - bd) + (ad + bc)i
\end{aligned}$$

This agrees with our definition of multiplication.

EXAMPLE 1 Write each expression in the form $a + bi$, where a and b are real numbers.

(a) $(3 + 4i) + (2 + 5i)$ (b) $(3 + 4i)(2 + 5i)$

SOLUTION

(a)
$$(3 + 4i) + (2 + 5i) = (3 + 2) + (4 + 5)i = 5 + 9i$$

(b)
$$\begin{aligned}
(3 + 4i)(2 + 5i) &= (3 + 4i)2 + (3 + 4i)(5i) \\
&= 6 + 8i + 15i + 20i^2 \\
&= 6 + 20(-1) + 23i \\
&= -14 + 23i
\end{aligned}$$
∎

The set $\mathbb{R}$ of real numbers may be identified with complex numbers of the form $a + 0i$. It is also convenient to denote the complex number $0 + bi$ by bi. Using these conventions we see that

$$(a + 0i) + (0 + bi) = (a + 0) + (0 + b)i = a + bi.$$

Thus, $a + bi$ may be regarded as the sum of two complex numbers a and bi (that is, $a + 0i$ and $0 + bi$).

The **identity element** relative to addition is 0 (or, equivalently, $0 + 0i$); that is,

$$(a + bi) + 0 = a + bi$$

for every complex number $a + bi$.

If $(-a) + (-b)i$ is added to $a + bi$, we obtain 0. This implies that $(-a) + (-b)i$ is the **additive inverse** of $a + bi$; that is,

$$-(a + bi) = (-a) + (-b)i.$$

Subtraction of complex numbers is defined using additive inverses:

$$(a + bi) - (c + di) = (a + bi) + \left[-(c + di)\right].$$

Since $-(c + di) = (-c) + (-d)i$, it follows that

$$(a + bi) - (c + di) = (a - c) + (b - d)i.$$

If c, d, and k are real numbers, then,

$$k(c + di) = (k + 0i)(c + di) = (kc - 0d) + (kd + 0c)i;$$

that is,

$$k(c + di) = kc + (kd)i.$$

One illustration of this formula is

$$3(5 + 2i) = 15 + 6i.$$

If, as in the next example, we are asked to write an expression in the form $a + bi$, we shall also accept the form $a - di$, since $a - di = a + (-d)i$.

EXAMPLE 2   Write each expression in the form $a + bi$.

(a) $4(2 + 5i) - (3 - 4i)$ (b) $(4 - 3i)(2 + i)$

(c) $i(3 - 2i)^2$ (d) i^{51}

SOLUTION

(a) $4(2 + 5i) - (3 - 4i) = 8 + 20i - 3 + 4i = 5 + 24i$

(b) $(4 - 3i)(2 + i) = 8 - 6i + 4i - 3i^2 = 11 - 2i$

(c) $i(3 - 2i)^2 = i(9 - 12i + 4i^2) = i(5 - 12i) = 5i - 12i^2 = 12 + 5i$

(d) Taking successive powers of i, we obtain $i^1 = i$, $i^2 = -1$, $i^3 = -i$, $i^4 = 1$, and then the cycle starts over: $i^5 = i$, $i^6 = i^2 = -1$, etc. In particular, $i^{51} = i^{48}i^3 = (i^4)^{12}i^3 = (1)^{12}i^3 = i^3 = -i$.   ■

The complex number $a - bi$ is called the **conjugate** of the complex number $a + bi$. Since

$$(a + bi) + (a - bi) = 2a$$

and $$(a + bi)(a - bi) = a^2 + b^2,$$

we see that *the sum and product of a complex number and its conjugate are real numbers.* Conjugates are useful for finding the multiplicative inverse $1/(a + bi)$ of $a + bi$, or for simplifying the quotient $(a + bi)/(c + di)$ of two complex numbers, as illustrated in the next example.

EXAMPLE 3   Express each fraction in the form $a + bi$.

(a) $\dfrac{1}{9 + 2i}$ (b) $\dfrac{7 - i}{3 - 5i}$

SOLUTION   We multiply numerator and denominator by the conjugate of the denominator, as follows:

(a) $$\frac{1}{9 + 2i} = \frac{1}{9 + 2i} \cdot \frac{9 - 2i}{9 - 2i} = \frac{9 - 2i}{81 + 4} = \frac{9}{85} - \frac{2}{85}i$$

(b) $$\frac{7 - i}{3 - 5i} = \frac{7 - i}{3 - 5i} \cdot \frac{3 + 5i}{3 + 5i} = \frac{21 - 3i + 35i - 5i^2}{9 - 25i^2}$$

$$= \frac{26 + 32i}{34} = \frac{13}{17} + \frac{16}{17}i$$ ■

It is easy to see that if p is any positive real number, then the equation $x^2 = -p$ has solutions in $\mathbb{C}$. As a matter of fact, one solution is $i\sqrt{p}$, since

$$(i\sqrt{p})^2 = i^2(\sqrt{p})^2 = (-1)p = -p.$$

Similarly, $-i\sqrt{p}$ is also a solution.

The next definition is motivated by the fact that $(i\sqrt{p})^2 = -p$.

DEFINITION

> If p is a positive real number, then the **principal square root** of $-p$ is denoted by $\sqrt{-p}$ and is defined by
>
> $$\sqrt{-p} = i\sqrt{p}.$$

We also express this as $\sqrt{-p} = \sqrt{p}\,i$; however, care must be taken so as *not* to write $\sqrt{pi}$ when $\sqrt{p}\,i$ is intended. Some examples of principal square roots are

$$\sqrt{-9} = i\sqrt{9} = i(3) = 3i, \qquad \sqrt{-5} = i\sqrt{5}, \qquad \sqrt{-1} = i\sqrt{1} = i.$$

The radical sign must be used with caution when the radicand is negative. For example, the formula $\sqrt{a}\sqrt{b} = \sqrt{ab}$, which holds for positive real numbers, is not true when a and b are both negative. To illustrate,

$$\sqrt{-3}\sqrt{-3} = (i\sqrt{3})(i\sqrt{3}) = i^2(\sqrt{3})^2 = (-1)3 = -3,$$

whereas

$$\sqrt{(-3)(-3)} = \sqrt{9} = 3.$$

Hence,

$$\sqrt{-3}\sqrt{-3} \neq \sqrt{(-3)(-3)}.$$

However, if only *one* of a or b is negative, then $\sqrt{a}\sqrt{b} = \sqrt{ab}$. In general, we shall not apply laws of radicals if radicands are negative. Instead, we shall change the form of radicals before performing any operations, as illustrated in the next example.

EXAMPLE 4   Express $(5 - \sqrt{-9})(-1 + \sqrt{-4})$ in the form $a + bi$.

SOLUTION

$$\begin{aligned}
(5 - \sqrt{-9})(-1 + \sqrt{-4}) &= (5 - i\sqrt{9})(-1 + i\sqrt{4}) \\
&= (5 - 3i)(-1 + 2i) \\
&= -5 + 3i + 10i - 6i^2 \\
&= -5 + 13i + 6 = 1 + 13i \quad \blacksquare
\end{aligned}$$

In the previous section we proved that if a, b, and c are real numbers such that $b^2 - 4ac \geq 0$, and if $a \neq 0$, then the solutions of the quadratic equation $ax^2 + bx + c = 0$ are

$$\frac{-b + \sqrt{b^2 - 4ac}}{2a} \quad \text{and} \quad \frac{-b - \sqrt{b^2 - 4ac}}{2a}.$$

We may now extend this fact to the case where $b^2 - 4ac < 0$. Indeed, the same manipulations used to obtain the Quadratic Formula, together with the developments in this section, show that if $b^2 - 4ac < 0$, then the solutions of $ax^2 + bx + c = 0$ are the two *complex* numbers given above. Notice that the solutions are conjugates of one another.

EXAMPLE 5 Find the solutions of the equation $5x^2 + 2x + 1 = 0$.

SOLUTION By the Quadratic Formula,

$$x = \frac{-2 \pm \sqrt{4 - 20}}{10} = \frac{-2 \pm \sqrt{-16}}{10} = \frac{-2 \pm 4i}{10}.$$

Dividing numerator and denominator by 2, we see that the solutions of the equation are $-\frac{1}{5} + \frac{2}{5}i$ and $-\frac{1}{5} - \frac{2}{5}i$. ∎

EXAMPLE 6 Find the solutions of the equation $x^3 - 1 = 0$.

SOLUTION The equation $x^3 - 1 = 0$ may be written

$$(x - 1)(x^2 + x + 1) = 0.$$

Setting each factor equal to zero and solving the resulting equations, we obtain the solutions

$$1, \quad \frac{-1 \pm \sqrt{1 - 4}}{2},$$

or, equivalently,

$$1, \quad -\frac{1}{2} + \frac{\sqrt{3}}{2}i, \quad -\frac{1}{2} - \frac{\sqrt{3}}{2}i. \quad ∎$$

The three solutions of $x^3 - 1 = 0$ are called the **cube roots of unity**. It can be shown that if n is any positive integer, the equation $x^n - 1 = 0$ has n distinct complex solutions, called the **nth roots of unity**.

Exercises 2.4

In Exercises 1–32 write the expression in the form $a + bi$.

1 $(3 + 2i) + (-5 + 4i)$

2 $(8 - 5i) + (2 - 3i)$

3 $(-4 + 5i) + (2 - i)$

4 $(5 + 7i) + (-8 - 4i)$

5 $(16 + 10i) - (9 + 15i)$

6 $(2 - 6i) - (7 + 2i)$

7 $7 - (3 - 7i)$

8 $-9 + (5 + 9i)$

9 $5i - (6 + 2i)$

10 $(10 + 7i) - 12i$

11 $(4 + 3i)(-1 + 2i)$

12 $(3 - 6i)(2 + i)$

13 $(-7 + i)(-3 + i)$

14 $(5 + 2i)^2$

15 $(2 + 3i)^2$

16 $7i(13 + 8i)$

17 $-9i(4 - 8i)$

18 $(1 + i)^3$

19 $(3 + i)^2(3 - i)^2$

20 $-3(-6 + 12i)$

21 i^{30}

22 i^{25}

23 $\dfrac{1}{3 + 2i}$

24 $\dfrac{1}{5 + 8i}$

25 $\dfrac{7}{5 - 6i}$

26 $\dfrac{-3}{2 - 5i}$

27 $\dfrac{4 - 3i}{2 + 4i}$

28 $\dfrac{4 + 3i}{-1 + 2i}$

29 $\dfrac{6 + 4i}{1 - 5i}$

30 $\dfrac{7 - 6i}{-5 - i}$

31 $\dfrac{21 - 7i}{i}$

32 $\dfrac{10 + 9i}{-3i}$

In Exercises 33 and 34 find x and y.

33 $(x - y) + 3i = 7 + yi$

34 $8 + (3x + y)i = 2x - 4i$

Find the solutions of the equations in Exercises 35–48.

35 $x^2 - 3x + 10 = 0$

36 $x^2 - 5x + 20 = 0$

37 $x^2 + 2x + 5 = 0$

38 $x^2 + 3x + 6 = 0$

39 $4x^2 + x + 3 = 0$

40 $-3x^2 + x - 5 = 0$

41 $x^3 - 125 = 0$

42 $x^3 + 27 = 0$

43 $x^6 - 64 = 0$

44 $x^4 = 81$

45 $4x^4 + 25x^2 + 36 = 0$

46 $27x^4 + 21x^2 + 4 = 0$

47 $x^3 + 3x^2 + 4x = 0$

48 $8x^3 - 12x^2 + 2x - 3 = 0$

49 If $z = a + bi$ is any complex number, its conjugate is often denoted by $\bar{z}$, that is, $\bar{z} = a - bi$. Prove each of the following.

(a) $\overline{z + w} = \bar{z} + \bar{w}$

(b) $\overline{z \cdot w} = \bar{z} \cdot \bar{w}$

(c) $\overline{z^2} = (\bar{z})^2$, $\overline{z^3} = (\bar{z})^3$, and $\overline{z^4} = (\bar{z})^4$

(d) $\bar{z} = z$ if and only if z is real.

50 Refer to Exercise 49. Prove that $\overline{z - w} = \bar{z} - \bar{w}$ and $\bar{\bar{z}} = z$.

2.5 Miscellaneous Equations

If an equation can be expressed in factored form *with zero on one side*, then solutions may often be obtained by setting each factor equal to zero, as illustrated in the next two examples.

EXAMPLE 1 Solve $x^3 + 2x^2 - x - 2 = 0$.

SOLUTION The left side may be factored by grouping, as follows:

$$x^3 + 2x^2 - x - 2 = 0$$

$$x^2(x + 2) - (x + 2) = 0$$

$$(x^2 - 1)(x + 2) = 0$$

$$(x + 1)(x - 1)(x + 2) = 0$$

Setting each factor equal to 0 gives us

$$x + 1 = 0, \qquad x - 1 = 0, \qquad x + 2 = 0,$$

or, equivalently,

$$x = -1, \qquad x = 1, \qquad x = -2.$$

We may check that these three numbers are solutions by substitution in the original equation. ∎

EXAMPLE 2 Solve $x^{3/2} = x^{1/2}$.

SOLUTION We may proceed as follows:

$$x^{3/2} - x^{1/2} = 0$$

$$x^{1/2}(x - 1) = 0$$

Hence, $x^{1/2} = 0$ or $x - 1 = 0$, from which we obtain the solutions $x = 0$ and $x = 1$. ∎

In Example 2 it would have been *incorrect* to divide both sides of the equation $x^{3/2} = x^{1/2}$ by $x^{1/2}$, obtaining $x = 1$, since this leads to the loss of the solution $x = 0$. In general, *do not divide both sides of an equation by an expression that contains variables.*

If equations involve radicals or fractional exponents, the method of raising both sides to a positive power is often used to find solutions. When this is done, the solutions of the new equation always contain the solutions of the original equation. For example, the solutions of

$$2x - 3 = \sqrt{x + 6}$$

are also solutions of

$$(2x - 3)^2 = (\sqrt{x + 6})^2.$$

In some cases the new equation has *more* solutions than the original equation. To illustrate, if we start with the equation $x = 3$ and square both sides, we obtain $x^2 = 9$. Note that the given equation has only one solution, 3, whereas the new equation has two solutions, 3 and -3. Any solution of the new equation that is not a solution of the original equation is called an **extraneous solution.** Since extraneous solutions may arise, it is *absolutely essential* to check all solutions obtained after raising both sides of an equation to some power.

EXAMPLE 3   Solve the equation

$$\sqrt[3]{x^2 - 1} = 2.$$

SOLUTION   Cubing both sides, we obtain

$$(\sqrt[3]{x^2 - 1})^3 = 2^3$$
$$x^2 - 1 = 8$$
$$x^2 = 9.$$

Hence, the only possible solutions of $\sqrt[3]{x^2 - 1} = 2$ are 3 or -3.

Let us check each of these numbers by substitution in $\sqrt[3]{x^2 - 1} = 2$. Substituting 3 for x in the equation, we obtain $\sqrt[3]{3^2 - 1} = 2$, or $\sqrt[3]{8} = 2$, which is a true statement. Thus, 3 is a solution. Similarly, -3 is a solution. Hence, the solutions of the given equation are 3 and -3.   ■

EXAMPLE 4   Solve the equation

$$3 + \sqrt{3x + 1} = x.$$

SOLUTION   We begin by isolating the radical on one side as follows:

$$\sqrt{3x + 1} = x - 3.$$

Next we square both sides and simplify, obtaining

$$(\sqrt{3x + 1})^2 = (x - 3)^2$$
$$3x + 1 = x^2 - 6x + 9$$
$$x^2 - 9x + 8 = 0$$
$$(x - 1)(x - 8) = 0.$$

Since the last equation has solutions 1 and 8, it follows that 1 and 8 are the only possible solutions of the original equation.

We now check each of these by substitution in $3 + \sqrt{3x + 1} = x$. Letting $x = 1$ gives us

$$3 + \sqrt{4} = 1, \quad \text{or} \quad 5 = 1$$

which is false. Consequently, 1 is not a solution. Letting $x = 8$ in the given equation we obtain

$$3 + \sqrt{25} = 8, \quad \text{or} \quad 3 + 5 = 8$$

which is true. Hence, the equation $3 + \sqrt{3x + 1} = x$ has only one solution, $x = 8$. ∎

For equations involving several radicals it may be necessary to raise sides to powers several times, as illustrated in the next example.

EXAMPLE 5   Solve the equation

$$\sqrt{2x - 3} - \sqrt{x + 7} + 2 = 0.$$

SOLUTION   Let us begin by writing

$$\sqrt{2x - 3} = \sqrt{x + 7} - 2.$$

Squaring both sides we obtain

$$2x - 3 = (x + 7) - 4\sqrt{x + 7} + 4,$$

which simplifies to $x - 14 = -4\sqrt{x + 7}.$

Squaring both sides of the last equation and simplifying gives us

$$x^2 - 28x + 196 = 16(x + 7)$$
$$x^2 - 28x + 196 = 16x + 112$$
$$x^2 - 44x + 84 = 0$$
$$(x - 42)(x - 2) = 0.$$

Hence, the only possible solutions of the given equation are 42 and 2.

We next check each of these by substitution in the original equation. Substituting $x = 42$ gives us

$$\sqrt{84 - 3} - \sqrt{42 + 7} + 2 = 0$$

or $9 - 7 + 2 = 0$

which is false. Hence, 42 is not a solution. If we substitute $x = 2$ we obtain

$$\sqrt{4 - 3} - \sqrt{2 + 7} + 2 = 0$$

or

$$1 - 3 + 2 = 0$$

which is true. Hence, the given equation has one solution, $x = 2$. ∎

An equation is said to be of **quadratic type** if it can be written in the form

$$au^2 + bu + c = 0$$

where $a \neq 0$ and u is an expression in some variable. If we find the solutions in terms of u, then the solutions of the original equation can be obtained by referring to the specific form of u. The technique is illustrated in the following example.

EXAMPLE 6   Solve $x^{2/3} + x^{1/3} - 6 = 0$.

SOLUTION   If we let $u = x^{1/3}$, then the equation can be written

$$u^2 + u - 6 = 0, \quad \text{or} \quad (u + 3)(u - 2) = 0.$$

This has solutions $u = -3$ and $u = 2$.

Since $u = x^{1/3}$, $x^{1/3} = -3$ or $x^{1/3} = 2$.

Cubing gives us $x = -27$ or $x = 8$.

We next check each of these by substitution. Letting $x = -27$ in the given equation, we obtain

$$[(-27)^{1/3}]^2 + (-27)^{1/3} - 6 = 9 - 3 - 6 = 0.$$

Thus, -27 is a solution. Similarly, it can be shown that 8 is a solution. Hence, the solutions of the given equation are -27 and 8. ∎

EXAMPLE 7   Solve $x^4 - 3x^2 + 1 = 0$.

SOLUTION   Letting $u = x^2$ gives us

$$u^2 - 3u + 1 = 0.$$

Using the Quadratic Formula,

$$u = \frac{3 \pm \sqrt{9 - 4}}{2} = \frac{3 \pm \sqrt{5}}{2}.$$

Since $u = x^2$,

$$x^2 = \frac{3 \pm \sqrt{5}}{2}, \quad \text{or} \quad x = \pm \sqrt{\frac{3 \pm \sqrt{5}}{2}}.$$

Thus, there are four solutions:

$$\sqrt{\frac{3 + \sqrt{5}}{2}}, \quad -\sqrt{\frac{3 + \sqrt{5}}{2}}, \quad \sqrt{\frac{3 - \sqrt{5}}{2}}, \quad -\sqrt{\frac{3 - \sqrt{5}}{2}}. \quad \blacksquare$$

Exercises 2.5

Solve the equations in Exercises 1–36.

1 $4x^3 + 12x^2 - 9x - 27 = 0$

2 $2x^3 + 5x^2 - 8x - 20 = 0$

3 $6x^5 + 10x^4 = 3x^3 + 5x^2$

4 $25z^4 + 5z = 125z^3 + z^2$

5 $y^{3/2} = 4y$ **6** $2x^3 = 5x^2$

7 $\dfrac{1}{x} = \dfrac{1}{x^2}$

8 $\sqrt{x}\sqrt[3]{x} - 3\sqrt[3]{x} - 2\sqrt{x} + 6 = 0$

9 $\sqrt{7 - 5x} = 8$ **10** $\sqrt{2x - 9} = 3^{-1}$

11 $2 + \sqrt[3]{1 - 5t} = 0$ **12** $\sqrt[3]{6 - s^2} + 5 = 0$

13 $\sqrt[5]{2x^2 + 1} - 2 = 0$ **14** $\sqrt[4]{2x^2 - 1} = x$

15 $3\sqrt{2x - 3} + 2\sqrt{7 - x} = 11$

16 $\sqrt{2x + 15} - 2 = \sqrt{6x + 1}$

17 $\sqrt{7 - 2x} - \sqrt{5 + x} = \sqrt{4 + 3x}$

18 $4\sqrt{1 + 3x} + \sqrt{6x + 3} = \sqrt{-6x - 1}$

19 $\sqrt{11 + 8x} + 1 = \sqrt{9 + 4x}$

20 $2\sqrt{x} - \sqrt{x - 3} = \sqrt{5 + x}$

21 $\sqrt{2\sqrt{x} + 1} = \sqrt{3x - 5}$

22 $\sqrt{5\sqrt{x}} = \sqrt{2x - 3}$ **23** $\sqrt{1 + 4\sqrt{x}} = \sqrt{x} + 1$

24 $\sqrt{x + 1} = \sqrt{x - 1}$ **25** $4x^4 - 37x^2 + 9 = 0$

26 $2x^4 - 9x^2 + 4 = 0$ **27** $3z^4 - 5z^2 + 1 = 0$

28 $2y^4 + y^2 - 5 = 0$

29 $3x^{2/3} + 8x^{1/3} - 3 = 0$

30 $2t^{1/3} - 5t^{1/6} + 2 = 0$

31 $3w - 19\sqrt{w} + 20 = 0$

32 $2x^{-2/3} + 7x^{-1/3} - 4 = 0$

33 $\dfrac{2}{(x - 1)^2} + \dfrac{3}{x - 1} - 2 = 0$

34 $\dfrac{4}{(x^2 - 1)^2} - \dfrac{5}{x^2 - 1} + 1 = 0$

35 $\left(\dfrac{t}{t + 1}\right)^2 + \dfrac{2t}{t + 1} - 15 = 0$

36 $6u^{-1/2} - 17u^{-1/4} + 5 = 0$

The formulas in Exercises 37–42 occur in mathematics and its applications. Solve for the indicated variable in terms of the remaining variables (all letters denote positive real numbers).

37 $S = \pi r \sqrt{r^2 + h^2}$ for h

38 $d = \frac{1}{2}\sqrt{4R^2 - C^2}$ for C

39 $y = \frac{b}{a}\sqrt{a^2 - x^2}$ for x

40 $T = 2\pi \sqrt{\dfrac{l}{g}}$ for l

41 $x^{2/3} + y^{2/3} = a^{2/3}$ for y

42 $y = (\sqrt[3]{a} - \sqrt[3]{x})^3$ for x

43 As sand leaks out of a certain container it forms a pile that has the shape of a right circular cone whose height is always one-half the diameter of the base. What is the diameter of the base at the instant that 144 cm^3 of sand has leaked out?

44 If the volume of a spherical balloon increases from $10\frac{2}{3}$ ft^3 to 36 ft^3, how much does the diameter increase?

45 A rectangular field has an area of 1200 ft^2, and the greatest distance between any two points in the field is 50 feet. What are the dimensions of the field?

46 A conical paper cup is to have a height of 3 inches. Find the radius of the cone that will result in a surface area of 6π in^2. (*Hint:* The lateral surface area of a cone is given by the equation $S = \pi r \sqrt{r^2 + h^2}$.)

Calculator Exercises 2.5

Approximate the solutions of the equations in Exercises 1–4 to the nearest hundredth.

1 $x^{6/7} - 28x^{1/7} = 0$

2 $45x^4 - 63x^2 + 14 = 0$

3 $62x^{2/3} - 75x^{1/3} - 23 = 0$

4 $12x^{-2/5} - 19x^{-1/5} - 11 = 0$

FIGURE FOR EXERCISE 5

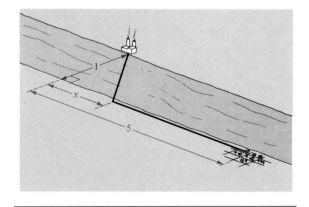

5 A power line is to be installed across a river that is 1 mile wide to a town that is 5 miles downstream (see figure). It costs \$7500 per mile to lay the cable underwater and \$6000 per mile over land. Determine how the cable should be installed if \$35,000 has been allocated for this project.

6 Adolphe Quetelet (1796–1874), the director of the Brussels Observatory from 1832–1874, was the first person to attempt to fit a mathematical expression to human growth data. Given that h denotes height in meters and t age in years, Quetelet's formula for males in Brussels can be expressed as

$$h + \frac{h}{100(h_m - h)} = at + \frac{h_0 + t}{1 + (4t/3)},$$

where $h_0 = 0.5$ is the height at birth, $h_m = 1.684$ is the final adult height, and $a = 0.545$.

(a) Find the expected height of a 12 year old.

(b) At what age should 50% of the adult height be reached?

2.6 Inequalities

Let us consider the inequality

$$x^2 - 3 < 2x + 4$$

where x is a variable. If certain numbers such as 4 or 5 are substituted for x, we obtain the false statements $13 < 12$ or $22 < 14$, respectively. Other numbers, such as 1 or 2, produce the true statements $-2 < 6$ or $1 < 8$. If a true statement is obtained when x is replaced by a real number a, then a is called a **solution** of the inequality. Thus 1 and 2 are solutions of the inequality $x^2 - 3 < 2x + 4$, whereas 4 and 5 are not solutions. To **solve** an inequality means to find all solutions. We say that two inequalities are **equivalent** if they have exactly the same solutions.

As with equations, to solve an inequality we replace it with a list of equivalent inequalities, ending with an inequality for which the solutions are obvious. The following properties are often useful, where a, b, and c denote real numbers.

PROPERTIES OF >

> (i) If $a > b$ and $b > c$, then $a > c$.
>
> (ii) If $a > b$, then $a + c > b + c$.
>
> (iii) If $a > b$, then $a - c > b - c$.
>
> (iv) If $a > b$ and $c > 0$, then $ac > bc$.
>
> (v) If $a > b$ and $c < 0$, then $ac < bc$.

PROOF   We will use the fact that both the sum and product of any two positive real numbers are positive. To prove (i) we first note that if $a > b$ and $b > c$, then $a - b$ and $b - c$ are both positive. Consequently, the sum $(a - b) + (b - c)$ is positive. Since the sum reduces to $a - c$, we see that $a - c$ is positive, which means that $a > c$.

To establish (ii) we again note that if $a > b$, then $a - b$ is positive. Since $(a + c) - (b + c) = a - b$, it follows that $(a + c) - (b + c)$ is positive; that is, $a + c > b + c$.

If $a > b$, then by (ii), $a + (-c) > b + (-c)$, or equivalently, $a - c > b - c$. This proves (iii).

To prove (iv) observe that if $a > b$ and $c > 0$, then $a - b$ and c are both positive and hence, so is the product $(a - b)c$. Consequently, $ac - bc$ is positive; that is, $ac > bc$.

Finally, to prove (v) we first note that if $c < 0$, then $0 - c$, or $-c$, is positive. In addition, if $a > b$, then $a - b$ is positive and hence, the product $(a - b)(-c)$ is positive. However, $(a - b)(-c) = -ac + bc$ and therefore $bc - ac$ is positive. This means that $bc > ac$, or $ac < bc$.   □

Similar results are true for the symbol $<$. Thus, if $a < b$, then we have $a + c < b + c$, and so on.

If x represents a real number, then by properties (ii) or (iii), adding or subtracting the same expression in x on both sides of an inequality leads to an equivalent inequality. By (iv) we may multiply both sides of an inequality by an expression containing x if we are certain that the expression is positive for all values of x under consideration. To illustrate, multiplication by $x^4 + 3x^2 + 5$ would be permissible since this expression is always positive. If we multiply both sides of an inequality by an expression that is always negative, such as $-7 - x^2$, then by (v) the inequality sign is reversed.

EXAMPLE 1   Solve the inequality $-3x + 4 > 11$.

SOLUTION   The following inequalities are equivalent (Supply reasons.).

$$-3x + 4 > 11$$
$$(-3x + 4) - 4 > 11 - 4$$
$$-3x > 7$$
$$(-\tfrac{1}{3})(-3x) < (-\tfrac{1}{3})(7)$$
$$x < -\tfrac{7}{3}$$

Thus, the solutions of $-3x + 4 > 11$ consist of all real numbers x such that $x < -\tfrac{7}{3}$. ∎

EXAMPLE 2   Solve the inequality $4x - 3 < 2x + 5$.

SOLUTION   The following is a list of equivalent inequalities:

$$4x - 3 < 2x + 5$$
$$(4x - 3) + 3 < (2x + 5) + 3$$
$$4x < 2x + 8$$
$$4x - 2x < (2x + 8) - 2x$$
$$2x < 8$$
$$\tfrac{1}{2}(2x) < \tfrac{1}{2}(8)$$
$$x < 4$$

Hence, the solutions of the given inequality consist of all real numbers x such that $x < 4$. ∎

FIGURE 2.5

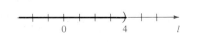

We can represent solutions of inequalities graphically. The **graph** of a set of real numbers is the collection of points on a coordinate line l that correspond to the numbers. To **sketch a graph** we darken an appropriate portion of l. The graph corresponding to the solutions of Example 2 consists of all points to the left of the point with coordinate 4 and is sketched in Figure 2.5, where it is understood that the black portion extends indefinitely to the left. The parenthesis in the figure indicates that the point corresponding to 4 is not part of the graph.

If $a < b$, the symbol (a, b) is sometimes used for all real numbers between a and b; this set is called an **open interval.**

DEFINITION OF OPEN INTERVAL

$$(a, b) = \{x : a < x < b\}$$

FIGURE 2.6

Open intervals (a, b), $(-1, 3)$, and $(2, 4)$.

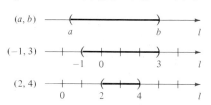

The expression to the right of the equal sign is translated "the set of all x such that $a < x < b$." The numbers a and b are called the **endpoints** of the interval. The graph of (a, b) consists of all points on a coordinate line that lie between the points corresponding to a and b. In Figure 2.6 we have sketched the graph of a general open interval (a, b) and also the special open intervals $(-1, 3)$ and $(2, 4)$. The parentheses in the figure indicate that the endpoints of the intervals are not to be included. For convenience, we shall use the terms *open interval* and *graph of an open interval* interchangeably.

The solutions of an inequality may often be described in terms of intervals, as illustrated in the next example.

EXAMPLE 3   Solve the inequality $-6 < 2x - 4 < 2$ and represent the solutions graphically.

SOLUTION   A real number x is a solution of the given inequality if and only if it is a solution of *both* of the inequalities

$$-6 < 2x - 4 \quad \text{and} \quad 2x - 4 < 2.$$

The first inequality is equivalent to each of the following:

$$-6 < 2x - 4$$
$$-6 + 4 < (2x - 4) + 4$$
$$-2 < 2x$$
$$\tfrac{1}{2}(-2) < \tfrac{1}{2}(2x)$$
$$-1 < x$$

The second inequality is equivalent to each of the following:

$$2x - 4 < 2$$

$$2x < 6$$

$$x < 3$$

Thus, x is a solution of the given inequality if and only if *both*

$$-1 < x \quad \text{and} \quad x < 3,$$

FIGURE 2.7

that is,

$$-1 < x < 3.$$

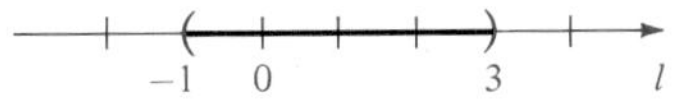

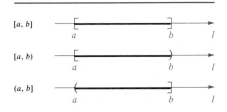

Hence, the solutions are all numbers in the open interval $(-1, 3)$. The graph is sketched in Figure 2.7.

An alternative (and shorter) method is to work with both inequalities simultaneously, as follows:

$$-6 < 2x - 4 < 2$$

$$-6 + 4 < 2x < 2 + 4$$

$$-2 < 2x < 6$$

$$-1 < x < 3 \qquad\qquad \blacksquare$$

If we wish to include an endpoint of an interval, a bracket is used instead of a parenthesis. If $a < b$, then **closed intervals,** denoted by $[a, b]$, and **half-open intervals,** denoted by $[a, b)$ or $(a, b]$, are defined as follows:

DEFINITION OF CLOSED AND HALF-OPEN INTERVALS

$$[a, b] = \{x : a \leq x \leq b\}$$

$$[a, b) = \{x : a \leq x < b\}$$

$$(a, b] = \{x : a < x \leq b\}$$

Typical graphs of intervals are sketched in Figure 2.8. A bracket indicates that the corresponding endpoint is part of the graph.

FIGURE 2.8

$[a, b]$

$[a, b)$

$(a, b]$

EXAMPLE 4 Solve the inequality

$$-5 \leq \frac{4 - 3x}{2} < 1$$

and sketch the graph corresponding to the solutions.

SOLUTION A number x is a solution if and only if it satisfies both of the inequalities

$$-5 \le \frac{4 - 3x}{2} \quad \text{and} \quad \frac{4 - 3x}{2} < 1.$$

We can either work with each inequality separately or proceed as in the alternative solution of Example 3, as follows:

$$-5 \le \frac{4 - 3x}{2} < 1$$

$$-10 \le 4 - 3x < 2$$

$$-10 - 4 \le -3x < 2 - 4$$

$$-14 \le -3x < -2$$

$$(-\tfrac{1}{3})(-14) \ge (-\tfrac{1}{3})(-3x) > (-\tfrac{1}{3})(-2)$$

$$\tfrac{14}{3} \ge x > \tfrac{2}{3}$$

$$\tfrac{2}{3} < x \le \tfrac{14}{3}.$$

FIGURE 2.9

Thus, the solutions of the inequality consist of all numbers in the interval $(\tfrac{2}{3}, \tfrac{14}{3}]$. The graph is sketched in Figure 2.9. ∎

EXAMPLE 5 The temperature readings on the Fahrenheit and Celsius scales are related by the equation $C = \tfrac{5}{9}(F - 32)$. What values of F correspond to $30 \le C \le 40$?

SOLUTION The following inequalities are equivalent:

$$30 \le C \le 40$$

$$30 \le \tfrac{5}{9}(F - 32) \le 40$$

$$\tfrac{9}{5}(30) \le F - 32 \le \tfrac{9}{5}(40)$$

$$54 \le F - 32 \le 72$$

$$86 \le F \le 104$$

Thus, a Celsius temperature range from $30\,°C$ to $40\,°C$ is the same as a Fahrenheit range from $86\,°F$ to $104\,°F$. ∎

To describe solutions of inequalities such as $x < 4$, $x > -2$, $x \le 7$, or $x \ge 3$, it is convenient to employ **infinite intervals.** If a is a real number,

we define

$$(-\infty, a) = \{x : x < a\}.$$

Thus, $(-\infty, a)$ denotes the set of all real numbers less than a. The symbol ∞ (**infinity**) is merely a notational device and does not represent a real number. To illustrate, the solutions in Example 1 are the numbers in the infinite interval $(-\infty, -\frac{7}{3})$, and those in Example 2 are in $(-\infty, 4)$. The graph of $(-\infty, 4)$ is sketched in Figure 2.5.

If we wish to include the point corresponding to a we write

$$(-\infty, a] = \{x : x \le a\}.$$

FIGURE 2.10

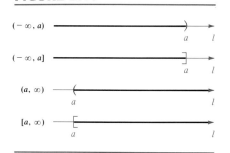

Other types of infinite intervals are defined by

$$(a, \infty) = \{x : x > a\}$$
$$[a, \infty) = \{x : x \ge a\}.$$

The set $\mathbb{R}$ of real numbers is sometimes denoted by $(-\infty, \infty)$.

Graphs of infinite intervals for an arbitrary real number a are sketched in Figure 2.10. The absence of a parentheses or bracket, on the left for the graphs of $(-\infty, a)$ and $(-\infty, a]$, and on the right for (a, ∞) and $[a, \infty)$, indicates that the black portions extend indefinitely.

EXAMPLE 6   Solve the inequality

$$\frac{1}{x - 2} > 0$$

FIGURE 2.11

and sketch the graph corresponding to the solutions.

SOLUTION   Since the numerator is positive, the fraction is positive if and only if $x - 2 > 0$, or equivalently, $x > 2$. Thus, the solutions are all numbers in the infinite interval $(2, \infty)$. See Figure 2.11.   ■

Exercises 2.6

1 Given $-6 < -4$, what inequality is obtained if:

(a) 5 is added to both sides?
(b) 5 is subtracted from both sides?
(c) both sides are multiplied by $\frac{1}{2}$?
(d) both sides are multiplied by $-\frac{1}{2}$?

2 Given $3 > -3$, what inequality is obtained if:

(a) 3 is added to both sides?
(b) -3 is added to both sides?
(c) both sides are divided by 3?
(d) both sides are divided by -3?

In Exercises 3–12, express the given inequality in interval notation and sketch a graph of the interval.

3 $2 < x < 5$

4 $-1 < x < 2$

5 $-1 < x \leq 3$

6 $-2 \geq x \geq -3$

7 $4 \geq x \geq 1$

8 $0 < x \leq 4$

9 $x > -1$

10 $x < 1$

11 $x \leq 2$

12 $x \geq -3$

Express the intervals in Exercises 13–18 as an inequality in the variable x.

13 $(-1, 7)$

14 $[5, 12]$

15 $(8, 9]$

16 $[-5, 0)$

17 $(5, \infty)$

18 $(-\infty, 7]$

Solve the inequalities in Exercises 19–42 and express the solutions in terms of intervals.

19 $5x - 6 > 11$

20 $3x - 5 < 10$

21 $2 - 7x \leq 16$

22 $7 - 2x \geq -3$

23 $3x + 1 < 5x - 4$

24 $6x - 5 > 9x + 1$

25 $4 - \frac{1}{2}x \geq -7 + \frac{1}{4}x$

26 $\frac{1}{3}x - 4 \leq \frac{1}{4}x - 6$

27 $-4 < 3x + 5 < 8$

28 $6 > 2x - 6 > 4$

29 $3 \geq \dfrac{7 - x}{2} \geq 1$

30 $-2 \leq \dfrac{5 - 3x}{4} \leq \dfrac{1}{2}$

31 $0 < 2 - \frac{3}{4}x \leq \frac{1}{2}$

32 $-3 < \frac{1}{2}x - 4 \leq 0$

33 $(3x + 1)(2x - 4) < (6x - 2)(x + 5)$

34 $(x + 2)(x - 5) < (x - 1)^2$

35 $(x + 2)^2 > x(x - 2)$

36 $2x(8x - 1) \geq (4x + 1)(4x - 3)$

37 $\dfrac{5}{3x + 7} > 0$

38 $\dfrac{1}{6 - 2x} < 0$

39 $(1 - 4x)^{-1} < 0$

40 $(x + 4)^{-1} > 0$

41 $\dfrac{1}{(x - 1)^2} > 0$

42 $\dfrac{1}{x^2 + 1} < 0$

43 The relationship between the Fahrenheit and Celsius temperature scales is given by $C = \frac{5}{9}(F - 32)$. If $60 \leq F \leq 80$, express the corresponding range for C in terms of an inequality.

44 In the study of electricity, Ohm's Law states that if R denotes the resistance of an object (in ohms), E the potential difference across the object (in volts), and I the current that flows through it (in amperes), then $R = E/I$ (see figure). If the voltage is 110, what values of the resistance will result in a current that does not exceed 10 amperes?

FIGURE FOR EXERCISE 44

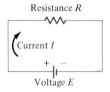

Resistance R

Current I

Voltage E

45 According to Hooke's Law, the force F (in pounds) required to stretch a certain spring x inches beyond its natural length is given by the formula $F = (4.5)x$ (see figure). If $10 \leq F \leq 18$, what are the corresponding values for x?

FIGURE FOR EXERCISE 45

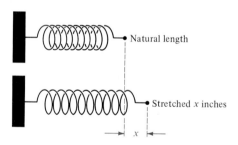

Natural length

Stretched x inches

x

46 If two resistors R_1 and R_2 are connnected in parallel in an electrical circuit, the net resistance R is given by $1/R = (1/R_1) + (1/R_2)$. (Compare Example 3 of Section 2.2.) If $R_1 = 10$ ohms, what values of R_2 will result in a net resistance of less than 5 ohms?

47 A convex lens has focal length $f = 5$ cm. If an object is placed at a distance of p cm from the lens, the distance q of the image from the lens is related to p and f by the formula $(1/p) + (1/q) = 1/f$ (see figure). How close must the object be to the lens for the image to be more than 12 cm from the lens?

FIGURE FOR EXERCISE 47

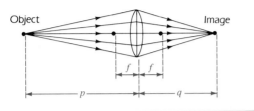

48 A construction firm is trying to decide which of two models of a type of crane to purchase. Model A costs $50,000 and requires $4000 per year to maintain. The corresponding figures for model B are $40,000 initial cost and maintenance costs of $5500 per year. For how many years must model A be used before it becomes more economical than B?

49 If $0 < a < b$, prove that $(1/a) > (1/b)$. Why is the restriction $0 < a$ necessary?

50 If $0 < a < b$, prove that $a^2 < b^2$. Why is the restriction $0 < a$ necessary?

2.7 More on Inequalities

If a is a real number, then the inequality $|a| < 1$ is equivalent to $-1 < a < 1$. This, in turn, means that a is in the open interval $(-1, 1)$. In general, if b is any positive real number, the following can be proved.

PROPERTIES OF ABSOLUTE VALUES ($b > 0$)

(i) $|a| < b$ if and only if $-b < a < b$.

(ii) $|a| > b$ if and only if either $a > b$ or $a < -b$.

(iii) $|a| = b$ if and only if $a = b$ or $a = -b$.

Properties (i) and (ii) are also true if $b = 0$. Thus, if $b \geq 0$, then

$$|a| \leq b \quad \text{if and only if} \quad -b \leq a \leq b$$

and $\qquad |a| \geq b \quad \text{if and only if} \quad a \geq b \quad \text{or} \quad a \leq -b.$

EXAMPLE 1 Solve the following inequalities:

(a) $|x| < 4$ (b) $|x| > 4$ (c) $|x| = 4$

SOLUTION

(a) Using property (i) with $a = x$ and $b = 4$, we see that

$$|x| < 4 \quad \text{if and only if} \quad -4 < x < 4.$$

Hence, the solutions are all real numbers in the open interval $(-4, 4)$.

(b) By property (ii), $|x| > 4$ means that either $x > 4$ or $x < -4$. Thus, the solutions consist of all real numbers in the two infinite intervals $(-\infty, -4)$ and $(4, \infty)$.

(c) By property (iii)

$$|x| = 4 \quad \text{if and only if} \quad x = 4 \quad \text{or} \quad x = -4. \qquad \blacksquare$$

For the solutions obtained in (b) of Example 1, we may use the **union symbol,** $\cup$, and write

$$(-\infty, -4) \cup (4, \infty)$$

to denote all real numbers that are either in $(-\infty, -4)$ or in $(4, \infty)$.

EXAMPLE 2   Solve the inequality $|x - 3| < 0.1$.

SOLUTION   By property (i), with $a = x - 3$ and $b = 0.1$, the inequality is equivalent to

$$-0.1 < x - 3 < 0.1$$

and hence to $-0.1 + 3 < (x - 3) + 3 < 0.1 + 3$

or to $2.9 < x < 3.1$.

Consequently, the solutions are the real numbers in the interval $(2.9, 3.1)$.

$$\blacksquare$$

EXAMPLE 3   If a and δ denote real numbers and $\delta > 0$, solve the inequality

$$|x - a| < \delta$$

and represent the solutions graphically.

SOLUTION   We may proceed as in Example 2. Thus,

$$-\delta < x - a < \delta$$
$$a - \delta < a + (x - a) < a + \delta$$
$$a - \delta < x < a + \delta.$$

FIGURE 2.12

$|x - a| < \delta$

The solutions of the inequality consist of all real numbers in the interval $(a - \delta, a + \delta)$. A typical graph is sketched in Figure 2.12.

$$\blacksquare$$

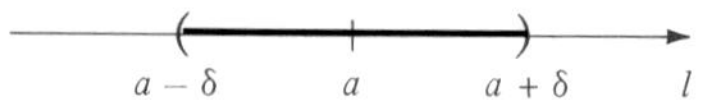

Note that Example 2 is the special case of Example 3 with $a = 3$ and $\delta = 0.1$.

EXAMPLE 4   Solve $|2x + 3| > 9$ and illustrate the solutions graphically.

SOLUTION   By property (ii) of absolute values with $a = 2x + 3$ and $b = 9$, the solutions of the inequality are the solutions of the following *two* inequalities:

(i) $2x + 3 > 9$ (ii) $2x + 3 < -9$

FIGURE 2.13

Inequality (i) is equivalent to $2x > 6$, or $x > 3$. This gives us the infinite interval $(3, \infty)$. Inequality (ii) is equivalent to $2x < -12$, or $x < -6$, which leads to the interval $(-\infty, -6)$. Consequently, the solutions of the inequality $|2x + 3| > 9$ consist of the numbers in the union $(-\infty, -6) \cup (3, \infty)$. The graph is sketched in Figure 2.13.   ■

To solve inequalities involving polynomials of degree greater than 1, we shall use the following theorem, which will be discussed further in Section 4.2. In the statement of the theorem, the phrase *successive solutions c and d* means that there are no other solutions between c and d.

THEOREM

> Let $a_n x^n + \cdots + a_1 x + a_0$ be a polynomial. If the real numbers c and d are successive solutions of the equation
> $$a_n x^n + \cdots + a_1 x + a_0 = 0,$$
> then when x is in the open interval (c, d) either all values of the polynomial are positive or all values are negative.

This theorem implies that if we choose *any* number k, such that $c < k < d$, and if the value of the polynomial is positive for $x = k$, then the polynomial is positive for *every* x in (c, d). Similarly, if the polynomial is negative for $x = k$, then it is negative throughout (c, d). We shall call the value of the polynomial at $x = k$ a **test value** for the interval (c, d). Test values may also be used on infinite intervals of the form $(-\infty, a)$ or (a, ∞), provided the polynomial equation has no solutions on these intervals. The use of test values is demonstrated in the following examples.

EXAMPLE 5   Solve $2x^2 - x < 3$.

SOLUTION   To use the method that will be described here, *it is essential to have all nonzero terms on one side* of the inequality sign. Thus, we begin by writing

$$2x^2 - x - 3 < 0.$$

Factoring gives us $(x + 1)(2x - 3) < 0.$

We see from the factored form that the equation $2x^2 - x - 3 = 0$ has solutions -1 and $\frac{3}{2}$. For reference, let us plot the corresponding points on a real axis, as in Figure 2.14. These points divide the axis into three parts and determine the following intervals:

FIGURE 2.14

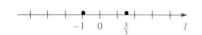

$$(-\infty, -1), \qquad (-1, \tfrac{3}{2}), \qquad (\tfrac{3}{2}, \infty).$$

The sign of the polynomial $2x^2 - x - 3$ in each interval can be found using a suitable test value.

If we choose -2 in $(-\infty, -1)$, then the polynomial $2x^2 - x - 3$ has the value

$$2(-2)^2 - (-2) - 3 = 8 + 2 - 3 = 7.$$

Since 7 is positive, it follows from the preceding theorem that the polynomial $2x^2 - x - 3$ is positive for every x in $(-\infty, -1)$.

If we choose 0 in $(-1, \frac{3}{2})$, then the polynomial has the value

$$2(0)^2 - (0) - 3 = -3.$$

Since -3 is negative, $2x^2 - x - 3 < 0$ for every x in $(-1, \frac{3}{2})$.

Finally, choosing 2 in $(\frac{3}{2}, \infty)$, we obtain

$$2(2)^2 - 2 - 3 = 8 - 2 - 3 = 3,$$

and since 3 is positive, $2x^2 - x - 3 > 0$ throughout $(\frac{3}{2}, \infty)$. It is convenient to tabulate these facts as follows:

Interval	$(-\infty, -1)$	$(-1, \frac{3}{2})$	$(\frac{3}{2}, \infty)$
k	-2	0	2
Test value of $2x^2 - x - 3$ at k	7	-3	3
Sign of $2x^2 - x - 3$ in interval	$+$	$-$	$+$

FIGURE 2.15

Figure 2.15 graphically illustrates the intervals in which $2x^2 - x - 3$ is positive or negative. Thus, the solutions $2x^2 - x - 3 < 0$, or equivalently $2x^2 - x < 3$, are the real numbers in the interval $(-1, \frac{3}{2})$. ∎

EXAMPLE 6 Solve $x^2 > 7x - 10$ and represent the solutions graphically.

SOLUTION As in Example 5, we bring all terms to one side of the inequality sign, obtaining

$$x^2 - 7x + 10 > 0.$$

FIGURE 2.16

Factoring gives us $(x - 2)(x - 5) > 0.$

Points corresponding to the solutions 2 and 5 of $x^2 - 7x + 10 = 0$ are plotted in Figure 2.16. Referring to the figure, we obtain the following intervals:

$$(-\infty, 2), \quad (2, 5), \quad (5, \infty).$$

We next use test values to determine the sign of $x^2 - 7x + 10$ in each interval. The following table summarizes results. (*Check each entry.*)

Interval	$(-\infty, 2)$	$(2, 5)$	$(5, \infty)$
k	0	3	6
Test value of $x^2 - 7x + 10$ at k	$0^2 - 7(0) + 10 = 10$	$3^2 - 7(3) + 10 = -2$	$6^2 - 7(6) + 10 = 4$
Sign of $x^2 - 7x + 10$ in interval	+	−	+

FIGURE 2.17

Figure 2.17 graphically illustrates where $x^2 - 7x + 10$ is positive or negative. Thus, $x^2 - 7x + 10 > 0$ if x is in either $(-\infty, 2)$ or $(5, \infty)$. The solutions of the inequality are given by $(-\infty, 2) \cup (5, \infty)$. The graph is sketched in Figure 2.18. ∎

FIGURE 2.18

EXAMPLE 7

Solve the inequality $\dfrac{x + 1}{x + 3} \le 2$ and represent the solutions graphically.

SOLUTION   We first take all nonzero terms to one side of the inequality symbol and then proceed as follows:

$$\frac{x+1}{x+3} \le 2$$

$$\frac{x+1}{x+3} - 2 \le 0$$

$$\frac{x+1-2x-6}{x+3} \le 0$$

$$\frac{-x-5}{x+3} \le 0$$

$$\frac{x+5}{x+3} \ge 0$$

FIGURE 2.19

The numerator and denominator of $(x+5)/(x+3)$ equal zero at $x = -5$ and $x = -3$, respectively. For reference, we plot these points as in Figure 2.19. Note that -5 is a solution of $(x+5)/(x+3) \ge 0$, but -3 is *not* a solution since a zero denominator occurs if -3 is substituted for x. The points in the figure determine the following intervals:

$$(-\infty, -5), \qquad (-5, -3), \qquad (-3, \infty)$$

Since $(x+5)/(x+3)$ is a quotient of two polynomials, it is always positive or always negative throughout each interval. (The sign of this quotient is the same as the sign of the product $(x+5)(x+3)$.) As before, we may use test values to determine the sign in each interval. The following table summarizes results:

Interval	$(-\infty, -5)$	$(-5, -3)$	$(-3, \infty)$
k	-6	-4	0
Test value of $(x+5)/(x+3)$ at k	$\frac{1}{3}$	-1	$\frac{5}{3}$
Sign of $(x+5)/(x+3)$ in interval	$+$	$-$	$+$

FIGURE 2.20

FIGURE 2.21

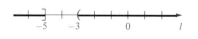

Figure 2.20 indicates where $(x+5)/(x+3)$ is positive or negative. Thus, the solutions of $(x+5)/(x+3) > 0$ are given by $(-\infty, -5) \cup (-3, \infty)$. The solutions of $(x+5)/(x+3) \ge 0$ are given by $(-\infty, -5] \cup (-3, \infty)$. The graph is sketched in Figure 2.21. ∎

2.7 **More on Inequalities**

FIGURE 2.22

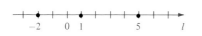

FIGURE 2.23

EXAMPLE 8 Solve $(x + 2)(x - 1)(x - 5) > 0$ and represent the solutions graphically.

SOLUTION The expression $(x + 2)(x - 1)(x - 5)$ is zero at -2, 1, and 5. The corresponding points are plotted on a real axis in Figure 2.22. These points determine the four intervals:

$$(-\infty, -2), \qquad (-2, 1), \qquad (1, 5), \qquad (5, \infty)$$

We next use test values, as shown in the following table.

Interval	$(-\infty, -2)$	$(-2, 1)$	$(1, 5)$	$(5, \infty)$
k	-3	0	2	6
Test value of $(x + 2)(x - 1)(x - 5)$	$(-1)(-4)(-8)$ $= -32$	$(2)(-1)(-5)$ $= 10$	$(4)(1)(-3)$ $= -12$	$(8)(5)(1)$ $= 40$
Sign of $(x + 2)(x - 1)(x - 5)$ in interval	$-$	$+$	$-$	$+$

FIGURE 2.24

Figure 2.23 shows where $(x + 2)(x - 1)(x - 5)$ is positive or negative. Thus, the solutions of $(x + 2)(x - 1)(x - 5) > 0$ are given by $(-2, 1) \cup (5, \infty)$. The graph is sketched in Figure 2.24. ∎

Exercises 2.7

Solve the inequalities in Exercises 1–38 and express the solutions in terms of intervals.

1 $|x| < 2$

2 $|x| < 25$

3 $|x| > 6$

4 $|x| > \frac{1}{2}$

5 $|-x| \le 10$

6 $|x| \ge 7$

7 $|x - 10| < 0.05$

8 $|x - 5| < 0.001$

9 $\left| \dfrac{x + 4}{3} \right| \le 2$

10 $\left| \dfrac{x + 1}{5} \right| \le 1$

11 $|5 - 3x| < 7$

12 $|7x + 4| < 10$

13 $|2x + 4| > 8$

14 $|6x - 7| \ge 4$

15 $\dfrac{3}{|5 - 2x|} \le 1$

16 $\left| \dfrac{4 - x}{2} \right| > 9$

17 $x^2 - x - 6 < 0$

18 $x^2 + 6x + 5 < 0$

19 $x(2x + 3) > 5$

20 $x(3x - 1) > 4$

21 $x^2 \le 10x$

22 $4x^2 \ge x$

23 $(2x + 1)(10 - 3x) < 0$

24 $(9 - 4x)(x + 1) \ge 0$

25 $x^2 < 9$

26 $25x^2 - 16 > 0$

27 $\dfrac{7x}{x^2 - 16} \geq 0$

28 $\dfrac{x - 2}{x^2 - 9} < 0$

29 $\dfrac{x^2 - 25}{4x^2 - 9} \leq 0$

30 $\dfrac{10}{x^4 - 16} > 0$

31 $\dfrac{x + 4}{2x - 1} < 3$

32 $\dfrac{1}{x - 2} > \dfrac{3}{x + 1}$

33 $x^3 < x$

34 $x^4 + 3x^2 > 4$

35 $x^3 - x^2 - 4x + 4 \geq 0$

36 $(x^2 - 4x + 4)(3x - 7) < 0$

37 $\dfrac{x^2 - x - 2}{x^2 - 4x + 3} \geq 0$

38 $\dfrac{x^2 + 2x - 3}{x^2 - 4x} \leq 0$

39 A boy shoots a toy rocket straight upward with an initial velocity of 72 feet per second. Its altitude s (in feet) after t seconds is given by $s = -16t^2 + 72t$. During what time interval will the rocket be at least 32 feet above the ground?

40 If the length of the pendulum in a grandfather clock is l cm, then its period T (seconds) is given by $T = 2\pi\sqrt{l/g}$ where g is a physical constant. If, under certain conditions, $g = 980$ and $98 \leq l \leq 100$, what is the corresponding range for T?

41 The braking distance d (in feet) of a car traveling v mph is approximated by $d = v + (v^2/20)$. Determine those velocities that result in braking distances of less than 75 feet.

42 For a satellite to maintain an orbit of height h km, its velocity (in km/sec) must equal $626.4/\sqrt{h + R}$ where $R = 6372$ km is the radius of the earth. What velocities will result in orbits of height more than 100 km from the earth's surface?

43 For a particular salmon population, the relationship between the number S of spawners and the number R of offspring that survive to maturity is given by the formula $R = 4500S/(S + 500)$. Under what conditions is $R > S$?

44 Determine when a cube with a side of x units will have the numerical value of its volume larger than the numerical value of its surface area.

45 For a drug to have a beneficial effect, its concentration in the bloodstream must exceed a certain value called the *minimum therapeutic level*. Suppose that the concentration c of a drug t hours after it is taken orally is given by $c = 20t/(t^2 + 4)$ mg/liter. If the minimum therapeutic level is 4 mg/liter, determine when this level is exceeded.

46 As a jar containing 10 moles of gas A is heated, the velocity of the gas molecules increases and a second gas, B, is formed. When two molecules of gas A collide, two molecules of gas B are formed. If the number of moles of gas B after t minutes is $10t/(t + 4)$, when will there be more of gas B than gas A?

2.8 Review

Define or discuss each of the following.

1 Solution of an equation

2 Root of an equation

3 Identity

4 Conditional equation

5 Equivalent equations

6 Linear equation

7 Quadratic equation

8 The Quadratic Formula

9 Discriminant of a quadratic equation

10 Complex numbers

11 Conjugate of a complex number

12 Extraneous solution

13 Properties of inequalities

14 Solution of an inequality

15 Equivalent inequalites

16 The graph of a set of real numbers

17 Open interval

18 Closed interval

19 Half-open interval

20 Infinite interval

Exercises 2.8

Solve the equations and inequalities in Exercises 1–34.

1 $\dfrac{3x+1}{5x+7} = \dfrac{6x+11}{10x-3}$

2 $2 - \dfrac{1}{x} = 1 + \dfrac{4}{x}$

3 $\dfrac{2}{x+5} - \dfrac{3}{2x+1} = \dfrac{5}{6x+3}$

4 $\dfrac{7}{x-2} - \dfrac{6}{x^2-4} = \dfrac{3}{2x+4}$

5 $\dfrac{1}{\sqrt{x}} - 2 = \dfrac{1-2\sqrt{x}}{\sqrt{x}}$

6 $2x^2 + 5x - 12 = 0$

7 $x(3x+4) = 5$

8 $\dfrac{x}{3x+1} = \dfrac{x-1}{2x+3}$

9 $(x-2)(x+1) = 3$

10 $4x^4 - 33x^2 + 50 = 0$

11 $x^{2/3} - 2x^{1/3} - 15 = 0$

12 $20x^3 + 8x^2 - 35x - 14 = 0$

13 $5x^2 = 2x - 3$

14 $x^2 + \frac{1}{3}x + 2 = 0$

15 $6x^4 + 29x^2 + 28 = 0$

16 $x^4 + x^2 + 1 = 0$

17 $\dfrac{1}{x} + 6 = \dfrac{5}{\sqrt{x}}$

18 $\sqrt[3]{4x-5} - 2 = 0$

19 $\sqrt{7x+2} + x = 6$

20 $\sqrt{x+4} = \sqrt[4]{6x+19}$

21 $\sqrt{3x+1} - \sqrt{x+4} = 1$

22 $10 - 7x < 4 + 2x$

23 $-\dfrac{1}{2} < \dfrac{2x+3}{5} < \dfrac{3}{2}$

24 $(3x-1)(10x+4) \ge (6x-5)(5x-7)$

25 $\dfrac{6}{10x+3} < 0$

26 $|4x+7| < 21$

27 $|16 - 3x| \ge 5$

28 $2 < |x-6| < 4$

29 $10x^2 + 11x > 6$

30 $x(x-3) \le 10$

31 $\dfrac{3}{2x+3} < \dfrac{1}{x-2}$

32 $\dfrac{x+1}{x^2-25} \le 0$

33 $x^3 > x^2$

34 $(x^2 - x)(x^2 - 5x + 6) < 0$

In Exercises 35–40 solve for the indicated variable in terms of the remaining variables.

35 $S = \pi(r+R)s$ for R

36 $S = 2(ab + bc + ac)$ for a

37 $n = \dfrac{\pi P R^4}{8VL}$ for R

38 $V = \frac{4}{3}\pi r^3$ for r

39 $\dfrac{1}{R} = (n-1)\left(\dfrac{1}{R_1} + \dfrac{1}{R_2}\right)$ for R_2

40 $V = \frac{1}{3}\pi h(R_1^2 + R_2^2 + R_1 R_2)$ for R_1

Express the complex numbers in Exercises 41–46 in the form $a + bi$ where a and b are real numbers.

41 $(7+5i) + (-8+3i)$

42 $(4+2i)(-5+4i)$

43 $(3+8i)^2$

44 $\dfrac{1}{9-2i}$

45 $\dfrac{6-3i}{2+7i}$

46 $\dfrac{20-8i}{4i}$

47 The diagonal of a square is 50 cm long. What change in the length of a side will increase the area by 46 cm²?

48 The surface area S of a sphere of radius r is given by the formula $S = 4\pi r^2$. If $r = 6$ inches, what change in radius will increase the surface area by 36π in²?

49 An airplane flew with the wind for 30 minutes and returned the same distance in 45 minutes. If the cruising speed of the airplane is 320 mph, what is the speed of the wind?

50 A merchant wishes to mix peanuts costing $1.50 per pound with cashews costing $4.00 per pound, obtaining 50 pounds of a mixture costing $2.40 per pound. How many pounds of each should be used?

51 A solution of ethyl alcohol that is 75% alcohol by weight is to be used as a bactericide. The solution is to be made by adding water to a 95% ethyl alcohol solution. How many grams of each should be used to prepare 400 grams of the bactericide?

52 A large solar heating panel requires 120 gallons of a fluid that is 30% antifreeze. The fluid comes in either a 50% solution or a 20% solution. How many gallons of each should be used to prepare the 120-gallon solution?

53 When two resistors R_1 and R_2 are connected in parallel, the net resistance R is given by $1/R = (1/R_1) + (1/R_2)$. If $R_1 = 5$ ohms, what is the value of R_2 such that the net resistance is 2 ohms?

54 A tugboat can bring a large ship into port in 2 hours, while a smaller tug can do the job in 3 hours. Predict the number of hours needed for both tugs to tow the ship to port.

55 A boat carries 10 gallons of gasoline and, when operated at full throttle in still water, travels at 20 mph and gets 16 miles per gallon. The boat is moving upstream into a 5 mph current. How far upstream can the boat travel and return on the 10 gallons of gasoline if it will be operated at full throttle during the entire trip?

56 A motorist averaged 45 mph driving outside the city limits and 25 mph within the city limits. If an 80-mile trip took the motorist 2 hours, how much time was spent driving within the city limits?

57 The width of a page in a book is 2 inches smaller than its length. The printed area is 72 in², with 1-inch margins at the top and bottom and $\frac{1}{2}$-inch margins at each side. What are the dimensions of the page?

58 A man puts a fence around a rectangular field and then subdivides the field into three smaller rectangular plots by placing two fences parallel to one of the sides. If the area of the field is 31,250 yd², and 1000 yards of fencing were used, what are the dimensions of the field?

59 A North-South highway intersects an East-West highway at a point P. An automobile crosses P at 10:00 A.M., traveling east at a constant rate of 20 mph. At that same instant another automobile is 2 miles north of P, traveling south at 50 mph. Find a formula that expresses the distance d between the automobiles at time t (hours) after 10:00 A.M. At what time will the automobiles be 104 miles apart?

60 A sales representative for a company estimates that gasoline consumption for her automobile averages 28 miles per gallon on the highway and 22 miles per gallon in the city. On a recent trip she covered 627 miles and used 24 gallons of gasoline. How much of the trip was spent driving in the city?

61 A large aquarium is to be constructed with sides 6 feet long and square ends.

(a) Find the length of the square end if the volume is to be 48 ft³.

(b) Find the length of the square end if 44 ft² of glass is to be used in its construction.

62 The length of a rectangular reflecting pool is to be four times its width, and a sidewalk of width 6 feet will surround the pool. If a total of 1440 ft² have been set aside for construction, what are the dimensions of the pool?

63 If the population P (in thousands) of a city is expected to increase according to the formula $P = 15 + \sqrt{3t + 2}$ where t (time) is in years, determine when the population is expected to reach 20,000 people.

64 A recent college graduate has two different job offers for a sales position in a computer firm. Job A pays $20,000 per year plus a 5% commission. Job B pays only $15,000 per year but the commission rate is 10%. How much yearly business must the salesman do for the second job to be more lucrative?

65 Boyle's Law for a certain gas states that $pv = 200$ where p denotes the pressure (lb/in²) and v denotes the volume (in³). If $25 \leq v \leq 50$, what is the corresponding range for p?

66 The *Lorentz contraction formula* in relativity theory relates the length L of an object moving at a velocity of V miles per second with respect to an observer to its length L_0 at rest. If c is the speed of light, then

$$L^2 = L_0^2 \left(1 - \frac{V^2}{c^2} \right).$$

For what velocities will L be less than $\frac{1}{2}L_0$? State your answer in terms of c.

CHAPTER 3

Functions and Graphs

One of the most important concepts in mathematics is that of *function*. ■ Indeed, without the notion of function, little progress could be made in mathematics or in any area of science. ■ In the first two sections of this chapter we consider coordinate systems and graphs in two dimensions. ■ The remainder of the chapter contains a discussion of functions and their graphs.

3.1 Coordinate Systems in Two Dimensions

In Section 1.2 we discussed a method of assigning coordinates to points on a line. Coordinate systems can also be introduced in planes by means of *ordered pairs*. The term **ordered pair** refers to two real numbers, where one is designated as the "first" number and the other as the "second." The symbol (a, b) is used to denote the ordered pair consisting of the real numbers a and b, where a is first and b is second. There are many uses for ordered pairs. We used them in Section 2.6 to denote open intervals. In this section they will represent points in a plane. Although ordered pairs are employed in different situations, there is little chance that we will confuse them, since it should always be clear from our discussion whether the symbol (a, b) represents an interval, a point, or some other mathematical object. We consider two ordered pairs (a, b) and (c, d) equal, and write

$$(a, b) = (c, d) \quad \text{if and only if} \quad a = c \quad \text{and} \quad b = d.$$

This implies, in particular, that $(a, b) \neq (b, a)$ if $a \neq b$.

A **rectangular,** or **Cartesian* coordinate system** may be introduced in a plane by considering two perpendicular coordinate lines in the plane that intersect in the origin O on each line. Unless specified otherwise, the same unit of length is chosen on each line. Usually one of the lines is horizontal with positive direction to the right, and the other line is vertical with positive direction upward, as indicated by the arrowheads in Figure 3.1. The two lines are called **coordinate axes,** and the point O is called the **origin.** The horizontal line is usually referred to as the **x-axis** and the vertical line as the **y-axis,** and they are labeled x and y, respectively. The plane is then called a **coordinate plane,** or an **xy-plane.** In certain applications different labels such as d or t are used for the coordinate lines. The coordinate axes divide the plane into four parts called the **first, second, third,** and **fourth quadrants** and labeled I, II, III, and IV, respectively. (See Figure 3.1(i).)

Each point P in an xy-plane may be assigned a unique ordered pair (a, b) as shown in Figure 3.1(ii). The number a is called the **x-coordinate** (or **abscissa**) of P, and b is called the **y-coordinate** (or **ordinate**). We say that P *has coordinates* (a, b). Conversely, every ordered pair (a, b) deter-

---

*The term "Cartesian" is used in honor of the French mathematician and philosopher René Descartes (1596–1650), who was one of the first to employ such coordinate systems.

FIGURE 3.1

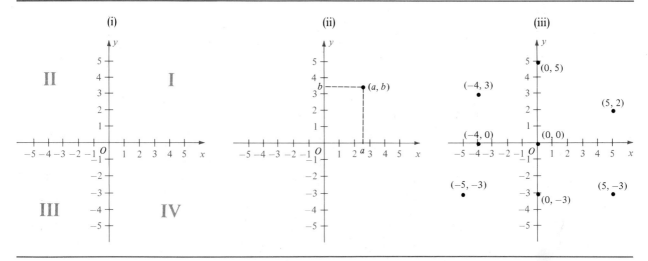

(i)   (ii)   (iii)

mines a point P in the xy-plane with coordinates a and b. We often refer to the *point* (a, b), or $P(a, b)$, meaning the point P with x-coordinate a and y-coordinate b. To **plot a point** $P(a, b)$ means to locate P in a coordinate plane and represent it by a dot, as illustrated by some points in Figure 3.1(iii).

The next statement provides a formula for finding the distance between two points in a coordinate plane.

DISTANCE FORMULA

> The distance $d(P_1, P_2)$ between any two points $P_1(x_1, y_1)$ and $P_2(x_2, y_2)$ in a coordinate plane is
> $$d(P_1, P_2) = \sqrt{(x_2 - x_1)^2 + (y_2 - y_1)^2}.$$

FIGURE 3.2

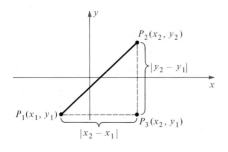

PROOF If $x_1 \neq x_2$ and $y_1 \neq y_2$, then, as illustrated in Figure 3.2, the points P_1, P_2, and $P_3(x_2, y_1)$ are vertices of a right triangle. By the Pythagorean Theorem,

$$[d(P_1, P_2)]^2 = [d(P_1, P_3)]^2 + [d(P_3, P_2)]^2.$$

We see from the figure that

$$d(P_1, P_3) = |x_2 - x_1| \quad \text{and} \quad d(P_3, P_2) = |y_2 - y_1|.$$

Since $|a|^2 = a^2$ for every real number a, we may write

$$[d(P_1, P_2)]^2 = (x_2 - x_1)^2 + (y_2 - y_1)^2.$$

Taking the square root of each side of the last equation gives us the desired formula.

If $y_1 = y_2$, the points P_1 and P_2 lie on the same horizontal line and

$$d(P_1, P_2) = |x_2 - x_1| = \sqrt{(x_2 - x_1)^2}.$$

Similarly, if $x_1 = x_2$, the points are on the same vertical line and

$$d(P_1, P_2) = |y_2 - y_1| = \sqrt{(y_2 - y_1)^2}.$$

These are special cases of the Distance Formula.

Although we referred to Figure 3.2, the argument used in our proof is independent of the positions of the points P_1 and P_2. ☐

In applying the Distance Formula, note that $d(P_1, P_2) = d(P_2, P_1)$ and, hence, the order in which we subtract the x-coordinates and the y-coordinates of the points is immaterial.

FIGURE 3.3

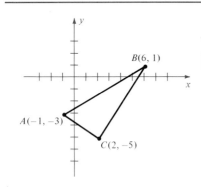

$A(-1, -3)$

$B(6, 1)$

$C(2, -5)$

EXAMPLE 1 Plot the points $A(-1, -3)$, $B(6, 1)$, and $C(2, -5)$. Prove that the triangle with vertices A, B, and C is a right triangle and find its area.

SOLUTION The points and the triangle are shown in Figure 3.3. From plane geometry, a triangle is a right triangle if and only if the sum of the squares of two of its sides is equal to the square of the remaining side. Using the Distance Formula,

$$d(A, B) = \sqrt{(-1 - 6)^2 + (-3 - 1)^2} = \sqrt{49 + 16} = \sqrt{65}$$
$$d(B, C) = \sqrt{(6 - 2)^2 + (1 + 5)^2} = \sqrt{16 + 36} = \sqrt{52}$$
$$d(A, C) = \sqrt{(-1 - 2)^2 + (-3 + 5)^2} = \sqrt{9 + 4} = \sqrt{13}.$$

Since $[d(A, B)]^2 = [d(B, C)]^2 + [d(A, C)]^2$, the triangle is a right triangle with hypotenuse AB. The area is $\frac{1}{2}\sqrt{52}\sqrt{13} = 13$ square units. ∎

It is easy to obtain a formula for the midpoint of a line segment. Let $P_1(x_1, y_1)$ and $P_2(x_2, y_2)$ be two points in a coordinate plane and let M be the midpoint of the segment $P_1 P_2$. The lines through P_1 and P_2 parallel

FIGURE 3.4

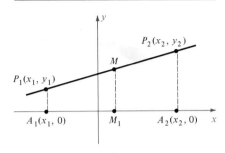

to the y-axis intersect the x-axis at $A_1(x_1, 0)$ and $A_2(x_2, 0)$ and, from plane geometry, the line through M parallel to the y-axis bisects the segment $A_1 A_2$ (see Figure 3.4). If $x_1 < x_2$, then $x_2 - x_1 > 0$, and hence $d(A_1, A_2) = x_2 - x_1$. Since M_1 is halfway from A_1 to A_2, the x-coordinate of M_1 is

$$x_1 + \tfrac{1}{2}(x_2 - x_1) = x_1 + \tfrac{1}{2}x_2 - \tfrac{1}{2}x_1$$
$$= \tfrac{1}{2}x_1 + \tfrac{1}{2}x_2$$
$$= \frac{x_1 + x_2}{2}.$$

It follows that the x-coordinate of M is also $(x_1 + x_2)/2$. Similarly, the y-coordinate of M is $(y_1 + y_2)/2$. Moreover, these formulas hold for all positions of P_1 and P_2. This gives us the following result.

MIDPOINT FORMULA

> The midpoint of the line segment from $P_1(x_1, y_1)$ to $P_2(x_2, y_2)$ is
>
> $$\left(\frac{x_1 + x_2}{2}, \frac{y_1 + y_2}{2} \right).$$

EXAMPLE 2   Find the midpoint M of the line segment from $P_1(-2, 3)$ to $P_2(4, -2)$. Plot the points P_1, P_2, and M and verify that

$$d(P_1, M) = d(P_2, M).$$

SOLUTION   By the Midpoint Formula, the coordinates of M are

FIGURE 3.5

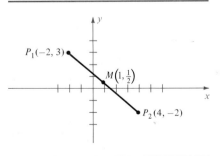

$$\left(\frac{-2 + 4}{2}, \frac{3 + (-2)}{2} \right) \quad \text{or} \quad \left(1, \frac{1}{2} \right).$$

The three points P_1, P_2, and M are plotted in Figure 3.5. Using the Distance Formula,

$$d(P_1, M) = \sqrt{(-2 - 1)^2 + (3 - \tfrac{1}{2})^2} = \sqrt{9 + \tfrac{25}{4}}$$
$$d(P_2, M) = \sqrt{(4 - 1)^2 + (-2 - \tfrac{1}{2})^2} = \sqrt{9 + \tfrac{25}{4}}.$$

Hence, $d(P_1, M) = d(P_2, M)$.

Exercises 3.1

1 Plot the following points on a rectangular coordinate system: $A(5, -2)$, $B(-5, -2)$, $C(5, 2)$, $D(-5, 2)$, $E(3, 0)$, $F(0, 3)$.

2 Plot the points $A(-3, 1)$, $B(3, 1)$, $C(-2, -3)$, $D(0, 3)$, and $E(2, -3)$ on a rectangular coordinate system and then draw the line segments AB, BC, CD, DE, and EA.

3 Plot $A(0, 0)$, $B(1, 1)$, $C(3, 3)$, $D(-1, -1)$, and $E(-2, -2)$. Describe the set of all points of the form (x, x) where x is a real number.

4 Plot $A(0, 0)$, $B(1, -1)$, $C(2, -2)$, $D(-1, 1)$, and $E(-3, 3)$. Describe the set of all points of the form $(a, -a)$ where a is a real number.

5 Describe the set of all points $P(x, y)$ in a coordinate plane such that:

(a) $x = 3$ (b) $y = -1$

(c) $x \geq 0$ (d) $xy > 0$

(e) $y < 0$.

6 Describe the set of all points $P(x, y)$ in a coordinate plane such that:

(a) $y = 0$ (b) $x = -5$

(c) $x/y < 0$ (d) $xy = 0$

(e) $y > 1$.

In Exercises 7–12 find (a) the distance $d(A, B)$ between the given points A and B; (b) the midpoint of the segment AB.

7 $A(4, -3)$, $B(6, 2)$ **8** $A(-2, -5)$, $B(4, 6)$

9 $A(-5, 0)$, $B(-2, -2)$ **10** $A(6, 2)$, $B(6, -2)$

11 $A(7, -3)$, $B(3, -3)$ **12** $A(-4, 7)$, $B(0, -8)$

In Exercises 13 and 14 prove that the triangle with the indicated vertices is a right triangle and find its area.

13 $A(8, 5)$, $B(1, -2)$, $C(-3, 2)$

14 $A(-6, 3)$, $B(3, -5)$, $C(-1, 5)$

15 Prove that the following points are vertices of a square: $A(-4, 2)$, $B(1, 4)$, $C(3, -1)$, $D(-2, -3)$.

16 Prove that the following points are vertices of a parallelogram: $A(-4, -1)$, $B(0, -2)$, $C(6, 1)$, $D(2, 2)$.

17 Given $A(-3, 8)$, find the coordinates of the point B such that $M(5, -10)$ is the midpoint of AB.

18 Given $A(5, -8)$ and $B(-6, 2)$, find the point on AB that is three-fourths of the way from A to B.

19 Given $A(-4, -3)$ and $B(6, 1)$, prove that $P(5, -11)$ is on the perpendicular bisector of AB.

20 Given $A(-4, -3)$, and $B(6, 1)$, find a formula that expresses the fact that $P(x, y)$ is on the perpendicular bisector of AB.

21 Find a formula that expresses the fact that $P(x, y)$ is a distance 5 from the origin. Describe the totality of all such points.

22 If r is a positive real number, find a formula that states that $P(x, y)$ is a distance r from a fixed point $C(h, k)$. Describe the totality of all such points.

23 Find all points on the y-axis that are 6 units from the point $(5, 3)$.

24 Find all points on the x-axis that are 5 units from $(-2, 4)$.

25 Let S denote the set of points of the form $(2x, x)$ where x is a real number. Find the point in S that is in the third quadrant and is a distance 5 from the point $(1, 3)$.

26 Let S denote the set of points of the form (x, x), where x is a real number. Find all points in S that are a distance 3 from the point $(-2, 1)$.

27 For what values of a is the distance between $(a, 3)$ and $(5, 2a)$ greater than $\sqrt{26}$?

28 Given the points $A(-2, 0)$ and $B(2, 0)$, find a formula not containing radicals that expresses the fact that the sum of the distances from $P(x, y)$ to A and to B, respectively, is 5.

29 Prove that the midpoint of the hypotenuse of any right triangle is equidistant from the vertices. (*Hint:* Label the vertices of the triangle $O(0, 0)$, $A(a, 0)$, and $B(0, b)$.)

30 Prove that the diagonals of any parallelogram bisect each other. (*Hint:* Label three of the vertices of the parallelogram $O(0, 0)$, $A(a, b)$, and $C(0, c)$.)

3.2 Graphs

If W is a set of ordered pairs, we may consider the point $P(x, y)$ in a coordinate plane that corresponds to the ordered pair (x, y) in W. The **graph** of W is the set of all such points. The phrase "sketch the graph of W" means to illustrate the significant features of the graph geometrically on a coordinate plane.

FIGURE 3.6

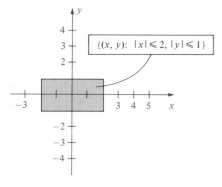

EXAMPLE 1   Sketch the graph of $W = \{(x, y) : |x| \leq 2, |y| \leq 1\}$.

SOLUTION   The notation describing W may be translated "the set of all ordered pairs (x, y) such that $|x| \leq 2$ and $|y| \leq 1$." These inequalities are equivalent to $-2 \leq x \leq 2$ and $-1 \leq y \leq 1$. Thus, the graph of W consists of all points within and on the boundary of the rectangular region shown in Figure 3.6.   ∎

EXAMPLE 2   Sketch the graph of $W = \{(x, y) : y = 2x - 1\}$.

SOLUTION   We wish to find the points (x, y), where the ordered pair (x, y) is in W. It is convenient to list coordinates of several such points in the following tabular form, where for each x, the value for y is obtained from $y = 2x - 1$.

x	-2	-1	0	1	2	3
y	-5	-3	-1	1	3	5

After we plot the points with these coordinates, it appears that they lie on a line and we sketch the graph (see Figure 3.7). Ordinarily, the few points we have plotted would not be enough to illustrate the graph; however, in this elementary case we can be reasonably sure that the graph is a line. In Section 3.5 we will prove this fact.   ∎

FIGURE 3.7

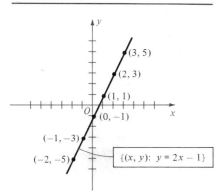

The x-coordinates of points at which a graph intersects the x-axis are called the **x-intercepts** of the graph. The y-coordinates of points at which a graph intersects the y-axis are called the **y-intercepts.** The graph in Figure 3.7 has one x-intercept, $\frac{1}{2}$, and one y-intercept, -1.

It is impossible to sketch the entire graph in Example 2, since x may be assigned values that are numerically as large as desired. Nevertheless, we call the drawing in Figure 3.7 *the graph of W* or *a sketch of the graph.* It is understood that the drawing is only a device for visualizing the actual graph and the line does not terminate as shown in the figure. In general, the sketch of a graph should illustrate enough of the graph so that the remaining parts are evident.

Given an equation in x and y, we say that an ordered pair (a, b) is a **solution** of the equation if equality is obtained when a is substituted for x and b for y. For example, $(2, 3)$ is a solution of $y = 2x - 1$ since substitution of 2 for x and 3 for y leads to $3 = 4 - 1$, or $3 = 3$. Two equations in x and y are **equivalent** if they have exactly the same solutions. The solutions of an equation in x and y determine a set W of ordered pairs, and we define the **graph of the equation** as the graph of W. Note that the graph of the equation $y = 2x - 1$ is the same as the graph of the set W in Example 2 (see Figure 3.7).

To sketch the graph of an equation we may plot points until a pattern emerges, and then sketch the graph accordingly. This is obviously a crude (and often inaccurate) way to arrive at the graph; however, it is a method often employed in elementary courses. As we progress through this text, techniques will be introduced that will enable us to sketch accurate graphs without plotting many points. To sketch graphs when complicated expressions are involved, it is usually necessary to employ methods introduced in calculus.

FIGURE 3.8

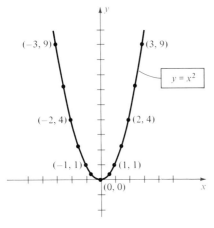

EXAMPLE 3 Sketch the graph of the equation $y = x^2$.

SOLUTION To sketch the graph, we must plot more points than in the previous example. Increasing successive x-coordinates by $\frac{1}{2}$, we obtain the following table.

x	-3	$-\frac{5}{2}$	-2	$-\frac{3}{2}$	-1	$-\frac{1}{2}$	0	$\frac{1}{2}$	1	$\frac{3}{2}$	2	$\frac{5}{2}$	3
y	9	$\frac{25}{4}$	4	$\frac{9}{4}$	1	$\frac{1}{4}$	0	$\frac{1}{4}$	1	$\frac{9}{4}$	4	$\frac{25}{4}$	9

Larger values of x produce larger values of y. For example, the points $(4, 16)$, $(5, 25)$, and $(6, 36)$ are on the graph, as are $(-4, 16)$, $(-5, 25)$, and $(-6, 36)$. Plotting the points given by the table and drawing a smooth curve through these points gives us the sketch in Figure 3.8, in which several points are labeled. ■

The graph in Example 3 is a **parabola.** The y-axis is called the **axis of the parabola.** The lowest point $(0, 0)$ is the **vertex** of the parabola and we say that the parabola **opens upward.** If the graph were inverted, as would be the case for $y = -x^2$, then the parabola **opens downward** and the vertex $(0, 0)$ is the highest point on the graph. In general, the graph of *any* equation of the form $y = ax^2$ for $a \neq 0$ is a parabola with vertex $(0, 0)$. Parabolas may also open to the right or to the left (see Example 4). In Chapter 4 we shall consider parabolas with axes *parallel* to the x- or y-axes.

If the coordinate plane in Figure 3.8 is folded along the y-axis, then the graph that lies in the left half of the plane coincides with that in the right half. We say that **the graph is symmetric with respect to the y-axis.** As in Figure 3.9(i), a graph is symmetric with respect to the y-axis provided that the point $(-x, y)$ is on the graph whenever (x, y) is on the graph. As in Figure 3.9(ii), **a graph is symmetric with respect to the x-axis** if whenever a point (x, y) is on the graph, then $(x, -y)$ is also on the graph. Certain graphs possess a type of symmetry, called **symmetry with respect to the origin.** In this situation, whenever a point (x, y) is on the graph, then $(-x, -y)$ is also on the graph, as illustrated in Figure 3.9(iii).

FIGURE 3.9

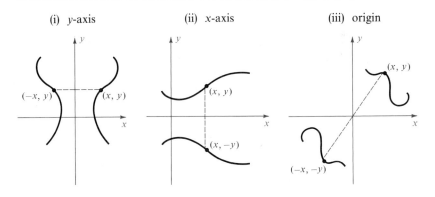

(i) y-axis   (ii) x-axis   (iii) origin

The following tests are useful for investigating these three types of symmetry for graphs of equations in x and y.

TESTS FOR SYMMETRY

(i) The graph of an equation is symmetric with respect to the y-axis if substitution of $-x$ for x leads to an equivalent equation.

(ii) The graph of an equation is symmetric with respect to the x-axis if substitution of $-y$ for y leads to an equivalent equation.

(iii) The graph of an equation is symmetric with respect to the origin if the simultaneous substitution of $-x$ for x and $-y$ for y leads to an equivalent equation.

If, in the equation of Example 3, we substitute $-x$ for x, we obtain $y = (-x)^2$, which is equivalent to $y = x^2$. Hence, by Symmetry Test (i), the graph is symmetric with respect to the y-axis.

If symmetry with respect to an axis exists, it is sufficient to determine the graph in half of the coordinate plane, since the remainder of the graph may be sketched by taking a mirror image, or reflection, of that half.

FIGURE 3.10

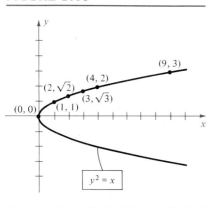

$y^2 = x$

EXAMPLE 4   Sketch the graph of the equation $y^2 = x$.

SOLUTION   Since substitution of $-y$ for y does not change the equation, the graph is symmetric with respect to the x-axis. (See Symmetry Test (ii).) Thus, it is sufficient to plot points with nonnegative y-coordinates and then reflect through the x-axis. Since $y^2 = x$, the y-coordinates of points above the x-axis are given by $y = \sqrt{x}$. Coordinates of some points on the graph are listed in the following table.

x	0	1	2	3	4	9
y	0	1	$\sqrt{2} \approx 1.4$	$\sqrt{3} \approx 1.7$	2	3

A portion of the graph is sketched in Figure 3.10. The graph is a parabola that opens to the right, with its vertex at the origin. In this case the x-axis is the axis of the parabola. ∎

FIGURE 3.11

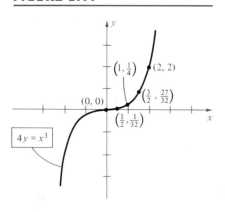

$4y = x^3$

EXAMPLE 5   Sketch the graph of the equation $4y = x^3$.

SOLUTION   If we substitute $-x$ for x and $-y$ for y, then

$$4(-y) = (-x)^3 \quad \text{or} \quad -4y = -x^3.$$

Multiplying both sides by -1, we see that the last equation has the same solutions as the equation $4y = x^3$. Hence, from Symmetry Test (iii), the graph is symmetric with respect to the origin. The following table lists some points on the graph.

x	0	$\frac{1}{2}$	1	$\frac{3}{2}$	2	$\frac{5}{2}$
y	0	$\frac{1}{32}$	$\frac{1}{4}$	$\frac{27}{32}$	2	$\frac{125}{32}$

By symmetry (or substitution) we see that the points $(-1, -\frac{1}{4})$, $(-2, -2)$, and so on, are on the graph. Plotting points leads to the sketch in Figure 3.11. ∎

If $C(h, k)$ is a point in a coordinate plane, then a circle with center C and radius $r > 0$ consists of all points in the plane that are r units from C. As shown in Figure 3.12(i), a point $P(x, y)$ is on the circle if and only if

FIGURE 3.12

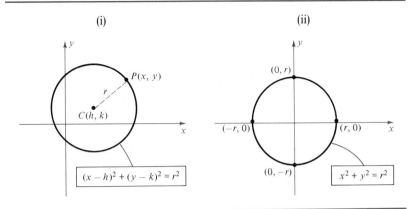

(i)   (ii)

$d(C, P) = r$ or, by the Distance Formula, if and only if

$$\sqrt{(x - h)^2 + (y - k)^2} = r.$$

The following equivalent equation is called the **equation of a circle of radius r and center (h, k).**

EQUATION OF A CIRCLE

$$(x - h)^2 + (y - k)^2 = r^2, \qquad r > 0$$

If $h = 0$ and $k = 0$, this equation reduces to $x^2 + y^2 = r^2$, which is an equation of a circle of radius r with center at the origin (see Figure 3.12(ii)). If $r = 1$, the graph is called a **unit circle.**

FIGURE 3.13

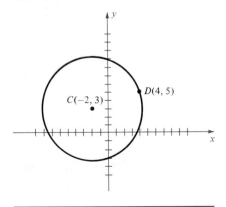

EXAMPLE 6   Find an equation of the circle that has center $C(-2, 3)$ and contains the point $D(4, 5)$.

SOLUTION   The circle is illustrated in Figure 3.13. Since D is on the circle, the radius r is $d(C, D)$. By the Distance Formula,

$$r = \sqrt{(-2 - 4)^2 + (3 - 5)^2} = \sqrt{36 + 4} = \sqrt{40}.$$

Using the equation of a circle with $h = -2$, $k = 3$, and $r = \sqrt{40}$, we obtain

$$(x + 2)^2 + (y - 3)^2 = 40$$

or

$$x^2 + y^2 + 4x - 6y - 27 = 0. \qquad \blacksquare$$

Squaring terms of $(x - h)^2 + (y - k)^2 = r^2$ and simplifying leads to an equation of the form

$$x^2 + y^2 + ax + by + c = 0$$

for some real numbers a, b, and c. Conversely, if we begin with the last equation, it is always possible, by *completing squares* (see page 78) to obtain an equation of the form

$$(x - h)^2 + (y - k)^2 = d.$$

The method will be illustrated in Example 7. If $d > 0$ the graph is a circle with center (h, k) and radius $r = \sqrt{d}$. If $d = 0$ the graph consists of only one point (h, k). Finally, if $d < 0$ the equation has no real solutions and hence there is no graph.

EXAMPLE 7 Find the center and radius of the circle with equation

$$x^2 + y^2 - 4x + 6y - 3 = 0.$$

SOLUTION We begin by arranging the equation as follows:

$$(x^2 - 4x) + (y^2 + 6y) = 3.$$

Next we complete the squares for the expressions within parentheses. Of course, to obtain equivalent equations, we must add the numbers to *both* sides of the equation. To complete the square for $x^2 - 4x$ we add $(-\frac{4}{2})^2 = (-2)^2 = 4$ to both sides, whereas for $y^2 + 6y$ we add $(\frac{6}{2})^2 = 3^2 = 9$. The resulting equation is

$$(x^2 - 4x + 4) + (y^2 + 6y + 9) = 3 + 4 + 9$$

or $$(x - 2)^2 + (y + 3)^2 = 16.$$

It follows that the center is $(2, -3)$ and the radius is 4. ∎

Exercises 3.2

In Exercises 1–8 sketch the graph of the set W of ordered pairs.

1 $W = \{(x, y): x = 4\}$

2 $W = \{(x, y): y = -3\}$

3 $W = \{(x, y): xy < 0\}$

4 $W = \{(x, y): xy = 0\}$

5 $W = \{(x, y): |x| < 2, |y| > 1\}$

6 $W = \{(x, y): |x| > 1, |y| \leq 2\}$

7 $W = \{(x, y):|x - 2| < 3, |y| \le 1\}$

8 $W = \{(x, y):|x - 1| < 5, |y + 1| < 2\}$

In Exercises 9–32 sketch the graph of the equation after plotting a sufficient number of points.

9 $y = 3x + 1$

10 $y = 4x - 3$

11 $y = -2x + 3$

12 $y = 2 - 3x$

13 $2y + 3x + 6 = 0$

14 $3y - 2x = 6$

15 $y = 2x^2$

16 $y = -x^2$

17 $y = 2x^2 - 1$

18 $y = -x^2 + 2$

19 $4y = x^2$

20 $3y + x^2 = 0$

21 $y = -\frac{1}{2}x^3$

22 $y = \frac{1}{2}x^3$

23 $y = x^3 - 2$

24 $y = 2 - x^3$

25 $y = \sqrt{x}$

26 $y = \sqrt{x} - 1$

27 $y = \sqrt{-x}$

28 $y = \sqrt{x - 1}$

29 $y = \sqrt{9 - x^2}$

30 $y = -\sqrt{4 - x^2}$

31 $x = -\sqrt{9 - y^2}$

32 $x = \sqrt{4 - y^2}$

Describe the graphs of the equations in Exercises 33–40.

33 $x^2 + y^2 = 16$

34 $x^2 + y^2 = 25$

35 $9x^2 + 9y^2 = 1$

36 $2x^2 + 2y^2 = 1$

37 $(x - 2)^2 + (y + 1)^2 = 4$

38 $(x + 3)^2 + (y - 4)^2 = 1$

39 $x^2 + (y - 3)^2 = 9$

40 $(x + 5)^2 + y^2 = 16$

In Exercises 41–50 find an equation of a circle satisfying the stated conditions.

41 Center $C(3, -2)$, radius 4

42 Center $C(-5, 2)$, radius 5

43 Center $C(\frac{1}{2}, -\frac{3}{2})$, radius 2

44 Center $C(\frac{1}{3}, 0)$, radius $\sqrt{3}$

45 Center at the origin, passing through $P(-3, 5)$

46 Center $C(-4, 6)$, passing through $P(1, 2)$

47 Center $C(-4, 2)$, tangent to the x-axis

48 Center $C(3, -5)$, tangent to the y-axis

49 Endpoints of a diameter $A(4, -3)$ and $B(-2, 7)$

50 Tangent to both axes, center in the first quadrant, radius 2

In Exercises 51–60 find the center and radius of the circle with the given equation.

51 $x^2 + y^2 + 2x - 10y + 10 = 0$

52 $x^2 + y^2 - 8x + 4y + 15 = 0$

53 $x^2 + y^2 - 6y + 5 = 0$

54 $x^2 + y^2 + 14x + 46 = 0$

55 $x^2 + y^2 - 10x + 8y = 0$

56 $x^2 + y^2 + 2y = 0$

57 $4x^2 + 4y^2 + 8x - 8y + 7 = 0$

58 $2x^2 + 2y^2 + 8x + 7 = 0$

59 $3x^2 + 3y^2 - 3x + 2y + 1 = 0$

60 $9x^2 + 9y^2 + 12x - 6y + 4 = 0$

Calculator Exercises 3.2

Sketch the graphs of the following equations after approximating the coordinates of a sufficient number of points.

1 $y = x^{2/3}$

2 $y = x^{3/2}$

3 $y = x^{5/2}$

4 $y = 2x^{3/5} - 1$

5 $y = \sqrt[5]{x} - 1$

6 $y = \sqrt[4]{x} + 1$

7 $x^{2/3} + y^{2/3} = 1$

8 $y = x/(x^2 + 1)$

3.3 Definition of Function

The notion of **correspondence** occurs frequently in everyday life. For example, to each book in a library there corresponds the number of pages in the book. As another example, to each human being there corresponds a birth date. To cite a third example, if the temperature of the air is recorded throughout a day, then at each instant of time there is a corresponding temperature. These examples of correspondence involve two sets D and E. In our first example D denotes the set of books in a library and E the set of positive integers. For each book x in D there corresponds a positive integer y in E, namely the number of pages in the book.

We sometimes depict correspondences by diagrams of the type shown in Figure 3.14, where the sets D and E are represented by points within regions in a plane. The curved arrow indicates that the element y of E corresponds to the element x of D. The two sets may have elements in common. As a matter of fact, we often have $D = E$.

Our examples indicate that *to each x in D there corresponds one and only one y in E*; that is, y *is unique* for a given x. However, the same element of E may correspond to different elements of D. For example, two different books may have the same number of pages, two different people may have the same birthday, and so on.

In most of our work D and E will be sets of numbers. To illustrate, let D and E both denote the set $\mathbb{R}$ of real numbers, and to each real number x let us assign its square x^2. Thus, to 3 we assign 9, to -5 we assign 25, and to $\sqrt{2}$, the number 2. This gives us a correspondence from $\mathbb{R}$ to $\mathbb{R}$.

Each of the preceding examples of a correspondence is a *function*.

FIGURE 3.14

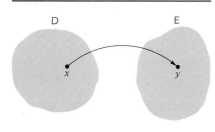

DEFINITION

> A **function** f from a set D to a set E is a correspondence that assigns to each element x of D a unique element y of E.

The element y of E is called the **value** of f at x and is denoted by $f(x)$ (read "f of x"). The set D is called the **domain** of the function. The **range** of f consists of all possible values $f(x)$, where x is in D.

We may now sketch the diagram in Figure 3.15. The curved arrows indicate that the elements $f(x)$, $f(w)$, $f(z)$, and $f(a)$ of E correspond to the elements x, w, z, and a of D. It is important to remember that *to each x in D there is assigned precisely one value $f(x)$ in E;* however, different elements of D, such as w and z in Figure 3.15, may have the same value in E.

FIGURE 3.15

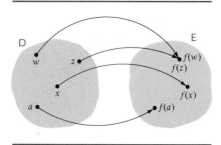

The symbols

$$D \xrightarrow{f} E, \quad f: D \to E \quad \text{or}$$

signify that f is a function from D to E. Beginning students are sometimes confused by the notations f and $f(x)$. Remember that f is used to represent the function. It is neither in D nor in E. However, $f(x)$ is an element of E, namely the element that f assigns to x.

Equality of two functions f and g from D to E is defined as follows:

$$f = g \quad \text{if and only if} \quad f(x) = g(x) \quad \text{for every } x \text{ in } D.$$

For example, if $g(x) = \frac{1}{2}(2x^2 - 6) + 3$ and $f(x) = x^2$ for all x in $\mathbb{R}$, then $g = f$ since $\frac{1}{2}(2x^2 - 6) + 3 = x^2$ for every x.

EXAMPLE 1   Let f be the function with domain $\mathbb{R}$ such that $f(x) = x^2$ for every x in $\mathbb{R}$. Find $f(-6)$, $f(\sqrt{3})$, and $f(a)$ for any real number a. What is the range of f?

SOLUTION   We may find values of f by substituting for x in the equation $f(x) = x^2$. Thus,

$$f(-6) = (-6)^2 = 36, \quad f(\sqrt{3}) = (\sqrt{3})^2 = 3, \quad \text{and} \quad f(a) = a^2.$$

By definition, the range of f consists of all numbers of the form $f(a) = a^2$, where a is in $\mathbb{R}$. Since the square of every real number is nonnegative, the range is contained in the set of all nonnegative real numbers. Moreover, every nonnegative real number c is a value of f since $f(\sqrt{c}) = (\sqrt{c})^2 = c$. Hence, the range of f is the set of all nonnegative real numbers.   ∎

It is often difficult to find the range of a function algebraically, as in the preceding example. In the next section we shall see how the range can be obtained from a graph.

If a function is defined as in Example 1, the symbols used for the function and variable are immaterial; that is, expressions such as $f(x) = x^2$, $f(s) = s^2$, $g(t) = t^2$, and $k(r) = r^2$ all define the same function. This is true because if a is any number in the domain, then the same value a^2 is obtained no matter which expression is employed.

In the remainder of our work the phrase "f is a function" will mean that the domain and range are sets of real numbers. If a function is defined by means of an expression, as in Example 1, and the domain D is not stated explicitly, then D is considered to be the totality of real numbers x such that $f(x)$ is real. To illustrate, if $f(x) = \sqrt{x - 2}$, then the domain is assumed to be the set of real numbers x such that $\sqrt{x - 2}$ is real; that is,

$x - 2 \geq 0$, or $x \geq 2$. Thus, the domain is the infinite interval $[2, \infty)$. If x is in the domain, we say that f **is defined at** x, or that $f(x)$ **exists.** If a set S is contained in the domain, we often say that f **is defined on** S. The terminology f **is undefined at** x means that x is not in the domain of f.

EXAMPLE 2

If $g(x) = \dfrac{\sqrt{4 + x}}{1 - x}$, find

(a) the domain of g.   (b) $g(5)$, $g(-2)$, $g(-a)$, $-g(a)$.

SOLUTION

(a) The fractional expression is a real number if and only if the radicand is nonnegative and the denominator is different from 0. Thus, $g(x)$ exists if and only if

$$4 + x \geq 0 \quad \text{and} \quad 1 - x \neq 0,$$

or equivalently, $x \geq -4 \quad \text{and} \quad x \neq 1.$

The domain may be expressed in terms of intervals as $[-4, 1) \cup (1, \infty)$.

(b) To find values of g, we substitute for x as follows:

$$g(5) = \frac{\sqrt{4 + 5}}{1 - 5} = \frac{\sqrt{9}}{-4} = -\frac{3}{4}$$

$$g(-2) = \frac{\sqrt{4 + (-2)}}{1 - (-2)} = \frac{\sqrt{2}}{3}$$

$$g(-a) = \frac{\sqrt{4 + (-a)}}{1 - (-a)} = \frac{\sqrt{4 - a}}{1 + a}$$

$$-g(a) = -\frac{\sqrt{4 + a}}{1 - a} = \frac{\sqrt{4 + a}}{a - 1}$$

Manipulations of the type given in the next example occur in calculus.

EXAMPLE 3   Suppose $f(x) = x^2 + 3x - 2$. If a and h are real numbers, and $h \neq 0$, find and simplify

$$\frac{f(a + h) - f(a)}{h}.$$

SOLUTION Since

$$f(a + h) = (a + h)^2 + 3(a + h) - 2$$
$$= (a^2 + 2ah + h^2) + (3a + 3h) - 2$$

and

$$f(a) = a^2 + 3a - 2,$$

we see that

$$\frac{f(a + h) - f(a)}{h} = \frac{[(a^2 + 2ah + h^2) + (3a + 3h) - 2] - (a^2 + 3a - 2)}{h}$$

$$= \frac{2ah + h^2 + 3h}{h}$$

$$= 2a + h + 3.$$ ∎

Many formulas that occur in mathematics and the sciences determine functions. As an illustration, the formula $A = \pi r^2$ for the area A of a circle of radius r assigns to each positive real number r a unique value of A. This determines a function f such that $f(r) = \pi r^2$, and we may write $A = f(r)$. The letter r, which represents an arbitrary number from the domain of f, is often called an **independent variable.** The letter A, which represents a number from the range of f, is called a **dependent variable,** since its value depends on the number assigned to r. If two variables r and A are related in this manner, it is customary to use the phrase "A is a function of r." To cite another example, if an automobile travels at a uniform rate of 50 miles per hour, then the distance d (miles) traveled in time t (hours) is given by $d = 50t$, and hence, the distance d is a function of time t.

EXAMPLE 4 Express the radius of a circle as a function of its area.

SOLUTION If A and r denote the area and radius, then

$$A = \pi r^2 \quad \text{and} \quad r^2 = \frac{A}{\pi}.$$

Since both A and r are positive,

$$r = \sqrt{\frac{A}{\pi}}.$$

This shows that r is a function of A, since to each value of A there corresponds a unique value $\sqrt{A/\pi}$ of r. ∎

EXAMPLE 5   A steel storage tank for propane gas is to be constructed in the shape of a right circular cylinder of altitude 10 feet with a hemisphere attached to each end. The radius r is yet to be determined. Express the volume V of the tank as a function of r.

FIGURE 3.16

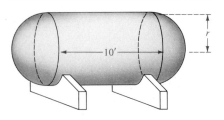

SOLUTION   The tank is sketched in Figure 3.16. The volume of the cylindrical part of the tank may be found by multiplying the altitude 10 by the area πr^2 of the base of the cylinder. This gives us

$$\text{volume of cylinder} = 10(\pi r^2) = 10\pi r^2.$$

The two hemispherical ends, taken together, form a sphere of radius r. Using the formula for the volume of a sphere, we obtain

$$\text{volume of the two ends} = \tfrac{4}{3}\pi r^3.$$

Thus, the volume V of the tank is

$$V = \tfrac{4}{3}\pi r^3 + 10\pi r^2.$$

We may write this in the factored form

$$V = \tfrac{1}{3}\pi r^2(4r + 30) = \tfrac{2}{3}\pi r^2(2r + 15). \qquad \blacksquare$$

FIGURE 3.17

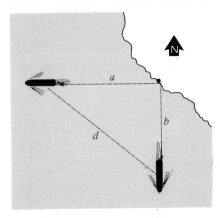

EXAMPLE 6   Two ships leave port at the same time, one sailing west at a rate of 17 miles per hour and the other sailing south at 12 miles per hour. If t is the time (in hours) after their departure, express the distance d between the ships as a function of t.

SOLUTION   To help visualize the problem we begin by drawing a picture and labeling it, as in Figure 3.17. Using the Pythagorean Theorem, we obtain

$$d^2 = a^2 + b^2$$

or

$$d = \sqrt{a^2 + b^2}.$$

Since distance = (rate)(time), and since the rates are 17 and 12, respectively,

$$a = 17t \quad \text{and} \quad b = 12t.$$

Substitution in $d = \sqrt{a^2 + b^2}$ gives us

$$d = \sqrt{(17t)^2 + (12t)^2} = \sqrt{289t^2 + 144t^2} = \sqrt{433t^2}\,.$$

Thus the functional relationship is

$$d = \sqrt{433}\,t\,.$$

An *approximate* formula for expressing d as a function of t is $d \approx (20.8)t$.

∎

If $f(x) = x$ for every x in the domain D of f, then f is called the **identity function** on D. A function f is a **constant function** if there is some (fixed) element c in the range such that $f(x) = c$ for every x in the domain. If a constant function is represented by a diagram of the type shown in Figure 3.15, *every* arrow from D terminates at the same point in E.

The types of functions described in the next definition occur frequently.

DEFINITION

> A function f with domain D is
> (i) **even** if $f(-x) = f(x)$ for every x in D,
> (ii) **odd** if $f(-x) = -f(x)$ for every x in D.

EXAMPLE 7

(a) If $f(x) = 3x^4 - 2x^2 + 5$, show that f is an even function.

(b) If $g(x) = 2x^5 - 7x^3 + 4x$, show that f is an odd function.

SOLUTION If x is any real number, then

(a)
$$\begin{aligned}
f(-x) &= 3(-x)^4 - 2(-x)^2 + 5 \\
&= 3x^4 - 2x^2 + 5 = f(x)
\end{aligned}$$

and hence f is even.

(b)
$$\begin{aligned}
g(-x) &= 2(-x)^5 - 7(-x)^3 + 4(-x) \\
&= -2x^5 + 7x^3 - 4x \\
&= -(2x^5 - 7x^3 + 4x) = -g(x)
\end{aligned}$$

Thus, g is odd.

∎

Exercises 3.3

1 If $f(x) = 2x^2 - 3x + 4$, find $f(1), f(-1), f(0)$, and $f(2)$.

2 If $f(x) = x^3 + 5x^2 - 1$, find $f(2), f(-2), f(0)$, and $f(-1)$.

3 If $f(x) = \sqrt{x - 1} + 2x$, find $f(1), f(3), f(5)$, and $f(10)$.

4 If $f(x) = \dfrac{x}{x - 2}$, find $f(1), f(3), f(-2)$, and $f(0)$.

In Exercises 5–8 find each of the following, where a, b, and h are real numbers:

(a) $f(a)$ (b) $f(-a)$

(c) $-f(a)$ (d) $f(a + h)$

(e) $f(a) + f(h)$

(f) $\dfrac{f(a + h) - f(a)}{h}$,   provided $h \neq 0$

5 $f(x) = 5x - 2$ **6** $f(x) = 3 - 4x$

7 $f(x) = 2x^2 - x + 3$ **8** $f(x) = x^3 - 2x$

In Exercises 9–12 find the following:

(a) $g(1/a)$ (b) $1/g(a)$ (c) $g(a^2)$

(d) $(g(a))^2$ (e) $g(\sqrt{a})$ (f) $\sqrt{g(a)}$

9 $g(x) = 3x^2$ **10** $g(x) = 3x - 8$

11 $g(x) = \dfrac{2x}{x^2 + 1}$ **12** $g(x) = \dfrac{x^2}{x + 1}$

In Exercises 13–20 find the domain of the function f.

13 $f(x) = \sqrt{3x - 5}$ **14** $f(x) = \sqrt{7 - 2x}$

15 $f(x) = \sqrt{4 - x^2}$ **16** $f(x) = \sqrt{x^2 - 9}$

17 $f(x) = \dfrac{x + 1}{x^3 - 9x}$ **18** $f(x) = \dfrac{4x + 7}{6x^2 + 13x - 5}$

19 $f(x) = \dfrac{\sqrt{x}}{2x^2 - 11x + 12}$ **20** $f(x) = \dfrac{x^3 - 1}{x^2 - 1}$

In Exercises 21–26 find the number x such that $f(x) = 4$. If $a > 0$, find x such that $f(x) = a$. Find the range of f.

21 $f(x) = 7x - 5$ **22** $f(x) = 3x$

23 $f(x) = \sqrt{x - 3}$ **24** $f(x) = 1/x$

25 $f(x) = x^3$ **26** $f(x) = \sqrt[3]{x - 4}$

In Exercises 27–36 determine whether f is even, odd, or neither even nor odd.

27 $f(x) = 3x^3 - 4x$ **28** $f(x) = 7x^6 - x^4 + 7$

29 $f(x) = 9 - 5x^2$ **30** $f(x) = 2x^5 - 4x^3$

31 $f(x) = 2$ **32** $f(x) = 2x^3 + x^2$

33 $f(x) = 2x^2 - 3x + 4$ **34** $f(x) = \sqrt{x^2 + 1}$

35 $f(x) = \sqrt[3]{x^3 - 4}$ **36** $f(x) = |x| + 5$

37 An open box is to be made from a rectangular piece of cardboard having dimensions 20 inches × 30 inches by cutting out identical squares of area x^2 from each corner and turning up the sides (see figure). Express the volume V of the box as a function of x.

FIGURE FOR EXERCISE 37

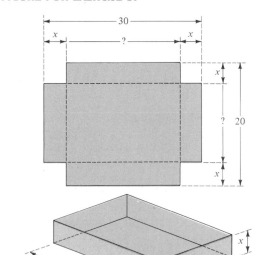

38 An aquarium of height 1.5 feet is to have a volume of 6 ft³. Let x denote the length of the base, and let y denote the width (see figure).

(a) Express y as a function of x.

(b) Express the total number of square feet of glass needed as a function of x.

FIGURE FOR EXERCISE 38

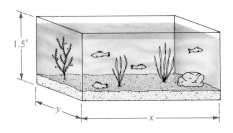

39 A small office building is to contain 500 ft² of floor space. The simple floor plans are shown in the figure.

   (a) Express the length y of the building as a function of the width x.

   (b) If the walls cost \$100 per running foot, express the cost C of the walls as a function of the width x. (Disregard the wall space above the door.)

FIGURE FOR EXERCISE 39

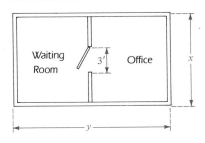

40 A meteorologist is inflating a spherical weather balloon with helium gas.

   (a) Express the surface area of the balloon as a function of the volume of gas it contains.

   (b) What is the surface area if the volume is 12 ft³?

41 A hot-air balloon is released at 1:00 P.M. and rises vertically at a rate of 2 meters per second. An observation point is situated 100 meters from a point on the ground directly below the balloon (see figure). If t denotes the time (in seconds) after 1:00 P.M., express the distance d between the balloon and the observation point as a function of t.

FIGURE FOR EXERCISE 41

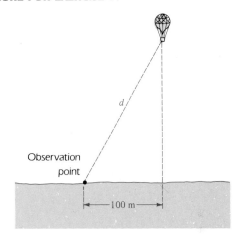

42 Triangle ABC is inscribed in a semicircle of diameter 15 (see figure).

   (a) If x denotes the length of side AC, express the length y of side BC as a function of x.

   (b) Express the area of triangle ABC as a function of x, and find the domain of this function.

FIGURE FOR EXERCISE 42

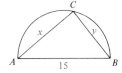

43 From an exterior point P that is h units from a circle of radius r, a tangent line is drawn to the circle (see figure).

FIGURE FOR EXERCISE 43

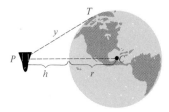

Let y denote the distance from the point P to the point of tangency T.

(a) Express y as a function of h. (*Hint:* If C is the center of the circle, then PT is perpendicular to CT.)

(b) If r is the radius of the earth, and h is the altitude of a space shuttle, then we can derive a formula for the maximum distance (to the earth) that an astronaut can see from the shuttle. In particular, if $h = 200$ miles and $r \approx 4000$ miles, approximate y.

44 The accompanying figure illustrates the apparatus for a tightrope walker. Two poles are set 50 feet apart, but the point of attachment P for the rope is yet to be determined.

(a) Express the length L of the rope as a function of the distance x from point P to the ground.

(b) If the total walk is be 75 feet, determine the height of the point of attachment P.

FIGURE FOR EXERCISE 44

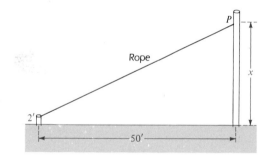

45 A steel storage tank for propane gas is to be constructed in the shape of a right circular cylinder of altitude 10 feet with a hemisphere attached to each end. The radius r is yet to be determined. Express the surface area S of the tank as a function of r. (Compare Example 5.)

46 A company sells running shoes to dealers at a rate of $20 per pair if less than 50 pairs are ordered. If a dealer orders 50 or more pairs (up to 600), the price per pair is reduced at a rate of 2 cents times the number ordered. Let A denote the amount of money received when x pairs are ordered. Express A as a function of x.

47 The relative positions of an aircraft runway and a 20-foot tall control tower are shown in the figure. The beginning of the runway is at a perpendicular distance of 300 feet from the base of the tower. If x denotes the distance an airplane has moved down the runway, express the distance d between the airplane and the control booth as a function of x.

FIGURE FOR EXERCISE 47

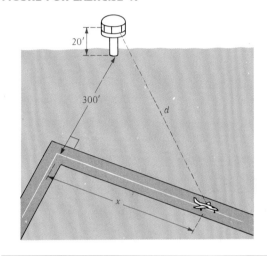

48 A man in a rowboat that is 2 miles from the nearest point A on a straight shoreline wishes to reach a house located at a point B that is 6 miles further downshore (see figure). He plans to row to a point P that is between A and B and is x miles from the house, and then walk the remainder of the distance. Suppose he can row at a rate of 3 mph and can walk at a rate of 5 mph. If T is the total time required to reach the house, express T as a function of x.

FIGURE FOR EXERCISE 48

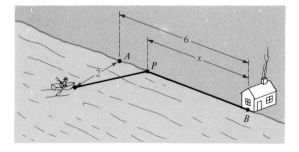

49 A right circular cylinder of radius r and height h is inscribed in a cone of altitude 12 and base radius 4, as illustrated in the figure at the right.

    (a) Express h as a function of r. (*Hint:* Use similar triangles.)

    (b) Express the volume V of the cylinder as a function of r.

50 Suppose a point $P(x, y)$ is moving along the parabola $y = x^2$. Let d denote the distance from the point $A(3, 1)$ to P. Express d as a function of x.

FIGURE FOR EXERCISE 49

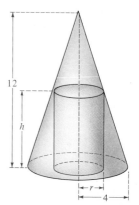

3.4 Graphs of Functions

FIGURE 3.18

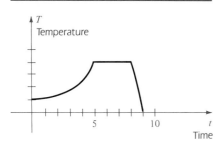

FIGURE 3.19

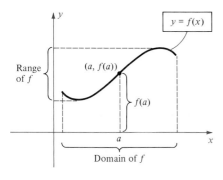

Graphs are often used to describe the variation of physical quantities. For example, a scientist may use the graph in Figure 3.18 to indicate the temperature T of a certain solution at various times t during an experiment. The sketch shows that the temperature increased gradually from time $t = 0$ to time $t = 5$, did not change between $t = 5$ and $t = 8$, and then decreased rapidly from $t = 8$ to $t = 9$. This visual aid reveals the variation of T more clearly than a long table of numerical values would.

    If f is a function, we may use a graph to show the change in $f(x)$ as x varies through the domain of f. By definition, the **graph of a function** f is the graph of the equation $y = f(x)$ for x in the domain of f. As shown in Figure 3.19, we often attach the label $y = f(x)$ to a sketch of the graph. Note that if $P(a, b)$ is a point on the graph, then the y-coordinate b is the function value $f(a)$. The figure exhibits the domain of f (the set of possible values of x) and the range of f (the corresponding values of y). Although we have pictured the domain and range as closed intervals, they may be infinite intervals or other sets of real numbers.

    It is important to note that since there is a unique value $f(a)$ for each a in the domain, only *one* point on the graph has x-coordinate a. Thus, *every vertical line intersects the graph of a function in at most one point.* Consequently, the graph of a function cannot be a figure such as a circle, in which a vertical line may intersect the graph in several points.

    The x-intercepts of the graph of a function f are the solutions of the equation $f(x) = 0$. These numbers are called the **zeros** of the function. The y-intercept of the graph is $f(0)$, if it exists.

FIGURE 3.20

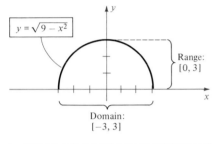

$y = \sqrt{9 - x^2}$

Range: [0, 3]

Domain: [−3, 3]

FIGURE 3.21

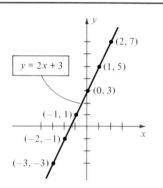

$y = 2x + 3$

(2, 7)
(1, 5)
(0, 3)
(−1, 1)
(−2, −1)
(−3, −3)

EXAMPLE 1 If $f(x) = \sqrt{9 - x^2}$, sketch the graph of f and find the domain and range of f.

SOLUTION By definition, the graph of f is the graph of the equation $y = \sqrt{9 - x^2}$. We know from our work with circles in Section 3.2 that the graph of $x^2 + y^2 = 9$ is a circle of radius 3 with center at the origin. Solving the equation $x^2 + y^2 = 9$ for y gives us $y = \pm\sqrt{9 - x^2}$. It follows that the graph of f is the *upper half* of the circle, as illustrated in Figure 3.20. Referring to the figure, we see that the domain of f is the closed interval $[-3, 3]$, and the range of f is the interval $[0, 3]$. ∎

EXAMPLE 2 If $f(x) = 2x + 3$, sketch the graph of f and find the domain and range of f.

SOLUTION The graph of f is the graph of the equation $y = 2x + 3$. The following table lists coordinates of several points on the graph.

x	−3	−2	−1	0	1	2
y	−3	−1	1	3	5	7

Plotting, it appears that the points lie on a line, and we sketch the graph as in Figure 3.21. (In the next section we shall *prove* that the graph is a line.)

Since the values of x and y may be any real numbers, both the domain and range of f are $\mathbb{R}$. ∎

If f is the function in Example 2 and if $x_1 < x_2$, then $2x_1 + 3 < 2x_2 + 3$; that is, $f(x_1) < f(x_2)$. Thus, as x-coordinates of points increase, y-coordinates also increase and the function f is then said to be *increasing*. If f is increasing, then the graph rises as x increases. For certain functions $f(x_1) > f(x_2)$ whenever $x_1 < x_2$. In this case the graph of f falls as x increases, and the function is called a *decreasing* function. In general, we shall speak of functions that increase or decrease on certain intervals, as in the following definition.

DEFINITION

Let a function f be defined on an interval I and let x_1, x_2 be numbers in I.

(i) f is **increasing** on I if $f(x_1) < f(x_2)$ whenever $x_1 < x_2$.
(ii) f is **decreasing** on I if $f(x_1) > f(x_2)$ whenever $x_1 < x_2$.
(iii) f is **constant** on I if $f(x_1) = f(x_2)$ for every x_1 and x_2.

Geometric interpretations of this definition are given in Figure 3.22, where the interval I is not indicated. Note that if f is constant, then the graph is part of a horizontal line.

FIGURE 3.22

(i) Increasing function

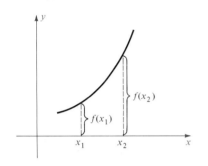

(ii) Decreasing function

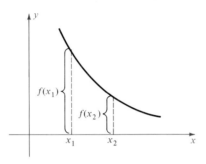

(iii) Constant function

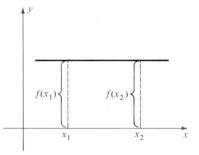

We shall use the phrases "f is increasing" and "$f(x)$ is increasing" interchangeably. This will also be done for the term "decreasing."

FIGURE 3.23

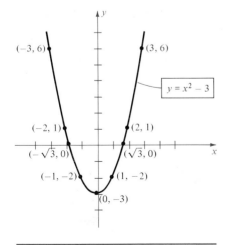

EXAMPLE 3 If $f(x) = x^2 - 3$, sketch the graph of f and find the domain and range of f. Determine the intervals on which f is increasing or decreasing.

SOLUTION Coordinates (x, y) of some points on the graph of f are listed in the following table, where $y = x^2 - 3$.

x	-3	-2	-1	0	1	2	3
y	6	1	-2	-3	-2	1	6

The x-intercepts are the solutions of the equation $f(x) = 0$, that is, of $x^2 - 3 = 0$. Thus the x-intercepts are $\pm\sqrt{3}$. The y-intercept is $f(0) = -3$. Plotting points and using the x-intercepts leads to the sketch in Figure 3.23.

Since x can have any value, the domain of f is $\mathbb{R}$. The range of f consists of all real numbers y such that $y \geq -3$, and hence is the interval $[-3, \infty)$.

We see from the graph that f is decreasing on $(-\infty, 0]$ and that f is increasing on $[0, \infty)$. ■

Referring to Figure 3.23, we see that $f(x) = x^2 - 3$ takes on its least value at $x = 0$. This smallest value, -3, is called the **minimum value** of f. The corresponding point $(0, -3)$ is the lowest point on the graph. Clearly, $f(x)$ does not attain a **maximum value,** that is, a *largest* value.

The solution to Example 3 could have been shortened by observing that since $(-x)^2 - 3 = x^2 - 3$, the graph of $y = x^2 - 3$ is symmetric with respect to the y-axis. This fact also follows from (i) of the next theorem.

THEOREM ON SYMMETRY

(i) The graph of an even function is symmetric with respect to the y-axis.

(ii) The graph of an odd function is symmetric with respect to the origin.

PROOF If f is even, then $f(-x) = f(x)$, and hence the equation $y = f(x)$ is not changed if $-x$ is substituted for x. Statement (i) now follows from Symmetry Test (i) of Section 3.2. The proof of (ii) is left to the reader. □

FIGURE 3.24

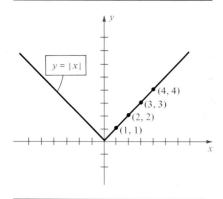

EXAMPLE 4 Sketch the graph of f for $f(x) = |x|$, and find the domain and range of f. Determine the intervals on which f is increasing or decreasing.

SOLUTION If $x \geq 0$, then $f(x) = x$ and hence the points (x, x) in the first quadrant are on the graph of f. Some special cases are $(0, 0)$, $(1, 1)$, $(2, 2)$, $(3, 3)$, and $(4, 4)$. Since $|-x| = |x|$ we see that f is an even function and hence, by the preceding theorem, the graph is symmetric with respect to the y-axis. Plotting points and using symmetry leads to the sketch in Figure 3.24.

The domain of f is $\mathbb{R}$, and the range is $[0, \infty)$. As in Example 3, this function is decreasing on $(-\infty, 0]$ and increasing on $[0, \infty)$. ∎

EXAMPLE 5 Sketch the graph of f if $f(x) = \sqrt{x - 1}$ and find the domain and range of f. Determine where f is increasing or decreasing.

SOLUTION The domain of f consists of all real numbers x such that $x \geq 1$ and hence is the interval $[1, \infty)$. The following table lists some points (x, y) on the graph, where $y = \sqrt{x - 1}$.

FIGURE 3.25

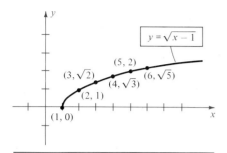

x	1	2	3	4	5	6
y	0	1	$\sqrt{2}$	$\sqrt{3}$	2	$\sqrt{5}$

Plotting points leads to the sketch in Figure 3.25.

We see from the graph that the range of f is $[0, \infty)$. The function is increasing throughout its domain. The x-intercept is 1, and there is no y-intercept. ■

EXAMPLE 6 Given $f(x) = x^2 + c$, sketch the graph of f if $c = 4$; if $c = -2$.

SOLUTION We shall sketch both graphs on the same coordinate axes. The graph of $y = x^2$ was sketched in Figure 3.8, and, for reference, is represented in color in Figure 3.26. To find the graph of $y = x^2 + 4$ we may simply add 4 to the y-coordinate of each point on the graph of $y = x^2$. This amounts to *shifting* the graph of $y = x^2$ *upward* 4 units as shown in the figure. For $c = -2$ we decrease y-coordinates by 2 and, hence, the graph of $y = x^2 - 2$ may be obtained by shifting the graph of $y = x^2$ *downward* 2 units. Each graph is a parabola symmetric with respect to the y-axis. To determine the correct position of each graph, it is advisable to plot several points. ■

FIGURE 3.26

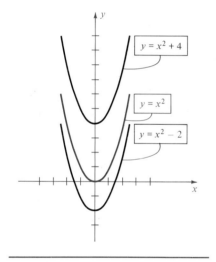

The graphs in the preceding example illustrate **vertical shifts** of the graph of $y = x^2$ and are special cases of the following general rules.

VERTICAL SHIFTS OF GRAPHS ($c > 0$)

To obtain the graph of:	shift the graph of $y = f(x)$:
$y = f(x) - c$	c units downward
$y = f(x) + c$	c units upward

FIGURE 3.27

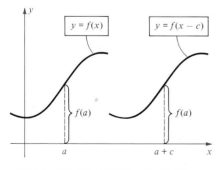

Similar rules can be stated for **horizontal shifts**. Specifically, if $c > 0$, consider the graphs of $y = f(x)$ and $y = f(x - c)$ sketched on the same coordinate axes, as illustrated in Figure 3.27.

Since $f(a) = f(a + c - c)$, we see that the point with x-coordinate a on the graph of $y = f(x)$ has the same y-coordinate as the point with x-coordinate $a + c$ on the graph of $y = f(x - c)$. This implies that the graph

FIGURE 3.28

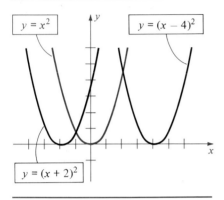

$y = x^2$

$y = (x - 4)^2$

$y = (x + 2)^2$

FIGURE 3.29

(i)

(ii)

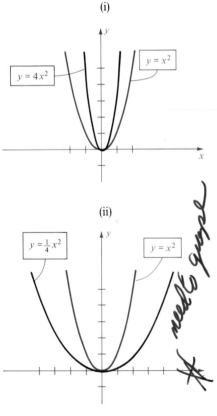

$y = 4x^2$

$y = x^2$

$y = \frac{1}{4}x^2$

$y = x^2$

of $y = f(x - c)$ can be obtained by shifting the graph of $y = f(x)$ to the right c units. Similarly, the graph of $y = f(x + c)$ can be obtained by shifting the graph of f to the left c units. These rules are listed for reference in the next box.

HORIZONTAL SHIFTS OF GRAPHS ($c > 0$)

To obtain the graph of:	shift the graph of $y = f(x)$:
$y = f(x - c)$	c units to the right
$y = f(x + c)$	c units to the left

EXAMPLE 7   Sketch the graph of f if

(a) $f(x) = (x - 4)^2$ (b) $f(x) = (x + 2)^2$

SOLUTION   The graph of $y = x^2$ is sketched in color in Figure 3.28. According to the rules for horizontal shifts, shifting this graph to the right 4 units gives us the graph of $y = (x - 4)^2$, whereas shifting to the left 2 units leads to the graph of $y = (x + 2)^2$. Students who are not convinced of the validity of this technique are urged to plot several points on each graph.   ∎

To obtain the graph of $y = cf(x)$ for some real number c, we may *multiply* the y-coordinates of points on the graph of $y = f(x)$ by c. For example, if $y = 2f(x)$, we double y-coordinates, or if $y = \frac{1}{2}f(x)$, we multiply each y-coordinate by $\frac{1}{2}$. If $c > 0$ (and $c \neq 1$) we shall refer to this procedure as **stretching** the graph of $y = f(x)$.

EXAMPLE 8   Sketch the graph of (a) $y = 4x^2$; (b) $y = \frac{1}{4}x^2$.

SOLUTION

(a) To sketch the graph of $y = 4x^2$ we may refer to the graph of $y = x^2$ (shown in color in Figure 3.29(i)) and multiply the y-coordinate of each point by 4. This gives us a narrower parabola that is sharper at the vertex, as illustrated in (i) of the figure. To obtain the correct shape, several points such as $(0, 0)$, $(\frac{1}{2}, 1)$, and $(1, 4)$ should be plotted.

(b) The graph of $y = \frac{1}{4}x^2$ may be sketched by multiplying y-coordinates of points on the graph of $y = x^2$ by $\frac{1}{4}$. The graph is a wider parabola that is flatter at the vertex, as shown in Figure 3.29(ii). ∎

The graph of $y = -f(x)$ may be obtained by multiplying the y-coordinate of each point on the graph of $y = f(x)$ by -1. Thus, every point (a, b) on the graph of $y = f(x)$ that lies above the x-axis determines a point $(a, -b)$ on the graph of $y = -f(x)$ that lies below the x-axis. Similarly, if (c, d) lies below the x-axis (that is, $d < 0$), then $(c, -d)$ lies above the x-axis. The graph of $y = -f(x)$ is called a **reflection** of the graph of $y = f(x)$ through the x-axis.

FIGURE 3.30

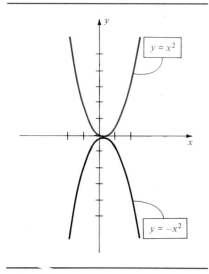

EXAMPLE 9 Sketch the graph of $y = -x^2$.

SOLUTION The graph may be found by plotting points; however, since the graph of $y = x^2$ is well known, we sketch it in color, as in Figure 3.30, and then multiply y-coordinates of points by -1. This gives us the reflection through the x-axis indicated in the figure. ∎

Sometimes functions are described in terms of more than one expression, as in the next examples. We shall call such functions **piecewise-defined functions.**

EXAMPLE 10 Sketch the graph of the function f that is defined as follows:

FIGURE 3.31

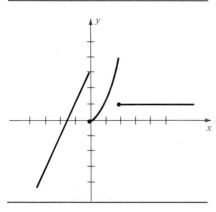

$$f(x) = \begin{cases} 2x + 3 & \text{if } x < 0 \\ x^2 & \text{if } 0 \leq x < 2 \\ 1 & \text{if } x \geq 2 \end{cases}$$

SOLUTION If $x < 0$, then $f(x) = 2x + 3$. This means that if x is negative, the expression $2x + 3$ should be used to find function values. Consequently, if $x < 0$, then the graph of f coincides with the graph in Figure 3.21 and we sketch that portion of the graph to the left of the y-axis as indicated in Figure 3.31.

If $0 \leq x < 2$, we use x^2 to find values of f, and therefore this part of the graph of f coincides with the graph of the equation $y = x^2$. We then sketch the part of the graph of f between $x = 0$ and $x = 2$, as indicated in Figure 3.31.

Finally, if $x \geq 2$, the values of f are always 1. That part of the graph is the horizontal half line illustrated in Figure 3.31. ∎

EXAMPLE 11 If x is any real number, then there exist consecutive integers n and $n + 1$ such that $n \leq x < n + 1$. Let f be the function defined as follows: If $n \leq x < n + 1$, then $f(x) = n$. Sketch the graph of f.

SOLUTION The x- and y-coordinates of some points on the graph may be listed as follows:

FIGURE 3.32

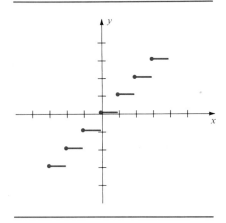

Values of x	$f(x)$
$\cdots$	$\cdots$
$-2 \leq x < -1$	-2
$-1 \leq x < 0$	-1
$0 \leq x < 1$	0
$1 \leq x < 2$	1
$2 \leq x < 3$	2
$\cdots$	$\cdots$

Since $f(x)$ does not change when x is between successive integers, the corresponding part of the graph is a segment of a horizontal line. Part of the graph is sketched in Figure 3.32. The graph continues indefinitely to the right and to the left. ∎

The symbol $[\![x]\!]$ is often used to denote the largest integer n such that $n \leq x$. For example $[\![1.6]\!] = 1$, $[\![\sqrt{5}]\!] = 2$, $[\![\pi]\!] = 3$, and $[\![-3.5]\!] = -4$. Using this notation, the function f of Example 11 may be defined by $f(x) = [\![x]\!]$. It is customary to refer to f as the **greatest integer function.**

Exercises 3.4

In Exercises 1–20 sketch the graph of f, find the domain and range of f, and describe the intervals in which f is increasing, decreasing, or constant.

1 $f(x) = 5x$

2 $f(x) = -3x$

3 $f(x) = -4x + 2$

4 $f(x) = 4x - 2$

5 $f(x) = 3 - x^2$

6 $f(x) = 4x^2 + 1$

7 $f(x) = 2x^2 - 4$

8 $f(x) = \frac{1}{100}x^2$

9 $f(x) = \sqrt{x + 4}$

10 $f(x) = \sqrt{4 - x}$

11 $f(x) = \sqrt{x} + 2$

12 $f(x) = 2 - \sqrt{x}$

13 $f(x) = 4/x$

14 $f(x) = 1/x^2$

15 $f(x) = |x - 2|$

16 $f(x) = |x + 2|$

17 $f(x) = |x| - 2$

18 $f(x) = |x| + 2$

19 $f(x) = \dfrac{x}{|x|}$

20 $f(x) = x + |x|$

In Exercises 21–30 sketch, on the same coordinate axes, the graphs of f for the stated values of c. (Make use of vertical shifts, horizontal shifts, stretching, or reflecting).

21 $f(x) = 3x + c$; $c = 0$, $c = 2$, $c = -1$

22 $f(x) = -2x + c$; $c = 0$, $c = 1$, $c = -3$

23 $f(x) = x^3 + c$; $c = 0$, $c = 1$, $c = -2$

24 $f(x) = -x^3 + c$; $c = 0$, $c = 2$, $c = -1$

25 $f(x) = \sqrt{4 - x^2} + c$; $c = 0$, $c = 4$, $c = -3$

26 $f(x) = c - |x|$; $c = 0$, $c = 5$, $c = -2$

27 $f(x) = 3(x - c)$; $c = 0$, $c = 2$, $c = 3$

28 $f(x) = -2(x - c)^2$; $c = 0$, $c = 3$, $c = 1$

29 $f(x) = (x + c)^3$; $c = 0$, $c = 2$, $c = -2$

30 $f(x) = c\sqrt{9 - x^2}$; $c = 0$, $c = 2$, $c = 3$.

31 The graph of a function f with domain $0 \le x \le 4$ is shown in the accompanying figure. Sketch the graph of each of the following:

(a) $y = f(x + 2)$

(b) $y = f(x - 2)$

(c) $y = f(x) + 2$

(d) $y = f(x) - 2$

(e) $y = 2f(x)$

(f) $y = \frac{1}{2}f(x)$

(g) $y = -2f(x)$

(h) $y = f(x - 3) + 1$

FIGURE FOR EXERCISE 31

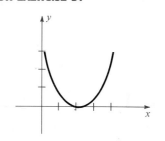

32 The graph of a function f with domain $0 \le x \le 4$ is shown in the accompanying figure. Sketch the graph of

each of the following:

(a) $y = f(x - 1)$

(b) $y = f(x + 3)$

(c) $y = f(x) - 2$

(d) $y = f(x) + 1$

(e) $y = 3f(x)$

(f) $y = \frac{1}{2}f(x)$

(g) $y = f(x + 2) - 2$

(h) $y = f(2x)$

FIGURE FOR EXERCISE 32

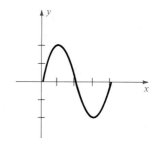

In Exercises 33–40 sketch the graph of the piecewise-defined function f.

33 $f(x) = \begin{cases} 2 & \text{if } x < 0 \\ -1 & \text{if } x \ge 0 \end{cases}$

34 $f(x) = \begin{cases} -1 & \text{if } x \text{ is an integer} \\ 1 & \text{if } x \text{ is not an integer} \end{cases}$

35 $f(x) = \begin{cases} 3 & \text{if } x < -3 \\ -x & \text{if } -3 \le x \le 3 \\ -3 & \text{if } x > 3 \end{cases}$

36 $f(x) = \begin{cases} x & \text{if } x < 0 \\ -2 & \text{if } 0 \le x < 1 \\ x^2 & \text{if } x \ge 1 \end{cases}$

37 $f(x) = \begin{cases} x^2 & \text{if } x \le -1 \\ x^3 & \text{if } |x| < 1 \\ 2x & \text{if } x \ge 1 \end{cases}$

38 $f(x) = \begin{cases} x & \text{if } x \le 1 \\ -x^2 & \text{if } 1 < x < 2 \\ x & \text{if } x \ge 2 \end{cases}$

39 $f(x) = \begin{cases} \dfrac{x^2 - 4}{x - 2} & \text{if } x \ne 2 \\ 3 & \text{if } x = 2 \end{cases}$

40 $f(x) = \begin{cases} \dfrac{x^2 - 1}{1 - x} & \text{if } x \neq 1 \\ 2 & \text{if } x = 1 \end{cases}$

41 If $[\![x]\!]$ denotes values of the greatest integer function, sketch the graph of f in each of the following:

(a) $f(x) = [\![x - 3]\!]$ (b) $f(x) = 2[\![x]\!]$

(c) $f(x) = -[\![x]\!]$

42 Explain why the graph of the equation $y^2 = x$ is not the graph of a function.

3.5 Linear Functions

The following type of function is very important in mathematics and its applications.

DEFINITION

> A function f is a **linear function** if
>
> $$f(x) = ax + b$$
>
> where a and b are real numbers and $a \neq 0$.

The reason for the term "linear" is that the graph of f is a line, as we shall see later in this section. Let us first introduce several fundamental concepts pertaining to lines. All lines referred to are considered to be in a coordinate plane.

DEFINITION

> Let l be a line that is not parallel to the y-axis, and let $P_1(x_1, y_1)$ and $P_2(x_2, y_2)$ be distinct points on l. The **slope** m of l is
>
> $$m = \frac{y_2 - y_1}{x_2 - x_1}.$$
>
> If l is parallel to the y-axis, then the slope is not defined.

Typical points P_1 and P_2 on a line l are shown in Figure 3.33. The numerator $y_2 - y_1$ in the formula for m is the vertical change in direction in proceeding from P_1 to P_2 and may be positive, negative, or zero. The denominator $x_2 - x_1$ is the amount of horizontal change in going from P_1 to P_2, and it may be positive or negative, but never zero, because l is not parallel to the y-axis.

FIGURE 3.33

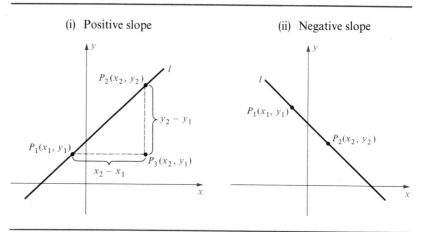

(i) Positive slope   (ii) Negative slope

In finding the slope of a line, it is immaterial which point is labeled P_1 and which is labeled P_2, since

$$\frac{y_2 - y_1}{x_2 - x_1} = \frac{y_1 - y_2}{x_1 - x_2}.$$

Consequently, we may assume that the points are labeled so that $x_1 < x_2$, as in Figure 3.33. In this situation $x_2 - x_1 > 0$, and hence the slope is positive, negative, or zero, depending on whether $y_2 > y_1$, $y_2 < y_1$, or $y_2 = y_1$. The slope of the line shown in Figure 3.33(i) is positive, whereas the slope of the line shown in (ii) of the figure is negative.

FIGURE 3.34

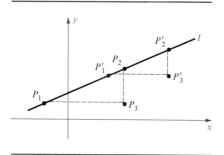

A **horizontal line** is a line that is parallel to the x-axis. A line is horizontal if and only if its slope is 0. A **vertical line** is a line that is parallel to the y-axis. The slope of a vertical line is undefined.

It is important to note that the definition of slope is independent of the two points that are chosen on l, for if other points $P_1'(x_1', y_1')$ and $P_2'(x_2', y_2')$ are used, then as in Figure 3.34, the triangle with vertices P_1', P_2', and $P_3'(x_2', y_1')$ is similar to the triangle with vertices P_1, P_2, and $P_3(x_2, y_1)$. Since the ratios of corresponding sides are equal, it follows that

$$m = \frac{y_2 - y_1}{x_2 - x_1} = \frac{y_2' - y_1'}{x_2' - x_1'}.$$

EXAMPLE 1   Sketch the lines through the following pairs of points and find their slopes.

(a) $A(-1, 4)$ and $B(3, 2)$ (b) $A(2, 5)$ and $B(-2, -1)$

(c) $A(4, 3)$ and $B(-2, 3)$ (d) $A(4, -1)$ and $B(4, 4)$

SOLUTION   The lines are sketched in Figure 3.35.

FIGURE 3.35

 (i) $m = -\frac{1}{2}$  (ii) $m = \frac{3}{2}$ (iii) $m = 0$ (iv) m undefined

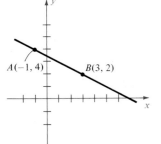   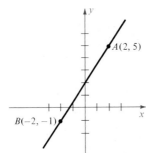   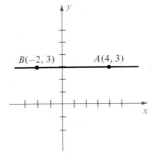   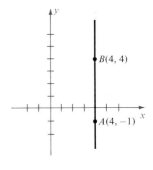

Using the definition of slope gives us

(a)
$$m = \frac{2 - 4}{3 - (-1)} = \frac{-2}{4} = -\frac{1}{2}$$

(b)
$$m = \frac{5 - (-1)}{2 - (-2)} = \frac{6}{4} = \frac{3}{2}$$

(c)
$$m = \frac{3 - 3}{-2 - 4} = \frac{0}{-6} = 0$$

(d)   The slope is undefined since the line is vertical. This is also seen by noting that, if the formula for m is used, the denominator is zero.   ■

EXAMPLE 2   Construct a line through $P(2, 1)$ that has slope $\frac{5}{3}$; $-\frac{5}{3}$.

SOLUTION   If the slope of a line is a/b and b is positive, then for every change of b units in the horizontal direction, the line rises or falls $|a|$ units, depending on whether a is positive or negative, respectively. If $P(2, 1)$ is on the line and $m = \frac{5}{3}$, we can obtain another point on the line by starting at P and moving 3 units to the right and 5 units upward. This gives us the point $Q(5, 6)$, and the line is determined (see Figure 3.36(i)). Similarly, if $m = -\frac{5}{3}$, we move 3 units to the right and 5 units downward, obtaining $Q(5, -4)$ as in Figure 3.36(ii).

FIGURE 3.36

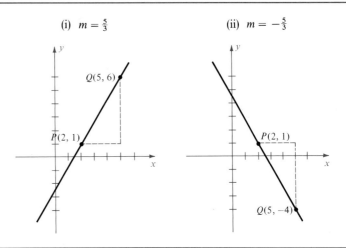

(i) $m = \frac{5}{3}$ (ii) $m = -\frac{5}{3}$

THEOREM

(i) The graph of the equation $x = a$ is a vertical line that has x-intercept a.

(ii) The graph of the equation $y = b$ is a horizontal line that has y-intercept b.

FIGURE 3.37

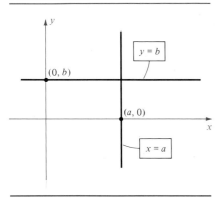

PROOF The equation $x = a$ may be written in the form $x + (0)y = a$. Some typical solutions of this equation are $(a, -2)$, $(a, 1)$, and $(a, 3)$. Evidently, every solution has the form (a, y), where y may have any value and a is fixed. It follows that the graph of $x = a$ is a line parallel to the y-axis with x-intercept a, as illustrated in Figure 3.37. This proves (i). Part (ii) is proved in similar fashion. □

Let us next find an equation of the line l through a point $P_1(x_1, y_1)$ with slope m (only one such line exists). If $P(x, y)$ is any point with $x \neq x_1$

FIGURE 3.38

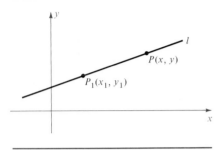

(see Figure 3.38), then P is on l if and only if the slope of the line through P_1 and P is m; that is,

$$\frac{y - y_1}{x - x_1} = m.$$

This equation may be written in the form

$$y - y_1 = m(x - x_1).$$

Note that (x_1, y_1) is also a solution of the last equation and hence the points on l are precisely the points that correspond to the solutions. This equation for l is referred to as the **Point-Slope Form.**

POINT-SLOPE FORM FOR THE EQUATION OF A LINE

> An equation of the line through the point (x_1, y_1) with slope m is
>
> $$y - y_1 = m(x - x_1).$$

EXAMPLE 3   Find an equation of the line through the points $A(1, 7)$ and $B(-3, 2)$.

SOLUTION   The slope m of the line is

$$m = \frac{7 - 2}{1 - (-3)} = \frac{5}{4}.$$

We may use the coordinates of either A or B for (x_1, y_1) in the Point-Slope Form. Using $A(1, 7)$ gives us

$$y - 7 = \tfrac{5}{4}(x - 1),$$

which is equivalent to

$$4y - 28 = 5x - 5, \quad \text{or} \quad 5x - 4y + 23 = 0. \qquad \blacksquare$$

The Point-Slope Form may be rewritten as $y = mx - mx_1 + y_1$, which is of the form

$$y = mx + b$$

with $b = -mx_1 + y_1$. The real number b is the y-intercept of the graph, as may be seen by setting $x = 0$. Since the equation $y = mx + b$ displays the slope m and y-intercept b of l, it is called the **Slope-Intercept Form** for the equation of a line. Conversely, if we start with $y = mx + b$, we may write

$$y - b = m(x - 0).$$

Comparing this last equation with the Point-Slope Form, we see that the graph is a line that has slope m and passes through the point $(0, b)$. This gives us the next result.

SLOPE-INTERCEPT FORM FOR THE EQUATION OF A LINE

> The graph of the equation $y = mx + b$ is a line having slope m and y-intercept b.

We have shown that every line is the graph of an equation of the form

$$ax + by + c = 0$$

where a, b, and c are real numbers, and a and b are not both zero. We call such an equation a **linear equation** in x and y. Let us show, conversely, that the graph of $ax + by + c = 0$ is always a line provided a and b are not both zero. On the one hand, if $b \neq 0$, we may solve for y, obtaining

$$y = \left(-\frac{a}{b} \right) x + \left(-\frac{c}{b} \right),$$

which, by the Slope-Intercept Form, is an equation of a line with slope $-a/b$ and y-intercept $-c/b$. On the other hand, if $b = 0$ but $a \neq 0$, then we may solve $ax + by + c = 0$ for x, obtaining $x = -c/a$, which is the equation of a vertical line with x-intercept $-c/a$. This establishes the following important theorem.

THEOREM

> The graph of a linear equation $ax + by + c = 0$ is a line and, conversely, every line is the graph of a linear equation.

For simplicity, we shall use the terminology *the line $ax + by + c = 0$* instead of the more accurate phrase *the line with equation $ax + by + c = 0$*.

FIGURE 3.39

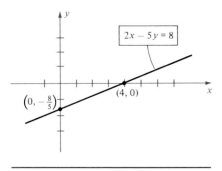

EXAMPLE 4 Sketch the graph of $2x - 5y = 8$.

SOLUTION From the preceding theorem we know that the graph is a line, and hence it is sufficient to find two points on the graph. Let us find the x- and y-intercepts. Substituting $y = 0$ in the given equation, we obtain the x-intercept 4. Substituting $x = 0$, we see that the y-intercept is $-\frac{8}{5}$. This leads to the graph in Figure 3.39.

Another method of solution is to express the given equation in Slope-Intercept Form. To do this we begin by isolating the term involving y on one side of the equal sign, obtaining

$$5y = 2x - 8.$$

Next, dividing both sides by 5 gives us

$$y = \frac{2}{5}x + \left(\frac{-8}{5}\right),$$

which is in the form $y = mx + b$. Hence, the slope is $m = \frac{2}{5}$, and the y-intercept is $b = -\frac{8}{5}$. We may then sketch a line through the point $(0, -\frac{8}{5})$ with slope $\frac{2}{5}$. ∎

The following theorem can be proved.

THEOREM

> Two nonvertical lines are parallel if and only if they have the same slope.

We shall use this fact in the next example.

EXAMPLE 5 Find an equation of the line that passes through the point $(5, -7)$ and is parallel to the line $6x + 3y - 4 = 0$.

SOLUTION Let us express the given equation in Slope-Intercept Form. We begin by writing

$$3y = -6x + 4$$

and then divide both sides by 3, obtaining

$$y = -2x + \tfrac{4}{3}.$$

The last equation is in Slope-Intercept Form with $m = -2$, and hence the slope is -2. Since parallel lines have the same slope, the required line also has slope -2. Applying the Point-Slope Form gives us

$$y + 7 = -2(x - 5).$$

This is equivalent to

$$y + 7 = -2x + 10 \quad \text{or} \quad 2x + y - 3 = 0. \qquad \blacksquare$$

The next theorem specifies conditions for perpendicular lines.

THEOREM

Two lines with slopes m_1 and m_2 are perpendicular if and only if $m_1 m_2 = -1$.

FIGURE 3.40

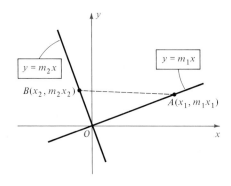

PROOF For simplicity, let us consider the special case of two lines that intersect at the origin O, as illustrated in Figure 3.40. In this case equations of the lines are $y = m_1 x$ and $y = m_2 x$. If, as in the figure, we choose points $A(x_1, m_1 x_1)$ and $B(x_2, m_2 x_2)$ different from O on the lines, then the lines are perpendicular if and only if angle AOB is a right angle. By the Pythagorean Theorem, angle AOB is a right angle if and only if

$$[d(A, B)]^2 = [d(O, B)]^2 + [d(O, A)]^2$$

or, by the Distance Formula,

$$(m_2 x_2 - m_1 x_1)^2 + (x_2 - x_1)^2 = (m_2 x_2)^2 + x_2^2 + (m_1 x_1)^2 + x_1^2.$$

Squaring the indicated terms and simplifying gives us

$$-2m_1 m_2 x_1 x_2 - 2x_1 x_2 = 0.$$

Dividing both sides by $-2x_1 x_2$, we see that the lines are perpendicular if and only if $m_1 m_2 + 1 = 0$, or $m_1 m_2 = -1$.

The same type of proof may be given if the lines intersect at *any* point (a, b). $\qquad \square$

A convenient way to remember the conditions for perpendicularity is to note that m_1 and m_2 must be *negative reciprocals* of one another, that is, $m_1 = -1/m_2$ and $m_2 = -1/m_1$.

EXAMPLE 6   Find an equation of the line that passes through the point $(5, -7)$ and is perpendicular to the line $6x + 3y - 4 = 0$.

SOLUTION   The line $6x + 3y - 4 = 0$ was considered in Example 5, and we found that its slope is -2. Hence, the slope of the required line is the negative reciprocal $-[1/(-2)]$, or $\frac{1}{2}$. Applying the Point-Slope Form gives us

$$y + 7 = \tfrac{1}{2}(x - 5),$$

which is equivalent to

$$2y + 14 = x - 5, \quad \text{or} \quad x - 2y - 19 = 0. \qquad \blacksquare$$

EXAMPLE 7   Find an equation for the perpendicular bisector of the line segment from $A(1, 7)$ to $B(-3, 2)$.

SOLUTION   By the Midpoint Formula, the midpoint M of the segment AB is $(-1, \frac{9}{2})$. Since the slope of AB is $\frac{5}{4}$ (see Example 3), it follows from the preceding theorem that the slope of the perpendicular bisector is $-\frac{4}{5}$. Applying the Point-Slope Form,

$$y - \tfrac{9}{2} = -\tfrac{4}{5}(x + 1).$$

Multiplying both sides by 10 and simplifying leads to $8x + 10y - 37 = 0$.
$\blacksquare$

The following theorem justifies the terminology used in the definition at the beginning of this section.

THEOREM

> The graph of a linear function is a line.

PROOF   If f is linear we may write

$$f(x) = ax + b$$

for some constants a and b. Since the graph of f is the graph of the equation $y = ax + b$, the conclusion follows at once from the Slope-Intercept Form.
$\square$

If y is a linear function of x, then $y = ax + b$ for some constants a and b, and we say that x and y are **linearly related.** Linear relationships between variables occur frequently in applications. The following example gives one illustration. For other applications see Exercises 41–52.

EXAMPLE 8   The relationship between the air temperature T (in °F) and the altitude h (in feet above sea level) is approximately linear. When the temperature at sea level is 60°, an increase of 5000 feet in altitude lowers the air temperature about 18°.

(a) Express T as a function of h.

(b) What is the air temperature at an altitude of 15,000 feet?

SOLUTION

(a) If T is a linear function of h, then

$$T = ah + b$$

for some constants a and b. Since $T = 60$ when $h = 0$, we see that

$$60 = a(0) + b \quad \text{or} \quad b = 60.$$

Thus, $\qquad\qquad\qquad\qquad T = ah + 60.$

In addition, we are given that if $h = 5000$, then $T = 60 - 18 = 42$. Substituting these values into the formula $T = ah + 60$, we obtain

$$42 = a(5000) + 60 \quad \text{or} \quad 5000a = -18.$$

Hence, $\qquad\qquad\qquad\qquad a = -\dfrac{18}{5000} = -\dfrac{9}{2500},$

and the (approximate) formula for T is

$$T = -\frac{9}{2500} h + 60.$$

(b) Using the formula for T obtained in part (a), the (approximate) temperature when $h = 15,000$ is

$$T = -\frac{9}{2500}(15,000) + 60 = -54 + 60 = 6\,°\text{F}. \qquad \blacksquare$$

Exercises 3.5

In Exercises 1–4 plot the points A and B and find the slope of the line through A and B.

1 $A(-4, 6)$, $B(-1, 18)$ **2** $A(6, -2)$, $B(-3, 5)$

3 $A(-1, -3)$, $B(-1, 2)$ **4** $A(-3, 4)$, $B(2, 4)$

5 Show that $A(-3, 1)$, $B(5, 3)$, $C(3, 0)$, and $D(-5, -2)$ are vertices of a parallelogram.

6 Show that $A(2, 3)$, $B(5, -1)$, $C(0, -6)$, and $D(-6, 2)$ are vertices of a trapezoid.

7 Prove that the points $A(6, 15)$, $B(11, 12)$, $C(-1, -8)$, and $D(-6, -5)$ are vertices of a rectangle.

8 Prove that the points $A(1, 4)$, $B(6, -4)$, and $C(-15, -6)$ are vertices of a right triangle.

9 If three consecutive vertices of a parallelogram are $A(-1, -3)$, $B(4, 2)$, and $C(-7, 5)$, find the fourth vertex.

10 Let $A(x_1, y_1)$ $B(x_2, y_2)$, $C(x_3, y_3)$, and $D(x_4, y_4)$ denote the vertices of an arbitrary quadrilateral. Prove that the line segments joining midpoints of adjacent sides form a parallelogram.

In Exercises 11–20 find an equation of the line satisfying the given conditions.

11 Through $A(2, -6)$, slope $\frac{1}{2}$

12 Slope -3, y-intercept 5

13 Through $A(-5, -7)$, $B(3, -4)$

14 x-intercept -4, y-intercept 8

15 Through $A(8, -2)$, y-intercept -3

16 Slope 6, x-intercept -2

17 Through $A(10, -6)$, parallel to (a) the y-axis; (b) the x-axis

18 Through $A(-5, 1)$, perpendicular to (a) the y-axis; (b) the x-axis

19 Through $A(7, -3)$, perpendicular to the line with equation $2x - 5y = 8$

20 Through $(-\frac{3}{4}, -\frac{1}{2})$, parallel to the line with equation $x + 3y = 1$

21 Given $A(3, -1)$ and $B(-2, 6)$, find an equation for the perpendicular bisector of the line segment AB.

22 Find an equation for the line that bisects the second and fourth quadrants.

23 Find equations for the altitudes of the triangle with vertices $A(-3, 2)$, $B(5, 4)$, $C(3, -8)$, and find the point at which the altitudes intersect.

24 Find equations for the medians of the triangle in Exercise 23, and find their point of intersection.

In Exercises 25–34 use the Slope-Intercept Form to find the slope and y-intercept of the line with the given equation, and sketch the graph of each equation.

25 $3x - 4y + 8 = 0$ **26** $2y - 5x = 1$

27 $x + 2y = 0$ **28** $8x = 1 - 4y$

29 $y = 4$ **30** $x + 2 = \frac{1}{2}y$

31 $5x + 4y = 20$ **32** $y = 0$

33 $x = 3y + 7$ **34** $x - y = 0$

35 Find a real number k such that the point $P(-1, 2)$ is on the line $kx + 2y - 7 = 0$.

36 Find a real number k such that the line $5x + ky - 3 = 0$ has y-intercept -5.

37 If a line l has nonzero x- and y-intercepts a and b, respectively, prove that an equation for l is of the form $(x/a) + (y/b) = 1$. (This is called the **intercept form** for the equation of a line.) Express the equation $4x - 2y = 6$ in intercept form.

38 Prove that an equation of the line through $P_1(x_1, y_1)$ and $P_2(x_2, y_2)$ is

$$(y - y_1)(x_2 - x_1) = (y_2 - y_1)(x - x_1).$$

(This is called the **two-point form** for the equation of a line.) Use the two-point form to find an equation of the line through $A(7, -1)$ and $B(4, 6)$.

39 Find all values of r such that the slope of the line through the points $(r, 4)$ and $(1, 3 - 2r)$ is less than 5.

40 Find all values of t such that the slope of the line through $(t, 3t + 1)$ and $(1 - 2t, t)$ is greater than 4.

41 Six years ago a house was purchased for $59,000. This year it was appraised at $95,000. Assuming that the value of the house is linearly related to time, find a formula that specifies the value at any time after the purchase date. When was the house worth $73,000?

42 Charles' Law for gases states that if the pressure remains constant, then the relationship between the volume V that a gas occupies and its temperature T in degrees Celsius is given by $V = V_0(1 + \frac{1}{273}T)$.

(a) What is the significance of V_0?

(b) What increase in temperature is needed to increase the volume from V_0 to $2V_0$?

(c) Sketch the graph of the equation on a TV-plane for the case $V_0 = 100$ and $T \geq 273$.

43 The electrical resistance R (in ohms) for a pure metal wire is linearly related to its temperature T (in °C) by the formula

$$R = R_0(1 + aT)$$

for some constants a and $R_0 > 0$.

(a) What is the significance of R_0?

(b) At absolute zero ($T = -273\,°C$), $R = 0$. Find a.

(c) At $0\,°C$, silver wire has a resistance of 1.25 ohms. At what temperature is the resistance doubled?

44 The freezing point of water is $0\,°C$ or $32\,°F$. The boiling point is $100\,°C$ or $212\,°F$. Using this information, find a linear relationship between the temperature in $°F$ and the temperature in $°C$. What temperature increase in $°F$ corresponds to an increase in temperature of $1\,°C$?

45 The IRS allows taxpayers to depreciate rental property over a 15-year period. Thus, if y denotes the value of the property after x years, and the property has an initial value of $60,000, then the points $(0, 60,000)$ and $(15, 0)$ are on the graph of $y = f(x)$. One common method of depreciation is *linear depreciation*, in which a line is fitted through these two points. (This method is also called *straight-line depreciation*.)

(a) Express y as a linear function of x.

(b) What is the yearly depreciation?

46 A piece of farm machinery, initially valued at $40,000, is depreciated over its useful lifetime of ten years. At the end of ten years, the equipment has a *scrap value* of $2000.

(a) Using *linear depreciation* (see Exercise 45) express the value y (in dollars) as a function of the age x (in years).

(b) What is the yearly depreciation?

47 The expected weight W (in long tons) of an adult humpback whale is related to its length L (in feet) by the linear equation

$$W = 1.70L - 42.8.$$

(a) A 30-foot humpback has been spotted by marine researchers. Estimate its weight.

(b) If the error in estimating the length could be as large as 2 feet, what is the corresponding error for the weight estimate?

48 Newborn blue whales measure approximately 24 feet and weigh 3 tons. Young whales are nursed for 7 months and, after weaning, measure an amazing 53 feet and weigh 23 tons.

(a) If t denotes the age of the whale (in months) and L the length (in feet), express L as a linear function of t. According to this linear model, what is the daily increase in length? (Use 1 month = 30 days.)

(b) Express the weight W as a linear function of t. What is the daily increase in weight?

(c) Express W as a linear function of L.

49 The owner of an ice-cream franchise must pay the parent company $500 per month plus 10% of the profits. In addition, there are fixed costs of $400 per month for items such as utilities and labor. Suppose that 50% of the monthly revenue is pure profit.

(a) Express the monthly take-home pay M of the owner as a function of the monthly revenue R.

(b) How much business must be done to merely break even?

50 Several rules of thumb have been suggested for modifying adult drug dosage levels for young children. Let a denote the adult dose (in mg) and let t be the age of the child (in years). Some typical rules are

$$y = \frac{t + 1}{24}a \qquad \text{(Cowling's Rule)}$$

and $\quad y = \dfrac{2}{25}ta \qquad$ (Friend's Rule)

(a) Graph these two linear equations on the same set of axes for $0 \le t \le 12$.

(b) At approximately what age do the two rules give the same recommendation?

51 In a simple video game for young children, airplanes fly from left to right along the curve $y = 1 + (1/x)$ and can shoot their bullets in the tangent direction at creatures placed along the x-axis at $x = 1, 2, 3, 4$, and 5. Using calculus it can be shown that the slope of the tangent line at $P(1, 2)$ is $m = -1$, while the tangent line at $Q(\tfrac{3}{2}, \tfrac{5}{3})$ has slope $m = -\tfrac{4}{9}$. If a player shoots when the plane is at P, will a target be hit? What if the player shoots at Q?

52 A hammer thrower is working on his form in a small practice area. The hammer spins, generating a circle with a radius of 5 feet, and when released, hits a tall screen that is 50 feet from the center of the throwing area. Let coordinate axes be introduced as shown in the figure (not to scale).

(a) If the hammer is released at $(-4, -3)$ and travels in the tangent direction, where will it hit the screen?

(b) If the hammer is to hit at $(0, -50)$, where on the circle should it be released?

FIGURE FOR EXERCISE 52

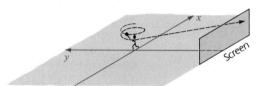

FIGURE FOR EXERCISE 51

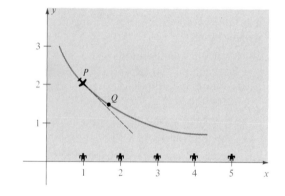

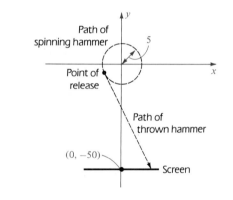

3.6 Operations on Functions

Functions are often defined in terms of sums, differences, products, and quotients of various expressions. For example, if

$$h(x) = x^2 + \sqrt{5x + 1},$$

we may regard $h(x)$ as a sum of values of the simpler functions f and g

defined by

$$f(x) = x^2 \quad \text{and} \quad g(x) = \sqrt{5x + 1}.$$

It is natural to refer to the function h as the *sum* of f and g.

In general, suppose f and g are *any* functions. Let I be *the intersection of their domains;* that is, the numbers *common* to both domains. The **sum** of f and g is the function h defined by

$$h(x) = f(x) + g(x)$$

where x is in I.

It is convenient to denote h by the symbol $f + g$. Since f and g are functions, not numbers, the $+$ used between f and g is not to be considered as addition of real numbers. It is used to indicate that the value of x for $f + g$ is $f(x) + g(x)$; that is,

$$(f + g)(x) = f(x) + g(x).$$

Similarly, the **difference** $f - g$ and the **product** fg of f and g are defined by

$$(f - g)(x) = f(x) - g(x) \quad \text{and} \quad (fg)(x) = f(x)g(x)$$

where x is in I. Finally, the **quotient** f/g of f by g is given by

$$\left(\frac{f}{g}\right)(x) = \frac{f(x)}{g(x)}$$

where x is in I and $g(x) \neq 0$.

EXAMPLE 1   If $f(x) = \sqrt{4 - x^2}$ and $g(x) = 3x + 1$, find the sum, difference, and product of f and g, and the quotient of f by g.

SOLUTION   The domain of f is the closed interval $[-2, 2]$ and the domain of g is $\mathbb{R}$. Consequently, the intersection of their domains is $[-2, 2]$, and the required functions are given by

$$(f + g)(x) = \sqrt{4 - x^2} + (3x + 1), \qquad -2 \leq x \leq 2$$
$$(f - g)(x) = \sqrt{4 - x^2} - (3x + 1), \qquad -2 \leq x \leq 2$$
$$(fg)(x) = \sqrt{4 - x^2}\,(3x + 1), \qquad -2 \leq x \leq 2$$
$$\left(\frac{f}{g}\right)(x) = \frac{\sqrt{4 - x^2}}{3x + 1}, \qquad -2 \leq x \leq 2, \quad x \neq -\tfrac{1}{3}. \qquad \blacksquare$$

A function f is a **polynomial function** if $f(x)$ is a polynomial (see page 27). A polynomial function may be thought of as a sum of functions whose values are of the form cx^k, where c is a real number and k is a nonnegative integer.

A function f is called **algebraic** if it can be expressed in terms of sums, differences, products, quotients, or roots of polynomial functions. For example, if

$$f(x) = 5x^4 - 2\sqrt[3]{x} + \frac{x(x^2 + 5)}{\sqrt{x^3 + \sqrt{x}}},$$

then f is an algebraic function. Functions that are not algebraic are termed **transcendental.** The exponential and logarithmic functions considered in Chapter 5 are examples of transcendental functions.

We shall conclude this section by describing an important method of using two functions f and g to obtain a third function. Suppose D, E, and K are sets of real numbers, and let f be a function from D to E and g a function from E to K. Using arrow notation we have

$$D \xrightarrow{f} E \xrightarrow{g} K$$

FIGURE 3.41

We shall use f and g to define a function from D to K.

For every x in D, the number $f(x)$ is in E. Since the domain of g is E, we may then find the number $g(f(x))$ in K. By associating $g(f(x))$ with x, we obtain a function from D to K called the *composite function* of g by f. This is illustrated pictorially in Figure 3.41, where the dashes indicate the correspondence we have defined from D to K.

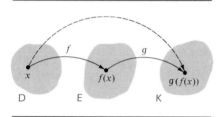

We sometimes use an operational symbol ∘ and denote a composite function by $g \circ f$. The following definition summarizes our discussion.

DEFINITION

Let f be a function from D to E and let g be a function from E to K. The **composite function** $g \circ f$ is the function from D to K defined by

$$(g \circ f)(x) = g(f(x))$$

for every x in D.

If the domain of g is a *subset* E' of E, then the domain of $g \circ f$ consists of all x in D such that $f(x)$ is in E'. (See Example 3.)

EXAMPLE 2   If $f(x) = x^3$ and $g(x) = 5x^2 + 2x + 1$, find $(g \circ f)(x)$.

SOLUTION   By definition,

$$(g \circ f)(x) = g(f(x)) = g(x^3).$$

Since $g(x^3)$ means that x^3 should be substituted for x in the expression for $g(x)$, we have

$$g(x^3) = 5(x^3)^2 + 2(x^3) + 1.$$

Consequently, $(g \circ f)(x) = 5x^6 + 2x^3 + 1.$ ∎

EXAMPLE 3   If the functions f and g are defined by $f(x) = x - 2$ and $g(x) = 5x + \sqrt{x}$, find $(g \circ f)(x)$ and the domain of $g \circ f$.

SOLUTION   Formal substitutions give us the following:

$$
\begin{aligned}
(g \circ f)(x) &= g(f(x)) && \text{(definition of } g \circ f) \\
&= g(x - 2) && \text{(definition of } f) \\
&= 5(x - 2) + \sqrt{x - 2} && \text{(definition of } g) \\
&= 5x - 10 + \sqrt{x - 2} && \text{(simplifying)}
\end{aligned}
$$

The domain of f is the set of all real numbers; however, the last equality implies that $(g \circ f)(x)$ is a real number only if $x \geq 2$. Thus, the domain of the composite function $g \circ f$ is the interval $[2, \infty)$.   ∎

Given f and g, it may also be possible to find $(f \circ g)(x) = f(g(x))$ as illustrated in the next example.

EXAMPLE 4   If $f(x) = x^2 - 1$ and $g(x) = 3x + 5$, find $(f \circ g)(x)$ and $(g \circ f)(x)$.

SOLUTION

$$
\begin{aligned}
(f \circ g)(x) &= f(g(x)) && \text{(definition of } f \circ g) \\
&= f(3x + 5) && \text{(definition of } g) \\
&= (3x + 5)^2 - 1 && \text{(definition of } f) \\
&= 9x^2 + 30x + 24 && \text{(simplifying)}
\end{aligned}
$$

Similarly,

$$
\begin{aligned}
(g \circ f)(x) &= g(f(x)) && \text{(definition of } g \circ f) \\
&= g(x^2 - 1) && \text{(definition of } f) \\
&= 3(x^2 - 1) + 5 && \text{(definition of } g) \\
&= 3x^2 + 2 && \text{(simplifying)} \quad\blacksquare
\end{aligned}
$$

We see from Example 4 that $f(g(x))$ and $g(f(x))$ are not always the same, that is, $f \circ g \neq g \circ f$. In certain cases it may happen that equality *does* occur. Of major importance is the case where $f(g(x))$ and $g(f(x))$ are not only identical, but both are equal to x. Of course, f and g must be very special functions for this to happen. In the next section we shall indicate the manner in which these functions will be restricted.

In certain applications, it is necessary to express a quantity y as a function of time t. The following example illustrates that it is often easier to introduce a third variable x, then express x as a function of t—that is, $x = g(t)$. Next express y as a function of x,—that is, $y = f(x)$—and finally form the composite function given by $y = f(x) = f(g(t))$.

EXAMPLE 5   A spherical toy balloon is being inflated with helium gas. If the radius of the balloon is changing at a rate of 1.5 cm/sec, express the volume V of the balloon as a function of time t (seconds).

SOLUTION   Let x denote the radius of the balloon. If we assume that the radius is 0 initially, then after t seconds

$$x = 1.5t \qquad \text{(radius of balloon after } t \text{ seconds).}$$

To illustrate, after 1 second the radius is 1.5 cm; after 2 seconds it is 3.0 cm; after 3 seconds it is 4.5 cm; and so on.

Next we write

$$V = \tfrac{4}{3}\pi x^3 \qquad \text{(volume of a sphere of radius } x).$$

This gives us a composite function relationship in which V is a function of x, and x is a function of t. By substitution, we obtain

$$V = \tfrac{4}{3}\pi x^3 = \tfrac{4}{3}\pi(1.5t)^3 = \tfrac{4}{3}\pi(\tfrac{3}{2}t)^3 = \tfrac{4}{3}\pi(\tfrac{27}{8}t^3).$$

Simplifying, we obtain the following formula for V as a function of t:

$$V = \tfrac{9}{2}\pi t^3. \qquad\qquad \blacksquare$$

Exercises 3.6

In Exercises 1–6, find the sum, difference, and product of f and g, and the quotient of f by g.

1 $f(x) = 3x^2$, $g(x) = 1/(2x - 3)$

2 $f(x) = \sqrt{x + 3}$, $g(x) = \sqrt{x + 3}$

3 $f(x) = x + (1/x)$, $g(x) = x - (1/x)$

4 $f(x) = x^3 + 3x$, $g(x) = 3x^2 + 1$

5 $f(x) = 2x^3 - x + 5$, $g(x) = x^2 + x + 2$

6 $f(x) = 7x^4 + x^2 - 1$, $g(x) = 7x^4 - x^3 + 4x$

In Exercises 7–24 find $(f \circ g)(x)$ and $(g \circ f)(x)$.

7 $f(x) = 3x + 2$, $g(x) = 2x - 1$

8 $f(x) = x + 4$, $g(x) = 5x - 3$

9 $f(x) = 4x^2 - 5$, $g(x) = 3x$

10 $f(x) = 7x + 1$, $g(x) = 2x^2$

11 $f(x) = 3x^2 + 2x$, $g(x) = 2x - 1$

12 $f(x) = 5x + 7$, $g(x) = 4 - x^2$

13 $f(x) = x - 1$, $g(x) = x^3$

14 $f(x) = x^3 + 5x$, $g(x) = 4x$

15 $f(x) = x^2 + 9x$, $g(x) = \sqrt{x + 9}$

16 $f(x) = \sqrt[3]{x^2 + 1}$, $g(x) = x^3 + 1$

17 $f(x) = \dfrac{1}{2x - 5}$, $g(x) = \dfrac{1}{x^2}$

18 $f(x) = \dfrac{x}{x + 1}$, $g(x) = x - 1$

19 $f(x) = |x|$, $g(x) = -5$

20 $f(x) = 7$, $g(x) = 10$

21 $f(x) = x^2$, $g(x) = 1/x^2$

22 $f(x) = \dfrac{1}{x + 1}$, $g(x) = x + 1$

23 $f(x) = 2x - 3$, $g(x) = \dfrac{x + 3}{2}$

24 $f(x) = x^3 - 1$, $g(x) = \sqrt[3]{x + 1}$

Use the method of Example 5 to solve Exercises 25–30.

25 A fire has started in a dry open field and spreads in the form of a circle. If the radius of this circle increases at the rate of 6 feet per minute, express the total fire area as a function of time t (in minutes).

26 A 100-foot-long cable of diameter 4 inches is submerged in seawater. Due to corrosion, the surface area of the cable decreases at the rate of 750 in^2 per year. Express the diameter of the cable as a function of time. (Ignore corrosion at the ends of the cable.)

27 A hot-air balloon rises vertically as a rope attached to the base of the balloon is released at the rate of 5 feet per second. The pulley that releases the rope is 20 feet from a platform where passengers board the balloon. Express the height of the balloon as a function of time.

FIGURE FOR EXERCISE 27

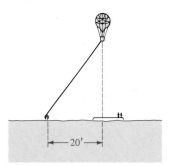

---

28 The diameter d of a cube is the distance between two opposite vertices. Express d as a function of the edge x of the cube. (*Hint:* First express the diagonal y of a face as a function of x.)

29 Refer to Exercise 44 of Section 3.3. The tightrope walker moves up the rope at a steady rate of 1 foot per second. If the rope is attached 30 feet up the pole, express the height h of the walker above the ground as a function of time t. (*Hint:* Let d denote the total distance traveled along the wire. First express d as a function of t, and then h as a function of d.)

30 Refer to Exercise 47 of Section 3.3. When the airplane is 500 feet down the runway, it has reached and will maintain a speed of 68 feet per second (or about 100 mph) until takeoff. Express the distance of the plane from the control tower as a function of time t (in seconds). (*Hint:* In the figure, first write x as a function of t.)

3.7 Inverse Functions

A function f may have the same value for different numbers in its domain. For example, if $f(x) = x^2$, then $f(2) = 4$ and $f(-2) = 4$, but $2 \neq -2$. To define *the inverse of a function*, it is essential that different numbers in the domain *always* give different values of f. Such functions are called *one-to-one*.

DEFINITION

> A function f with domain D and range E is a **one-to-one function** if whenever $a \neq b$ in D, then $f(a) \neq f(b)$ in E.

EXAMPLE 1

(a) If $f(x) = 3x + 2$, prove that f is one-to-one.

(b) If $g(x) = x^4 + 2x^2$ prove that g is not one-to-one.

SOLUTION

(a) If $a \neq b$, then $3a \neq 3b$ and hence $3a + 2 \neq 3b + 2$, or $f(a) \neq f(b)$. Thus, f is one-to-one.

(b) The function g is not one-to-one, since different numbers in the domain may have the same value. For example, although $-1 \neq 1$, both $g(-1)$ and $g(1)$ are equal to 3. ∎

THEOREM

> (i) If f is an increasing function throughout its domain, then f is one-to-one.
>
> (ii) If f is a decreasing function throughout its domain, then f is one-to-one.

PROOF Suppose f is an increasing function. If a and b are in the domain of f and $a \neq b$, then either $a < b$ or $b < a$, and hence either

$f(a) < f(b)$ or $f(b) < f(a)$, respectively. Thus, $f(a) \neq f(b)$; that is, f is one-to-one. A similar proof may be given for (ii). □

FIGURE 3.42

(i) $y = f(x)$

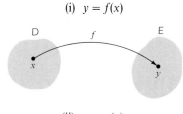

(ii) $x = g(y)$

Suppose f is a one-to-one function with domain D and range E. Thus, for each number y in E, there is *exactly one* number x in D such that $f(x) = y$, as illustrated by the arrow in Figure 3.42(i). Since x is *unique*, we may define a function g from E to D by means of the following rule:

$$g(y) = x.$$

(See Figure 3.42(ii)). In words, g *reverses the correspondence* given by f. Since

$$g(y) = x \quad \text{and} \quad f(x) = y$$

for x in D and y in E, we see, by substitution, that

$$g(f(x)) = x \quad \text{and} \quad f(g(y)) = y.$$

The last two equations may be written

$$(g \circ f)(x) = x \quad \text{and} \quad (f \circ g)(y) = y,$$

which means that $g \circ f$ and $f \circ g$ are the identity functions on D and E, respectively (see page 133); because of this, we call g the *inverse function* of f and denote it by f^{-1}, as indicated in the next definition.

DEFINITION

> Let f be a one-to-one function with domain D and range E. A function f^{-1} with domain E and range D is called the **inverse function of f** if
>
> $$f^{-1}(f(x)) = x \quad \text{for every } x \text{ in } D$$
>
> and $\qquad f(f^{-1}(y)) = y \quad \text{for every } y \text{ in } E.$

The -1 used in the notation f^{-1} should not be mistaken for an exponent; that is, $f^{-1}(y)$ *does not mean* $1/[f(y)]$. The reciprocal $1/[f(y)]$ may be denoted by $[f(y)]^{-1}$.

It is important to note that *a function f that either increases or decreases throughout its domain has an inverse function*, since by the previous theorem such functions are one-to-one.

When we considered functions in previous sections we usually let x denote an arbitrary number in the domain. Similarly, for the inverse function f^{-1}, we may wish to consider $f^{-1}(x)$, *where x is in the domain E of f^{-1}.* In this event, the two formulas in our definition are written

$$f^{-1}(f(x)) = x \quad \text{for every } x \text{ in } D$$

and

$$f(f^{-1}(x)) = x \quad \text{for every } x \text{ in } E.$$

The diagrams in Figure 3.42 contain a hint for finding the inverse of a one-to-one function in certain cases. If possible, we *solve the equation $y = f(x)$ for x in terms of y,* obtaining an equation of the form $x = g(y)$. If the two conditions $g(f(x)) = x$ and $f(g(x)) = x$ are true for all x in the domains of f and g, respectively, then g is the required inverse function f^{-1}. The following three guidelines summarize this procedure, where in guideline 2, in anticipation of finding f^{-1}, we write $x = f^{-1}(y)$ instead of $x = g(y)$.

GUIDELINES: **Finding f^{-1} in Simple Cases**

1 Verify that f is a one-to-one function (or that f is increasing or decreasing) throughout its domain.

2 Solve the equation $y = f(x)$ for x in terms of y, obtaining an equation of the form $x = f^{-1}(y)$.

3 Verify the two conditions

$$f^{-1}(f(x)) = x \quad \text{and} \quad f(f^{-1}(x)) = x$$

for all x in the domains of f and f^{-1}, respectively. ■ ■ ■

The success of this method depends on the nature of the equation $y = f(x)$, since we must be able to solve for x in terms of y. That is what we mean by "simple cases" in the heading of the guidelines.

EXAMPLE 2 If $f(x) = 3x - 5$, find the inverse function of f.

SOLUTION We shall follow the three guidelines. First, we note that the graph of the linear function f is a line of slope 3, and hence f is increasing throughout $\mathbb{R}$. Thus the inverse function f^{-1} exists. Moreover, since the domain and range of f is $\mathbb{R}$, the same is true for f^{-1}.

As in guideline 2, we consider

$$y = 3x - 5$$

and then solve for x in terms of y, obtaining

$$x = \frac{y + 5}{3}.$$

We now formally let

$$f^{-1}(y) = \frac{y + 5}{3}.$$

Since the symbol used for the variable is immaterial, we may also write

$$f^{-1}(x) = \frac{x + 5}{3}.$$

Finally, we verify that the two conditions

$$f^{-1}(f(x)) = x \quad \text{and} \quad f(f^{-1}(x)) = x$$

are fulfilled. Thus,

$$f^{-1}(f(x)) = f^{-1}(3x - 5) \qquad \text{(definition of } f\text{)}$$

$$= \frac{(3x - 5) + 5}{3} \qquad \text{(definition of } f^{-1}\text{)}$$

$$= x \qquad \text{(simplifying)}$$

Also,

$$f(f^{-1}(x)) = f\left(\frac{x + 5}{3}\right) \qquad \text{(definition of } f^{-1}\text{)}$$

$$= 3\left(\frac{x + 5}{3}\right) - 5 \qquad \text{(definition of } f\text{)}$$

$$= x \qquad \text{(simplifying)}$$

This proves that the inverse function of f is given by

$$f^{-1}(x) = \frac{x + 5}{3}.$$

■

FIGURE 3.43

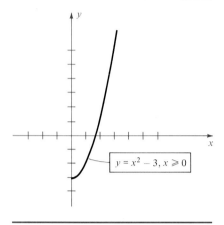

$y = x^2 - 3, \ x \geqslant 0$

EXAMPLE 3   Find the inverse function of f if $f(x) = x^2 - 3$ and $x \geq 0$.

SOLUTION   The graph of f is sketched in Figure 3.43. The domain D is $[0, \infty)$, and the range E is $[-3, \infty)$. Since f is increasing on D, it has an inverse function f^{-1} that has domain E and range D.

As in guideline 2, we consider the equation

$$y = x^2 - 3.$$

and solve for x, obtaining

$$x = \pm\sqrt{y + 3}.$$

Since x is nonnegative, we reject $x = -\sqrt{y + 3}$ and let

$$f^{-1}(y) = \sqrt{y + 3} \quad \text{or, equivalently,} \quad f^{-1}(x) = \sqrt{x + 3}.$$

Finally, we verify that $f^{-1}(f(x)) = x$ for x in $D = [0, \infty)$, and that $f(f^{-1}(x)) = x$ for x in $E = [-3, \infty)$. Thus,

$$f^{-1}(f(x)) = f^{-1}(x^2 - 3) = \sqrt{(x^2 - 3) + 3} = \sqrt{x^2} = x \quad \text{if } x \geq 0,$$

and

$$f(f^{-1}(x)) = f(\sqrt{x + 3}) = (\sqrt{x + 3})^2 - 3 = (x + 3) - 3 = x \quad \text{if } x \geq -3.$$

This proves that the inverse function is given by

$$f^{-1}(x) = \sqrt{x + 3} \quad \text{for } x \geq -3. \qquad \blacksquare$$

An interesting relationship exists between the graphs of a function f and its inverse function f^{-1}. We first note that $b = f(a)$ means the same thing as $a = f^{-1}(b)$. These equations imply that the point (a, b) is on the graph of f if and only if the point (b, a) is on the graph of f^{-1}. As an illustration, in Example 3 we found that the functions f and f^{-1} given by

$$f(x) = x^2 - 3 \quad \text{and} \quad f^{-1}(x) = \sqrt{x + 3}$$

are inverse functions of one another, provided that x is suitably restricted. Some points on the graph of f are $(0, -3), (1, -2), (2, 1)$, and $(3, 6)$. Corresponding points on the graph of f^{-1} are $(-3, 0), (-2, 1), (1, 2)$, and $(6, 3)$. The graphs of f and f^{-1} are sketched on the same coordinate axes in Figure 3.44. If the page is folded along the line l that bisects quadrants I and III (as indicated by the dashes in the figure), then the graphs of f and f^{-1} coincide. Note that an equation for l is $y = x$. The two graphs are said to be *reflections* of one another through the line l (or *symmetric* with respect to l). This is typical of the graph of every function f that has an inverse function f^{-1}. (See Exercise 34.)

FIGURE 3.44

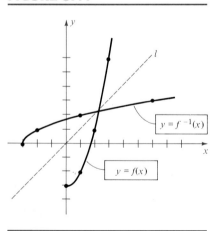

$y = f^{-1}(x)$

$y = f(x)$

Exercises 3.7

In Exercises 1–12 determine whether the function f is one-to-one.

1 $f(x) = 2x + 9$

2 $f(x) = 1/(7x + 9)$

3 $f(x) = 5 - 3x^2$

4 $f(x) = 2x^2 - x - 3$

5 $f(x) = \sqrt{x}$

6 $f(x) = x^3$

7 $f(x) = |x|$

8 $f(x) = 4$

9 $f(x) = x^2 - 3x + 2$

10 $f(x) = \sqrt[3]{x}$

11 $f(x) = 1/x$

12 $f(x) = \sqrt{9 - x^2}$

In Exercises 13–16 prove that f and g are inverse functions of one another and sketch the graphs of f and g on the same coordinate plane.

13 $f(x) = 7x + 5$; $g(x) = (x - 5)/7$

14 $f(x) = x^2 - 1$, $x \geq 0$; $g(x) = \sqrt{x + 1}$, $x \geq -1$

15 $f(x) = \sqrt{2x - 4}$, $x \geq 2$; $g(x) = \frac{1}{2}(x^2 + 4)$, $x \geq 0$

16 $f(x) = x^3 + 1$; $g(x) = \sqrt[3]{x - 1}$

In Exercises 17–30 find the inverse function of f.

17 $f(x) = 4x - 3$

18 $f(x) = 9 - 7x$

19 $f(x) = \dfrac{1}{2x + 5}$, $x > -\frac{5}{2}$

20 $f(x) = \dfrac{1}{3x - 1}$, $x > \frac{1}{3}$

21 $f(x) = 9 - x^2$, $x \geq 0$

22 $f(x) = 4x^2 + 1$, $x \geq 0$

23 $f(x) = 5x^3 - 2$

24 $f(x) = 7 - 2x^3$

25 $f(x) = \sqrt{3x - 5}$, $x \geq \frac{5}{3}$

26 $f(x) = \sqrt{4 - x^2}$, $0 \leq x \leq 2$

27 $f(x) = \sqrt[3]{x} + 8$

28 $f(x) = (x^3 + 1)^5$

29 $f(x) = x$

30 $f(x) = -x$

31 (a) Prove that the linear function defined by $f(x) = ax + b$ for $a \neq 0$ has an inverse function, and find $f^{-1}(x)$.

(b) Does a constant function have an inverse? Explain.

32 If f is a one-to-one function with domain D and range E, prove that f^{-1} is a one-to-one function with domain E and range D.

33 Prove that a one-to-one function has only one inverse function.

34 Establish the fact that the graph of f^{-1} is the reflection of the graph of f through the line $y = x$ by verifying each of the following:

(a) If $P(a, b)$ is on the graph of f, then $Q(b, a)$ is on the graph of f^{-1}.

(b) The midpoint of line segment PQ is on the line $y = x$.

(c) The line PQ is perpendicular to the line $y = x$.

The graphs of one-to-one functions are shown in Exercises 35–38. Use the reflection property to sketch the graph of f^{-1}. Determine the domain and range of (a) f; (b) f^{-1}.

35

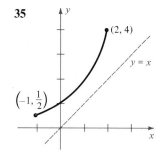

36

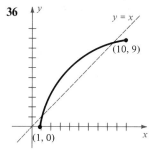

37

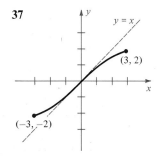

38

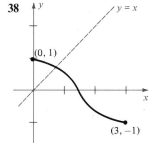

3.8 Variation

We shall next introduce terminology that is used in science to describe relationships among variable quantities. In the following definition the letters u, v and w are variables. The domains and ranges of the variables will not be specified. In any particular problem they should be evident.

DEFINITION

> (i) The phrase u **varies directly as** v, or u **is directly proportional to** v, means that $u = kv$ for a some real number k.
>
> (ii) If n is a positive real number, then the phrase u **varies directly as the nth power of** v, or u **is directly proportional to the nth power of** v, means that $u = kv^n$ for some real number k.
>
> (iii) The phrase u **varies inversely as** v, or u **is inversely proportional to** v, means that $u = k/v$, where k is a real number. If $u = k/v^n$ for some positive real number n, then u **varies inversely as the nth power of** v.
>
> (iv) The phrase u **varies jointly as w and** v, or u **is jointly proportional to w and** v, means $u = kwv$ for some real number k. If $u = kw^n v^m$ for positive real numbers n and m, then u **varies jointly as the nth power of w and the mth power of** v.

Note that in (i)–(iii) the variable u is a function of the variable v. In (iv) the variable depends on *both* w and v and determines what is called a function of *two* variables. These functions arise in many applications and are studied in advanced courses in mathematics.

The number k in the definition is sometimes called the **constant of variation** or the **constant of proportionality.** For example, if an automobile is moving at a rate of 50 miles per hour, then the distance d it travels in t hours is given by $d = 50t$. Hence, as in (i), the distance d is directly proportional to the time t, and the constant of proportionality is 50.

To illustrate (ii) of the definition, the formula $A = \pi r^2$ for the area of a circle states that the area A varies directly as the square of the radius r. The constant of proportionality is π. As another illustration, the formula $V = \frac{4}{3}\pi r^3$ for the volume of a sphere of radius r states that the volume V is directly proportional to the cube of the radius. The constant of proportionality in this case is $\frac{4}{3}\pi$.

In direct variation (i) the dependent variable increases numerically as the independent variable increases, whereas in inverse variation (iii) the absolute value of the dependent variable decreases as the independent variable increases.

In many applied problems the constant of proportionality can be determined by examining experimental facts, as illustrated in the following example.

EXAMPLE 1 If the temperature remains constant, then the pressure of an enclosed gas is inversely proportional to the volume. The pressure of a certain gas within a spherical balloon of radius 9 inches is 20 pounds per square inch. If the radius of the balloon increases to 12 inches, find the new pressure of the gas.

SOLUTION The original volume is $\frac{4}{3}\pi(9)^3 = 972\pi$ cubic inches. If we denote the pressure by P and the volume by V, then by (iii) of the definition,

$$P = \frac{k}{V}$$

for some real number k. Since $P = 20$ when $V = 972\pi$,

$$20 = \frac{k}{972\pi}$$

and hence, $k = 20(972\pi) = 19440\pi$. Consequently, a formula for P is

$$P = \frac{19440\pi}{V}.$$

If the radius is 12 inches, then $V = \frac{4}{3}(12)^3\pi = 2304\pi$ cubic inches. Substituting this number for V in the previous equation gives us

$$P = \frac{19440\pi}{2304\pi} = \frac{135}{16} = 8.4375.$$

Thus, the pressure is 8.4375 pounds per square inch when the radius is 12 inches. ■

Combinations of the types of variation may occur. It is inconvenient to discuss all possible situations, but perhaps the following illustration will suffice. If a variable s varies jointly as u and the cube of v, and inversely as the square of w, then

$$s = k\frac{uv^3}{w^2},$$

where k is some real number.

EXAMPLE 2 The weight that can be safely supported by a beam with a rectangular cross section varies jointly as the width and square of the depth of the cross section, and inversely as the length of the beam. If a 2-by-4-inch beam that is 8 feet long safely supports a load of 500 pounds, what weight can be safely supported by a 2-by-8-inch beam that is 10 feet long? (Assume that the width is the *shorter* dimension of the cross section.)

SOLUTION If the width, depth, length, and weight are denoted by w, d, l, and W, respectively, then

$$W = k\,\frac{wd^2}{l}.$$

According to the given data,

$$500 = k\,\frac{2(4^2)}{8}.$$

Solving for k we obtain $k = 125$, and hence the formula for W is

$$W = 125\left(\frac{wd^2}{l}\right).$$

To answer the question we substitute $w = 2$, $d = 8$, and $l = 10$, obtaining

$$W = 125\left(\frac{2 \cdot 8^2}{10}\right) = 1600 \text{ pounds.} \qquad \blacksquare$$

Exercises 3.8

In Exercises 1–8 express each statement as a formula and determine the constant of proportionality from the given conditions.

1 a is directly proportional to v. If $v = 30$, then $a = 12$.

2 s varies directly as t. If $t = 10$, then $s = 18$.

3 r varies directly as s and inversely as t. If $s = -2$ and $t = 4$, then $r = 7$.

4 w varies directly as z and inversely as the square root of u. If $z = 2$ and $u = 9$, then $w = 6$.

5 y is directly proportional to the square of x and inversely proportional to the cube of z. If $x = 5$ and $z = 3$, then $y = 25$.

6 q is inversely proportional to the sum of x and y. If $x = 0.5$ and $y = 0.7$, then $q = 1.4$.

7 c varies jointly as the square of a and the cube of b. If $a = 7$ and $b = -2$, then $c = 16$.

8 r is jointly proportional to s and v and inversely proportional to the cube of p. If $s = 2$, $v = 3$, and $p = 5$, then $r = 40$.

9 The pressure acting at a point in a liquid is directly proportional to the distance from the surface of the liquid to the point. In a certain oil tank the pressure at a depth of 2 feet is 118 pounds per ft^2. Find the pressure at a depth of 5 feet.

10 Hooke's Law states that the force required to stretch a spring x units beyond its natural length is directly proportional to x. If a weight of 4 pounds stretches a spring from its natural length of 10 inches to a length of 10.3 inches, what weight will stretch it to a length of 11.5 inches?

11 The electrical resistance of a wire varies directly as its length and inversely as the square of its diameter. If a wire 100 feet long of diameter 0.01 inches has a resistance of 25 ohms, find the resistance in a wire made of the same material that has a diameter of 0.015 inches and is 50 feet long.

12 The intensity of illumination I from a source of light varies inversely as the square of the distance d from the source. If a searchlight has an intensity of 1,000,000 candlepower at 50 feet, what is the intensity at a distance of 1 mile?

13 The period of a simple pendulum, that is, the time required for one complete oscillation, varies directly as the square root of its length. If a pendulum 2 feet long has a period of 1.5 seconds, find the period of a pendulum 6 feet long.

14 A circular cylinder is often used in physiology as a simple representation of a human limb.

(a) Show that the volume V of a cylinder varies jointly as the length L and the square of the circumference C.

(b) The formula obtained in (a) can be used to approximate the volume of a limb from length and circumference measurements. Suppose the (average) circumference of a man's forearm is 22 cm and the length is 27 cm. Approximate the volume of the forearm.

15 Kepler's Third Law states that the period T of a planet, that is, the time needed to make one complete revolution about the sun, is directly proportional to the $\frac{3}{2}$ power of the average distance d from the sun. For the planet earth, $T = 365$ days and $d = 93$ million miles. Venus is 67 million miles from the sun. Estimate the period of Venus.

16 A motorcycle daredevil has made a jump of 150 feet. His speed coming off the ramp was 70 mph. It is known from physics that the range of a projectile is directly proportional to the square of the velocity. If he can at-

tain 80 mph coming off the ramp and maintain proper balance, estimate the possible length of such a jump.

17 Police can sometimes estimate the speed V at which an automobile was traveling before the brakes were applied from the length L of the skid marks. Suppose that on a dry surface $L = 50$ feet when $V = 35$ mph. Assuming that the speed is directly proportional to the square root of the length of the skid marks, estimate the initial speed if the skid measures 150 feet.

18 Coulomb's Law in electricity asserts that the force F of attraction between two oppositely charged particles varies jointly as the magnitudes Q_1 and Q_2 of the charges, and inversely as the square of the distance d between the particles.

(a) Find a formula for F.

(b) What is the effect of reducing the distance between the particles by a factor of $\frac{1}{4}$?

19 Threshold weight W is defined to be that unhealthy weight beyond which mortality increases significantly. For middle-aged males, this weight is directly proportional to the third power of the height h. For a 6-foot male, W is about 200 pounds. Estimate the threshold weight for an individual who is 5 feet, 6 inches tall.

20 The Ideal Gas Law asserts that the volume V that a gas occupies is jointly proportional to the number n of moles of gas and to the temperature T (in degrees Kelvin), but is inversely proportional to the pressure P (in atmospheres). What is the effect on the volume if the number of moles is doubled but the temperature and pressure are reduced by a factor of $\frac{1}{2}$?

21 Poiseuille's Law asserts that the blood flow rate F (in liters/minute) through major arteries is jointly proportional to the fourth power of the radius R and to the blood pressure P. During heavy exercise, normal flow rates sometimes triple. If the radius increases by 10%, approximately how much harder must the heart pump?

22 Suppose 200 trout are caught, tagged, and released in a lake's general population. Let y denote the number of tagged fish that are recovered later, when a sample of n trout are caught. The validity of the *mark-recapture* method for estimating the lake's total trout population is based on the assumption that y is directly proportional to n. If 10 tagged trout are recovered from a sample of 300, estimate the total trout population of the lake.

3.9 Review

Define or discuss each of the following.

1 Ordered pair
2 Rectangular coordinate system in a plane
3 Coordinate axes
4 Quadrants
5 Coordinates of a point
6 Distance formula
7 Midpoint formula
8 Graph of an equation in x and y
9 Tests for symmetry
10 Equation of a circle
11 Unit circle
12 Slope of a line
13 Point-Slope Form
14 Slope-Intercept Form
15 Linear equation in x and y
16 Function
17 Domain and range of a function

18 One-to-one function
19 Identity function
20 Constant function
21 Even function
22 Odd function
23 Graph of a function
24 Linear function
25 Increasing function
26 Decreasing function
27 Vertical shifts of graphs
28 Horizontal shifts of graphs
29 Stretching of graphs
30 Reflections of graphs
31 Composite function of two functions
32 Inverse function
33 Variation

Exercises 3.9

1 Plot the points $A(3, 1)$, $B(-5, -3)$, and $C(4, -1)$, and prove that they are vertices of a right triangle. What is the area of the triangle?

2 Given points $P(-5, 9)$ and $Q(-8, -7)$, find (a) the midpoint of the segment PQ, and (b) a point T such that Q is the midpoint of PT.

3 Describe the set of all points (x, y) in a coordinate plane such that $y/x < 0$.

4 Find the slope of the line through $C(11, -5)$ and $D(-8, 6)$.

5 Prove that the points $A(-3, 1)$, $B(1, -1)$, $C(4, 1)$, and $D(3, 5)$ are vertices of a trapezoid.

6 Find an equation of the circle that has center $C(7, -4)$ and passes through the point $Q(-3, 3)$.

7 Find an equation of the circle that has center $C(-5, -1)$ and is tangent to the line $x = 4$.

8 Express the equation $8x + 3y - 24 = 0$ in Slope-Intercept Form.

9 Find an equation of the line through $A(\frac{1}{2}, -\frac{1}{3})$ that is (a) parallel to the line $6x + 2y + 5 = 0$; (b) perpendicular to the line $6x + 2y + 5 = 0$.

10 Find an equation of the line that has x-intercept -3 and passes through the center of the circle that has the equation $x^2 + y^2 - 4x + 10y + 26 = 0$.

Sketch the graphs of the equations in Exercises 11–21.

11 $2y + 5x - 8 = 0$ **12** $x = 3y + 4$

13 $x + 5 = 0$ **14** $2y - 7 = 0$

15 $y = \sqrt{1 - x}$ **16** $3x - 7y^2 = 0$

17 $9y^2 + 2x = 0$ **18** $y^2 = 16 - x^2$

19 $x^2 + y^2 + 4x - 16y + 64 = 0$

20 $x^2 + y^2 - 8x = 0$ **21** $y - x^2 = 1$

22 Find the domain and range of f if:

   (a) $f(x) = \sqrt{3x - 4}$ (b) $f(x) = 1/(x + 3)^2$

23 If $f(x) = x/\sqrt{x + 3}$ find (a)–(g):

   (a) $f(1)$ (b) $f(-1)$ (c) $f(0)$
   (d) $f(-x)$ (e) $-f(x)$ (f) $f(x^2)$
   (g) $(f(x))^2$

In Exercises 24–30 sketch the graph of f. Find the domain, the range, and the intervals in which f is increasing or decreasing.

24 $f(x) = |x + 3|$ **25** $f(x) = \dfrac{1 - 3x}{2}$

26 $f(x) = \sqrt{2 - x}$ **27** $f(x) = 1 - \sqrt{x + 1}$

28 $f(x) = 9 - x^2$ **29** $f(x) = 1000$

30 $f(x) = \begin{cases} x^2 & \text{if } x < 0 \\ 3x & \text{if } 0 \le x < 2 \\ 6 & \text{if } x \ge 2 \end{cases}$

31 Sketch the graphs of the following equations, making use of shifting, stretching, or reflecting.

   (a) $y = \sqrt{x}$ (b) $y = \sqrt{x + 4}$
   (c) $y = \sqrt{x} + 4$ (d) $y = 4\sqrt{x}$
   (e) $y = \frac{1}{4}\sqrt{x}$ (f) $y = -\sqrt{x}$

32 Determine whether f is even, odd, or neither even nor odd:

   (a) $f(x) = \sqrt[3]{x^3 + 4x}$
   (b) $f(x) = \sqrt[3]{3x^2 - x^3}$
   (c) $f(x) = \sqrt[3]{x^4 + 3x^2 + 5}$

In Exercises 33 and 34 find $(f \circ g)(x)$ and $(g \circ f)(x)$.

33 $f(x) = 2x^2 - 5x + 1,\ g(x) = 3x + 2$

34 $f(x) = \sqrt{3x + 2},\ g(x) = 1/x^2$

In Exercises 35 and 36 find $f^{-1}(x)$ and sketch the graphs of f and f^{-1} on the same coordinate plane.

35 $f(x) = 10 - 15x$

36 $f(x) = 9 - 2x^2,\ x \le 0$

37 If the altitude and radius of a right circular cylinder are equal, express the volume V as a function of the circumference C of the base.

38 A company plans to manufacture a container having the shape of a right circular cylinder, open at the top, and having a capacity of 24π in^3. If the cost of the material for the bottom is 30 cents per in^2 and that for the curved sides is 10 cents per in^2, express the total cost C of the material as a function of the radius r of the base of the container.

39 An open rectangular storage shelter consisting of two vertical sides, 4 feet wide and a flat roof is to be attached to an existing structure as illustrated in the figure. The flat roof is made of tin and costs \$5 per square foot while the other two sides are made of plywood costing \$2 per square foot.

   (a) If \$400 is available for construction, express the length y as a function of the height x.

   (b) Express the volume V inside the shelter as a function of x.

FIGURE FOR EXERCISE 39

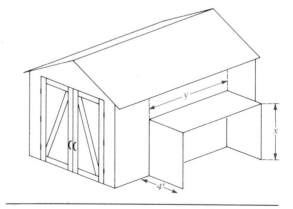

40 Five years ago the population of a small community was 15,000 people. Due to industrial development, the population has grown to 21,000. Assuming that the population P is linearly related to time t, express P as a function of t. When will the population number 30,000?

41 An automobile presently gets 20 miles per gallon but is in need of a tune-up that will cost $50. The tune-up will improve gasoline mileage by 10%.

(a) If gasoline costs $1.25 per gallon, find a linear function that gives the cost C of driving x miles without the tune-up.

(b) Find a linear function that gives the cost C of driving x miles with the tune-up.

(c) How many miles must the automobile be driven before the tune-up saves money?

42 Water in a paper conical filter drips into a cup as shown in the figure. Suppose 5 in^3 of water is poured into the filter. Let x denote the height of the water in the filter and let y denote the height of the water in the cup.

(a) Express the radius r (see figure) as a function of x. (*Hint:* Use similar triangles.)

(b) Express the height y of the water in the cup as a function of x. (*Hint:* What is the sum of the two volumes shown in the figure?)

FIGURE FOR EXERCISE 42

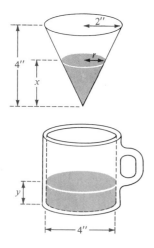

43 A cross section of a rectangular pool of dimensions 80 feet by 40 feet is shown in the figure. The pool is being filled with water at the rate of 10 ft^3 per minute.

(a) Express the volume V as a function of depth h at the deep end for $0 \le h \le 6$ and then for $6 \le h \le 9$.

(b) Express V as a function of time t. Then express h as a function of t.

FIGURE FOR EXERCISE 43

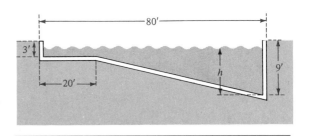

44 Find a formula that expresses the fact that w varies jointly as x and the square of y, and inversely as z if $w = 30$ when $x = 3$, $y = 2$, and $z = 5$.

45 The power P generated by a wind rotor is jointly proportional to the area A swept out by the blades and the third power of the wind velocity v. Suppose the diameter of the circular area swept out by the blades is 10 feet, and $P = 3000$ watts when $v = 20$ mph. Find the power generated when the wind velocity is 30 mph.

46 In a certain county, the average number of telephone calls per day between two cities is directly proportional to each of their populations and is inversely proportional to the square of the distance between them. Crestview and Cedar City are 25 miles apart and have populations of 10,000 and 5000, respectively. Telephone records indicate an average of 2000 calls per day between the two cities. Estimate the average number of calls between Crestview and Palms, a city of 15,000 people that is 100 miles from Crestview.

CHAPTER 4

Polynomial Functions, Rational Functions, and Conic Sections

Polynomial functions are the most basic functions in algebra.

■ Techniques for sketching their graphs are discussed in the first two sections. ■ Then we turn our attention to division and study methods for finding zeros of polynomial functions. ■ Next, we consider quotients of polynomial functions, that is, *rational functions*.

■ The chapter concludes with a brief survey of conic sections.

4.1 Quadratic Functions

Among the most important functions in mathematics are those defined as follows:

DEFINITION

> A function f is a **polynomial function** if
>
> $$f(x) = a_n x^n + a_{n-1} x^{n-1} + \cdots + a_1 x + a_0$$
>
> where the coefficients $a_0, a_1, \ldots, a_n$ are real numbers and the exponents are nonnegative integers.

If $a_n \neq 0$ in the preceding definition, we say that f has **degree n.** Note that a polynomial function of degree 1 is a linear function. If the degree is 2, then, as in the next definition, f is called a *quadratic function.*

DEFINITION

> A function f is a **quadratic function** if
>
> $$f(x) = ax^2 + bx + c$$
>
> where a, b, and c are real numbers and $a \neq 0$.

If $b = c = 0$ in the preceding definition, then $f(x) = ax^2$, and the graph is a parabola with vertex at the origin, opening upward if $a > 0$ or downward if $a < 0$ (see Figures 3.29 and 3.30). If $b = 0$ and $c \neq 0$, then

$$f(x) = ax^2 + c,$$

and from the discussion of vertical shifts in Section 3.4, the graph is a parabola with vertex at the point $(0, c)$ on the y-axis. Some typical graphs are illustrated in Figure 3.26. Another is given in the next example.

EXAMPLE 1   Sketch the graph of f if $f(x) = -\frac{1}{2}x^2 + 4$.

SOLUTION   The graph of $y = -\frac{1}{2}x^2$ is similar in shape to the graph of $y = -x^2$, sketched in Figure 3.30, but is somewhat wider. The graph of $y = -\frac{1}{2}x^2 + 4$ may be found by shifting the graph of $y = -\frac{1}{2}x^2$ upward 4 units. Coordinates of several points on the graph are listed in the following table.

FIGURE 4.1

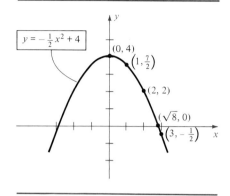

$y = -\frac{1}{2}x^2 + 4$

$(0, 4)$

$\left(1, \frac{7}{2}\right)$

$(2, 2)$

$(\sqrt{8}, 0)$

$\left(3, -\frac{1}{2}\right)$

x	0	1	2	$\sqrt{8}$	3
y	4	$\frac{7}{2}$	2	0	$-\frac{1}{2}$

Plotting and using symmetry with respect to the y-axis gives us the sketch in Figure 4.1. ■

If $f(x) = ax^2 + bx + c$ and $b \neq 0$, then by completing the square (see page 78), we can change the form of $f(x)$ to

$$f(x) = a(x - h)^2 + k$$

for some real numbers h and k. This technique is illustrated in the next example. As we shall see, the graph of f can be readily obtained from this new form.

EXAMPLE 2 If $f(x) = 3x^2 + 24x + 50$, express $f(x)$ in the form $a(x - h)^2 + k$.

SOLUTION Before completing the square *it is essential that we factor out the coefficient of x^2 from the first two terms* of $f(x)$ as follows:

$$f(x) = 3x^2 + 24x + 50$$
$$= 3(x^2 + 8x \quad) + 50.$$

We may now complete the square for the expression $x^2 + 8x$ by adding the square of one-half the coefficient of x, that is, $(\frac{8}{2})^2$, or 16. However, if we add 16 to the expression within parentheses, then, because of the factor 3, we are actually adding 48 to $f(x)$. Hence, we must compensate by subtracting 48:

$$f(x) = 3(x^2 + 8x \quad) + 50$$
$$= 3(x^2 + 8x + 16) + 50 - 48$$
$$= 3(x + 4)^2 + 2,$$

which has the desired form, with $a = 3$, $h = -4$, and $k = 2$. ■

If $f(x) = ax^2 + bx + c$, then by completing the square as in Example 2, we see that the graph of f is the same as the graph of an equation of the form

$$y = a(x - h)^2 + k.$$

From the discussion of horizontal shifts in Section 3.4, we can find the graph of $y = a(x - h)^2$ by shifting the graph of $y = ax^2$ either to the left or right, depending on the sign of h. Thus, $y = a(x - h)^2$ is an equation of a parabola that has vertex $(h, 0)$ and a vertical axis. A typical graph is sketched in Figure 4.2(i) for the case in which both a and h are positive. Since

FIGURE 4.2

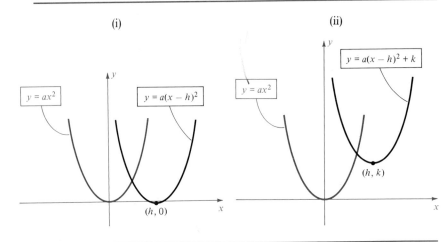

(i)

(ii)

$y = ax^2$

$y = a(x - h)^2$

$(h, 0)$

$y = ax^2$

$y = a(x - h)^2 + k$

(h, k)

the graph of $y = a(x - h)^2 + k$ can be obtained from that of $y = a(x - h)^2$ by a *vertical* shift of $|k|$ units, it follows that *the graph of a quadratic function f is a parabola* that has vertex (h, k) and a vertical axis. The sketch in Figure 4.2(ii) illustrates one possible graph. Observe that since (h, k) is the lowest (or highest) point on the parabola, $f(x)$ has its minimum (or maximum) value at $x = h$. This value is $f(h) = k$.

We have derived the following equation of a parabola that has vertex (h, k) and a vertical axis.

STANDARD EQUATION OF A PARABOLA (VERTICAL AXIS)

$$y - k = a(x - h)^2$$

This parabola opens upward if $a > 0$ or downward if $a < 0$.

EXAMPLE 3   Sketch the graph of f if $f(x) = 2x^2 - 6x + 4$ and find the minimum value of $f(x)$.

SOLUTION   The graph of f is a parabola and is the same as the graph of $y = 2x^2 - 6x + 4$. We begin by completing the square:

$$y = 2x^2 - 6x + 4$$
$$= 2(x^2 - 3x \quad) + 4$$
$$= 2(x^2 - 3x + \tfrac{9}{4}) + (4 - \tfrac{9}{2})$$
$$= 2(x - \tfrac{3}{2})^2 - \tfrac{1}{2}$$

FIGURE 4.3

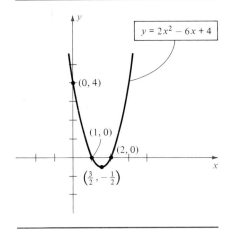

If we write the last equation as

$$y + \tfrac{1}{2} = 2(x - \tfrac{3}{2})^2$$

and compare it with the standard equation of a parabola, we see that $h = \frac{3}{2}$ and $k = -\frac{1}{2}$. Consequently, the vertex (h, k) of the parabola is $(\frac{3}{2}, -\frac{1}{2})$. Since $a = 2 > 0$, the parabola opens upward. The y-intercept is $f(0) = 4$. To find the x-intercepts we solve $2x^2 - 6x + 4 = 0$, or the equivalent equation $(2x - 2)(x - 2) = 0$, obtaining $x = 1$ and $x = 2$. Plotting the vertex together with the x- and y-intercepts provides enough points for a reasonably accurate sketch (see Figure 4.3). The minimum value of f occurs at the vertex, and hence is $f(\frac{3}{2}) = -\frac{1}{2}$. ∎

EXAMPLE 4 Sketch the graph of f if $f(x) = 8 - 2x - x^2$ and find the maximum value of $f(x)$.

SOLUTION We know that the graph is a parabola with a vertical axis. Let us set $y = f(x)$ and find the vertex by completing the square:

FIGURE 4.4

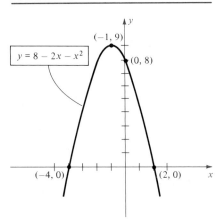

$$
\begin{aligned}
y &= -x^2 - 2x + 8 \\
&= -(x^2 + 2x) + 8 \\
&= -(x^2 + 2x + 1) + 8 + 1 \\
&= -(x + 1)^2 + 9.
\end{aligned}
$$

If we now write $\qquad y - 9 = -(x + 1)^2$

and compare this equation with the standard equation of a parabola, we see that $h = -1, k = 9$, and hence the vertex is $(-1, 9)$. Since $a = -1 < 0$, the parabola opens downward. To find the x-intercepts we solve the equation $8 - 2x - x^2 = 0$, or, equivalently, $x^2 + 2x - 8 = 0$. Factoring gives us $(x + 4)(x - 2) = 0$, and hence the intercepts are $x = -4$ and $x = 2$. The y-intercept is 8. Using this information gives us the sketch in Figure 4.4. The maximum value of f occurs at the vertex and hence is $f(-1) = 9$. ∎

The x-intercepts of the graph of $y = ax^2 + bx + c$ are the solutions of the quadratic equation $ax^2 + bx + c = 0$ and hence are $(-b \pm \sqrt{b^2 - 4ac})/2a$. If $b^2 - 4ac > 0$, the equation has two real and unequal solutions and the graph has two x-intercepts. If $b^2 - 4ac = 0$, the equation has one (double)

FIGURE 4.5

$y = ax^2 + bx + c$

$b^2 - 4ac > 0$

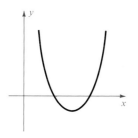

$b^2 - 4ac = 0$

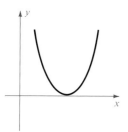

$b^2 - 4ac < 0$

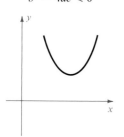

FIGURE 4.6

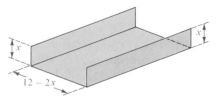

solution, and the graph is tangent to the x-axis. If $b^2 - 4ac < 0$, the equation has no real solutions, and the graph has no x-intercepts. We have illustrated the three cases in Figure 4.5 for $a > 0$. A similar situation occurs if $a < 0$, but in this case the parabolas open downward.

We can solve certain applied problems by finding maximum or minimum values of quadratic functions. The next example is one illustration.

EXAMPLE 5 A long rectangular sheet of metal, 12 inches wide, is to be made into a rain gutter by turning up two sides so that they are perpendicular to the sheet. How many inches should be turned up to give the gutter its greatest capacity?

SOLUTION The gutter is illustrated in Figure 4.6, where since x denotes the number of inches turned up on each side, the width of the base of the gutter is expressed as $12 - 2x$ inches. The capacity will be greatest when the area of the rectangle with sides of lengths x and $12 - 2x$ has its greatest value. Letting $f(x)$ denote this area, we have

$$f(x) = x(12 - 2x)$$
$$= 12x - 2x^2$$
$$= -2x^2 + 12x.$$

We may find the maximum value of $f(x)$ by completing a square:

$$f(x) = -2(x^2 - 6x)$$
$$= -2(x^2 - 6x + 9) + 18$$
$$= -2(x - 3)^2 + 18.$$

The graph of f is the graph of the equation

$$y = -2(x - 3)^2 + 18 \quad \text{or} \quad y - 18 = -2(x - 3)^2$$

and hence is a parabola that opens downward. The largest value of $y = f(x)$ occurs at the vertex (3, 18); that is at $x = 3$. Thus, 3 inches should be turned up to achieve maximum capacity. ■

Parabolas (and hence quadratic functions) occur in applications of mathematics to the physical world. For example, it can be shown that if a projectile is fired, and we assume that it is acted upon only by the force of gravity (that is, air resistance and other outside factors are ignored), then the path of the projectile is parabolic. Properties of parabolas are used in the design of mirrors for telescopes and searchlights and in the construction of radar antenna.

Exercises 4.1

1 If $f(x) = ax^2 + 2$, sketch the graph of f for each value of a.

(a) $a = 2$ (b) $a = 5$

(c) $a = \frac{1}{2}$ (d) $a = -3$

2 If $f(x) = 4x^2 + c$, sketch the graph of f for each value of c.

(a) $c = 2$ (b) $c = 5$

(c) $c = \frac{1}{2}$ (d) $c = -3$

In Exercises 3–6 sketch the graph of f and find the vertex.

3 $f(x) = 4x^2 - 9$ **4** $f(x) = 2x^2 + 3$

5 $f(x) = 9 - 4x^2$ **6** $f(x) = 16 - 9x^2$

In Exercises 7–10 use the Quadratic Formula to find the zeros of f.

7 $f(x) = 4x^2 - 11x - 3$

8 $f(x) = 25x^2 + 10x + 1$

9 $f(x) = 9x^2 - 12x + 4$

10 $f(x) = 10x^2 + x - 21$

In Exercises 11–14 use the technique of completing the square to express $f(x)$ in the form $a(x - h)^2 + k$, and then find the maximum or minimum value of $f(x)$.

11 $f(x) = 2x^2 - 16x + 23$

12 $f(x) = 3x^2 - 12x + 7$

13 $f(x) = -5x^2 - 10x + 3$

14 $f(x) = -2x^2 - 12x - 12$

In Exercises 15–24 sketch the graph of f and then find the maximum or minimum value of $f(x)$.

15 $f(x) = x^2 + 5x + 4$ **16** $f(x) = x^2 - 6x$

17 $f(x) = 8x - 12 - x^2$ **18** $f(x) = 10 + 3x - x^2$

19 $f(x) = x^2 + x + 3$ **20** $f(x) = x^2 + 2x + 5$

21 $f(x) = 3x^2 - 12x + 16$ **22** $f(x) = 2x^2 + 12x + 13$

23 $f(x) = -5x^2 - 10x - 4$

24 $f(x) = -3x^2 + 24x - 42$

25 Flights of leaping animals typically form parabolas. The figure illustrates a frog jump superimposed on a rectangular coordinate system. The length of the leap is 9 feet and the maximum height off the ground is 3 feet. Find a quadratic function f that specifies the path of the frog.

FIGURE FOR EXERCISE 25

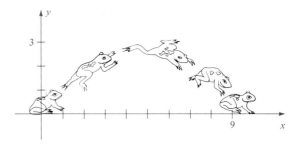

26 In the 1940s, the human cannonball stunt was performed regularly by Emmanuel Zacchini for The Ringling Brothers and Barnum & Bailey Circus. The tip of the cannon rose 15 feet off the ground, and the total horizontal distance traveled was 175 feet. When the cannon is aimed at a 45-degree angle, the equation of the parabolic flight has the form $y = ax^2 + x + c$ (see figure).

(a) Find values for a and c that correspond to the given information.

(b) Find the maximum height attained by the human cannonball.

FIGURE FOR EXERCISE 26

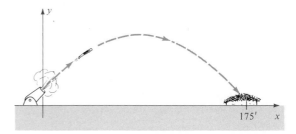

27 One section of a suspension bridge has its weight uniformly distributed between twin towers that are 400 feet apart and rise 90 feet above the horizontal roadway (see figure). A cable strung between the tops of the towers has the shape of a parabola, and its center point is 10 feet above the roadway. Suppose coordinate axes are introduced as shown in the figure.

(a) Find an equation for the parabola.

(b) Nine equally spaced vertical cables are used to support the bridge (see figure). Find the total length of these supports.

FIGURE FOR EXERCISE 27

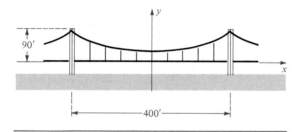

28 Traffic engineers are designing a stretch of highway that will connect a horizontal highway with one having a 20% grade (i.e., slope $\frac{1}{5}$) as illustrated in the figure. The smooth transition is to take place over a horizontal distance of 800 feet using a parabolic piece of highway to connect points A and B. If the equation of the parabolic segment is $y = ax^2 + bx + c$, it can be shown that the slope of the tangent line at the point $P(x, y)$ on the parabola is given by $m = 2ax + b$.

FIGURE FOR EXERCISE 28

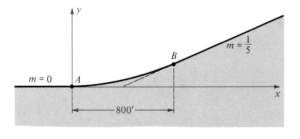

(a) Find an equation of the parabola that has a tangent line of slope 0 at A and $\frac{1}{5}$ at B.

(b) Find the coordinates of B.

29 Find an equation of a parabola that has a vertical axis and passes through the points $A(2, 3)$, $B(-1, 6)$, and $C(1, 0)$.

30 Prove that there is exactly one line of a given slope m that intersects the parabola $x^2 = 4py$ in exactly one point, and that its equation is $y = mx - pm^2$.

31 A person standing on the top of a building projects an object directly upward with a velocity of 144 feet per second. Its height $s(t)$ in feet above the ground after t seconds is given by $s(t) = -16t^2 + 144t + 100$. What is its maximum height? What is the height of the building?

32 A toy rocket is shot straight up into the air with an initial velocity of v_0 feet per second, and its height $s(t)$ in feet above the ground after t seconds is given by $s(t) = -16t^2 + v_0 t$.

(a) The rocket hits the ground after 12 seconds. What is the initial velocity v_0?

(b) What is the maximum height attained by the rocket?

33 The growth rate y (in pounds per month) of infants is related to their present weight x (in pounds) by the formula $y = cx(21 - x)$ for some constant $c > 0$. At what weight is the rate of growth a maximum?

34 The gasoline mileage y (in miles per gallon) of a small economy car is related to the velocity v (in mph) by the formula

$$y = -\tfrac{1}{30}v^2 + 2.5v \quad \text{for } 0 < v < 70.$$

What is the most economical speed for a long trip?

35 One thousand feet of chain-link fence is to be used to construct six cages for a zoo exhibit. The design is shown in the figure.

(a) Express the width y as a function of the length x.

(b) Express the total enclosed area A of the exhibit as a function of x.

(c) Find the dimensions that maximize the enclosed area.

FIGURE FOR EXERCISE 35

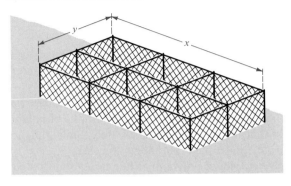

36 A man wishes to put a fence around a rectangular field and then subdivide the field into three smaller rectangular plots by placing two fences parallel to one of the sides. If he can afford only 1000 yards of fencing, what dimensions will give the maximum rectangular area?

37 A piece of wire 24 inches long is bent into the shape of a rectangle having width x and length y.

(a) Express y as a function of x.

(b) Express the area A of the rectangle as a function of x.

(c) Prove that the area A is greatest if the rectangle is a square.

38 A company sells running shoes to dealers at a rate of $20 per pair if less than 50 pairs are ordered. If a dealer orders 50 or more pairs (up to 600), the price per pair is reduced at a rate of 2 cents times the number ordered. What size order will produce the maximum amount of money for the company?

39 A boy tosses a baseball from the edge of a plateau down a hill as illustrated in the figure. The ball, thrown at an angle of 45 degrees, lands 50 feet down the hill, which is defined by the line $4y + 3x = 0$. Using calculus, it can be shown that the path of the baseball is given by the equation $y = ax^2 + x + c$ for some constants a and c.

(a) Ignoring the height of the boy, find an equation for the path.

(b) What is the maximum height of the ball *off the ground*?

FIGURE FOR EXERCISE 39

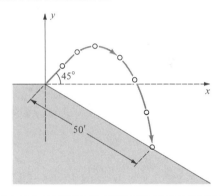

40 A cable television firm presently serves 5000 households and charges $20 per month. A marketing survey indicates that each decrease of $1 in the monthly charge will result in 500 new customers. Let $R(x)$ denote the total monthly revenue when the monthly charge is x dollars.

(a) Determine the revenue function R.

(b) Sketch the graph of R and find the value of x that results in maximum monthly revenue.

4.2 Graphs of Polynomial Functions of Degree Greater Than 2

Let f be a polynomial function of degree n; that is,

$$f(x) = a_n x^n + a_{n-1} x^{n-1} + \cdots + a_1 x + a_0$$

for some $a_n \neq 0$. The domain of f is $\mathbb{R}$. If the degree is odd, then the range of f is also $\mathbb{R}$; however, if the degree is even, then the range is an infinite interval of the form $(-\infty, a]$ or $[a, \infty)$. These facts are illustrated by the graphs in this section.

Recall that if $f(c) = 0$, then c is a **zero** of f, or of $f(x)$. We also call c a **solution,** or **root,** of the equation $f(x) = 0$. The zeros of f are the x-intercepts of the graph of f.

If a polynomial function has degree 0, then $f(x) = a$ for some nonzero real number a, and the graph is a horizontal line. Graphs of polynomial functions of degree 1 (linear functions) are lines. Polynomial functions of degree 2 (quadratic functions) have parabolas for their graphs. In this section we shall study graphs of polynomial functions of degree greater than 2.

If f has degree n, and all the coefficients except a_n are zero, then

$$f(x) = ax^n \quad \text{for some } a = a_n \neq 0.$$

FIGURE 4.7

(i)

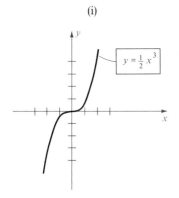

In this case, if $n = 1$, the graph of f is a line that passes through the origin, whereas if $n = 2$, the graph is a parabola with vertex at the origin. Several illustrations with $n = 3$ are given in the next example.

EXAMPLE 1 Sketch the graph of f if (a) $f(x) = \frac{1}{2}x^3$; (b) $f(x) = -\frac{1}{2}x^3$.

SOLUTION
(a) The following table lists several points on the graph of $y = \frac{1}{2}x^3$.

x	0	$\frac{1}{2}$	1	$\frac{3}{2}$	2	$\frac{5}{2}$
y	0	$\frac{1}{16} \approx 0.06$	$\frac{1}{2}$	$\frac{27}{16} \approx 1.7$	4	$\frac{125}{16} \approx 7.8$

(ii)

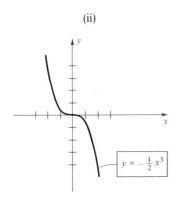

Since f is an odd function, the graph of f is symmetric with respect to the origin (see Section 3.2), and hence the points $(-\frac{1}{2}, -\frac{1}{16})$, $(-1, -\frac{1}{2})$, and so on are also on the graph. The graph is sketched in Figure 4.7(i).

(b) If $y = -\frac{1}{2}x^3$, the graph can be obtained from that in part (a) by multiplying all y-coordinates by -1. Reflecting the graph in part (a) through the x-axis, we obtain the sketch shown in Figure 4.7(ii). ∎

In general, if $f(x) = ax^3$, then increasing the absolute value of the coefficient a results in a graph that rises or falls more sharply. For example, if $f(x) = 10x^3$, then $f(1) = 10$, $f(2) = 80$, and $f(-2) = -80$. The effect of using a larger exponent and holding a fixed as, for example, in $f(x) = \frac{1}{2}x^5$, also leads to a graph that rises or falls more rapidly if $|x| > 1$.

If $f(x) = ax^n$ and n is an *even* integer, then the graph of f is symmetric with respect to the y-axis as illustrated in Figure 4.8 for the case $|a| = 1$. Note that as the exponent increases, the graph becomes flatter at the origin. It also rises (or falls) more rapidly if we let $|x|$ increase through values greater than 1.

FIGURE 4.8

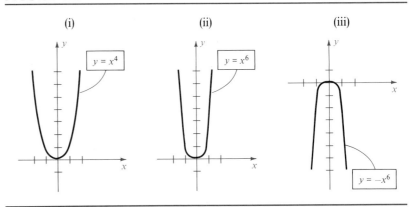

(i) $y = x^4$ (ii) $y = x^6$ (iii) $y = -x^6$

FIGURE 4.9

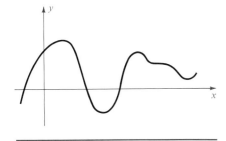

A complete analysis of graphs of polynomial functions of degree greater than 2 requires methods that are used in calculus. As the degree increases, the graphs usually become more complicated. However, they always have a smooth appearance with a number of *peaks* (high points) and *valleys* (low points) as illustrated in Figure 4.9. The points at which peaks and valleys occur are sometimes called **turning points** for the graph of f. At a turning point, f changes from an increasing function to a decreasing function, or vice versa.

A crude method for obtaining a rough sketch of the graph of a polynomial function is to plot many points and then fit a curve to the resulting configuration; however, this is usually an extremely tedious procedure. This method is based on a property of polynomial functions called **continuity.** Continuity, which is studied extensively in calculus, implies that a small change in x produces a small change in $f(x)$. The next theorem specifies another important property of polynomial functions. The proof requires advanced mathematical methods.

INTERMEDIATE VALUE THEOREM FOR POLYNOMIAL FUNCTIONS

If f is a polynomial function and $f(a) \neq f(b)$ for $a < b$, then f takes on every value between $f(a)$ and $f(b)$ in the interval $[a, b]$.

FIGURE 4.10

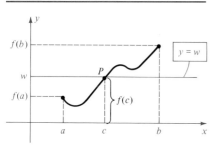

The Intermediate Value Theorem states that if w is any number between $f(a)$ and $f(b)$, then there is a number c between a and b such that $f(c) = w$. If the graph of the polynomial function f is regarded as extending continuously from the point $(a, f(a))$ to the point $(b, f(b))$, as illustrated in Figure 4.10, then for any number w between $f(a)$ and $f(b)$ it appears that a horizontal line with y-intercept w should intersect the graph in at least one point P. The x-coordinate c of P is a number such that $f(c) = w$.

A corollary to the Intermediate Value Theorem is that if $f(a)$ and $f(b)$ have opposite signs, then there is at least one number c between a and b such that $f(c) = 0$; that is, f has a zero at c. This implies that if the point $(a, f(a))$ on the graph of a polynomial function lies below the x-axis, and the point $(b, f(b))$ lies above the x-axis, or vice versa, then the graph crosses the x-axis at least once between the points $(a, 0)$ and $(b, 0)$. A by-product of this fact is that if c and d are *successive* zeros of $f(x)$, that is, there are no other zeros between c and d, then $f(x)$ *does not change sign on the interval* (c, d). Thus, if we choose any number k such that $c < k < d$, and if $f(k)$ is positive, then $f(x)$ is positive throughout (c, d). Similarly, if $f(k)$ is negative, then $f(x)$ is negative throughout (c, d). We shall call $f(k)$ a **test value** for $f(x)$ on the interval (c, d). Test values may also be used on infinite intervals of the form $(-\infty, a)$ or (a, ∞), provided that $f(x)$ has no zeros on these intervals. The use of test values in graphing is similar to that used for inequalities in Section 2.7.

EXAMPLE 2 If $f(x) = x^3 + x^2 - 4x - 4$, determine all values of x such that $f(x) > 0$, and all values of x such that $f(x) < 0$. Use this information to help sketch the graph of f.

SOLUTION The graph of f lies above the x-axis for values of x such that $f(x) > 0$, whereas the graph lies below the x-axis if $f(x) < 0$. We may factor $f(x)$ by grouping terms as follows:

FIGURE 4.11

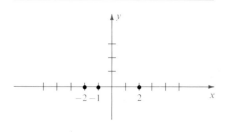

$$f(x) = (x^3 + x^2) - (4x + 4)$$
$$= x^2(x + 1) - 4(x + 1)$$
$$= (x^2 - 4)(x + 1)$$
$$= (x + 2)(x - 2)(x + 1)$$

We see from the last equation that the zeros of $f(x)$ (the x-intercepts of the graph) are -2, -1, and 2. The corresponding points on the graph (see Figure 4.11) divide the x-axis into four parts, and we consider the

open intervals

$$(-\infty, -2), \quad (-2, -1), \quad (-1, 2), \quad \text{and} \quad (2, \infty).$$

The sign of $f(x)$ in each of these intervals can be determined by finding a suitable test value. Thus, if we choose -3 in $(-\infty, -2)$, then

$$f(-3) = (-3)^3 + (-3)^2 - 4(-3) - 4$$
$$= -27 + 9 + 12 - 4 = -10.$$

Since the test value $f(-3) = -10$ is negative, $f(x)$ is negative throughout the interval $(-\infty, -2)$.

If we choose $-\frac{3}{2}$ in the interval $(-2, -1)$, then the corresponding test value is

FIGURE 4.12

$$f(-\tfrac{3}{2}) = (-\tfrac{3}{2})^3 + (-\tfrac{3}{2})^2 - 4(-\tfrac{3}{2}) - 4$$
$$= -\tfrac{27}{8} + \tfrac{9}{4} + 6 - 4 = \tfrac{7}{8}.$$

Since $f(-\frac{3}{2}) = \frac{7}{8}$ is positive, $f(x)$ is positive throughout $(-2, -1)$. The following table summarizes these facts and lists suitable test values for the remaining two intervals.

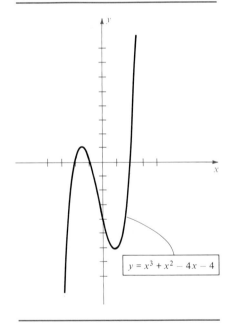

$y = x^3 + x^2 - 4x - 4$

Interval	$(-\infty, -2)$	$(-2, -1)$	$(-1, 2)$	$(2, \infty)$
Test value	$f(-3) = -10$	$f(-\tfrac{3}{2}) = \tfrac{7}{8}$	$f(0) = -4$	$f(3) = 20$
Sign of $f(x)$	$-$	$+$	$-$	$+$
Position of graph	Below x-axis	Above x-axis	Below x-axis	Above x-axis

Using the information from the table and plotting several points gives us the graph in Figure 4.12. To find the turning points of the graph it is necessary to use methods developed in calculus. ∎

The graph of every polynomial function of degree 3 has an S-shaped appearance similar to that shown in Figure 4.12, or it has an inverted version of that graph if the coefficient of x^3 is negative. However, sometimes the graph may have only one x-intercept or the S shape may be elongated, as in Figure 4.7.

EXAMPLE 3 If $f(x) = x^4 - 4x^3 + 3x^2$, determine all values of x such that $f(x) > 0$, and all x such that $f(x) < 0$. Sketch the graph of f.

SOLUTION We shall follow the same steps used in the solution of Example 2. Thus, we begin by factoring $f(x)$:

$$f(x) = x^2(x^2 - 4x + 3)$$
$$= x^2(x - 1)(x - 3).$$

The zeros of $f(x)$, that is, the x-intercepts of the graph are, in *increasing* order,

$$0, \quad 1, \quad \text{and} \quad 3.$$

The corresponding points on the graph divide the x-axis into four parts, and we consider the open intervals

$$(-\infty, 0), \quad (0, 1), \quad (1, 3), \quad (3, \infty).$$

We next determine the sign of $f(x)$ in each interval by using suitable test values. The following table summarizes results. You should check each entry.

FIGURE 4.13

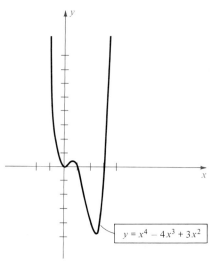

$y = x^4 - 4x^3 + 3x^2$

Interval	$(-\infty, 0)$	$(0, 1)$	$(1, 3)$	$(3, \infty)$
Test value	$f(-1) = 8$	$f(\frac{1}{2}) = \frac{5}{16}$	$f(2) = -4$	$f(4) = 48$
Sign of $f(x)$	$+$	$+$	$-$	$+$
Position of graph	Above x-axis	Above x-axis	Below x-axis	Above x-axis

Making use of the information in the table and plotting several points gives us the sketch in Figure 4.13. ∎

EXAMPLE 4 Show that $f(x) = x^5 + 2x^4 - 6x^3 + 2x - 3$ has a zero between 1 and 2.

SOLUTION Substitution for x gives us

$$f(1) = 1 + 2 - 6 + 2 - 3 = -4$$
$$f(2) = 32 + 32 - 48 + 4 - 3 = 17.$$

Since $f(1)$ and $f(2)$ have opposite signs, it follows from the Intermediate Value Theorem that $f(c) = 0$ for some real number c between 1 and 2.

∎

The preceding example illustrates a scheme for locating zeros of polynomials. By using a method of *successive approximation*, each zero can be approximated to any degree of accuracy (see, for example, Calculator Exercises 1–10).

Exercises 4.2

1 If $f(x) = ax^3 + 2$, sketch the graph of f for each value of a.

(a) $a = 2$ (b) $a = 4$

(c) $a = \frac{1}{4}$ (d) $a = -2$

2 If $f(x) = 2x^3 + c$, sketch the graph of f for each value of c.

(a) $c = 2$ (b) $c = 4$

(c) $c = \frac{1}{4}$ (d) $c = -2$

In Exercises 3–18 determine all x such that $f(x) > 0$, and all x such that $f(x) < 0$. Sketch the graph of f.

3 $f(x) = \frac{1}{2}x^3 - 4$ **4** $f(x) = -\frac{1}{4}x^3 - 16$

5 $f(x) = \frac{1}{8}x^4 + 2$ **6** $f(x) = 1 - x^5$

7 $f(x) = x^3 - 9x$ **8** $f(x) = 16x - x^3$

9 $f(x) = -x^3 - x^2 + 2x$

10 $f(x) = x^3 + x^2 - 12x$

11 $f(x) = (x + 4)(x - 1)(x - 5)$

12 $f(x) = (x + 2)(x - 3)(x - 4)$

13 $f(x) = x^4 - 16$ **14** $f(x) = 16 - x^4$

15 $f(x) = -x^4 - 3x^2 + 4$

16 $f(x) = x^4 - 7x^2 - 18$

17 $f(x) = x(x - 2)(x + 1)(x + 3)$

18 $f(x) = x(x + 1)^2(x - 3)(x - 5)$

19 If $f(x)$ is a polynomial, and if the coefficients of all odd powers of x are 0, show that f is an even function.

20 If $f(x)$ is a polynomial, and if the coefficients of all even powers of x are 0, show that f is an odd function.

21 If $f(x) = 3x^3 - kx^2 + x - 5k$, find a number k such that the graph of f contains the point $(-1, 4)$.

22 If one zero of $f(x) = x^3 - 2x^2 - 16x + 16k$ is 2, find two other zeros.

In Exercises 23–28 show that f has a zero between a and b.

23 $f(x) = x^3 - 4x^2 + 3x - 2$; $a = 3$, $b = 4$

24 $f(x) = 2x^3 + 5x^2 - 3$; $a = -3$, $b = -2$

25 $f(x) = -x^4 + 3x^3 - 2x + 1$; $a = 2$, $b = 3$

26 $f(x) = 2x^4 + 3x - 2$; $a = \frac{1}{2}$, $b = \frac{3}{4}$

27 $f(x) = x^5 + x^3 + x^2 + x + 1$; $a = -\frac{1}{2}$, $b = -1$

28 $f(x) = x^5 - 3x^4 - 2x^3 + 3x^2 - 9x - 6$; $a = 3$, $b = 4$

29 The third-degree Legendre polynomial

$$P(x) = \frac{1}{2}(5x^3 - 3x)$$

occurs in the solution of heat transfer problems in physics and engineering. Sketch the graph of P after first determining where $P(x) > 0$ and $P(x) < 0$.

30 The fourth-degree Chebyshev polynomial

$$f(x) = 8x^4 - 8x^2 + 1$$

occurs in statistical studies. Determine where $f(x) > 0$. (*Hint:* Let $z = x^2$ and use the Quadratic Formula.)

31 An open box is to be made from a rectangular piece of cardboard having dimensions 20 inches × 30 inches by cutting out identical squares of area x^2 from each corner and turning up the sides (see Exercise 37 of Section 3.3). Show that the volume of the box is given by the third-degree polynomial

$$V(x) = x(20 - 2x)(30 - 2x).$$

For what positive values of x is $V(x) > 0$? Sketch the graph of V for $x > 0$.

32 The frame for a shipping crate is to be constructed from 24 feet of 2 × 2 lumber.

(a) If the crate is to have square ends of side x feet, express the volume V of the crate as a function of x. (See figure.)

(b) Sketch the graph of V for $x > 0$.

FIGURE FOR EXERCISE 32

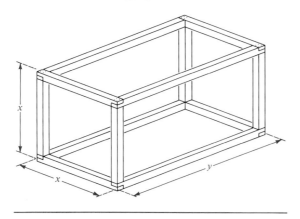

33 A meteorologist determines that the temperature T (in °F) on a certain cold winter day was given by

$$T = 0.05t(t - 12)(t - 24)$$

where t is the time (in hours) and $t = 0$ corresponds to 6 A.M.

(a) When was $T > 0$ and when was $T < 0$? Sketch a graph that depicts the temperature T for $0 \le t \le 24$.

(b) Show that the temperature was 32 °F sometime between 12 noon and 1 P.M. (*Hint:* Use the Intermediate Value Theorem.)

34 A diver stands on the very end of a diving board before beginning a dive. It is known that the deflection d of the board at a position s feet from the stationary end is given by the third-degree polynomial

$$d = cs^2(3L - s) \quad \text{for} \quad 0 \le s \le L,$$

where L is the length of the board and c is a positive constant that depends on the weight of the diver and on the physical properties of the board (see figure). Suppose the board is 10 feet long.

(a) If the deflection at the end of the board is 1 foot, find c.

(b) Show that the deflection is $\frac{1}{2}$ foot somewhere between $s = 6.5$ and $s = 6.6$.

FIGURE FOR EXERCISE 34

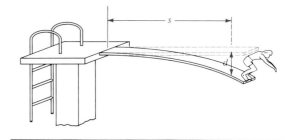

35 A herd of 100 deer is introduced to a small island. The herd at first increases rapidly, but eventually the food resources of the island dwindle and the population declines. Suppose that the number $N(t)$ of deer after t years is given by

$$N(t) = -t^4 + 21t^2 + 100.$$

(a) Determine the positive values of t for which $N(t) > 0$. Does the population become extinct? If so, when?

(b) Sketch the graph of N for $t > 0$.

36 It can be shown by means of calculus that the rate R at which the deer population in Exercise 35 grows (or declines) at time t is given by

$$R = -4t^3 + 42t \quad \text{(deer per year)}$$

(a) When does the population cease to grow?

(b) Determine the positive values of t for which $R > 0$.

Calculator Exercises 4.2

If a zero of a polynomial lies between two integers, then a method of *successive approximation* may be used to approximate the zero to any degree of accuracy. One such method is illustrated in Exercise 1.

1 If $f(x) = x^3 - 3x + 1$:

(a) Show that f has a zero between 1 and 2.

(b) By increasing values of x by tenths, show that f has a zero between 1.5 and 1.6.

(c) By increasing values of x by hundredths, show that f has a zero between 1.53 and 1.54.

In (c) of Exercise 1, a zero of $x^3 - 3x + 1$ is said to be *isolated between successive hundredths* in the interval $[1, 2]$. In Exercises 2–6 isolate a zero of $f(x)$ between successive hundredths in the interval $[a, b]$.

2 $f(x) = x^3 + 5x - 3; \quad a = 0, \ b = 1$

3 $f(x) = 2x^3 - 4x^2 - 3x + 1; \quad a = 2, \ b = 3$

4 $f(x) = x^4 - 4x^3 + 3x^2 - 8x + 2; \quad a = 3, \ b = 4$

5 $f(x) = x^5 - 2x^2 + 4; \quad a = -2, \ b = -1$

6 $f(x) = x^4 - 2x^3 + 10x - 25; \quad a = 2, \ b = 3$

In Exercises 7–10 isolate a zero of $f(x)$ between successive thousandths in the interval $[a, b]$.

7 $f(x) = x^4 + 2x^3 - 5x^2 + 1; \quad a = 1, \ b = 2$

8 $f(x) = x^4 - 5x^2 + 2x - 5; \quad a = 2, \ b = 3$

9 $f(x) = x^5 + x^2 - 9x - 3; \quad a = -2, \ b = -1$

10 $f(x) = x^5 + 3x^4 - x^3 + 2x^2 + 6x - 2;$
$a = -4, \ b = -3$

4.3 Properties of Division

In the following discussion, symbols such as $f(x)$ and $g(x)$ will be used to denote polynomials in x. If a polynomial $g(x)$ is a factor of a polynomial $f(x)$, then $f(x)$ is said to be **divisible** by $g(x)$. For example, the polynomial $x^4 - 16$ is divisible by $x^2 - 4$, by $x^2 + 4$, by $x + 2$, and by $x - 2$; but $x^2 + 3x + 1$ is not a factor of $x^4 - 16$. However, by *long division*, we can write

$$
\begin{array}{r}
x^2 - 3x + 8 \\
x^2 + 3x + 1 \overline{\smash{\big)}\ x^4 \qquad\qquad\qquad - 16} \\
\underline{x^4 + 3x^3 + \ x^2} \\
-3x^3 - \ x^2 \\
\underline{-3x^3 - 9x^2 - \ 3x} \\
8x^2 + \ 3x - 16 \\
\underline{8x^2 + 24x + \ 8} \\
-21x - 24
\end{array}
$$

The polynomial $x^2 - 3x + 8$ is called the **quotient** and $-21x - 24$ is the **remainder.**

Note that the long division process ends when we arrive at a polynomial (the remainder) that either is 0 or has smaller degree than the divisor. The result of this division is often written

$$\frac{x^4 - 16}{x^2 + 3x + 1} = (x^2 - 3x + 8) + \left(\frac{-21x - 24}{x^2 + 3x + 1}\right).$$

Multiplying both sides of the equation by $x^2 + 3x + 1$, we obtain

$$x^4 - 16 = (x^2 + 3x + 1)(x^2 - 3x + 8) + (-21x - 24).$$

This example illustrates the following theorem, which we state without proof.

DIVISION ALGORITHM FOR POLYNOMIALS

If $f(x)$ and $g(x)$ are polynomials and if $g(x) \neq 0$, then there exist unique polynomials $q(x)$ and $r(x)$ such that

$$f(x) = g(x)q(x) + r(x)$$

where either $r(x) = 0$, or the degree of $r(x)$ is less than the degree of $g(x)$. The polynomial $q(x)$ is called the **quotient,** and $r(x)$ is the **remainder** in the division of $f(x)$ by $g(x)$.

An interesting special case occurs if $f(x)$ is divided by a polynomial of the form $x - c$ where c is a real number. If $x - c$ is a factor of $f(x)$, then

$$f(x) = (x - c)q(x)$$

for some polynomial $q(x)$; that is, the remainder $r(x)$ is 0. If $x - c$ is not a factor of $f(x)$, then the degree of the remainder $r(x)$ is less than the degree of $x - c$, and hence $r(x)$ must have degree 0. This, in turn, means that the remainder is a nonzero number. Consequently, in all cases we have

$$f(x) = (x - c)q(x) + d$$

where the remainder d is some real number (possibly $d = 0$). If c is substituted for x in the equation $f(x) = (x - c)q(x) + d$, we obtain

$$f(c) = (c - c)q(c) + d,$$

which reduces to $f(c) = d$. This proves the following theorem.

REMAINDER THEOREM

If a polynomial $f(x)$ is divided by $x - c$, then the remainder is $f(c)$.

EXAMPLE 1 If $f(x) = x^3 - 3x^2 + x + 5$, use the Remainder Theorem to find $f(2)$.

SOLUTION According to the Remainder Theorem, $f(2)$ is the remainder when $f(x)$ is divided by $x - 2$. By long division,

$$
\begin{array}{r}
x^2 - x - 1 \\
x - 2 \overline{\smash{\big)}\ x^3 - 3x^2 + x + 5} \\
\underline{x^3 - 2x^2} \\
-x^2 + x \\
\underline{-x^2 + 2x} \\
-x + 5 \\
\underline{-x + 2} \\
3
\end{array}
$$

Hence, $f(2) = 3$. We may check this fact by direct substitution. Thus, $f(2) = 2^3 - 3(2)^2 + 2 + 5 = 3$. ∎

FACTOR THEOREM

A polynomial $f(x)$ has a factor $x - c$ if and only if $f(c) = 0$.

PROOF By the Remainder Theorem, $f(x) = (x - c)q(x) + f(c)$ for some quotient $q(x)$. If $f(c) = 0$, then $f(x) = (x - c)q(x)$; that is, $x - c$ is a factor of $f(x)$. Conversely, if $x - c$ is a factor, then the remainder upon division of $f(x)$ by $x - c$ must be 0, and hence, by the Remainder Theorem, $f(c) = 0$. □

The Factor Theorem is useful for finding factors of polynomials, as illustrated in the next example.

EXAMPLE 2 Show that $x - 2$ is a factor of the polynomial

$$f(x) = x^3 - 4x^2 + 3x + 2.$$

SOLUTION Since $f(2) = 8 - 16 + 6 + 2 = 0$, it follows from the Factor Theorem that $x - 2$ is a factor of $f(x)$. Of course, another method of solution would be to divide $f(x)$ by $x - 2$ and show that the remainder is 0. The quotient in the division would be another factor of $f(x)$. ∎

EXAMPLE 3 Find a polynomial $f(x)$ of degree 3 that has zeros $2, -1$, and 3.

SOLUTION By the Factor Theorem, $f(x)$ has factors $x - 2$, $x + 1$, and $x - 3$. We may then write

$$f(x) = a(x - 2)(x + 1)(x - 3)$$

where any nonzero value may be assigned to a. If we let $a = 1$ and multiply, we obtain

$$f(x) = x^3 - 4x^2 + x + 6.$$ ∎

To apply the Remainder Theorem it is necessary to divide a given polynomial by $x - c$. A method called **synthetic division** may be used to simplify this work. The following rules state how to proceed. The method can be justified by a careful (and lengthy) comparison with the method of long division.

Synthetic Division of $a_n x^n + a_{n-1} x^{n-1} + \cdots + a_1 x + a_0$ **by** $x - c$

1 Begin with the following display, supplying zeros for any missing coefficients in the given polynomial.

$$\begin{array}{c|ccccc} c & a_n & a_{n-1} & a_{n-2} & \cdots & a_1 & a_0 \\ \hline & a_n \end{array}$$

2 Multiply a_n by c and place the product ca_n underneath a_{n-1} as indicated by the arrow in the following display. (This arrow, and others, is used only to help clarify these rules, and will not appear in *specific* synthetic divisions.) Next find the sum $b_1 = a_{n-1} + ca_n$ and place it below the line as shown.

$$\begin{array}{c|ccccccc} c & a_n & a_{n-1} & a_{n-2} & \cdots & & a_1 & a_0 \\ & & ca_n & cb_1 & cb_2 & \cdots & cb_{n-2} & cb_{n-1} \\ \hline & a_n & b_1 & b_2 & \cdots & & b_{n-2} & b_{n-1} & r \end{array}$$

3 Multiply b_1 by c and place the product cb_1 underneath a_{n-2} as indicated by another arrow. Next find the sum $b_2 = a_{n-2} + cb_1$ and place it below the line as shown.

4 Continue this process, as indicated by the arrows, until the final sum $r = a_0 + cb_{n-1}$ is obtained. The numbers

$$a_n, \quad b_1, \quad b_2, \quad \ldots, \quad b_{n-2}, \quad b_{n-1}$$

are the coefficients of the quotient $q(x)$; that is,

$$q(x) = a_n x^{n-1} + b_1 x^{n-2} + \cdots + b_{n-2} x + b_{n-1},$$

and r is the remainder. ∎ ∎ ∎

The following examples illustrate synthetic division for some special cases.

EXAMPLE 4 Use synthetic division to find the quotient and remainder if $2x^4 + 5x^3 - 2x - 8$ is divided by $x + 3$.

SOLUTION Since the divisor is $x + 3$, the c in the expression $x - c$ is -3. Hence, the synthetic division takes this form:

$$
\begin{array}{r|rrrrr}
-3 & 2 & 5 & 0 & -2 & -8 \\
   &   & -6 & 3 & -9 & 33 \\
\hline
   & 2 & -1 & 3 & -11 & 25
\end{array}
$$

The first four numbers in the third row are the coefficients of the quotient $q(x)$ and the last number is the remainder r. Thus,

$$q(x) = 2x^3 - x^2 + 3x - 11 \quad \text{and} \quad r = 25.$$ ∎

Synthetic division can be used to find values of polynomial functions, as illustrated in the next example.

EXAMPLE 5 If $f(x) = 3x^5 - 38x^3 + 5x^2 - 1$, use synthetic division to find $f(4)$.

SOLUTION By the Remainder Theorem, $f(4)$ is the remainder when $f(x)$ is divided by $x - 4$. Dividing synthetically, we obtain

$$
\begin{array}{r|rrrrrr}
4 & 3 & 0 & -38 & 5 & 0 & -1 \\
  &   & 12 & 48 & 40 & 180 & 720 \\
\hline
  & 3 & 12 & 10 & 45 & 180 & 719
\end{array}
$$

Consequently, $f(4) = 719$. ∎

Synthetic division may be employed to help find zeros of polynomials. By the method illustrated in the preceding example, $f(c) = 0$ if and only if the remainder in the synthetic division by $x - c$ is 0.

EXAMPLE 6 Show that -11 is a zero of the polynomial

$$f(x) = x^3 + 8x^2 - 29x + 44.$$

SOLUTION Dividing synthetically by $x - (-11) = x + 11$ gives us

$$
\begin{array}{r|rrrr}
-11 & 1 & 8 & -29 & 44 \\
    &   & -11 & 33 & -44 \\
\hline
    & 1 & -3 & 4 & 0
\end{array}
$$

Thus, $f(-11) = 0$. ∎

Example 6 shows that the number -11 is a solution of the equation $x^3 + 8x^2 - 29x + 44 = 0$. In Section 4.5 we shall use synthetic division to find rational solutions of equations.

Exercises 4.3

In Exercises 1–6 find the quotient $q(x)$ and the remainder $r(x)$ if $f(x)$ is divided by $g(x)$.

1 $f(x) = x^4 + 3x^3 - 2x + 5$, $g(x) = x^2 + 2x - 4$

2 $f(x) = 4x^3 - x^2 + x - 3$, $g(x) = x^2 - 5x$

3 $f(x) = 5x^3 - 2x$, $g(x) = 2x^2 + 1$

4 $f(x) = 3x^4 - x^3 - x^2 + 3x + 4$, $g(x) = 2x^3 - x + 4$

5 $f(x) = 7x^3 - 5x + 2$, $g(x) = 2x^4 - 3x^2 + 9$

6 $f(x) = 10x - 4$, $g(x) = 8x^2 - 5x + 17$

In Exercises 7–16 use synthetic division to find the quotient and remainder assuming the first polynomial is divided by the second.

7 $2x^3 - 3x^2 + 4x - 5$, $x - 2$

8 $3x^3 - 4x^2 - x + 8$, $x + 4$

9 $x^3 - 8x - 5$, $x + 3$

10 $5x^3 - 6x^2 + 15$, $x - 4$

11 $3x^5 + 6x^2 + 7$, $x + 2$

12 $-2x^4 + 10x - 3$, $x - 3$

13 $4x^4 - 5x^2 + 1$, $x - \frac{1}{2}$

14 $9x^3 - 6x^2 + 3x - 4$, $x - \frac{1}{3}$

15 $x^n - 1$, $x - 1$ where n is any positive integer

16 $x^n + 1$, $x + 1$ where n is any positive integer

In Exercises 17–28 use the Remainder Theorem to find $f(c)$.

17 $f(x) = 2x^3 - x^2 - 5x + 3$, $c = 4$

18 $f(x) = 4x^3 - 3x^2 + 7x + 10$, $c = 3$

19 $f(x) = x^4 + 5x^3 - x^2 + 5$, $c = -2$

20 $f(x) = x^4 - 7x^2 + 2x - 8$, $c = -3$

21 $f(x) = x^6 - 3x^4 + 4$, $c = \sqrt{2}$

22 $f(x) = x^5 - x^4 + x^3 - x^2 + x - 1$, $c = -1$

23 $f(x) = x^4 - 4x^3 + x^2 - 3x - 5$, $c = 2$

24 $f(x) = 0.3x^3 + 0.04x - 0.034$, $c = -0.2$

25 $f(x) = x^6 - x^5 + x^4 - x^3 + x^2 - x + 1$, $c = 4$

26 $f(x) = 8x^5 - 3x^2 + 7$, $c = \frac{1}{2}$

27 $f(x) = x^2 + 3x - 5$, $c = 2 + \sqrt{3}$

28 $f(x) = x^3 - 3x^2 - 8$, $c = 1 + \sqrt{2}$

In Exercises 29–32 use synthetic division to show that c is a zero of $f(x)$.

29 $f(x) = 3x^4 + 8x^3 - 2x^2 - 10x + 4, \ c = -2$

30 $f(x) = 4x^3 - 9x^2 - 8x - 3, \ c = 3$

31 $f(x) = 4x^3 - 6x^2 + 8x - 3, \ c = \frac{1}{2}$

32 $f(x) = 27x^4 - 9x^3 + 3x^2 + 6x + 1, \ c = -\frac{1}{3}$

33 Determine k so that $f(x) = x^3 + kx^2 - kx + 10$ is divisible by $x + 3$.

34 Determine all values of k such that $f(x) = k^2x^3 - 4kx - 3$ is divisible by $x - 1$.

35 Use the Factor Theorem to show that $x - 2$ is a factor of $f(x) = x^4 - 3x^3 - 2x^2 + 5x + 6$.

36 Show that $x + 2$ is a factor of $f(x) = x^{12} - 4096$.

37 Prove that $f(x) = 3x^4 + x^2 + 5$ has no factor of the form $x - c$ where c is a real number.

38 Find the remainder if the polynomial $3x^{100} + 5x^{85} - 4x^{38} + 2x^{17} - 6$ is divided by $x + 1$.

39 Use the Factor Theorem to prove that $x - y$ is a factor of $x^n - y^n$ for all positive integers n. Assuming n is even, show that $x + y$ is also a factor of $x^n - y^n$.

40 Assuming n is an odd positive integer, prove that $x + y$ is a factor of $x^n + y^n$.

4.4 The Zeros of a Polynomial

The Factor and Remainder Theorems can be extended to the system of complex numbers. Thus, a complex number $c = a + bi$ is a zero of a polynomial $f(x)$; that is, $f(c) = 0$ if and only if $x - c$ is a factor of $f(x)$. Except in special cases, zeros of polynomials are very difficult to find. For example, $f(x) = x^5 - 3x^4 + 4x^3 + 4x - 10$ has no obvious zeros. Moreover, there is no formula that can be used to find the zeros. In spite of the practical difficulty of determining zeros of polynomials, it is possible to make some headway concerning the *theory* of such zeros. The results in this section form the basis for work in what is known as *The Theory of Equations*.

FUNDAMENTAL THEOREM OF ALGEBRA

If a polynomial $f(x)$ has positive degree and complex coefficients, then $f(x)$ has at least one complex zero.

The usual proof of this theorem requires results from the field of mathematics called *functions of a complex variable*. In turn, a prerequisite for studying this field is a strong background in calculus. The first proof of the Fundamental Theorem of Algebra was given by the German mathematician Carl Friedrich Gauss (1777–1855), who is considered by many to be the greatest mathematician of all time.

As a special case of the Fundamental Theorem, if all the coefficients of $f(x)$ are real, then $f(x)$ has at least one complex zero. If $a + bi$ is a complex zero, it may happen that $b = 0$, in which case we refer to the number as a **real zero.** If the Fundamental Theorem is combined with the Factor Theorem, the following useful corollary is obtained.

COROLLARY

> Every polynomial of positive degree has a factor of the form $x - c$ where c is a complex number.

The corollary enables us, at least in theory, to express every polynomial $f(x)$ of positive degree as a product of polynomials of degree 1. If $f(x)$ has degree $n > 0$, then applying the corollary gives us

$$f(x) = (x - c_1)f_1(x)$$

where c_1 is a complex number and $f_1(x)$ is a polynomial of degree $n - 1$. If $n - 1 > 0$, we may apply the corollary again, obtaining

$$f_1(x) = (x - c_2)f_2(x)$$

where c_2 is a complex number and $f_2(x)$ is a polynomial of degree $n - 2$. Hence,

$$f(x) = (x - c_1)(x - c_2)f_2(x).$$

Continuing this process, after n steps we arrive at a polynomial $f_n(x)$ of degree 0. Thus, $f_n(x) = a$ for some nonzero number a, and we may write

$$f(x) = a(x - c_1)(x - c_2) \cdots (x - c_n)$$

where each complex number c_j is a zero of $f(x)$. Evidently, the leading coefficient of the polynomial on the right in the last equation is a. It follows that a is the leading coefficient of $f(x)$. We have proved the following theorem.

THEOREM

> If $f(x)$ is a polynomial of degree $n > 0$, then there exist n complex numbers $c_1, c_2, \ldots, c_n$ such that
>
> $$f(x) = a(x - c_1)(x - c_2) \cdots (x - c_n)$$
>
> where a is the leading coefficient of $f(x)$. Each number c_j is a zero of $f(x)$.

COROLLARY

> A polynomial of degree $n > 0$ has at most n different complex zeros.

PROOF We shall give an indirect proof. Suppose $f(x)$ has *more* than n different complex zeros. Let us choose $n + 1$ of these zeros and label them $c_1, c_2, \ldots, c_n$, and c. We may use the c_j to obtain the factorization indicated in the statement of the theorem. Substituting c for x and using the fact that $f(c) = 0$, we obtain

$$0 = a(c - c_1)(c - c_2) \cdots (c - c_n).$$

However, each factor on the right side is different from zero because $c \neq c_j$ for every j. Since the product of nonzero numbers cannot equal zero, we have a contradiction. □

EXAMPLE 1 Find a polynomial $f(x)$ of degree 3 with zeros 2, -1, and 3 that has the value 5 at $x = 1$.

SOLUTION By the Factor Theorem $f(x)$ has factors $x - 2$, $x + 1$, and $x - 3$. No other factors of degree 1 exist, since, by the Factor Theorem, another linear factor $x - c$ would produce a fourth zero of $f(x)$ in violation of the last corollary. Hence, $f(x)$ has the form

$$f(x) = a(x - 2)(x + 1)(x - 3)$$

for some number a. If $f(x)$ has the value 5 at $x = 1$, then $f(1) = 5$; that is,

$$a(1 - 2)(1 + 1)(1 - 3) = 5 \quad \text{or} \quad 4a = 5.$$

Consequently, $a = \frac{5}{4}$ and

$$f(x) = \tfrac{5}{4}(x - 2)(x + 1)(x - 3).$$

Multiplying the four factors we obtain

$$f(x) = \tfrac{5}{4}x^3 - 5x^2 + \tfrac{5}{4}x + \tfrac{15}{2}. \qquad \blacksquare$$

The numbers $c_1, c_2, \ldots, c_n$ in the preceding theorem are not necessarily all different. To illustrate, the polynomial $f(x) = x^3 + x^2 - 5x + 3$ has the factorization

$$f(x) = (x + 3)(x - 1)(x - 1).$$

If a factor $x - c$ occurs m times in the factorization, then c is called a **zero of multiplicity** m of $f(x)$, or a **root of multiplicity** m of the equation

$f(x) = 0$. In the preceding illustration, 1 is a zero of multiplicity 2 and -3 is a zero of multiplicity 1.

As another illustration, if

$$f(x) = (x - 2)(x - 4)^3(x + 1)^2,$$

then $f(x)$ has degree 6 and possesses three distinct zeros 2, 4, and -1, where 2 has multiplicity 1, 4 has multiplicity 3, and -1 has multiplicity 2.

If $f(x)$ is a polynomial of degree n and $f(x) = a(x - c_1)(x - c_2) \cdots (x - c_n)$, then the n complex numbers $c_1, c_2, \ldots, c_n$ are zeros of $f(x)$. If a zero of multiplicity m is counted as m zeros, this tells us that $f(x)$ has at least n zeros (not necessarily all different). Combining this with the fact that $f(x)$ has at most n zeros gives us the next result.

THEOREM

> If $f(x)$ is a polynomial of degree $n > 0$ and if a zero of multiplicity m is counted m times, then $f(x)$ has precisely n zeros.

EXAMPLE 2 Express $f(x) = x^5 - 4x^4 + 13x^3$ as a product of linear factors, and list the five zeros of $f(x)$.

SOLUTION We begin by writing

$$f(x) = x^3(x^2 - 4x + 13).$$

By the Quadratic Formula, the zeros of the polynomial $x^2 - 4x + 13$ are

$$\frac{4 \pm \sqrt{16 - 52}}{2} = \frac{4 \pm \sqrt{-36}}{2} = \frac{4 \pm 6i}{2} = 2 \pm 3i.$$

Hence, by the Factor Theorem, $x^2 - 4x + 13$ has factors $x - (2 + 3i)$ and $x - (2 - 3i)$, and we obtain the desired factorization

$$f(x) = x \cdot x \cdot x \cdot (x - 2 - 3i)(x - 2 + 3i).$$

Since $x - 0$ occurs as a factor three times, the number 0 is zero of multiplicity three, and the five zeros of $f(x)$ are 0, 0, 0, $2 + 3i$, and $2 - 3i$. ∎

The next theorem may be used to obtain information about the zeros of a polynomial $f(x)$ with real coefficients. In the statement of the theorem it is assumed that the terms of $f(x)$ are arranged in order of decreasing powers of x, and that terms with zero coefficients are deleted. It is also assumed that the **constant term,** that is, the term that does not contain x, is different from 0. We say there is a **variation of sign** in $f(x)$ if two consecutive coefficients have opposite signs. To illustrate, the polynomial

$$f(x) = 2x^5 - 7x^4 + 3x^2 + 6x - 5$$

has three variations in sign, since there is one variation from $2x^5$ to $-7x^4$, a second from $-7x^4$ to $3x^2$, and a third from $6x$ to -5.

The theorem also refers to the variations of sign in $f(-x)$. Using the previous illustration, note that

$$f(-x) = 2(-x)^5 - 7(-x)^4 + 3(-x)^2 + 6(-x) - 5$$
$$= -2x^5 - 7x^4 + 3x^2 - 6x - 5.$$

Hence, there are two variations of sign in $f(-x)$: one from $-7x^4$ to $3x^2$, and a second from $3x^2$ to $-6x$.

DESCARTES' RULE OF SIGNS

Let $f(x)$ be a polynomial with real coefficients and nonzero constant term.

(i) The number of positive real solutions of the equation $f(x) = 0$ either is equal to the number of variations of sign in $f(x)$ or is less than that number by an even integer.

(ii) The number of negative real solutions of the equation $f(x) = 0$ either is equal to the number of variations of sign in $f(-x)$ or is less than that number by an even integer.

The proof of Descartes' Rule is rather technical and will not be given in this text.

EXAMPLE 3   Discuss the number of possible positive and negative real solutions, and nonreal complex solutions of the equation

$$2x^5 - 7x^4 + 3x^2 + 6x - 5 = 0.$$

SOLUTION   The polynomial $f(x)$ on the left side of the equation is the same as the one given in the illustration preceding the statement of Descartes' Rule. Since there are three variations of sign in $f(x)$, the equation has either three positive real solutions or one positive real solution.

Since $f(-x) = -2x^5 - 7x^4 + 3x^2 - 6x - 5$ has two variations of sign, the given equation has either two negative solutions or no negative solutions. The solutions that are not real numbers are complex numbers of the form $a + bi$ where a and b are real with $b \neq 0$. The following table summarizes the various possibilities that can occur for solutions of the equation.

Number of positive real solutions	1	1	3	3
Number of negative real solutions	0	2	0	2
Number of nonreal, complex solutions	4	2	2	0
Total number of solutions	5	5	5	5

Descartes' Rule stipulates that the constant term of the polynomial is different from 0. If the constant term is 0, as in the equation

$$x^4 - 3x^3 + 2x^2 - 5x = 0,$$

then we may write

$$x(x^3 - 3x^2 + 2x - 5) = 0.$$

In this case $x = 0$ is a solution, and Descartes' Rule may be applied to $x^3 - 3x^2 + 2x - 5 = 0$.

EXAMPLE 4   Discuss the nature of the roots of the equation

$$3x^5 + 4x^3 + 2x - 5 = 0.$$

SOLUTION   The polynomial $f(x)$ on the left side of the equation has one variation of sign and hence, by (i) of Descartes' Rule, the equation has precisely one positive real root. Since

$$f(-x) = 3(-x)^5 + 4(-x)^3 + 2(-x) - 5$$
$$= -3x^5 - 4x^3 - 2x - 5,$$

$f(-x)$ has no variations of sign and hence, by (ii) of Descartes' Rule, there are no negative real roots. Thus, the equation has one real root and four nonreal, complex roots.   ∎

When applying Descartes' Rule, we count roots of multiplicity k as k roots. For example, given $x^2 - 2x + 1 = 0$, the polynomial $x^2 - 2x + 1$ has two variations of sign and hence the equation has either two positive real roots or none. The factored form of the equation is $(x - 1)^2 = 0$, and hence 1 is a root of multiplicity 2.

We shall conclude this section with a discussion of *bounds* for the real solutions of an equation $f(x) = 0$ where $f(x)$ is a polynomial with real coefficients. By definition, a real number b is an **upper bound** for the solutions if no solution is greater than b. A real number a is a **lower bound** for the solutions if no solution is less than a. Thus, if r is a real solution of $f(x) = 0$, then $a \leq r \leq b$. Note that upper and lower bounds are not unique, since any number greater than b is also an upper bound, and any number less than a is a lower bound.

We may use synthetic division to find upper and lower bounds for the solutions of $f(x) = 0$. Recall that if $f(x)$ is divided synthetically by $x - c$, then the third row that appears in the division process consists of the coefficients of the quotient $q(x)$ together with the remainder $f(c)$. The following theorem indicates how this third row may be used to find upper and lower bounds for the real solutions.

BOUNDS FOR REAL ZEROS OF POLYNOMIALS

Suppose that $f(x)$ is a polynomial with real coefficients and positive leading coefficient and that $f(x)$ is divided synthetically by $x - c$.

(i) If $c > 0$ and if all numbers in the third row of the division process are either positive or zero, then c is an upper bound for the real solutions of the equation $f(x) = 0$.

(ii) If $c < 0$ and if the numbers in the third row of the division process are alternately positive and negative (where a 0 in the third row is considered to be either positive or negative), then c is a lower bound for the real solutions of the equation $f(x) = 0$.

A general proof of this theorem can be patterned after the solution given in the next example.

EXAMPLE 5   Find upper and lower bounds for the real solutions of the equation

$$2x^3 + 5x^2 - 8x - 7 = 0.$$

SOLUTION   If we divide synthetically by $x - 1$ and $x - 2$, we obtain

```
1│ 2    5   -8   -7        2│ 2    5   -8   -7
        2    7   -1                4   18   20
   ─────────────────           ─────────────────
   2    7   -1   -8           2    9   10   13
```

Since all numbers in the third row of the synthetic division by $x - 2$ are positive, it follows from (i) of the preceding theorem that 2 is an upper bound for the real solutions of the equation. This fact is also evident if we express the division by $x - 2$ in the Division Algorithm form

$$2x^3 + 5x^2 - 8x - 7 = (x - 2)(2x^2 + 9x + 10) + 13.$$

If $x > 2$, then it is obvious that the right side of the equation is positive, and hence is not zero. Consequently, $2x^3 + 5x^2 - 8x - 7$ is not zero if $x > 2$.

After some trial-and-error attempts using $x - (-1)$ and $x - (-2)$, we find that synthetic division by $x - (-3)$ and $x - (-4)$ gives us

$$
\begin{array}{r|rrrr}
-3 & 2 & 5 & -8 & -7 \\
   &   & -6 & 3 & 15 \\
\hline
   & 2 & -1 & -5 & 8 \\
\end{array}
\qquad
\begin{array}{r|rrrr}
-4 & 2 & 5 & -8 & -7 \\
   &   & -8 & 12 & -16 \\
\hline
   & 2 & -3 & 4 & -23 \\
\end{array}
$$

Since the numbers in the third row of the synthetic division by $x - (-4)$ are alternately positive and negative, it follows from (ii) of the preceding theorem that -4 is a lower bound for the real solutions. This can also be proved by expressing the division by $x + 4$ in the form

$$2x^3 + 5x^2 - 8x - 7 = (x + 4)(2x^2 - 3x + 4) - 23.$$

If $x < -4$, then the right side of this equation is negative (Why?), and therefore is not zero. Hence, $2x^3 + 5x^2 - 8x - 7$ is not zero if $x < -4$.

It follows that all real solutions of the given equation lie in the interval $[-4, 2]$. ■

Exercises 4.4

In Exercises 1–6 find a polynomial $f(x)$ of degree 3 with the indicated zeros and satisfying the given conditions.

1 $5, -2, -3;\quad f(2) = 4$ 2 $4, 1, -6;\quad f(5) = 2$

3 $2, 3, 1;\quad f(0) = 12$ 4 $\sqrt{2}, \pi, 0;\quad f(0) = 0$

5 $2 + i, 2 - i, -4;\quad f(1) = 3$

6 $1 + 2i, 1 - 2i, 5;\quad f(-2) = 1$

7 Find a polynomial of degree 4 such that both -2 and 3 are zeros of multiplicity 2.

8 Find a polynomial of degree 5 such that -2 is a zero of multiplicity 3 and 4 is a zero of multiplicity 2.

9 Find a polynomial $f(x)$ of degree 8 such that 2 is a zero of multiplicity 3, 0 is a zero of multiplicity 5, and $f(3) = 54$.

10 Find a polynomial $f(x)$ of degree 7 such that 1 is a zero of multiplicity 2, -1 is a zero of multiplicity 2, 0 is a zero of multiplicity 3, and $f(2) = 36$.

In Exercises 11–18 find the zeros of the polynomials and state the multiplicity of each zero.

11 $f(x) = (x + 4)^3(3x - 4)$

12 $f(x) = (x - 5)^2(4x + 7)^3$

13 $f(x) = 2x^5 - 8x^4 - 10x^3$

14 $f(x) = (4x^2 - 5)^2$

15 $f(x) = (9x^2 - 25)^4(x^2 + 16)$

16 $f(x) = (2x^2 + 13x - 7)^3$

17 $f(x) = (x^2 + x - 2)^2(x^2 - 4)$

18 $f(x) = 4x^6 + x^4$

19 Show that -3 is a zero of multiplicity 2 of the polynomial $f(x) = x^4 + 7x^3 + 13x^2 - 3x - 18$, and express $f(x)$ as a product of linear factors.

20 Show that 4 is a zero of multiplicity 2 of the polynomial $f(x) = x^4 - 9x^3 + 22x^2 - 32$, and express $f(x)$ as a product of linear factors.

21 Show that 1 is a zero of multiplicity 5 of the polynomial $f(x) = x^6 - 4x^5 + 5x^4 - 5x^2 + 4x - 1$, and express $f(x)$ as a product of linear factors.

22 Show that -1 is a zero of multiplicity 4 of the polynomial $f(x) = x^5 + x^4 - 6x^3 - 14x^2 - 11x - 3$, and express $f(x)$ as a product of linear factors.

In Exercises 23–30 use Descartes' Rule of Signs to determine the number of possible positive, negative, and nonreal complex solutions of the equation.

23 $4x^3 - 6x^2 + x - 3 = 0$

24 $5x^3 - 6x - 4 = 0$

25 $4x^3 + 2x^2 + 1 = 0$

26 $3x^3 - 4x^2 + 3x + 7 = 0$

27 $3x^4 + 2x^3 - 4x + 2 = 0$

28 $2x^4 - x^3 + x^2 - 3x + 4 = 0$

29 $x^5 + 4x^4 + 3x^3 - 4x + 2 = 0$

30 $2x^6 + 5x^5 + 2x^2 - 3x + 4 = 0$

In Exercises 31–36 find the smallest and largest integers that are upper and lower bounds, respectively, for the real solutions of the equation.

31 $x^3 - 4x^2 - 5x + 7 = 0$

32 $2x^3 - 5x^2 + 4x - 8 = 0$

33 $x^4 - x^3 - 2x^2 + 3x + 6 = 0$

34 $2x^4 - 9x^3 - 8x - 10 = 0$

35 $2x^5 - 13x^3 + 2x - 5 = 0$

36 $3x^5 + 2x^4 - x^3 - 8x^2 - 7 = 0$

37 Let $f(x)$ and $g(x)$ be polynomials of degree not greater than n, where n is a positive integer. Show that if $f(x)$ and $g(x)$ are equal in value for more than n distinct values of x, then $f(x)$ and $g(x)$ are identical; that is, coefficients of like powers are the same. (*Hint:* Write

$$f(x) = a_n x^n + a_{n-1} x^{n-1} + \cdots + a_1 x + a_0$$
$$g(x) = b_n x^n + b_{n-1} x^{n-1} + \cdots + b_1 x + b_0$$

and consider

$$h(x) = f(x) - g(x) = (a_n - b_n)x^n + \cdots + (a_0 - b_0).$$

Then show that $h(x)$ has more than n distinct zeros and conclude that $a_j = b_j$ for all j.)

4.5 Complex and Rational Zeros of Polynomials

Example 2 of the preceding section illustrates an interesting fact about polynomials with real coefficients: The two complex zeros $2 + 3i$ and $2 - 3i$ of $x^5 - 4x^4 + 13x^3$ are conjugates of one another. The relationship is not accidental, since the following general result is true.

THEOREM

If a polynomial $f(x)$ of degree $n > 0$ has real coefficients, and if z is a complex zero of $f(x)$, then the conjugate $\bar{z}$ of z is also a zero of $f(x)$.

PROOF   We may write

$$f(x) = a_n x^n + a_{n-1} x^{n-1} + \cdots + a_1 x + a_0$$

where each coefficient a_k is a real number and $a_n \neq 0$. If $f(z) = 0$, then

$$a_n z^n + a_{n-1} z^{n-1} + \cdots + a_1 z + a_0 = 0.$$

If two complex numbers are equal, then so are their conjugates. Hence, the conjugate of the left side of the last equation equals the conjugate of the right side; that is,

$$\overline{a_n z^n + a_{n-1} z^{n-1} + \cdots + a_1 z + a_0} = \bar{0} = 0$$

where the fact that $\bar{0} = 0$ follows from $\bar{0} = \overline{0 + 0i} = 0 - 0i = 0$.

It is not difficult to prove that if z and w are complex numbers, then $\overline{z + w} = \bar{z} + \bar{w}$. More generally, the conjugate of *any* sum of complex numbers is the sum of the conjugates. Consequently,

$$\overline{a_n z^n} + \overline{a_{n-1} z^{n-1}} + \cdots + \overline{a_1 z} + \overline{a_0} = 0.$$

It can also be shown that if z and w are complex numbers, then $\overline{z \cdot w} = \bar{z} \cdot \bar{w}$, $\overline{z^n} = \bar{z}^n$ for every positive integer n, and $\bar{z} = z$ if and only if z is real. (See Exercise 49 of Section 2.4.) Thus, for every j and k,

$$\overline{a_j z^k} = \overline{a_j} \cdot \overline{z^k} = \overline{a_j} \cdot \bar{z}^k = a_j \bar{z}^k,$$

and therefore,

$$a_n \bar{z}^n + a_{n-1} \bar{z}^{n-1} + \cdots + a_1 \bar{z} + a_0 = 0.$$

The last equation states that $f(\bar{z}) = 0$, which completes the proof.   □

EXAMPLE 1   Find a polynomial $f(x)$ of degree 4 that has real coefficients and zeros $2 + i$ and $-3i$.

SOLUTION   By the preceding theorem, $f(x)$ must also have zeros $2 - i$ and $3i$. Applying the Factor Theorem, $f(x)$ has factors $x - (2+i)$, $x - (2-i)$, $x - (-3i)$, and $x - (3i)$. Multiplying those factors gives us a polynomial of the required type:

$$f(x) = [x - (2 + i)][x - (2 - i)](x - 3i)(x + 3i)$$
$$= (x^2 - 4x + 5)(x^2 + 9)$$
$$= x^4 - 4x^3 + 14x^2 - 36x + 45.$$ ∎

If a polynomial with real coefficients is factored as on page 200, some of the factors $x - c_k$ may have a complex coefficient c_k. However, it is always possible to obtain a factorization into polynomials with real coefficients, as stated in the next theorem.

THEOREM

> Every polynomial with real coefficients and positive degree n can be expressed as a product of linear and quadratic polynomials with real coefficients, where the quadratic factors have no real zeros.

PROOF Since $f(x)$ has precisely n complex zeros $c_1, c_2, \ldots, c_n$, we may write

$$f(x) = a(x - c_1)(x - c_2) \cdots (x - c_n)$$

where a is the leading coefficient of $f(x)$. Of course, some of the zeros may be real. In such cases we obtain the linear factors referred to in the statement of the theorem. If a zero c_k is not real, then by the preceding theorem the conjugate $\overline{c_k}$ is also a zero of $f(x)$ and hence must be one of the numbers $c_1, c_2, \ldots, c_n$. This implies that both $x - c_k$ and $x - \overline{c_k}$ appear in the factorization of $f(x)$. If those factors are multiplied, we obtain

$$(x - c_k)(x - \overline{c_k}) = x^2 - (c_k + \overline{c_k})x + c_k\overline{c_k},$$

which has *real* coefficients, since $c_k + \overline{c_k}$ and $c_k\overline{c_k}$ are real numbers. (See page 88.) Thus, the complex zeros of $f(x)$ and their conjugates give rise to quadratic polynomials that are irreducible over $\mathbb{R}$. This completes the proof. □

EXAMPLE 2 Express $x^4 - 2x^2 - 3$ as a product (a) of linear polynomials, and (b) of linear and quadratic polynomials with real coefficients that are irreducible over $\mathbb{R}$.

SOLUTION
(a) We can find the zeros of the given polynomial by solving the equation $x^4 - 2x^2 - 3 = 0$, which may be regarded as quadratic in x^2. Solving for x^2 by means of the Quadratic Formula, we obtain

$$x^2 = \frac{2 \pm \sqrt{4 + 12}}{2} = \frac{2 \pm 4}{2}.$$

Thus, $x^2 = 3$ and $x^2 = -1$. Hence, the zeros are $\sqrt{3}$, $-\sqrt{3}$, i, and $-i$, and we obtain the factorization

$$x^4 - 2x^2 - 3 = (x - \sqrt{3})(x + \sqrt{3})(x - i)(x + i).$$

(b) Multiplying the last two factors in the preceding factorization gives us

$$x^4 - 2x^2 - 3 = (x - \sqrt{3})(x + \sqrt{3})(x^2 + 1),$$

which is of the form stated in the preceding theorem.

The solution of this example could also have been obtained by factoring the original expression without first finding the zeros. Thus,

$$x^4 - 2x^2 - 3 = (x^2 - 3)(x^2 + 1)$$
$$= (x + \sqrt{3})(x - \sqrt{3})(x + i)(x - i). \qquad \blacksquare$$

We have already pointed out that it is generally very difficult to find the zeros of a polynomial of high degree. However, if all the coefficients are integers or rational numbers, there is a method for finding the *rational* zeros, if they exist. The method is a consequence of the following theorem.

THEOREM ON RATIONAL ZEROS

> Suppose that $f(x) = a_n x^n + a_{n-1} x^{n-1} + \cdots + a_1 x + a_0$ is a polynomial with integral coefficients. If c/d is a rational zero of $f(x)$, where c and d have no common prime factors and $c > 0$, then c is a factor of a_0 and d is a factor of a_n.

PROOF Let us show that c is a factor of a_0. If $c = 1$, the theorem follows at once, since 1 is a factor of *any* number. Now suppose that $c \neq 1$. In this case $c/d \neq 1$, for if $c/d = 1$, we obtain $c = d$, and since c and d have no prime factor in common, this implies that $c = d = 1$, a contradiction. Hence, in the following discussion we have $c \neq 1$ and $c \neq d$.

Since $f(c/d) = 0$,

$$a_n(c^n/d^n) + a_{n-1}(c^{n-1}/d^{n-1}) + \cdots + a_1(c/d) + a_0 = 0.$$

Multiplying by d^n and then adding $-a_0 d^n$ to both sides, we obtain

$$a_n c^n + a_{n-1} c^{n-1} d + \cdots + a_1 c d^{n-1} = -a_0 d^n$$

or $\qquad c(a_n c^{n-1} + a_{n-1} c^{n-2} d + \cdots + a_1 d^{n-1}) = -a_0 d^n.$

This shows that c is a factor of the integer $a_0 d^n$. If c is factored into primes, such as $c = p_1 p_2 \cdots p_k$, then each prime p_j is also a factor of $a_0 d^n$. However, by hypothesis, none of the p_j is a factor of d. This implies that each p_j is a factor of a_0, that is, c is a factor of a_0. A similar argument may be used to prove that d is a factor of a_n. $\qquad\square$

The technique of using the preceding theorem for finding rational solutions of equations with integral coefficients is illustrated in the following example.

EXAMPLE 3 Find all rational solutions of the equation

$$3x^4 + 14x^3 + 14x^2 - 8x - 8 = 0.$$

SOLUTION The problem is equivalent to finding the rational zeros of the polynomial on the left side of the equation. According to the preceding theorem, if c/d is a rational zero and $c > 0$, then c is a divisor of -8 and d is a divisor of 3. Hence, the possible choices for c are 1, 2, 4, and 8, and the choices for d are ± 1 and ± 3. Consequently, any rational roots are included among the numbers ± 1, ± 2, ± 4, ± 8, $\pm \frac{1}{3}$, $\pm \frac{2}{3}$, $\pm \frac{4}{3}$, and $\pm \frac{8}{3}$. We can reduce the number of possibilities by finding upper and lower bounds for the real solutions; however, we shall not do so here. It is necessary to check to see which of the numbers, if any, are zeros. Synthetic division is the appropriate method for this task. After perhaps many trial-and-error attempts, we obtain:

$$
\begin{array}{r|rrrrr}
-2 & 3 & 14 & 14 & -8 & -8 \\
   &   & -6 & -16 & 4 & 8 \\
\hline
   & 3 & 8 & -2 & -4 & 0 \\
\end{array}
$$

This result shows that -2 is a zero. Moreover, the synthetic division provides the coefficients of the quotient in the division of the polynomial by $x + 2$. Hence, we have the following factorization of the given polynomial:

$$(x + 2)(3x^3 + 8x^2 - 2x - 4).$$

The remaining solutions of the equation must be zeros of the second factor, and therefore we may use that polynomial to check for solutions. Again proceeding by trial and error, we ultimately find that synthetic division by $x + \frac{2}{3}$ gives us the following:

$$
\begin{array}{r|rrrr}
-\frac{2}{3} & 3 & 8 & -2 & -4 \\
             &   & -2 & -4 & 4 \\
\hline
             & 3 & 6 & -6 & 0 \\
\end{array}
$$

Therefore, $-\frac{2}{3}$ is a zero.

The remaining zeros are solutions of the equation $3x^2 + 6x - 6 = 0$, or equivalently, $x^2 + 2x - 2 = 0$. By the Quadratic Formula, this equation has solutions

$$\frac{-2 \pm \sqrt{4 - 4(-2)}}{2} = \frac{-2 \pm \sqrt{12}}{2} = \frac{-2 \pm 2\sqrt{3}}{2}.$$

Hence, the given polynomial has two rational roots, -2 and $-\frac{2}{3}$, and two irrational roots, $-1 + \sqrt{3}$ and $-1 - \sqrt{3}$. ∎

The Theorem on Rational Zeros may be applied to equations with rational coefficients. We merely multiply both sides of the equation by the least common denominator of all the coefficients to obtain an equation with integral coefficients, and then proceed as in Example 3.

EXAMPLE 4 Find all rational solutions of the equation

$$\tfrac{2}{3}x^4 + \tfrac{1}{2}x^3 - \tfrac{5}{4}x^2 - x - \tfrac{1}{6} = 0.$$

SOLUTION Multiplying both sides of the equation by 12, we obtain the equivalent equation

$$8x^4 + 6x^3 - 15x^2 - 12x - 2 = 0.$$

If c/d is a rational solution, then the choices for c are 1 and 2, and the choices for d are ± 1, ± 2, ± 4, and ± 8. Hence, the only possible rational roots are ± 1, ± 2, $\pm\frac{1}{2}$, $\pm\frac{1}{4}$, and $\pm\frac{1}{8}$. Trying various possibilities we obtain, by synthetic division,

$$
\begin{array}{r|rrrrr}
-\frac{1}{2} & 8 & 6 & -15 & -12 & -2 \\
 & & -4 & -1 & 8 & 2 \\
\hline
 & 8 & 2 & -16 & -4 & 0
\end{array}
$$

Hence, $-\frac{1}{2}$ is a solution. Using synthetic division on the coefficients of the quotient gives us

$$
\begin{array}{r|rrrr}
-\frac{1}{4} & 8 & 2 & -16 & -4 \\
 & & -2 & 0 & 4 \\
\hline
 & 8 & 0 & -16 & 0
\end{array}
$$

Consequently, $-\frac{1}{4}$ is a solution. The last synthetic division gave us the quotient $8x^2 - 16$. Setting this equal to zero and solving, we obtain $x^2 = 2$, or $x = \pm\sqrt{2}$. Thus, the given equation has rational solutions $-\frac{1}{2}$, $-\frac{1}{4}$ and irrational solutions $\sqrt{2}$, $-\sqrt{2}$. ∎

EXAMPLE 5 A grain silo has the shape of a right circular cylinder with a hemisphere attached to the top. If the total height of the structure is 30 feet, what is the radius of the cylinder that will result in a total volume of 1008π cubic feet?

FIGURE 4.14

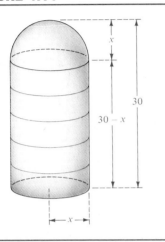

SOLUTION Let x denote the radius of the cylinder. A sketch of the silo, appropriately labeled, is shown in Figure 4.14. Since the volume of the cylinder is $\pi x^2(30 - x)$, and the volume of the hemisphere is $\frac{2}{3}\pi x^3$, we must solve the equation

$$\pi x^2(30 - x) + \tfrac{2}{3}\pi x^3 = 1008\pi.$$

Each of the following is equivalent to the preceding equation:

$$3x^2(30 - x) + 2x^3 = 3024$$
$$90x^2 - 3x^3 + 2x^3 = 3024$$
$$90x^2 - x^3 = 3024$$
$$x^3 - 90x^2 + 3024 = 0$$

To look for rational roots, we first factor 3024 into primes, obtaining $3024 = 2^4 \cdot 3^3 \cdot 7$. It follows that some of the positive factors of 3024 are

$$1, \quad 2, \quad 3, \quad 4, \quad 6, \quad 8, \quad 9, \quad 12, \ldots$$

Dividing synthetically, we eventually arrive at

$$\begin{array}{r|rrrr} 6 & 1 & -90 & 0 & 3024 \\ & & 6 & -504 & -3024 \\ \hline & 1 & -84 & -504 & 0 \end{array}$$

Thus, 6 is a solution of the equation $x^3 - 90x^2 + 3024 = 0$, and the desired radius is 6 feet.

The remaining two solutions of the equation can be found by solving $x^2 - 84x - 504 = 0$. It is not difficult to show that neither of the solutions of this quadratic equation satisfy the conditions of the problem. ∎

The discussion in this section gives no practical information about finding the irrational zeros of polynomials. The examples we have worked are not typical of problems encountered in applications. Indeed, polynomials with rational coefficients often have *no* rational zeros. A standard way to approximate irrational zeros is to use a calculus technique called Newton's Method. In practice, computers have taken over the task of approximating irrational solutions of equations.

Exercises 4.5

In Exercises 1–8 find a polynomial with real coefficients that has the given zero (or zeros) and degree.

1 $4 + i$; degree 2

2 $3 - 3i$; degree 2

3 $4, 3 - 2i$; degree 3

4 $-2, 1 + 5i$; degree 3

5 $1 + 4i, 2 - i$; degree 4

6 $3, 0, 8 - 7i$; degree 4

7 $5i, 1 + i, 0$; degree 5

8 $i, 2i, 3i$; degree 6

9 Does there exist a polynomial of degree 3 with real coefficients that has zeros $1, -1$, and i? Justify your answer.

10 The complex number i is a zero of the polynomial $f(x) = x^3 - ix^2 + 2ix + 2$; however, the conjugate $-i$ of i is not a zero. Why doesn't this contradict the first theorem of this section?

In Exercises 11–28 find all solutions of the given equations.

11 $x^3 - x^2 - 10x - 8 = 0$

12 $x^3 + x^2 - 14x - 24 = 0$

13 $2x^3 - 3x^2 - 17x + 30 = 0$

14 $12x^3 + 8x^2 - 3x - 2 = 0$

15 $x^4 + 3x^3 - 30x^2 - 6x + 56 = 0$

16 $3x^5 - 10x^4 - 6x^3 + 24x^2 + 11x - 6 = 0$

17 $2x^3 - 7x^2 - 10x + 24 = 0$

18 $2x^3 - 3x^2 - 8x - 3 = 0$

19 $6x^3 + 19x^2 + x - 6 = 0$

20 $6x^3 + 5x^2 - 17x - 6 = 0$

21 $8x^3 + 18x^2 + 45x + 27 = 0$

22 $3x^3 - x^2 + 11x - 20 = 0$

23 $x^4 - x^3 - 9x^2 - 3x - 36 = 0$

24 $3x^4 + 16x^3 + 28x^2 + 31x + 30 = 0$

25 $9x^4 + 15x^3 - 20x^2 - 20x + 16 = 0$

26 $15x^4 + 4x^3 + 11x^2 + 4x - 4 = 0$

27 $4x^5 + 12x^4 - 41x^3 - 99x^2 + 10x + 24 = 0$

28 $4x^5 + 24x^4 - 13x^3 - 174x^2 + 9x + 270 = 0$

Show that the equations in Exercises 29–36 have no rational roots.

29 $x^3 + 3x^2 - 4x + 6 = 0$

30 $3x^3 - 4x^2 + 7x + 5 = 0$

31 $x^5 - 3x^3 + 4x^2 + x - 2 = 0$

32 $2x^5 + 3x^3 + 7 = 0$

33 $3x^4 + 7x^3 + 3x^2 - 8x - 10 = 0$

34 $2x^4 - 3x^3 + 6x^2 - 24x + 5 = 0$

35 $8x^4 + 16x^3 - 26x^2 - 12x + 15 = 0$

36 $5x^5 + 2x^4 + x^3 - 10x^2 - 4x - 2 = 0$

37 If n is an odd positive integer, prove that a polynomial of degree n with real coefficients has at least one real zero.

38 Show that the theorem on page 208 is not necessarily true if $f(x)$ has complex coefficients.

39 Complete the proof of the Theorem on Rational Zeros by showing that d is a factor of a_n.

40 If a polynomial of the form

$$x^n + a_{n-1}x^{n-1} + \cdots + a_1x + a_0,$$

where each a_k is an integer, has a rational root r, show that r is an integer and is a factor of a_0.

41 An open box is to be made from a rectangular piece of cardboard having dimensions 20 inches $\times$ 30 inches by removing squares of area x^2 from each corner and turning up the sides. Show that there are two boxes that have a volume of 1000 in^3. Which box has the smaller surface area? (Compare Exercise 31 of Section 4.2.)

42 The frame for a shipping crate is to be constructed from 24 feet of 2 $\times$ 2 lumber. Assuming the crate is to have square ends of length x feet, determine the value(s) of x that result in a volume of 4 ft^3. (Compare Exercise 32 of Section 4.2.)

43 A meteorologist determines that the temperature T (in °F) on a certain cold winter day was given by $T = 0.05t(t - 12)(t - 24)$ for $0 \le t \le 24$, where t is time in hours and $t = 0$ corresponds to 6 A.M. At what time(s) of the day was the temperature 32°F? (Compare Exercise 33 of Section 4.2.)

44 A herd of 100 deer is introduced to a small island. Assuming the number $N(t)$ of deer after t years is given by $N(t) = -t^4 + 21t^2 + 100$ (for $t > 0$), determine when the herd size exceeds 180. (Compare Exercise 35 of Section 4.2.)

45 A right triangle has area 30 ft^2 and a hypotenuse that is 1 foot longer than one of the legs.

(a) If x denotes the length of this leg, show that $2x^3 + x^2 - 3600 = 0$.

(b) Show that there is a single positive root of the equation in part (a) and that this root is less than 13.

(c) Find the dimensions of the triangle.

46 A storage tank for propane gas is to be constructed in the shape of a right circular cylinder of altitude 10 feet with a hemisphere attached to each end. Determine the radius x, so that the resulting volume is 27π ft^3. (Compare Example 5 of Section 3.3.)

47 A storage shelter is to be constructed in the shape of a cube with a triangular prism forming the roof (see figure). The length x of a side of the cube is yet to be determined.

(a) If the total height of the structure is 6 feet, show that its volume V is given by $V = x^3 + \frac{1}{2}x^2(6 - x)$.

(b) Determine x so that the volume is 80 ft^3.

FIGURE FOR EXERCISE 47

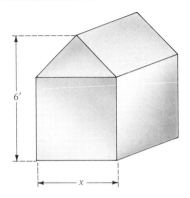

48 A canvas camping tent has the shape of a pyramid with a square base. An 8-foot pole will form the center support as illustrated in the figure. Find the length x of a side of the base so that the total canvas needed for the sides and bottom is 384 ft^2.

FIGURE FOR EXERCISE 48

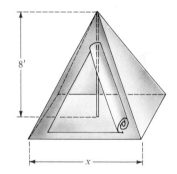

4.6 Rational Functions

A function f is a **rational function** if, for all x in its domain,

$$f(x) = \frac{g(x)}{h(x)}$$

where $g(x)$ and $h(x)$ are polynomials. The zeros of the numerator and denominator are of major importance. *Throughout this section we shall assume that $g(x)$ and $h(x)$ have no common factors* and hence no common zeros. If $g(c) = 0$, then $f(c) = 0$. However, if $h(c) = 0$, then $f(c)$ is undefined. As indicated in the next example, the behavior of $f(x)$ requires special attention when x is near a zero of the denominator $h(x)$.

EXAMPLE 1   Sketch the graph of f if

$$f(x) = \frac{1}{x - 2}.$$

SOLUTION   The numerator 1 is never zero and hence $f(x)$ has no zeros. This means that the graph has no x-intercepts.

The denominator $x - 2$ is zero at $x = 2$. If x is close to 2, and $x > 2$, then $f(x)$ is large. For example,

$$f(2.1) = \frac{1}{2.1 - 2} = \frac{1}{0.1} = 10$$

$$f(2.01) = \frac{1}{2.01 - 2} = \frac{1}{0.01} = 100$$

$$f(2.001) = \frac{1}{2.001 - 2} = \frac{1}{0.001} = 1000.$$

If x is close to 2, and $x < 2$, then $|f(x)|$ is large, but $f(x)$ is negative. Thus,

$$f(1.9) = \frac{1}{1.9 - 2} = \frac{1}{-0.1} = -10$$

$$f(1.99) = \frac{1}{1.99 - 2} = \frac{1}{-0.01} = -100$$

$$f(1.999) = \frac{1}{1.999 - 2} = \frac{1}{-0.001} = -1000.$$

Several other values of $f(x)$ are displayed in the following table:

x	-8	-1	0	1	3	4	12
$f(x)$	$-\frac{1}{10}$	$-\frac{1}{3}$	$-\frac{1}{2}$	-1	1	$\frac{1}{2}$	$\frac{1}{10}$

Observe that as $|x|$ increases, $f(x)$ approaches zero. Plotting points and paying attention to what happens near $x = 2$ gives us the sketch in Figure 4.15. ∎

FIGURE 4.15

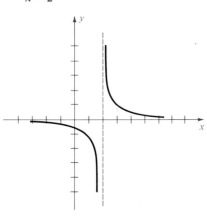

In Example 1, $f(x)$ can be made as large as desired by choosing x close to 2 (and $x > 2$). We denote this fact by writing

$$f(x) \to \infty \quad \text{as} \quad x \to 2^{+}.$$

We say that $f(x)$ *increases without bound* (or $f(x)$ *becomes positively infinite*) *as x approaches 2 from the right*. It is important to remember that the symbol ∞ (read "infinity") does not represent a real number, but is used merely as an abbreviation for certain types of functional behavior.

For the case $x < 2$ we write

$$f(x) \to -\infty \quad \text{as} \quad x \to 2^{-}$$

and say that $f(x)$ *decreases without bound* (or $f(x)$ *becomes negatively infinite*) *as x approaches 2 from the left*.

In general, the notation $x \to a^{+}$ will signify that x approaches a from the *right*, that is, through values *greater* than a. The symbol $x \to a^{-}$ will mean that x approaches a from the *left*, that is, through values *less* than a. Some illustrations of the manner in which a function f may increase or decrease without bound, together with the notation used, are shown in Figure 4.16. In the figure, a is pictured as positive, but we can also have $a \le 0$.

FIGURE 4.16

(i) $f(x) \to \infty$ as $x \to a^{-}$

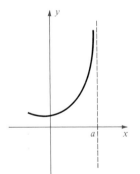

(ii) $f(x) \to \infty$ as $x \to a^{+}$

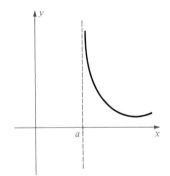

(iii) $f(x) \to -\infty$ as $x \to a^{-}$

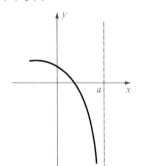

(iv) $f(x) \to -\infty$ as $x \to a^{+}$

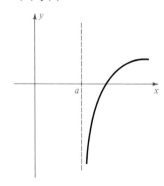

DEFINITION

The line $x = a$ is a **vertical asymptote** for the graph of a function f if

$$f(x) \to \infty \quad \text{or} \quad f(x) \to -\infty$$

as x approaches a from either the left or right.

In Figure 4.16 the dashed lines represent vertical asymptotes. Note that in Figure 4.15 the line $x = 2$ is a vertical asymptote for the graph of $y = 1/(x - 2)$.

Vertical asymptotes are common for graphs of rational functions. Indeed, *if the number a is a zero of the denominator h(x), then the graph of* $f(x) = g(x)/h(x)$ *has the vertical asymptote* $x = a$.

We are also interested in values of $f(x)$ when $|x|$ is large. As an illustration, consider Example 1, where $f(x) = 1/(x - 2)$. If we assign very large values to x, then $f(x)$ is close to 0. Thus,

$$f(1002) = \frac{1}{1000} = 0.001 \quad \text{and} \quad f(1{,}000{,}002) = \frac{1}{1{,}000{,}000} = 0.000001.$$

Moreover, we can make $f(x)$ as close to 0 as we desire by choosing x sufficiently large. This is expressed symbolically by

$$f(x) \to 0 \quad \text{as} \quad x \to \infty,$$

which is read $f(x)$ *approaches* 0 *as x increases without bound* (or *as x becomes positively infinite*).

Similarly, in Example 1, we write

$$f(x) \to 0 \quad \text{as} \quad x \to -\infty,$$

which is read $f(x)$ *approaches* 0 *as x decreases without bound* (or *as x becomes negatively infinite*).

In Example 1 the line $y = 0$, that is, the x-axis, is called a *horizontal asymptote* for the graph. In general, we have the following definition. The notation should be self-evident.

DEFINITION

> The line $y = c$ is a **horizontal asymptote** for the graph of a function f if
>
> $$f(x) \to c \quad \text{as} \quad x \to \infty \quad \text{or as} \quad x \to -\infty.$$

Some typical horizontal asymptotes (for $x \to \infty$) are illustrated in Figure 4.17. The manner in which the graph "approaches" the line $y = c$ may vary, depending on the nature of the function. Similar sketches may be made for the case $x \to -\infty$. Note that, as in the third sketch, the graph of f may cross a horizontal asymptote.

FIGURE 4.17

$f(x) \to c$ as $x \to \infty$

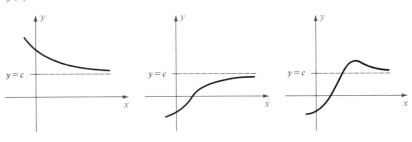

The following theorem is useful for locating horizontal asymptotes for the graph of a rational function.

THEOREM ON HORIZONTAL ASYMPTOTES

Let $f(x) = \dfrac{a_n x^n + a_{n-1}x^{n-1} + \cdots + a_1 x + a_0}{b_k x^k + b_{k-1}x^{k-1} + \cdots + b_1 x + b_0}$.

(i) If $n < k$, then the x-axis is a horizontal asymptote for the graph of f.

(ii) If $n = k$, then the line $y = a_n/b_k$ is a horizontal asymptote.

(iii) If $n > k$, the graph of f has no horizontal asymptote.

Proofs of (i) and (ii) of this theorem may be patterned after the solution to the following example. A similar argument can be given in case (iii).

EXAMPLE 2   Find the horizontal asymptotes for the graph of f if:

(a) $f(x) = \dfrac{3x - 1}{x^2 - x - 6}$ (b) $f(x) = \dfrac{5x^2 + 1}{3x^2 - 4}$

SOLUTION

(a) The degree of the numerator $3x - 1$ is less than the degree of the denominator $x^2 - x - 6$, and hence by (i) of the theorem, the x-axis is a horizontal asymptote. To verify this directly, we divide numerator and denominator of the quotient by x^2, obtaining

$$f(x) = \frac{\left(\dfrac{3x-1}{x^2}\right)}{\left(\dfrac{x^2-x-6}{x^2}\right)} = \frac{\dfrac{3}{x} - \dfrac{1}{x^2}}{1 - \dfrac{1}{x} - \dfrac{6}{x^2}}, \qquad x \neq 0.$$

If x is very large, then both $1/x$ and $1/x^2$ are close to 0, and hence,

$$f(x) \approx \frac{0-0}{1-0-0} = \frac{0}{1} = 0.$$

Thus, $\qquad\qquad\qquad f(x) \to 0 \quad \text{as} \quad x \to \infty.$

Since $f(x)$ is the y-coordinate of a point on the graph, this means that the x-axis is a horizontal asymptote.

(b) If $f(x) = (5x^2 + 1)/(3x^2 - 4)$, then the numerator and denominator have the same degree, and hence by (ii) of the theorem, the line $y = \frac{5}{3}$ is a horizontal asymptote. This may be proved directly by dividing numerator and denominator of $f(x)$ by x^2, obtaining

$$f(x) = \frac{5 + \dfrac{1}{x^2}}{3 - \dfrac{4}{x^2}}$$

Since $1/x^2 \to 0$ as $x \to \infty$, we see that

$$f(x) \to \frac{5+0}{3-0} = \frac{5}{3} \quad \text{as} \quad x \to \infty. \qquad\blacksquare$$

We shall next list some guidelines for sketching the graph of a rational function. Their use will be illustrated in Examples 3, 4, and 5.

GUIDELINES: **Sketching the Graph of $f(x) = \dfrac{g(x)}{h(x)}$ where $g(x)$ and $h(x)$ are Polynomials That Have No Common Factor**

STEP 1 Find the real zeros of the numerator $g(x)$ and use them to plot the points corresponding to the x-intercepts.

STEP 2 Find the real zeros of the denominator $h(x)$. For each zero a, the line $x = a$ is a vertical asymptote. Represent $x = a$ with dashes.

STEP 3   Find the sign of $f(x)$ in each of the intervals determined by the zeros of $g(x)$ and $h(x)$. Use these signs to determine whether the graph lies above or below the x-axis in each interval.

STEP 4   If $x = a$ is a vertical asymptote, use the information in Step 3 to determine whether $f(x) \to \infty$ or $f(x) \to -\infty$ for each case:

$$\text{(i)} \quad x \to a^-; \qquad \text{(ii)} \quad x \to a^+.$$

Make note of this by sketching a portion of the graph on each side of $x = a$.

STEP 5   Use the information in Step 3 to determine the manner in which the graph intersects the x-axis.

STEP 6   Apply the Theorem on Horizontal Asymptotes. If there is a horizontal asymptote, represent it with dashes.

STEP 7   Sketch the graph, using the information found in the preceding steps and plotting points wherever necessary.   ■ ■ ■

EXAMPLE 3   Sketch the graph of f if

$$f(x) = \frac{x - 1}{x^2 - x - 6}.$$

SOLUTION   We begin by factoring the denominator as follows:

$$f(x) = \frac{x - 1}{(x + 2)(x - 3)}.$$

We shall obtain the graph by following the steps listed in the guidelines.

Step 1:   The numerator $x - 1$ has the zero 1, and we plot the point $(1, 0)$ on the graph, as shown in Figure 4.18.

Step 2:   The denominator has zeros -2 and 3. Hence, the lines $x = -2$ and $x = 3$ are vertical asymptotes, and we represent them with dashes, as in Figure 4.18.

Step 3:   The zeros -2, 1, and 3 of the numerator and denominator of $f(x)$ determine the following intervals:

$$(-\infty, -2), \quad (-2, 1), \quad (1, 3), \quad \text{and} \quad (3, \infty).$$

Since $f(x)$ is a quotient of two polynomials, it follows from our work in Section 4.2 that $f(x)$ is always positive or always negative throughout each

FIGURE 4.18

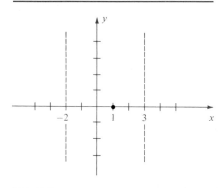

interval. Using test values to determine the sign of $f(x)$, we arrive at the following table.

Interval	$(-\infty, -2)$	$(-2, 1)$	$(1, 3)$	$(3, \infty)$
Test value	$f(-3) = -\frac{2}{3}$	$f(0) = \frac{1}{6}$	$f(2) = -\frac{1}{4}$	$f(4) = \frac{1}{2}$
Sign of $f(x)$	$-$	$+$	$-$	$+$
Position of graph	Below x-axis	Above x-axis	Below x-axis	Above x-axis

Step 4: We shall use the fourth row of the table in Step 3 to investigate the behavior of $f(x)$ near each vertical asymptote.

(a) Consider the vertical asymptote $x = -2$. Since the graph lies *below* the x-axis throughout the interval $(-\infty, -2)$, it follows that

$$f(x) \to -\infty \quad \text{as} \quad x \to -2^-.$$

Since the graph lies *above* the x-axis throughout the interval $(-2, 1)$, it follows that

$$f(x) \to \infty \quad \text{as} \quad x \to -2^+.$$

FIGURE 4.19

We note these facts in Figure 4.19 by sketching portions of the graph on each side of the line $x = -2$.

(b) Consider the vertical asymptote $x = 3$. The graph lies *below* the x-axis throughout the interval $(1, 3)$, and hence

$$f(x) \to -\infty \quad \text{as} \quad x \to 3^-.$$

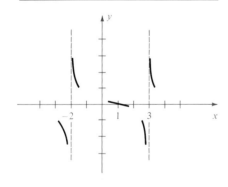

The graph lies *above* the x-axis throughout $(3, \infty)$, and hence

$$f(x) \to \infty \quad \text{as} \quad x \to 3^+.$$

We note these facts in Figure 4.19 by sketching portions of the graph on each side of $x = 3$.

Step 5: Referring to the fourth row of the table in Step 3, we see that the graph crosses the x-axis at $(1, 0)$ in a manner similar to that illustrated in Figure 4.19.

Step 6: The degree of the numerator $x - 1$ is less than the degree of the denominator $x^2 - x - 6$. Hence, by (i) of the Theorem on Horizontal Asymptotes, the x-axis is a horizontal asymptote.

FIGURE 4.20

$$y = \frac{x - 1}{x^2 - x - 6}$$

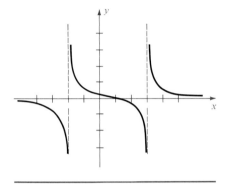

Step 7: Using the information found in Steps 4, 5, and 6, and plotting several points, we obtain the sketch in Figure 4.20. ■

EXAMPLE 4 Sketch the graph of f if

$$f(x) = \frac{x^2}{x^2 - x - 2}.$$

SOLUTION Factoring the denominator gives us

$$f(x) = \frac{x^2}{(x + 1)(x - 2)}.$$

We shall again follow the guidelines listed earlier.

Step 1: The numerator x^2 has 0 as a zero, and hence the graph intersects the x-axis at $(0, 0)$, as shown in Figure 4.21.

FIGURE 4.21

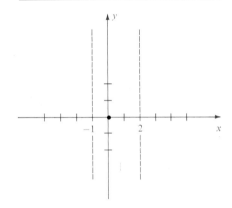

Step 2: Since the denominator has zeros -1 and 2, the lines $x = -1$ and $x = 2$ are vertical asymptotes, and we represent them with dashes, as in Figure 4.21.

Step 3: The intervals determined by the zeros in Steps 1 and 2 are

$$(-\infty, -1), \quad (-1, 0), \quad (0, 2), \quad \text{and} \quad (2, \infty).$$

Following the procedure used in Step 3 of Example 3, we arrive at the following table.

Interval	$(-\infty, -1)$	$(-1, 0)$	$(0, 2)$	$(2, \infty)$
Test value	$f(-2) = 1$	$f(-\frac{1}{2}) = -\frac{1}{5}$	$f(1) = -\frac{1}{2}$	$f(3) = \frac{9}{4}$
Sign of $f(x)$	+	−	−	+
Position of graph	Above x-axis	Below x-axis	Below x-axis	Above x-axis

Step 4: We refer to the fourth row of the table in Step 3 and proceed as follows:

(a) Consider the vertical asymptote $x = -1$. Since the graph is above the x-axis in $(-\infty, -1)$, it follows that

$$f(x) \to \infty \quad \text{as} \quad x \to -1^-.$$

FIGURE 4.22

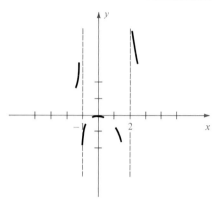

FIGURE 4.23

$$y = \frac{x^2}{x^2 - x - 2}$$

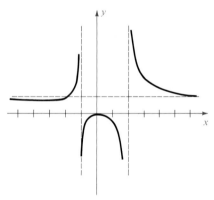

Since the graph is below the x-axis in $(-1, 0)$, we have

$$f(x) \to -\infty \quad \text{as} \quad x \to -1^+.$$

We note these facts in Figure 4.22 by sketching portions of the graph on each side of $x = -1$.

(b) Consider the vertical asymptote $x = 2$. The graph is below the x-axis in the interval $(0, 2)$, and hence

$$f(x) \to -\infty \quad \text{as} \quad x \to 2^-.$$

The graph is above the x-axis in $(2, \infty)$, and hence

$$f(x) \to \infty \quad \text{as} \quad x \to 2^+.$$

These facts are noted in Figure 4.22.

Step 5: Referring to the table in Step 3, we see that the graph lies below the x-axis in both of the intervals $(-1, 0)$ and $(0, 2)$. Consequently, the graph intersects, but does not cross, the x-axis at $(0, 0)$.

Step 6: The numerator and denominator of $f(x)$ have the same degree, and both leading coefficients are 1. Hence, by (ii) of the Theorem on Horizontal Asymptotes, the line $y = \frac{1}{1} = 1$ is a horizontal asymptote. We sketch this line with dashes, as in Figure 4.23.

Step 7: Using the information found in Steps 4, 5, and 6, and plotting several points, we obtain the graph sketched in Figure 4.23. The graph intersects the horizontal asymptote at $x = -2$; this may be verified by solving the equation $x^2/(x^2 - x - 2) = 1$. The fact that the graph lies below the horizontal asymptote if $x < -2$ and above it if $-2 < x < -1$ may be verified by plotting points. ∎

EXAMPLE 5 Sketch the graph of f if

$$f(x) = \frac{2x^4}{x^4 + 1}.$$

SOLUTION In this solution we shall not formally write down each step in the guidelines. Note that since $f(-x) = f(x)$, the function is even, and hence the graph is symmetric with respect to the y-axis.

The graph intersects the x-axis at $(0, 0)$. Since the denominator of $f(x)$ has no real zeros, the graph has no vertical asymptotes.

The numerator and denominator of $f(x)$ have the same degree. Since the leading coefficients are 2 and 1, respectively, it follows from (ii) of the

FIGURE 4.24

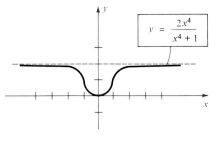

$$y = \frac{2x^4}{x^4 + 1}$$

Theorem on Horizontal Asymptotes, that the line $y = \frac{2}{1} = 2$ is a horizontal asymptote. We represent this line with dashes in Figure 4.24.

Plotting several points and making use of the symmetry with respect to the y-axis leads to the sketch in Figure 4.24.   ■

If $f(x) = g(x)/h(x)$ for polynomials $g(x)$ and $h(x)$, and *if the degree of $g(x)$ is one greater than the degree of $h(x)$*, then the graph of f has an **oblique asymptote** $y = ax + b$; that is, the graph approaches this line as $x \to \infty$ or as $x \to -\infty$. To find the oblique asymptote we may use division to express $f(x)$ in the form

$$f(x) = \frac{g(x)}{h(x)} = (ax + b) + \frac{r(x)}{h(x)}$$

where either $r(x) = 0$ or the degree of $r(x)$ is less than the degree of $h(x)$. It follows from (i) of the Theorem on Horizontal Asymptotes that

$$\frac{r(x)}{h(x)} \to 0 \quad \text{as} \quad x \to \infty \quad \text{or as} \quad x \to -\infty.$$

Consequently, $f(x)$ gets closer and closer to $ax + b$ as $|x|$ increases without bound. The next example illustrates a special case of this procedure.

EXAMPLE 6   Find all the asymptotes and sketch the graph of f if

$$f(x) = \frac{x^2 - 9}{2x - 4}.$$

SOLUTION   A vertical asymptote occurs if $2x - 4 = 0$, that is, if $x = 2$.

The degree of the numerator of $f(x)$ is greater than the degree of the denominator. Hence by (iii) of the Theorem on Horizontal Asymptotes, there is no horizontal asymptote. However, since the degree of the numerator $x^2 - 9$ is *one* greater than the degree of the denominator $2x - 4$, the graph has an oblique asymptote. Dividing, we obtain

$$\begin{array}{r} \frac{1}{2}x + 1 \\ 2x - 4 \overline{\smash{)}\ x^2 - 9} \\ \underline{x^2 - 2x } \\ 2x - 9 \\ \underline{2x - 4} \\ -5 \end{array}$$

Therefore,
$$\frac{x^2 - 9}{2x - 4} = \left(\frac{1}{2}x + 1\right) - \frac{5}{2x - 4}.$$

As we indicated in the discussion preceding this example, the line $y = \frac{1}{2}x + 1$ is an oblique asymptote. This line and the vertical asymptote $x = 2$ are sketched (with dashes) in (i) of Figure 4.25.

FIGURE 4.25

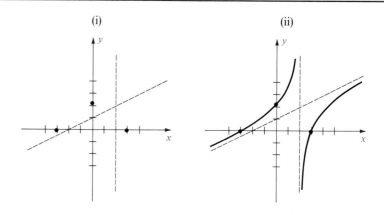

The x-intercepts of the graph are the solutions of the equation $x^2 - 9 = 0$, and hence are 3 and -3. The y-intercept is $f(0) = \frac{9}{4}$. The corresponding points are plotted in (i) of Figure 4.25. It is now easy to show that the graph has the shape indicated in (ii) of Figure 4.25. ∎

Graphs of rational functions may become increasingly complicated as the degrees of the polynomials in the numerator and denominator increase. Techniques developed in calculus must be employed for a thorough treatment of such graphs.

Exercises 4.6

Sketch the graph of f in Exercises 1–20.

1 $f(x) = \dfrac{1}{x + 2}$

2 $f(x) = \dfrac{1}{x + 3}$

3 $f(x) = \dfrac{-2}{x + 4}$

4 $f(x) = \dfrac{-3}{x - 1}$

5 $f(x) = \dfrac{x}{x - 5}$

6 $f(x) = \dfrac{x}{3x + 2}$

7 $f(x) = \dfrac{4}{(x - 1)^2}$

8 $f(x) = \dfrac{-1}{(x + 2)^2}$

9 $f(x) = \dfrac{1}{x^2 - 4}$

10 $f(x) = \dfrac{2}{x^2 + x - 2}$

11 $f(x) = \dfrac{5x}{4 - x^2}$

12 $f(x) = \dfrac{x^2}{x^2 - 4}$

13 $f(x) = \dfrac{x^2}{x^2 - 7x + 10}$

14 $f(x) = \dfrac{x}{x^2 - x - 6}$

15 $f(x) = \dfrac{3x + 2}{x}$

16 $f(x) = \dfrac{x^2 - 4}{x^2}$

17 $f(x) = \dfrac{4}{x^2 + 4}$

18 $f(x) = \dfrac{3x}{x^2 + 1}$

19 $f(x) = \dfrac{1}{x^3 + x^2 - 6x}$

20 $f(x) = \dfrac{x^2 - x}{16 - x^2}$

In Exercises 21–24 find the vertical and oblique asymptotes, and sketch the graph of f.

21 $f(x) = \dfrac{x^2 - x - 6}{x + 1}$

22 $f(x) = \dfrac{2x^2 - x - 3}{x - 2}$

23 $f(x) = \dfrac{8 - x^3}{2x^2}$

24 $f(x) = \dfrac{x^3 + 1}{x^2 - 9}$

25 A cylindrical container for storing radioactive waste is to be constructed from lead. This container must be 6 inches thick (see figure). The volume of the outside cylinder shown in the figure is to be 16π ft^3.

(a) Express the height h of the inside cylinder as a function of the inside radius r.

(b) Show that the inside volume is given by a rational function V such that

$$V(r) = \pi r^2 \left[\frac{16}{(r + 0.5)^2} - 1 \right].$$

FIGURE FOR EXERCISE 25

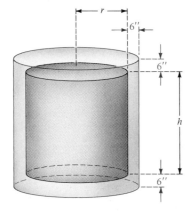

(c) What values of r must be excluded in the formula for $V(r)$?

26 Young's Rule is a formula that is used for modifying adult drug dosage levels for young children. (Compare Exercise 50 of Section 3.5.) If a denotes the adult dose (in mg), and if t is the age of the child (in years), then the child's dose y is given by $y = ta/(t + 12)$. Sketch the graph of this equation for $t > 0$.

27 Salt water of concentration 0.1 pounds of salt per gallon flows into a large tank that initially contains 50 gallons of pure water.

(a) If the flow rate of salt water into the tank is 5 gallons per minute, what is the volume $V(t)$ of water and the amount $A(t)$ of salt in the tank at time t?

(b) Show that the salt concentration at time t is given by $c(t) = t/(10t + 100)$.

(c) Discuss the behavior of $c(t)$ as $t \to \infty$.

28 An important problem in fishery science is predicting the next years' adult breeding population R (the recruits) from the number S that are presently spawning. For some species (such as North Sea herring), the relationship between R and S takes the form $R = aS/(S + b)$. What is the interpretation of the constant a? Conclude that for large values of S, recruitment is more or less constant.

29 Coulomb's Law in electricity asserts that the force of attraction F between two charged particles is inversely proportional to the square of the distance between the particles and directly proportional to the product of the charges. Suppose a particle of charge $+1$ is placed on a coordinate line between two particles of charge -1 as shown in the figure.

(a) Show that the net force acting on the particle of charge $+1$ is given by

$$F(x) = -\frac{k}{x^2} + \frac{k}{(x - 2)^2}$$

for some $k > 0$.

(b) Let $k = 1$ and sketch the graph of F for $0 < x < 2$.

FIGURE FOR EXERCISE 29

30 Biomathematicians have proposed many different functions for describing the effect of light on the rate at which photosynthesis can take place. If the function is to be realistic, then it must exhibit the *photoinhibition effect;* that is, the rate of production P of photosynthesis must decrease to 0 as the light intensity I reaches high levels (see figure). Which of the following functions might be used and which may not be used? Why?

(a) $P = \dfrac{aI}{b + I}$

(b) $P = \dfrac{aI}{b + I^2}$

FIGURE FOR EXERCISE 30

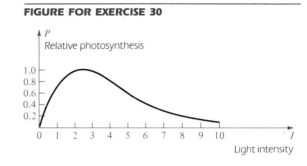

4.7 Conic Sections

The geometric figures considered in this section can be obtained by intersecting a double-napped right circular cone with a plane. For this reason they are called **conic sections** or simply **conics.** If, as in Figure 4.26(i), the plane cuts entirely across one nappe of the cone and is not perpendicular to the axis, then the curve of intersection is called an **ellipse.** If the plane is perpendicular to the axis of the cone, a **circle** results. If the plane does not cut across one entire nappe and does not intersect both nappes, as illustrated in Figure 4.26(ii), the curve of intersection is a **parabola.** If the plane cuts through both nappes of the cone, as in (iii) of the figure, we obtain a **hyperbola.**

FIGURE 4.26

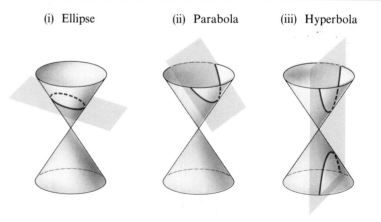

(i) Ellipse   (ii) Parabola   (iii) Hyperbola

The conic sections were studied extensively by the ancient Greeks, who used methods of Euclidean geometry. A remarkable fact about conic sections is that, although they were studied thousands of years ago, they are far from obsolete. Indeed, they are important tools for present-day investigations in outer space and for the study of the behavior of atomic particles. Several applications of parabolas were mentioned in Section 4.1. There are also numerous applications involving ellipses and hyperbolas. For example, orbits of planets are ellipses. If the ellipse is very flat, the curve resembles the path of a comet. Elliptic gears or cams are sometimes used in machines. The hyperbola is useful for describing the path of an alpha particle in the electric field of the nucleus of an atom. The interested person can find many other applications of conic sections.

We know from our work in Section 4.1 that if $a \neq 0$, then the graph of $y = ax^2 + bx + c$ is a parabola that opens upward if $a > 0$ or downward if $a < 0$. By interchanging the variables x and y we obtain

$$x = ay^2 + by + c.$$

If $b = c = 0$, then $x = ay^2$ and the graph is a parabola with vertex at the origin, symmetric with respect to the x-axis, and opening to the right if $a > 0$ or to the left if $a < 0$, as illustrated in Figure 4.27.

FIGURE 4.27

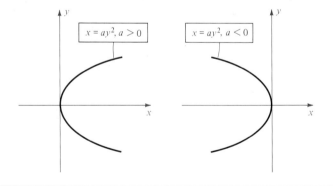

If $b = 0$ and $c \neq 0$, then $x = ay^2 + c$, and the graph may be obtained by shifting one of the graphs in Figure 4.27 to the right or left, depending on the sign of c.

If $b \neq 0$ we may complete the square, obtaining the following:

STANDARD EQUATION OF A PARABOLA (HORIZONTAL AXIS)

$$x - h = a(y - k)^2$$

This parabola has vertex (h, k) and opens to the right if $a > 0$ or to the left if $a < 0$.

EXAMPLE 1   Sketch the graph of the equation $x = 3y^2 + 8y - 3$.

SOLUTION   Since $a = 3 > 0$, the graph is a parabola that opens to the right. We complete the square as follows:

$$x = 3y^2 + 8y - 3$$
$$= 3(y^2 + \tfrac{8}{3}y) - 3$$
$$= 3(y^2 + \tfrac{8}{3}y + \tfrac{16}{9}) - 3 - \tfrac{16}{3}$$
$$= 3(y + \tfrac{4}{3})^2 - \tfrac{25}{3}.$$

FIGURE 4.28

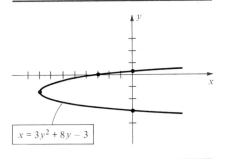

$x = 3y^2 + 8y - 3$

If we write the last equation as

$$x + \tfrac{25}{3} = 3(y + \tfrac{4}{3})^2$$

and compare with the standard equation we see that $h = -\tfrac{25}{3}$ and $k = -\tfrac{4}{3}$. Hence, the vertex is $(-\tfrac{25}{3}, -\tfrac{4}{3})$. To find the y-intercepts of the graph, we let $x = 0$ in the given equation and obtain

$$0 = 3y^2 + 8y - 3 = (3y - 1)(y + 3).$$

Thus, the y-intercepts are $\tfrac{1}{3}$ and -3. The x-intercept -3 is found by setting $y = 0$ in the equation of the parabola. Plotting the vertex and the points corresponding to the intercepts leads to the sketch in Figure 4.28.   ■

The graph of the equation

EQUATION OF AN ELLIPSE

$$\frac{x^2}{a^2} + \frac{y^2}{b^2} = 1$$

where a and b are positive real numbers, and $a \neq b$, is an **ellipse** with center at the origin. (If $a = b$ the graph is a circle.) To find the x-intercepts, we let $y = 0$, obtaining $\pm a$. Similarly, letting $x = 0$ gives us the y-intercepts $\pm b$. We may solve the equation for y in terms of x as follows:

$$\frac{y^2}{b^2} = 1 - \frac{x^2}{a^2} = \frac{a^2 - x^2}{a^2}$$

$$y^2 = \frac{b^2}{a^2}(a^2 - x^2)$$

$$y = \pm\frac{b}{a}\sqrt{a^2 - x^2}$$

To obtain points on the graph, the radicand $a^2 - x^2$ must be nonnegative. This will be true if $-a \le x \le a$. Consequently, the entire graph lies between the vertical lines $x = -a$ and $x = a$.

For each permissible value of x there correspond two values for y. Let us consider the nonnegative values given by

$$y = \frac{b}{a}\sqrt{a^2 - x^2}.$$

FIGURE 4.29

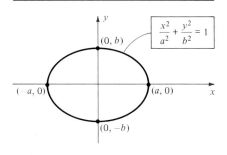

If we let x vary from $-a$ to 0, we see that y increases from 0 to b. As x varies from 0 to a, y decreases from b to 0. This gives us the upper half of the graph in Figure 4.29. The lower half is the graph of $y = (-b/a)\sqrt{a^2 - x^2}$. Note that the graph is symmetric with respect to the x-axis, the y-axis, and the origin.

The horizontal and vertical line segments that join the x- and y-intercepts in Figure 4.29 are called the **axes** of the ellipse. The longer axis is called the **major axis,** and the shorter axis is called the **minor axis** of the ellipse. The endpoints of the major axis are called the **vertices** of the ellipse.

Multiplying both sides of $(x^2/a^2) + (y^2/b^2) = 1$ by a^2b^2 we obtain

$$b^2x^2 + a^2y^2 = a^2b^2,$$

which may be written in the form

$$Ax^2 + By^2 = C$$

where A, B, and C are positive real numbers and $A \ne B$. Conversely, the graph of an equation of this type is an ellipse with center at the origin. Knowing this, the graph may be readily sketched by using the x- and y-intercepts and plotting several points.

EXAMPLE 2 Sketch the graph of $9x^2 + 16y^2 = 144$.

FIGURE 4.30

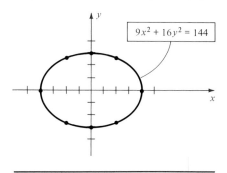

SOLUTION From the preceding discussion we know that the graph is an ellipse with center at the origin. The x-intercepts (found by letting $y = 0$) are ± 4, and the y-intercepts (found by letting $x = 0$) are ± 3. Let us next locate the points on the graph with x-coordinate 2. Substituting $x = 2$ in the given equation we obtain

$$9(2)^2 + 16y^2 = 144 \quad \text{or} \quad y^2 = \frac{108}{16} = \frac{27}{4}$$

and hence

$$y = \pm\frac{\sqrt{27}}{2} \approx 2.6.$$

Consequently, the points $(2, \pm\sqrt{27}/2)$ are on the graph. Similarly, $(-2, \pm\sqrt{27}/2)$ are solutions of the given equation. The graph is sketched in Figure 4.30. ∎

In Example 2 the major axis of the ellipse is on the x-axis. For the graph of $16x^2 + 9y^2 = 144$ sketched in Figure 4.31, the major axis is on the y-axis.

FIGURE 4.31

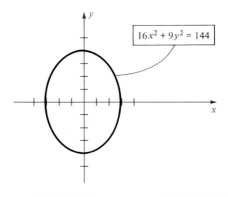

$16x^2 + 9y^2 = 144$

The graph of the equation

$$\frac{x^2}{a^2} - \frac{y^2}{b^2} = 1$$

EQUATION OF A HYPERBOLA

FIGURE 4.32

$$\frac{x^2}{a^2} - \frac{y^2}{b^2} = 1$$

$y = -\dfrac{b}{a}x$

$y = \dfrac{b}{a}x$

$W(0, b)$

$V'(-a, 0)$

$V(a, 0)$

$W'(0, -b)$

where a and b are positive real numbers, is a **hyperbola** with center at the origin. The graph has the general shape illustrated in Figure 4.32. (The significance of the dashes will be pointed out later.) Note that the x-intercepts are $\pm a$. The corresponding points $V(a, 0)$ and $V'(-a, 0)$ are called the **vertices**, and the line segment $V'V$ is known as the **transverse axis** of the hyperbola. There are no y-intercepts, since the equation $-y^2/b^2 = 1$ has no real solutions.

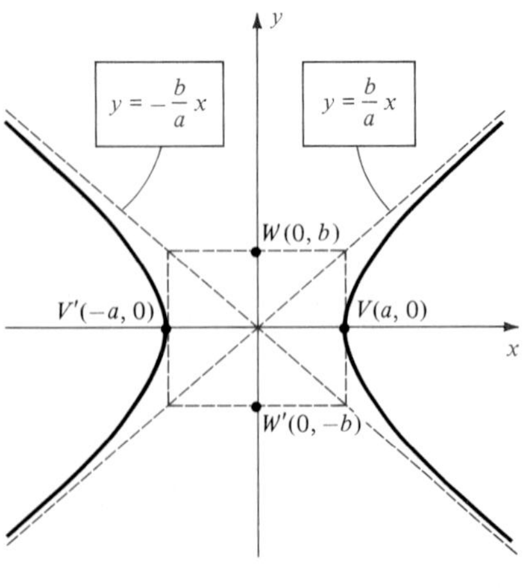

If the equation $(x^2/a^2) - (y^2/b^2) = 1$ is solved for y, we obtain

$$y = \pm \frac{b}{a}\sqrt{x^2 - a^2}.$$

There are no points (x, y) on the graph if $x^2 - a^2 < 0$, or equivalently, if $-a < x < a$. However, there *are* points $P(x, y)$ on the graph if $x \geq a$ or $x \leq -a$. If $x \geq a$, we may write the last equation in the form

$$y = \pm \frac{b}{a}\sqrt{x^2\left(1 - \frac{a^2}{x^2}\right)} = \pm \frac{b}{a}x\sqrt{1 - \frac{a^2}{x^2}}.$$

If x is large (in comparison to a), then $1 - (a^2/x^2) \approx 1$, and hence the y-coordinate of the point $P(x, y)$ on the hyperbola is close to either $(b/a)x$ or $-(b/a)x$. Thus, the point $P(x, y)$ is close to the line $y = (b/a)x$ when y is positive, or the line $y = -(b/a)x$ when y is negative. As x increases (or decreases), we say that the point $P(x, y)$ *approaches* one of these lines. A corresponding situation exists if $x \leq -a$. The lines with equations

$$y = \pm \frac{b}{a} x$$

are called the **asymptotes** of the hyperbola $(x^2/a^2) - (y^2/b^2) = 1$. The asymptotes serve as excellent guides for sketching the graph. This is illustrated in Figure 4.32 where we have represented the asymptotes by dashed lines. The two curves that make up the hyperbola are called the **branches** of the hyperbola.

A convenient way to sketch the asymptotes is first to plot the vertices $V(a, 0)$ and $V'(-a, 0)$ and the points $W(0, b)$ and $W'(0, -b)$ (see Figure 4.32). The line segment $W'W$ of length $2b$ is called the **conjugate axis** of the hyperbola. If horizontal and vertical lines are drawn through the endpoints of the conjugate annd transverse axes, respectively, then the diagonals of the resulting rectangle have slopes b/a and $-b/a$. Consequently, by extending these diagonals, we obtain the asymptotes. The hyperbola is then sketched using the asymptotes as a guide.

FIGURE 4.33

$$\frac{x^2}{4} - \frac{y^2}{9} = 1$$

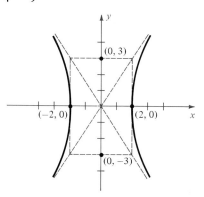

EXAMPLE 3   Discuss and sketch the graph of the equation

$$9x^2 - 4y^2 = 36.$$

SOLUTION   Dividing both sides by 36, we obtain

$$\frac{x^2}{4} - \frac{y^2}{9} = 1,$$

which is of the form shown in Figure 4.32 with $a^2 = 4$ and $b^2 = 9$. Hence, $a = 2$ and $b = 3$. The vertices $(\pm 2, 0)$ and the endpoints $(0, \pm 3)$ of the conjugate axis determine a rectangle whose diagonals (extended) give us the asymptotes. The graph is sketched in Figure 4.33. ∎

If we interchange x and y in the preceding discussion we obtain

$$\frac{y^2}{a^2} - \frac{x^2}{b^2} = 1.$$

In this case the graph is a hyperbola having vertices $(0, \pm a)$ and endpoints of conjugate axes $(\pm b, 0)$.

FIGURE 4.34

$4y^2 - 2x^2 = 1$

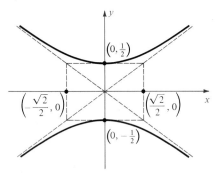

EXAMPLE 4 Discuss and sketch the graph of the equation

$$4y^2 - 2x^2 = 1.$$

SOLUTION We may rewrite the equation as

$$\frac{y^2}{\frac{1}{4}} - \frac{x^2}{\frac{1}{2}} = 1,$$

which is in the form $(y^2/a^2) - (x^2/b^2) = 1$ with $a^2 = \frac{1}{4}$, $b^2 = \frac{1}{2}$. Thus, $a = \frac{1}{2}$ and $b = \sqrt{2}/2 \approx 0.7$. The vertices are $(0, \pm\frac{1}{2})$ and the endpoints of the conjugate axes are $(\pm\sqrt{2}/2, 0)$. As in Example 3 we use these four points to construct a rectangle and then obtain the asymptotes from the diagonals. This leads to the sketch in Figure 4.34. ∎

Our discussion of ellipses and hyperbolas can be extended to the case where the center is any point (h, k) and the axes are parallel to the x- and y-axes. Two typical graphs are illustrated in Figure 4.35. Note that their equations may be obtained from those discussed earlier by replacing x by $x - h$ and y by $y - k$. We shall not prove these facts.

FIGURE 4.35

(i) $\dfrac{(x - h)^2}{a^2} + \dfrac{(y - k)^2}{b^2} = 1$ (ii) $\dfrac{(x - h)^2}{a^2} - \dfrac{(y - k)^2}{b^2} = 1$

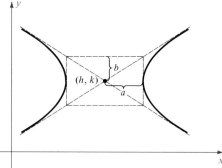

If $a > b$, as in Figure 4.35(i), the major axis of the ellipse is parallel to the x-axis; however, if $a < b$, the major axis is parallel to the y-axis. If we change the equation in (ii) to $(y - k)^2/a^2 - (x - h)^2/b^2 = 1$, we obtain a hyperbola with transverse axis parallel to the y-axis.

Squaring the terms $(x - h)^2$ and $(y - k)^2$ of the equations in Figure 4.35 and then simplifying leads to an equation of the form

$$Ax^2 + By^2 + Cx + Dy + F = 0.$$

For ellipses, A and B have the same sign, whereas for hyperbolas, A and B have opposite signs. Conversely, given such an equation, we may use the technique of completing squares to obtain the center (h, k) of the conic (whenever the graph exists).

EXAMPLE 5 Sketch the graph of the equation

$$16x^2 + 9y^2 + 64x - 18y - 71 = 0.$$

SOLUTION We begin by writing the equation in the form

$$16(x^2 + 4x) + 9(y^2 - 2y) = 71.$$

Next we complete the squares for the expressions within parentheses, obtaining

$$16(x^2 + 4x + 4) + 9(y^2 - 2y + 1) = 71 + 64 + 9.$$

Note that, by adding 4 to $x^2 + 4x$ we add 64 to the left side of the equation and hence must compensate by adding 64 to the right side. Similarly, adding 1 to the expression $y^2 - 2y$ adds 9 to the left side, and consequently 9 must also be added to the right side. The last equation may be written

$$16(x + 2)^2 + 9(y - 1)^2 = 144.$$

Dividing by 144 we obtain

$$\frac{(x + 2)^2}{9} + \frac{(y - 1)^2}{16} = 1,$$

which has the form shown in Figure 4.35(i) with $h = -2$ and $k = 1$. Thus, the graph is an ellipse with center $(-2, 1)$. Since $16 > 9$, the major axis is parallel to the y-axis. The lengths of *half* the major and minor axes are 4 and 3, respectively. This gives us the vertices indicated in Figure 4.36, and we obtain a rough sketch of the graph. If more accuracy is desired we could find the x- and y-intercepts, or plot several additional points. ∎

FIGURE 4.36

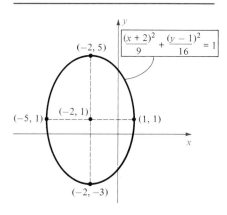

EXAMPLE 6 Discuss and sketch the graph of the equation

$$9x^2 - 4y^2 - 54x - 16y + 29 = 0.$$

SOLUTION We arrange our work as follows:

$$9(x^2 - 6x) - 4(y^2 + 4y) = -29$$

$$9(x^2 - 6x + 9) - 4(y^2 + 4y + 4) = -29 + 81 - 16$$

Note that, because of the -4 outside the second parentheses, adding 4 within that parentheses results in *subtracting* 16 from the left side of the equation. Hence, we must compensate by subtracting 16 from the right side. Next we write

$$9(x - 3)^2 - 4(y + 2)^2 = 36$$

$$\frac{(x - 3)^2}{4} - \frac{(y + 2)^2}{9} = 1,$$

which has the form shown in Figure 4.35(ii) with $h = 3$, $k = -2$, $a = 2$, and $b = 3$. Thus, the graph is a hyperbola with center $(3, -2)$ and transverse axis parallel to the x-axis.

Since $a = 2$ we may find the vertices by proceeding 2 units to the left and right of the center. The endpoints of the conjugate axis are found in like manner using $b = 3$. These four points may be used to construct a rectangle whose diagonals determine the asymptotes. We then sketch the graph as in Figure 4.37. ∎

FIGURE 4.37

$$\frac{(x - 3)^2}{4} - \frac{(y + 2)^2}{9} = 1$$

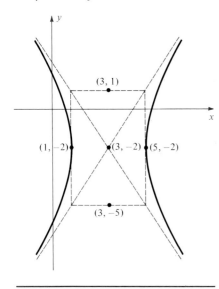

Exercises 4.7

Sketch the graphs of the equations in Exercises 1–48.

1 $x = 4y^2$

2 $x = -2y^2$

3 $x = 4y^2 - 3$

4 $x = 8 - 2y^2$

5 $x = 9 - y^2$

6 $x = y^2 + 4$

7 $x = y^2 + 4y + 1$

8 $x = y^2 + 6y + 5$

9 $x = -2y^2 + 12y - 13$

10 $x = 4y - y^2$

11 $y = x^2 - 4x - 6$

12 $y = 3x^2 - 30x + 75$

13 $x^2 + 5y^2 = 25$

14 $4x^2 + 25y^2 = 100$

15 $9x^2 + 4y^2 = 36$

16 $9x^2 + y^2 = 81$

17 $5x^2 + 3y^2 = 15$

18 $4x^2 + 3y^2 = 24$

19 $25x^2 + 9y^2 = 16$

20 $x^2 + 8y^2 = 4$

21 $\dfrac{x^2}{4} + \dfrac{y^2}{9} = 2$

22 $\dfrac{x^2}{4} + \dfrac{y^2}{4} = 1$

23 $9x^2 - 25y^2 = 225$

24 $16x^2 - 9y^2 = 144$

25 $25y^2 - 9x^2 = 225$

26 $16y^2 - 9x^2 = 144$

27 $y^2 - 36x^2 = 36$

28 $x^2 - 5y^2 = 10$

29 $x^2 - y^2 = 1$

30 $y^2 - x^2 = 4$

31 $4x^2 - 9y^2 = -1$

32 $y^2 - 25x^2 = 1$

33 $\dfrac{(x+5)^2}{16} + \dfrac{(y-4)^2}{9} = 1$

34 $8(x+3)^2 + (y-6)^2 = 32$

35 $\dfrac{(x-2)^2}{36} - \dfrac{(y+7)^2}{49} = 1$

36 $\dfrac{(y-2)^2}{9} - (x+4)^2 = 1$

37 $4(x+2)^2 + (y-2)^2 = 1$

38 $(x+1)^2 + 9y^2 = 36$

39 $16y^2 - 100(x-3)^2 = 1600$

40 $(x-7)^2 - (y+5)^2 = 1$

41 $4x^2 + y^2 + 24x - 10y + 45 = 0$

42 $9x^2 + 16y^2 + 36x + 96y + 36 = 0$

43 $9x^2 + y^2 - 108x - 4y + 319 = 0$

44 $x^2 + 4y^2 - 2x = 0$

45 $y^2 - 4x^2 + 6y - 40x - 107 = 0$

46 $25y^2 - 9x^2 - 100y - 54x + 10 = 0$

47 $9x^2 - y^2 - 36x + 12y - 9 = 0$

48 $4y^2 - x^2 + 32y - 8x + 49 = 0$

In Exercises 49 and 50 find an equation of an ellipse with center at the origin that has the given intercepts.

49 x-intercepts ± 10, y-intercepts ± 5

50 x-intercepts ± 8, y-intercepts $\pm\frac{1}{2}$

51 An arch of a bridge is semielliptical with its longer axis horizontal. The base of the arch is 30 feet across and the highest part of the arch is 10 feet above the horizontal roadway. Find the height of the arch 6 feet from the center of the base.

52 Determine A so that the point $(2, -3)$ is on the conic $Ax^2 + 2y^2 = 4$. Is the conic an ellipse or a hyperbola?

53 If a square with sides parallel to the coordinate axes is inscribed in the ellipse given by $(x^2/a^2) + (y^2/b^2) = 1$, express the area A of the square in terms of a and b.

54 The **eccentricity** e of the ellipse $(x^2/a^2) + (y^2/b^2) = 1$ is defined as the ratio $\sqrt{a^2 - b^2}/a$. Prove that $0 < e < 1$. If a is fixed and b varies, describe the general shape of the ellipse when the eccentricity is close to 1 and when it is close to 0.

55 The graphs of the equations

$$\frac{x^2}{a^2} - \frac{y^2}{b^2} = 1 \quad \text{and} \quad \frac{x^2}{a^2} - \frac{y^2}{b^2} = -1$$

are called **conjugate hyperbolas.** Sketch the graphs of both equations on the same coordinate plane if $a = 5$ and $b = 3$. Describe the relationship between the two graphs.

56 Find an equation of a hyperbola having x-intercepts ± 2 and asymptotes $y = \pm 3x$.

57 A line segment of length $a + b$ moves with its endpoints A and B attached to the coordinate axes, as illustrated in the figure. Prove that if $a \neq b$, then the point P traces an ellipse.

FIGURE FOR EXERCISE 57

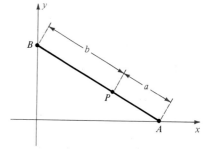

58 Consider the ellipse $px^2 + qy^2 = pq$ where $p > 0$ and $q > 0$. Prove that if m is any real number, there are exactly two lines of slope m that intersect the ellipse in precisely one point, and that equations of the lines are $y = mx \pm \sqrt{p + qm^2}$.

59 A point $P(x, y)$ is the same distance from the point $(4, 0)$ as it is from the circle $x^2 + y^2 = 4$. (Referring to the figure on page 238, $d_1 = d_2$.) Show that the collection of all such points forms a branch of a hyperbola and sketch its graph.

FIGURE FOR EXERCISE 59

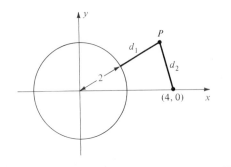

60 From a point P on the circle $x^2 + y^2 = 4$, a line segment is drawn perpendicular to the diameter AB, and the midpoint M is found (see figure). Find an equation of the collection of all such midpoints M and sketch the graph.

FIGURE FOR EXERCISE 60

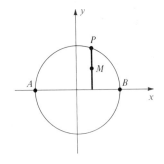

61 The physicist Ernest Rutherford discovered that when alpha particles are shot toward the nucleus of an atom, they are eventually repulsed away from the nucleus along hyperbolic paths. The figure illustrates the path of a particle that starts toward the origin along the line $y = \frac{1}{2}x$ and comes within 3 units of the nucleus. Find an equation of the path.

FIGURE FOR EXERCISE 61

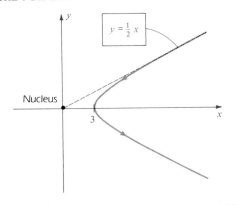

62 An Air Force jet is executing a high speed manuever along the path $2y^2 - x^2 = 8$. How close does the jet come to a town located at $(3, 0)$? (*Hint:* Let S denote the square of the distance from a point (x, y) on the path to $(3, 0)$ and find the minimum value of S. What is $\sqrt{S}$?)

FIGURE FOR EXERCISE 62

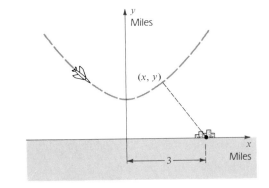

4.8 Review

Define or discuss each of the following.

1 Polynomial function

2 Quadratic function

3 Graph of a quadratic function

4 Maximum or minimum values of quadratic functions

Exercises 4.8

Sketch the graphs of the equations in Exercises 1–4.

1   $y = x^2 + 6x + 16$

2   $4x^2 + y = 10$

3   $4y = (x + 2)(x - 1)^2(3 - x)$

4   $y = \frac{1}{15}(x^5 - 20x^3 + 64x)$

In Exercises 5–16 sketch the graph of f.

5   $f(x) = (x - 4)^2$

6   $f(x) = 12x^2 + 5x - 3$

7   $f(x) = (x + 2)^3$

8   $f(x) = 2x^2 + x^3 - x^4$

9   $f(x) = x^3 + 2x^2 - 8x$

10 $f(x) = x^6 - 32$

11 $f(x) = \dfrac{-2}{(x + 1)^2}$

12 $f(x) = \dfrac{1}{(x - 1)^3}$

13 $f(x) = \dfrac{3x^2}{16 - x^2}$

14 $f(x) = \dfrac{x}{(x + 5)(x^2 - 5x + 4)}$

15 $f(x) = \dfrac{x^2 + 2x - 8}{x + 3}$

16 $f(x) = \dfrac{x^4 - 16}{x^3}$

In Exercises 17 and 18 find the maximum or minimum value of $f(x)$ by completing the square.

17 $f(x) = 5x^2 + 30x + 49$

18 $f(x) = -3x^2 + 30x - 82$

19 The interior of a half-mile race track consists of a rectangle with semicircles at two opposite ends. Find the dimensions that will maximize the area of the rectangle.

20 At 1:00 P.M. ship A is 30 miles due south of ship B and is sailing north at a rate of 15 mph. If ship B is sailing west at a rate of 10 mph, what is the time at which the distance between the ships is minimal (see figure)?

FIGURE FOR EXERCISE 20

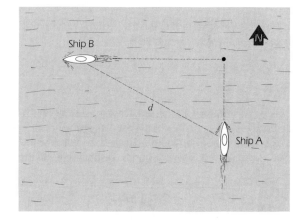

In Exercises 21–24 find the quotient and remainder if $f(x)$ is divided by $g(x)$.

21 $f(x) = 3x^5 - 4x^3 + x + 5$, $g(x) = x^3 - 2x + 7$

22 $f(x) = 7x^2 + 3x - 10$, $g(x) = x^3 - x^2 + 10$

23 $f(x) = 9x + 4$, $g(x) = 2x - 5$

24 $f(x) = 4x^3 - x^2 + 2x - 1$, $g(x) = x^2$

25 If $f(x) = -4x^4 + 3x^3 - 5x^2 + 7x - 10$, use the Remainder Theorem to find $f(-2)$.

26 Use the Remainder Theorem to prove that $x - 3$ is a factor of $2x^4 - 5x^3 - 4x^2 + 9$.

In Exercises 27 and 28 use synthetic division to find the quotient and remainder if $f(x)$ is divided by $g(x)$.

27 $f(x) = 6x^5 - 4x^2 + 8$, $g(x) = x + 2$

28 $f(x) = 2x^3 + 5x^2 - 2x + 1$, $g(x) = x - \sqrt{2}$

In Exercises 29 and 30 find polynomials with real coefficients that have the indicated zeros and degrees, and that satisfy the given conditions.

29 $-3 + 5i$, -1; degree 3; $f(1) = 4$

30 $1 - i$, 3, 0; degree 4; $f(2) = -1$

31 Find a polynomial of degree 7 such that -3 is a zero of multiplicity 2 and 0 is a zero of multiplicity 5.

32 Show that 2 is a zero of multiplicity 3 of the polynomial $x^5 - 4x^4 - 3x^3 + 34x^2 - 52x + 24$ and express this polynomial as a product of linear factors.

In Exercises 33 and 34 find the zeros of the polynomials and state the multiplicity of each zero.

33 $(x^2 - 2x + 1)^2(x^2 + 2x - 3)$

34 $x^6 + 2x^4 + x^2$

In Exercises 35 and 36 (a) use Descartes' Rule of Signs to determine the number of positive, negative, and nonreal complex solutions; and (b) find the smallest and largest integers that are upper and lower bounds, respectively, of the real solutions.

35 $2x^4 - 4x^3 + 2x^2 - 5x - 7 = 0$

36 $x^5 - 4x^3 + 6x^2 + x + 4 = 0$

37 Prove that $7x^6 + 2x^4 + 3x^2 + 10$ has no real zeros.

Find all solutions of the equations in Exercises 38–40.

38 $x^4 + 9x^3 + 31x^2 + 49x + 30 = 0$

39 $16x^3 - 20x^2 - 8x + 3 = 0$

40 $x^4 - 7x^2 + 6 = 0$

Sketch the graphs of the equations in Exercises 41–48.

41 $x^2 + 4y^2 = 64$

42 $4x^2 + y^2 = 64$

43 $x^2 - 4y^2 = 64$

44 $4y^2 - x^2 = 64$

45 $4x^2 + 9y^2 + 24x - 36y + 36 = 0$

46 $4x^2 + y^2 - 24x + 4y + 36 = 0$

47 $y^2 - 2x^2 + 6y + 8x - 3 = 0$

48 $4x^2 - y^2 - 40x - 8y + 88 = 0$

49 Find an equation of the ellipse that passes through the point (2, 3) and has $(\pm 5, 0)$ as endpoints of the major axis.

50 Find an equation of the hyperbola with vertices $(0, \pm 4)$ and asymptotes $y = \pm 2x$.

51 Show that the total volume V inside the shelter described in Exercise 39 of Section 3.9 is given by the equation $V = \frac{1}{3}x(400 - 16x)$ and find the design that maximizes the space inside the shelter.

52 A rocket is fired up a hillside, following a path given by $y = -0.016x^2 + 1.6x$. The hillside has slope $\frac{1}{5}$ as illustrated in the figure.

(a) Where does the rocket land?

(b) Find the maximum height of the rocket *above the ground*.

FIGURE FOR EXERCISE 52

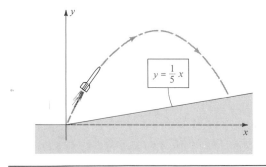

53 When a particular basketball player leaps straight up for a dunk, his distance $f(t)$ (in feet) off the ground after t

seconds is given by

$$f(t) = -\tfrac{1}{2}gt^2 + 16t.$$

(a) If $g = 32$, what is the player's *hang time;* that is, what is the total number of seconds that the player is in the air?

(b) Find the player's *vertical leap;* that is, find the maximum distance of his feet from the floor.

(c) On the moon $g = \tfrac{32}{6}$. Rework parts (a) and (b) for a player on the moon.

54 The cost $C(x)$ of cleaning up x percent of an oil spill that has washed ashore increases greatly as x approaches 100. Suppose that

$$C(x) = \frac{20x}{101 - x} \quad \text{(thousand dollars)}.$$

(a) Compare $C(100)$ to $C(90)$.

(b) Sketch the graph of C for $0 < x < 100$.

55 A bridge is to be constructed across a river that is 200 feet wide. The arch of the bridge is to be semielliptical and must be constructed so that a ship less than 50 feet wide and 30 feet high can pass safely through (see figure). Find an equation for the arch and calculate the height of the arch in the middle of the river.

FIGURE FOR EXERCISE 55

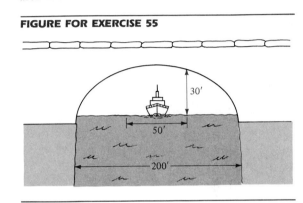

CHAPTER 5

Exponential and Logarithmic Functions

Exponential and logarithmic functions have applications in almost every field of human endeavor. ■ They are especially useful in the study of chemistry, biology, physics, and engineering to describe the manner in which quantities vary. ■ In this chapter we shall examine properties of these functions and consider many of their applications in everyday life.

5.1 Exponential Functions

Throughout this section the letter a will denote a positive real number. In Chapter 1 we defined a^r for every rational number r as follows: if m and n are integers with $n > 0$, then $a^{m/n} = \sqrt[n]{a^m}$. Using methods developed in calculus, we can define a^x for every *real* number x. To illustrate, for a^π we could use the nonterminating decimal representation $3.1415926\ldots$ for π and consider the following *rational* powers of a:

$$a^3, \quad a^{3.1}, \quad a^{3.14}, \quad a^{3.141}, \quad a^{3.1415}, \quad a^{3.14159}, \quad \ldots$$

If a^x is properly defined, then each successive power gets closer to a^π. In this chapter we shall assume that a^x can be obtained in similar fashion for every real number x and that the Laws of Exponents are valid in this more general setting.

It can be shown that the following result about rational exponents is also true for *real* exponents.

THEOREM

> If a is a real number such that $a > 1$, then:
>
> (i) $a^r > 1$ for every positive rational number r,
>
> (ii) if r and s are rational numbers such that $r < s$, then $a^r < a^s$.

PROOF

(i) Multiplying both sides of the inequality $a > 1$ by a, we obtain $a^2 > a$ and hence $a^2 > a > 1$. Thus, $a^2 > 1$. Multiplying both sides of $a^2 > 1$ by a gives us $a^3 > a$ and hence $a^3 > 1$. Continuing this process, we see that $a^p > 1$ for every positive integer p. (A rigorous proof of this fact requires the method of mathematical induction.) Similarly, if $0 < a \le 1$ it follows that $a^q \le 1$ for every positive integer q.

Now consider $r = p/q$ for positive integers p and q. If it were true that $a^{p/q} \le 1$, then from the previous discussion, $(a^{p/q})^q \le 1$ or $a^p \le 1$, which contradicts the fact that $a^p > 1$ for every positive integer p. Consequently, $a^r > 1$ for every positive rational number r.

(ii) If r and s are rational numbers such that $r < s$, then $s - r$ is a positive rational number, and hence from part (i), $1 < a^{s-r}$. Multiplying both sides of the last inequality by a^r,

$$a^r < (a^{s-r})a^r \quad \text{or} \quad a^r < a^s \qquad \square$$

Since to each real number x there corresponds a unique real number a^x, we can define a function as follows:

DEFINITION

> Let $a > 0$. The **exponential function** f **with base** a is defined by
>
> $$f(x) = a^x$$
>
> for every real number x.

FIGURE 5.1

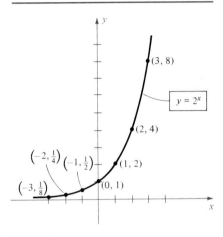

$y = 2^x$

If $a > 1$, and if x_1 and x_2 are real numbers such that $x_1 < x_2$, then $a^{x_1} < a^{x_2}$; that is, $f(x_1) < f(x_2)$. This means that if $a > 1$, then the exponential function f with base a is increasing throughout $\mathbb{R}$. It can also be shown that if $0 < a < 1$, then f is decreasing throughout $\mathbb{R}$.

EXAMPLE 1 Sketch the graph of f if $f(x) = 2^x$.

SOLUTION Coordinates of some points on the graph of $y = 2^x$ are listed in the following table.

x	-3	-2	-1	0	1	2	3	4
y	$\frac{1}{8}$	$\frac{1}{4}$	$\frac{1}{2}$	1	2	4	8	16

Plotting points and using the fact that f is increasing gives us the sketch in Figure 5.1. ∎

For any $a > 1$, the graph of the exponential function with base a has the general appearance of the graph in Figure 5.2(i); however, the exact

FIGURE 5.2

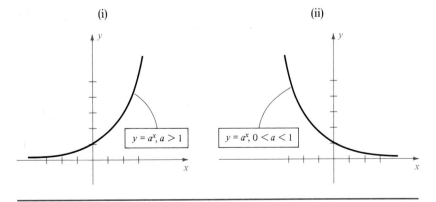

(i) $y = a^x, a > 1$

(ii) $y = a^x, 0 < a < 1$

shape depends on the value of a. If $0 < a < 1$, the graph has the appearance illustrated in (ii) of the figure. In both cases the domain of f is $\mathbb{R}$ and the range is the set of positive real numbers.

Since $a^0 = 1$, the y-intercept is always 1. If $a > 1$, then as x decreases through negative values, the graph approaches the x-axis but never intersects it, since $a^x > 0$ for all x. This means that the x-axis is a *horizontal asymptote* for the graph. As x increases through positive values, the graph rises very rapidly. Indeed, given $f(x) = 2^x$, if we begin with $x = 0$ and consider successive unit changes in x, then the corresponding changes in y are 1, 2, 4, 8, 16, 32, 64, and so on. This type of variation is characteristic of the **exponential law of growth**. In this case f is called a **growth function**. Figure 5.2(ii) illustrates **exponential decay**.

FIGURE 5.3

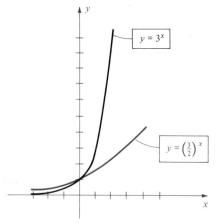

EXAMPLE 2 If $f(x) = \left(\tfrac{3}{2}\right)^x$ and $g(x) = 3^x$, sketch the graphs of f and g on the same coordinate plane.

SOLUTION The following table displays coordinates of several points on the graphs.

x	-2	-1	0	1	2	3	4
$\left(\tfrac{3}{2}\right)^x$	$\tfrac{4}{9} \approx .4$	$\tfrac{2}{3} \approx .7$	1	$\tfrac{3}{2}$	$\tfrac{9}{4} \approx 2.3$	$\tfrac{27}{8} \approx 3.4$	$\tfrac{81}{16} \approx 5.1$
3^x	$\tfrac{1}{9} \approx .1$	$\tfrac{1}{3} \approx .3$	1	3	9	27	81

Plotting points we obtain Figure 5.3, in which color has been used for the graph of f to distinguish it from the graph of g. ∎

Example 2 illustrates the fact that if $1 < a < b$, then $a^x < b^x$ for positive values of x and $b^x < a^x$ for negative values of x. In particular, since $\tfrac{3}{2} < 2 < 3$, the graph of $y = 2^x$ in Example 1 lies between the graphs of f and g in Example 2.

FIGURE 5.4

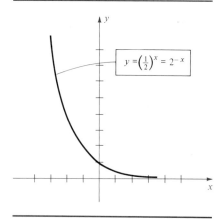

EXAMPLE 3 Sketch the graph of the equation $y = \left(\tfrac{1}{2}\right)^x$.

SOLUTION Some points on the graph may be obtained from the following table:

x	-3	-2	-1	0	1	2	3
$\left(\tfrac{1}{2}\right)^x$	8	4	2	1	$\tfrac{1}{2}$	$\tfrac{1}{4}$	$\tfrac{1}{8}$

The graph is sketched in Figure 5.4. Since $\left(\tfrac{1}{2}\right)^x = 2^{-x}$, the graph is the same as the graph of the equation $y = 2^{-x}$. ∎

In advanced mathematics and applications it is often necessary to consider functions such that $f(x) = a^p$, where p is some expression in x. The next example illustrates the case $p = -x^2$.

EXAMPLE 4 Sketch the graph of f if $f(x) = 2^{-x^2}$.

SOLUTION If we rewrite $f(x)$ as

$$f(x) = \frac{1}{2^{(x^2)}}$$

then it is evident that as $|x|$ increases, the point $(x, f(x))$ approaches the x-axis. Thus, the x-axis is a horizontal asymptote for the graph. The maximum value of $f(x)$ occurs at $x = 0$. Since f is an even function, the graph is symmetric with respect to the y-axis. Several points on the graph are $(0, 1)$, $(1, \frac{1}{2})$, and $(2, \frac{1}{16})$. Plotting and using symmetry gives us the sketch in Figure 5.5. Functions similar to f arise in the branch of mathematics called *probability*. ∎

FIGURE 5.5

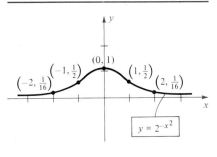

$y = 2^{-x^2}$

APPLICATION: **Bacterial Growth**

Exponential functions occur in the study of the growth of certain populations. As an illustration, it might be observed experimentally that the number of bacteria in a culture doubles every hour. If 1000 bacteria are present at the start of the experiment, then the experimenter would obtain the readings listed below, where t is the time in hours and $f(t)$ is the bacteria count at time t.

t (time)	0	1	2	3	4
$f(t)$ (bacteria count)	1000	2000	4000	8000	16,000

It appears that $f(t) = (1000)2^t$. With this formula we can predict the number of bacteria present at any time t. For example, at $t = 1.5 = \frac{3}{2}$,

$$f(t) = (1000)2^{3/2} \approx 2828.$$

The graph of f is sketched in Figure 5.6. ∎∎∎

FIGURE 5.6

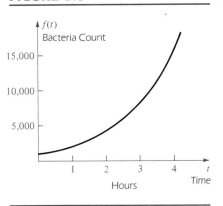

APPLICATION: **Radioactive Decay**

Certain physical quantities *decrease* exponentially. In such cases if a is the base of the exponential function, then $0 < a < 1$. One of the most

5 EXPONENTIAL AND LOGARITHMIC FUNCTIONS

FIGURE 5.7

Decay of polonium

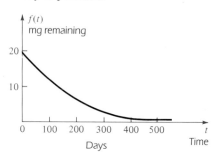

common examples is the decay of a radioactive substance. As an illustration, the polonium isotope ^{210}Po has a half-life of approximately 140 days; that is, given any amount, one-half of it will disintegrate in 140 days. If 20 mg of ^{210}Po is present initially, then the following table indicates the amount remaining after various intervals of time.

t (days)	0	140	280	420	560
Amount remaining (mg)	20	10	5	2.5	1.25

The sketch in Figure 5.7 illustrates the exponential nature of the disintegration. ■ ■ ■

APPLICATION: **Compound Interest**

Compound interest provides a good illustration of exponential growth. If a sum of money P, called the **principal,** is invested at a *simple* interest rate r, then the interest at the end of one interest period is the product Pr when r is expressed as a decimal. For example, if $P = \$1000$ and the interest rate is 9% per year, then $r = 0.09$, and the interest at the end of one year is $1000(0.09)$, or $90.

If the interest is reinvested at the end of this period, then the new principal is

$$P + Pr \quad \text{or} \quad P(1 + r).$$

Note that to find the new principal we multiply the original principal by $(1 + r)$. In the preceding illustration the new principal is $1000(1.09)$, or $1090.

After another time period has elapsed, the new principal may be found by multiplying $P(1 + r)$ by $(1 + r)$. Thus, the principal after two time periods is $P(1 + r)^2$. If we continue to reinvest, the principal after three periods is $P(1 + r)^3$; after four it is $P(1 + r)^4$; and in general, the amount A invested after k time periods is

$$A = P(1 + r)^k.$$

Interest accumulated by means of this formula is called **compound interest.** Note that A is expressed in terms of an exponential function with base $1 + r$. The time period may vary and may be measured in years, months, weeks, days, or any other suitable unit of time. When applying the formula for A, remember that r is the interest rate per time period expressed as a decimal. For example, if the rate is stated as 6% *per year compounded* *monthly*, then the rate per month is $\frac{6}{12}\%$, or equivalently, 0.5%. Thus, $r = 0.005$ and k is the number of months. If $100 is invested at this rate, then the formula for A is

$$A = 100(1 + 0.005)^k = 100(1.005)^k$$

Generally, suppose that r is the yearly interest rate (expressed as a decimal) and that interest is compounded n times per year. The interest rate per time period is r/n. If the principal P is invested for t years, then the number of interest periods is nt, and the amount A after t years is given by the following formula.

**COMPOUND
INTEREST
FORMULA**

$$A = P\left(1 + \frac{r}{n}\right)^{nt}$$

■ ■ ■

EXAMPLE 5   Suppose that $1000 is invested at an interest rate of 9% compounded monthly. Find the new amount of principal after 5 years; after 10 years; after 15 years. Illustrate graphically the growth of the investment.

FIGURE 5.8

Compound interest: $A = 1000(1.0075)^{12t}$

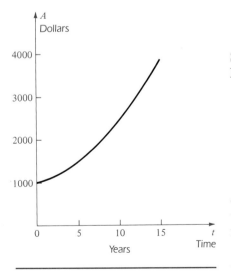

Years

Time

SOLUTION   Applying the Compound Interest Formula with $r = 0.09$, $n = 12$, and $P = \$1000$, the amount after t years is

$$A = 1000\left(1 + \frac{0.09}{12}\right)^{12t} = 1000(1.0075)^{12t}.$$

Substituting $t = 5$, 10, and 15, and using a calculator, we obtain the following amounts:

$$\text{After 5 years:} \quad A = 1000(1.0075)^{60} = \$1565.68$$
$$\text{After 10 years:} \quad A = 1000(1.0075)^{120} = \$2451.36$$
$$\text{After 15 years:} \quad A = 1000(1.0075)^{180} = \$3838.04$$

The exponential nature of the increase is indicated by the fact that during the first five years, the growth in the investment is $565.68; during the second five-year period, the growth is $885.68; and during the last five-year period, it is $1368.68.

The sketch in Figure 5.8 illustrates the growth of $1000 invested over a period of 15 years.

■

Exercises 5.1

In Exercises 1–24 sketch the graph of the function f.

1 $f(x) = 4^x$

2 $f(x) = 5^x$

3 $f(x) = 10^x$

4 $f(x) = 8^x$

5 $f(x) = 3^{-x}$

6 $f(x) = 4^{-x}$

7 $f(x) = -2^x$

8 $f(x) = -3^x$

9 $f(x) = 4 - 2^{-x}$

10 $f(x) = 2 + 3^{-x}$

11 $f(x) = (\frac{2}{3})^x$

12 $f(x) = (\frac{3}{4})^{-x}$

13 $f(x) = (\frac{5}{2})^{-x}$

14 $f(x) = (\frac{4}{3})^x$

15 $f(x) = 2^{|x|}$

16 $f(x) = 2^{-|x|}$

17 $f(x) = 2^{x+3}$

18 $f(x) = 3^{x+2}$

19 $f(x) = 2^{3-x}$

20 $f(x) = 3^{-2-x}$

21 $f(x) = 3^{1-x^2}$

22 $f(x) = 2^{-(x+1)^2}$

23 $f(x) = 3^x + 3^{-x}$

24 $f(x) = 3^x - 3^{-x}$

25 One hundred elk, each one year old, are introduced into a game preserve. The number $N(t)$ still alive after t years is predicted to be $N(t) = 100(0.9)^t$. Estimate the number of elk still alive after (a) 1 year; (b) 5 years; (c) 10 years.

26 A drug is eliminated from the body through urine. The initial dose is 10 mg and the amount $A(t)$ in the body t hours later is given by $A(t) = 10(0.8)^t$.

(a) Estimate the amount of the drug in the body 8 hours after the initial dose.

(b) What percentage of the drug still in the body is eliminated each hour?

27 The number of bacteria in a certain culture increased from 600 to 1800 between 7:00 A.M. and 9:00 A.M. Assuming exponential growth and using methods of calculus, it can be shown that the number $f(t)$ of bacteria t hours after 7:00 A.M. was given by $f(t) = 600(3)^{t/2}$.

(a) Estimate the number of bacteria in the culture at 8:00 A.M.; 10:00 A.M.; 11:00 A.M.

(b) Sketch the graph of f from $t = 0$ to $t = 4$.

28 According to Newton's Law of Cooling, the rate at which an object cools is directly proportional to the difference in temperature between the object and the surrounding

medium. If a certain object cools from $125°$ to $100°$ in 30 minutes when surrounded by air that has a temperature of $75°$, then its temperature $f(t)$ after t hours of cooling is given by $f(t) = 50(2)^{-2t} + 75$.

(a) Assuming $t = 0$ corresponds to 1:00 P.M., approximate to the nearest tenth of a degree the temperature at 2:00 P.M., 3:30 P.M., and 4:00 P.M.

(b) Sketch the graph of f from $t = 0$ to $t = 4$.

29 The radioactive isotope ^{210}Bi has a half-life of 5 days; that is, the number of radioactive particles will decrease to one-half the number in 5 days. If there are 100 mg of ^{210}Bi present at $t = 0$, then the amount $f(t)$ remaining after t days is given by $f(t) = 100(2)^{-t/5}$.

(a) How much ^{210}Bi remains after 5 days? 10 days? 12.5 days?

(b) Sketch the graph of f from $t = 0$ to $t = 30$.

30 An important problem in oceanography is to determine the amount of light that can penetrate to various ocean depths. The Beer-Lambert Law asserts that an exponential function I such that $I(x) = I_0 a^x$ should be used to model this phenomenon. Assuming $I(x) = 10(0.4)^x$ is the amount of light (in calories/cm^2/second) reaching a depth of x meters,

(a) What is the amount of light at a depth of 2 meters?

(b) Sketch the graph of I from $x = 0$ to $x = 5$.

31 The half-life of radium is 1600 years; that is, given any quantity, one-half of it will disintegrate in 1600 years. If the initial amount is q_0 milligrams, then the quantity $q(t)$ remaining after t years is given by $q(t) = q_0 2^{kt}$. Find k.

32 If 10 grams of salt are added to a quantity of water, then the amount $q(t)$ that is undissolved after t minutes is given by $q(t) = 10(\frac{4}{5})^t$. Sketch a graph that shows the value $q(t)$ at any time from $t = 0$ to $t = 10$.

33 If $1000 is invested at a rate of 12% per year compounded monthly, what is the principal after (a) 1 month? (b) 2 months? (c) 6 months? (d) 1 year?

34 If a savings fund pays interest at a rate of 10% compounded semiannually, how much money invested now will amount to $5000 after one year?

35 If a certain make of automobile is purchased for C dollars, then its trade-in value $v(t)$ at the end of t years is given by $v(t) = 0.78C(0.85)^{t-1}$. If the original cost is $10,000, calculate to the nearest dollar the value after (a) 1 year; (b) 4 years; (c) 7 years.

36 If the value of real estate increases at a rate of 10% per year, then after t years the value V of a house purchased for P dollars is given by $V = P(1.1)^t$. If a house is purchased for $80,000 in 1986, what will it be worth in 1990?

37 Why was $a < 0$ ruled out in the discussion of a^x?

38 Prove that if $0 < a < 1$ and r and s are rational numbers such that $r < s$, then $a^r > a^s$.

39 How does the graph of $y = a^x$ compare with the graph of $y = -a^x$?

40 If $a > 1$, how does the graph of $y = a^x$ compare with the graph of $y = a^{-x}$?

Calculator Exercises 5.1

In Exercises 1 and 2 sketch the graph for $-3 \le x \le 3$ by choosing values of x at intervals of length 0.5; that is, $x = -3$, -2.5, -2, -1.5, and so on.

1 $y = (1.8)^x$

2 $y = (2.3)^{-x^2}$

Solve Exercises 3–8 by using the Compound Interest Formula.

3 Assuming $1000 is invested at an interest rate of 6% per year compounded quarterly, find the principal at the end of (a) one year; (b) two years; (c) five years; (d) ten years.

4 Rework Exercise 3 for an interest rate of 6% per year compounded monthly.

5 If $10,000 is invested at a rate of 9% per year compounded semiannually, how long will it take for the principal to exceed (a) $15,000? (b) $20,000? (c) $30,000?

6 A certain department store requires its credit card customers to pay interest at the rate of 18% per year, compounded monthly, on any unpaid bills. If a man buys a television set for $500 on credit and then makes no payments for one year, how much does he owe at the end of the year?

7 A boy deposits $500 in a savings account that pays interest at a rate of 6% per year compounded weekly. How much is in the account after one year?

8 A savings fund pays interest at the rate of 9% compounded daily. How much should be invested in order to have $2000 at the end of 10 weeks?

5.2 The Natural Exponential Function

At the end of Section 5.1 we discussed the *Compound Interest Formula*

$$A = P\left(1 + \frac{r}{n}\right)^{nt}$$

where P is the principal invested, r is the interest rate (expressed as a decimal), n is the number of interest periods per year, and t is the number of years that the principal is invested. The next example illustrates what happens if the rate and total time invested are fixed, but the *time period* for compounding interest is varied.

EXAMPLE 1 Suppose $1000 is invested at a compound interest rate of 9%. Find the new amount of principal after one year if the interest is compounded monthly; weekly; daily; hourly; each minute.

SOLUTION If we let $P = \$1000$, $t = 1$, and $r = 0.09$ in the Compound Interest Formula, then

$$A = 1000\left(1 + \frac{0.09}{n}\right)^n$$

for n interest periods per year. To find the desired amounts, we let n have the following values:

$$12, \quad 52, \quad 365, \quad 8760, \quad 525{,}600.$$

We have assumed there are 365 days in a year and, hence, $(365)(24) = 8760$ hours, and $(8760)(60) = 525{,}600$ minutes. (Actually, in business transactions an investment year is considered to be 360 days). Using the Compound Interest Formula (and a calculator), we obtain the following table:

Time period for compounding interest	Amount of principal after one year
Month	$1000\left(1 + \dfrac{0.09}{12}\right)^{12} = \$1093.81$
Week	$1000\left(1 + \dfrac{0.09}{52}\right)^{52} = \$1094.09$
Day	$1000\left(1 + \dfrac{0.09}{365}\right)^{365} = \$1094.16$
Hour	$1000\left(1 + \dfrac{0.09}{8760}\right)^{8760} = \$1094.17$
Minute	$1000\left(1 + \dfrac{0.09}{525{,}600}\right)^{525{,}600} = \$1094.17$

Note that, in the preceding example, after a certain time period is reached, the number of interest periods per year has little effect on the final amount. If interest had been compounded each *second*, the result would still be $1094.17. Thus, the amount approaches a fixed value as n increases. Interest is said to be **compounded continuously** if the number n of time periods per year increases without bound. Evidently, if we allow this to happen in Example 1, the amount of principal after one year is the same as that obtained for a time period of one hour, or of one minute.

If we let $P = 1$, $r = 1$, and $t = 1$ in the Compound Interest Formula, we obtain

$$A = \left(1 + \frac{1}{n}\right)^n.$$

The expression on the right of the equation occurs in the study of calculus. In Example 1, in a similar situation, as n increased, A approached a limiting value. The same phenomenon occurs here, as illustrated by the following table, which was obtained using a calculator.

n	Approximation to $\left(1 + \dfrac{1}{n}\right)^n$
1	2.0000000
10	2.5937425
100	2.7048138
1000	2.7169238
10,000	2.7181459
100,000	2.7182546
1,000,000	2,7182818
10,000,000	2.7182818

It can be proved that as n increases, $[1 + (1/n)]^n$ gets closer and closer to a certain irrational number, denoted by e. As indicated by the values in the table, e can be assigned the following decimal approximation:

$$e \approx 2.71828$$

The number e arises naturally in the investigation of many physical phenomena. For this reason, the function f defined by $f(x) = e^x$ is called the **natural exponential function.** It is one of the most important functions that occurs in advanced mathematics and applications. Since $2 < e < 3$, the graph of $y = e^x$ lies "between" the graphs of $y = 2^x$ and $y = 3^x$, as shown in Figure 5.9.

A brief table of values of e^x and e^{-x} is given in Table 2 of Appendix II. Some calculators have an $\boxed{e^x}$ key for approximating values of the natural

FIGURE 5.9

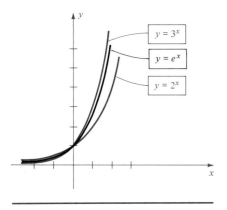

$y = 3^x$

$y = e^x$

$y = 2^x$

exponential function. There may also be a $\boxed{y^x}$ key that can be used for any positive base y. Approximations to e^x can then be found by calculating $(2.71828)^x$.

EXAMPLE 2   Verify the graph of $y = e^x$ sketched in Figure 5.9 by plotting a sufficient number of points.

SOLUTION   The following table may be obtained by using a calculator or Table 2 of Appendix II, and rounding off values of e^x to two decimal places.

x	-2.0	-1.5	-1.0	-0.5	0	0.5	1.0	1.5	2.0
e^x (approx.)	0.14	0.22	0.37	0.61	1.00	1.65	2.72	4.48	7.39

Plotting points and using the fact that the exponential function with base e is increasing leads to the graph of $y = e^x$ sketched in Figure 5.9.   ∎

EXAMPLE 3   Sketch the graph of f if

$$f(x) = \frac{e^x + e^{-x}}{2}.$$

SOLUTION   Note that f is an even function, because

$$f(-x) = \frac{e^{-x} + e^{-(-x)}}{2} = \frac{e^{-x} + e^x}{2} = f(x).$$

Thus, the graph is symmetric with respect to the y-axis (see page 123). Using a calculator or Table 2 of Appendix II, we obtain the following approximations to $f(x)$:

x	0	0.5	1.0	1.5	2.0
$f(x)$ (approx.)	1	1.13	1.54	2.35	3.76

Plotting points and using symmetry with respect to the y-axis gives us the sketch in Figure 5.10.   ∎

FIGURE 5.10

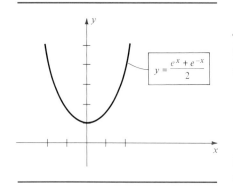

$y = \dfrac{e^x + e^{-x}}{2}$

APPLICATION: **Flexible Cables**

The function f of Example 3 is important in applied mathematics and engineering, where it is called the **hyperbolic cosine function.** This function

FIGURE 5.11

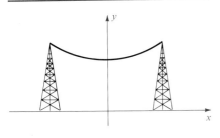

can be used to describe the shape of a uniform flexible cable, or chain, whose ends are supported from the same height. This is often the case for telephone or power lines, as illustrated in Figure 5.11. If we introduce a coordinate system as indicated in the figure, then it can be shown that an equation that corresponds to the shape of the cable is $y = (a/2)(e^{x/a} + e^{-x/a})$ where a is a real number. The graph is called a **catenary,** after the Latin word for *chain*. Note that the function in Example 3 is the special case in which $a = 1$. ■ ■ ■

APPLICATION: **Radiotherapy**

One of the many fields in which exponential functions with base e play an important role is *radiotherapy*, the treatment of tumors by radiation. Of major interest is the fraction of a tumor population that survives a treatment. This *surviving fraction* depends not only on the energy and nature of the radiation, but also on the depth, size, and characteristics of the tumor itself. The exposure to radiation may be thought of as a number of potentially damaging events, where only one "hit" is required to kill a tumor cell. Suppose that each cell has exactly one "target" that must be hit. If k denotes the average target size of a tumor cell, and if x is the number of damaging events (the *dose*), then the surviving fraction $f(x)$ is

$$f(x) = e^{-kx}.$$

This is called the *one-target–one-hit surviving fraction*.

Next, suppose that each cell has n targets and that hitting any one of the targets results in the death of a cell. In this case, the *n-target–one-hit surviving fraction* is

$$f(x) = 1 - (1 - e^{-kx})^n.$$

FIGURE 5.12

Surviving fraction of tumor cells after a radiation treatment

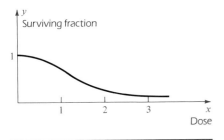

The graph of f may be analyzed to determine what effect increasing the dosage x will have on decreasing the surviving fraction of tumor cells. Note that $f(0) = 1$; that is, if there is no dose, then all cells survive. As a special case, if $k = 1$ and $n = 2$, then

$$f(x) = 1 - (1 - e^{-x})^2 = 1 - (1 - 2e^{-x} + e^{-2x})$$
$$= 2e^{-x} - e^{-2x}.$$

A complete analysis of the graph of f requires methods of calculus. It can be shown that the graph has the shape indicated in Figure 5.12. The "shoulder" on the curve near the point $(0, 1)$ represents the threshold nature of the treatment; that is, a small dose results in very little tumor elimination. Note that for a large x, an increase in dosage has little effect on the surviving fraction. To determine the ideal dose that should be administered to a patient, specialists in radiation therapy must also take into account the number of healthy cells that are killed during a treatment.

■ ■ ■

Problems of the type illustrated in the next example occur in the study of calculus. (See also Exercises 9–12.)

EXAMPLE 4 Find the zeros of f if $f(x) = x^2(-2e^{-2x}) + 2xe^{-2x}$.

SOLUTION We may factor $f(x)$ as follows:

$$f(x) = 2xe^{-2x} - 2x^2e^{-2x}$$
$$= 2xe^{-2x}(1 - x).$$

To find the zeros of f, we must solve the equation $f(x) = 0$. Since $e^{-2x} > 0$ for all x, it follows that $f(x) = 0$ if and only if $x = 0$ or $1 - x = 0$. Thus, the zeros of f are 0 and 1. ■

Exercises 5.2

In Exercises 1–7 use your knowledge of the graph of $y = e^x$ to help you sketch the graph of f.

1 (a) $f(x) = e^{-x}$ (b) $f(x) = -e^x$

2 (a) $f(x) = e^{2x}$ (b) $f(x) = 2e^x$

3 (a) $f(x) = e^{x+4}$ (b) $f(x) = e^x + 4$

4 (a) $f(x) = e^{-2x}$ (b) $f(x) = -2e^x$

5 $f(x) = \dfrac{e^{2x} + e^{-2x}}{2}$

6 $f(x) = \dfrac{e^x - e^{-x}}{2}$

7 $f(x) = \dfrac{2}{e^x + e^{-x}}$

(*Hint:* Take reciprocals of y-coordinates in Example 3.)

8 In statistics the **normal distribution function** is defined by

$$f(x) = \frac{1}{\sigma\sqrt{2\pi}} e^{(-1/2)[(x-\mu)/\sigma]^2}$$

for real numbers μ and $\sigma > 0$. (μ is called the *mean* and σ is called the *variance* of the distribution.) Sketch the graph of f for the case $\sigma = 1$ and $\mu = 0$.

In Exercises 9–12 find the zeros of f.

9 $f(x) = xe^x + e^x$

10 $f(x) = -x^2e^{-x} + 2xe^{-x}$

11 $f(x) = x^3(4e^{4x}) + 3x^2e^{4x}$

12 $f(x) = x^2(2e^{2x}) + 2xe^{2x} + e^{2x} + 2xe^{2x}$

Simplify the expressions in Exercises 13 and 14.

13 $\dfrac{(e^x + e^{-x})(e^x + e^{-x}) - (e^x - e^{-x})(e^x - e^{-x})}{(e^x + e^{-x})^2}$

14 $\dfrac{(e^x - e^{-x})^2 - (e^x + e^{-x})^2}{(e^x + e^{-x})^2}$

Calculator Exercises 5.2

1 An exponential function W such that $W(t) = W_0 e^{kt}$ (for $k > 0$) describes the first month of growth for crops such as maize, cotton, and soybeans. Here $W(t)$ is the total weight in mg, W_0 is the weight on the day of emergence, and t is the time in days. If, for a species of soybean, $k = 0.2$ and $W_0 = 68$ mg, predict the weight at the end of the month ($t = 30$).

2 Refer to Exercise 1. It is often difficult to measure the weight W_0 of the plant when it first emerges from the soil. If, for a species of cotton, $k = 0.21$ and $W(10) = 575$ mg, estimate W_0.

3 The 1980 population of the United States was approximately 227 million, and the population has been growing at a rate of 0.7% per year. It is possible to show, using calculus, that the population $N(t)$, t years later, may be approximated by $N(t) = 227e^{0.007t}$. If this growth trend continues, predict the population in the year 2000.

4 The 1980 population estimate for India was 651 million, and the population has been growing at a rate of about 2% per year. The population $N(t)$, t years later, may be approximated by $N(t) = 651e^{0.02t}$. Assuming that this rapid growth rate continues, estimate the population of India in the year 2000.

5 In fishery science, the collection of fish that results from one annual reproduction is referred to as a *cohort*. It is usually assumed that the number $N(t)$ still alive after t years is given by an exponential function. For Pacific halibut, $N(t) = N_0 e^{-0.2t}$ where N_0 is the initial size of the cohort. What percentage of the original number is still alive after 10 years?

6 The radioactive tracer ^{51}Cr can be used to locate the position of the placenta in a pregnant woman. Often the

tracer must be ordered from a medical lab. If A_0 units (microcuries) are shipped, then because of radioactive decay, the number of units $A(t)$ present after t days is given by $A(t) = A_0 e^{-0.0249t}$. If 35 units are shipped and it takes 2 days for the tracer to arrive, how many units are then available for the test? If 35 units are needed for the test, how many units should be shipped?

7 In 1966, the International Whaling Commission protected the world population of blue whales from hunting. In 1978, the population in the southern hemisphere was thought to number 5000. Now without predators and with an abundant food supply, the population $N(t)$ is expected to grow exponentially according to the formula $N(t) = 5000e^{0.047t}$ where t is in years. Predict the population in (a) 1990; (b) 2000.

8 The length (in cm) of many common commercial fish t years old is closely approximated by a von-Bertalanffy growth function $f(t) = a(1 - be^{-kt})$ where a, b, and k are constants.
  (a) For Pacific halibut, $a = 200$, $b = 0.956$, and $k = 0.18$. Estimate the length of a typical 10-year-old halibut.
  (b) What is the interpretation of the constant a in the formula?

9 Under certain conditions the atmospheric pressure p (in inches) at altitude h feet is given by $p = 29e^{-0.000034h}$. What is the pressure at an altitude of 40,000 feet?

10 Starting with c milligrams of the polonium isotope ^{210}Po, the amount remaining after t days may be approximated by $A = ce^{-0.00495t}$. If the initial amount is 50 milligrams, find, to the nearest hundredth, the amount remaining after (a) 30 days; (b) 180 days; (c) 365 days.

5.3 Logarithmic Functions

If $f(x) = a^x$ and $a > 1$, then f is increasing throughout $\mathbb{R}$, whereas if $0 < a < 1$, then f is decreasing (see Figure 5.2). Thus, if $a > 0$ and $a \neq 1$, then f is a one-to-one function and hence has an inverse function f^{-1}

(see Section 3.7). The inverse of the exponential function with base a is called the **logarithmic function with base a** and is denoted by $\log_a$. Its values are denoted by $\log_a (x)$ or $\log_a x$, read "**the logarithm of x with base a.**" Since

$$y = f^{-1}(x) \quad \text{if and only if} \quad x = f(y),$$

the definition of $\log_a$ may be expressed as follows:

DEFINITION OF $\log_a x$

$$y = \log_a x \quad \text{if and only if} \quad x = a^y.$$

Since the domain and range of the exponential function with base a are $\mathbb{R}$ and the positive real numbers, respectively, the domain of its inverse $\log_a$ is the positive real numbers and the range is $\mathbb{R}$. Thus, in the definition, $x > 0$ and y is in $\mathbb{R}$.

Note that

$$\text{if} \quad y = \log_a x, \quad \text{then} \quad x = a^y = a^{\log_a x}.$$

In words, $\log_a x$ *is the exponent to which a must be raised in order to obtain x.* As illustrations,

$$\log_2 8 = 3 \qquad \text{since} \qquad 2^3 = 8$$
$$\log_5 \tfrac{1}{25} = -2 \qquad \text{since} \qquad 5^{-2} = \tfrac{1}{25}$$
$$\log_{10} 10{,}000 = 4 \quad \text{since} \quad 10^4 = 10{,}000.$$

The next theorem is an immediate consequence of the definition of logarithm.

THEOREM

(i) $\quad a^{\log_a x} = x \quad$ for every $x > 0$

(ii) $\quad \log_a a = 1$

(iii) $\quad \log_a 1 = 0$

We have already proved (i). To prove (ii) and (iii) it is sufficient to note that $a^1 = a$ and $a^0 = 1$, respectively.

EXAMPLE 1 Find s if:

(a) $\log_4 2 = s$ (b) $\log_5 s = 2$ (c) $\log_s 8 = 3$

SOLUTION

(a) If $\log_4 2 = s$, then $4^s = 2$ and hence $s = \frac{1}{2}$.

(b) If $\log_5 s = 2$, then $5^2 = s$ and hence $s = 25$.

(c) If $\log_s 8 = 3$, then $s^3 = 8$ and hence $s = \sqrt[3]{8} = 2$. ■

EXAMPLE 2 Solve the equation $\log_4 (5 + x) = 3$.

SOLUTION If $\log_4 (5 + x) = 3$, then by the definition of logarithm,

$$5 + x = 4^3 \quad \text{or} \quad 5 + x = 64.$$

Hence, the solution is $x = 59$. ■

The following laws are fundamental for all work with logarithms of positive real numbers u and w.

LAWS OF LOGARITHMS

> (i) $\log_a (uw) = \log_a u + \log_a w$
>
> (ii) $\log_a (u/w) = \log_a u - \log_a w$
>
> (iii) $\log_a (u^c) = c \log_a u$ for every real number c

PROOF To prove (i) we begin by letting

$$r = \log_a u \quad \text{and} \quad s = \log_a w.$$

Applying the definition of logarithm, $a^r = u$ and $a^s = w$. Consequently,

$$a^r a^s = uw$$

and hence, $a^{r+s} = uw.$

By the definition of logarithm, the last equation is equivalent to

$$r + s = \log_a (uw).$$

Since $r = \log_a u$ and $s = \log_a w$, we obtain

$$\log_a u + \log_a w = \log_a uw.$$

This completes the proof of Law (i).

To prove (ii) we begin as in the proof of (i), but divide a^r by a^s, obtaining

$$\frac{a^r}{a^s} = \frac{u}{w} \quad \text{or} \quad a^{r-s} = \frac{u}{w}.$$

Using the definition of logarithm, we may write the last equation as

$$r - s = \log_a (u/w).$$

Substituting for r and s gives us

$$\log_a u - \log_a w = \log_a (u/w)$$

This proves Law (ii).

Finally, if c is any real number, then

$$(a^r)^c = u^c \quad \text{or} \quad a^{cr} = u^c.$$

By the definition of logarithm, the last equality implies that

$$cr = \log_a u^c.$$

Substituting for r, we obtain

$$c \log_a u = \log_a u^c.$$

This proves Law (iii).   □

The following examples illustrate uses of the Laws of Logarithms.

EXAMPLE 3   If $\log_a 3 = 0.4771$ and $\log_a 2 = 0.3010$, find:

(a) $\log_a 6$ (b) $\log_a \frac{3}{2}$ (c) $\log_a \sqrt{2}$ (d) $\dfrac{\log_a 3}{\log_a 2}$

SOLUTION

(a) Since $6 = 2 \cdot 3$, we may use Law (i) to obtain

$$\log_a 6 = \log_a (2 \cdot 3) = \log_a 2 + \log_a 3$$
$$= 0.4771 + 0.3010 = 0.7781.$$

(b) By Law (ii),

$$\log_a \tfrac{3}{2} = \log_a 3 - \log_a 2$$
$$= 0.4771 - 0.3010 = 0.1761.$$

(c) Using Law (iii),

$$\log_a \sqrt{2} = \log_a 2^{1/2} = \tfrac{1}{2} \log_a 2$$
$$= \tfrac{1}{2}(0.3010) = 0.1505.$$

(d) There is no law of logarithms that allows us to simplify $(\log_a 3)/(\log_a 2)$. Consequently, we *divide* 0.4771 by 0.3010, obtaining the approximation 1.585. It is important to notice the difference between this problem and part (b). ■

EXAMPLE 4   Solve the following equations:

(a) $\log_2 (2x + 3) = \log_2 11 + \log_2 3$

(b) $\log_4 (x + 6) - \log_4 10 = \log_4 (x - 1) - \log_4 2$

SOLUTION

(a) Using Law (i), we may write the equation as

$$\log_2 (2x + 3) = \log_2 (11 \cdot 3)$$

or $\qquad\qquad\qquad \log_2 (2x + 3) = \log_2 33.$

Since the bases are equal, we must have

$$2x + 3 = 33 \quad \text{or} \quad 2x = 30.$$

Hence, the solution is $x = 15$.

(b) The given equation is equivalent to

$$\log_4 (x + 6) - \log_4 (x - 1) = \log_4 10 - \log_4 2.$$

Applying Law (ii),

$$\log_4 \left(\frac{x + 6}{x - 1} \right) = \log_4 \frac{10}{2} = \log_4 5$$

and hence, $\qquad\qquad\qquad \dfrac{x + 6}{x - 1} = 5.$

The last equation implies that

$$x + 6 = 5x - 5 \quad \text{or} \quad 4x = 11.$$

Thus, the solution is $x = \tfrac{11}{4}$. ■

Extraneous solutions sometimes occur in the process of solving equations that involve logarithms, as illustrated in the next example.

EXAMPLE 5   Solve the equation $2 \log_7 x = \log_7 36$.

SOLUTION   Applying Law (iii), we obtain $2 \log_7 x = \log_7 x^2$, and substitution in the given equation leads to

$$\log_7 x^2 = \log_7 36.$$

Consequently, $x^2 = 36$ and, hence, either $x = 6$ or $x = -6$. However, $x = -6$ is not a solution of the original equation since x must be positive in order for $\log_7 x$ to exist. Thus, the only solution is $x = 6$.

The preceding difficulty could have been avoided by writing the given equation as

$$\log_7 x = \tfrac{1}{2} \log_7 36 = \log_7 36^{1/2} = \log_7 6$$

and, therefore, $x = 6$.   ∎

The Laws of Logarithms are often used as in the following two examples.

EXAMPLE 6

Express $\log_a \dfrac{x^3 \sqrt{y}}{z^2}$ in terms of the logarithms of x, y, and z.

SOLUTION   Writing $\sqrt{y}$ as $y^{1/2}$ and using the three Laws of Logarithms, we obtain

$$\log_a \frac{x^3 y^{1/2}}{z^2} = \log_a (x^3 y^{1/2}) - \log_a z^2$$
$$= \log_a x^3 + \log_a y^{1/2} - \log_a z^2$$
$$= 3 \log_a x + \tfrac{1}{2} \log_a y - 2 \log_a z.$$ ∎

EXAMPLE 7   Express in terms of one logarithm:

$$\tfrac{1}{3} \log_a (x^2 - 1) - \log_a y - 4 \log_a z.$$

SOLUTION   Using the Laws of Logarithms we have

$$\tfrac{1}{3} \log_a (x^2 - 1) - \log_a y - 4 \log_a z = \log_a (x^2 - 1)^{1/3} - \log_a y - \log_a z^4$$
$$= \log_a \sqrt[3]{x^2 - 1} - (\log_a y + \log_a z^4)$$
$$= \log_a \sqrt[3]{x^2 - 1} - \log_a yz^4$$
$$= \log_a \frac{\sqrt[3]{x^2 - 1}}{yz^4}.$$ ∎

It is important to note that there are no laws for expressing $\log_a(u + w)$ or $\log_a(u - w)$ in terms of simpler logarithms. It is evident that

$$\log_a(u + w) \neq \log_a u + \log_a w,$$

since the latter sum equals $\log_a(uw)$. Similarly,

$$\log_a(u - w) \neq \log_a u - \log_a w.$$

Logarithmic functions occur frequently in applications. Indeed, if two variables u and v are related such that u is an exponential function of v, then v is a logarithmic function of u.

EXAMPLE 8 The number N of bacteria in a certain culture after t hours is given by $N = (1000)2^t$. Express t as a logarithmic function of N with base 2.

SOLUTION If $N = (1000)2^t$, then $2^t = \dfrac{N}{1000}$. Changing to logarithmic form,

$$t = \log_2 \frac{N}{1000}. \qquad \blacksquare$$

Exercises 5.3

Change the equations in Exercises 1–8 to logarithmic form.

1 $4^3 = 64$

2 $3^5 = 243$

3 $2^7 = 128$

4 $5^3 = 125$

5 $10^{-3} = 0.001$

6 $10^{-2} = 0.01$

7 $t^r = s$

8 $v^w = u$

Change the equations in Exercises 9–16 to exponential form.

9 $\log_{10} 1000 = 3$

10 $\log_3 81 = 4$

11 $\log_3 \frac{1}{243} = -5$

12 $\log_4 \frac{1}{64} = -3$

13 $\log_7 1 = 0$

14 $\log_9 1 = 0$

15 $\log_t r = p$

16 $\log_v w = q$

Find the numbers in Exercises 17–22.

17 $\log_4 \frac{1}{16}$

18 $\log_2 32$

19 $\log_{10} 100$

20 $\log_8 64$

21 $10^{\log_{10} 5}$

22 $\log_{10} 0.0001$

Find the solutions of the equations in Exercises 23–34.

23 $\log_3(x - 4) = 2$

24 $\log_2(x - 5) = 4$

25 $\log_9 x = \frac{3}{2}$

26 $\log_4 x = -\frac{3}{2}$

27 $\log_5 x^2 = -2$

28 $\log_{10} x^2 = -4$

29 $\log_6(2x - 3) = \log_6 12 - \log_6 3$

30 $2 \log_3 x = 3 \log_3 5$

31 $\log_2 x - \log_2(x + 1) = 3 \log_2 4$

32 $\log_5 x + \log_5(x + 6) = \frac{1}{2} \log_5 9$

33 $\log_{10} x^2 = \log_{10} x$

34 $\frac{1}{2} \log_5(x - 2) = 3 \log_5 2 - \frac{3}{2} \log_5(x - 2)$

In Exercises 35–42 express the logarithm in terms of logarithms of x, y, and z.

35 $\log_a \dfrac{x^2 y}{z^3}$

36 $\log_a \dfrac{x^3 y^2}{z^5}$

37 $\log_a \dfrac{\sqrt{x}z^2}{y^4}$

38 $\log_a x \sqrt[3]{\dfrac{y^2}{z^4}}$

39 $\log_a \sqrt[3]{\dfrac{x^2}{yz^5}}$

40 $\log_a \dfrac{\sqrt{x}\,y^6}{\sqrt[3]{z^2}}$

41 $\log_a \sqrt{x\sqrt{yz^3}}$

42 $\log_a \sqrt[3]{x^2 y \sqrt{z}}$

In Exercises 43–46 write the expression as one logarithm.

43 $2 \log_a x + \frac{1}{3} \log_a (x - 2) - 5 \log_a (2x + 3)$

44 $5 \log_a x - \frac{1}{2} \log_a (3x - 4) + 3 \log_a (5x + 1)$

45 $\log_a (y^2 x^3) - 2 \log_a x \sqrt[3]{y} + 3 \log_a \left(\dfrac{x}{y} \right)$

46 $2 \log_a \dfrac{y^3}{x} - 3 \log_a y + \frac{1}{2} \log_a x^4 y^2$

47 Starting with q_0 milligrams of pure radium, the amount q remaining after t years is $q = q_0(2)^{-t/1600}$. Use logarithms with base 2 to solve for t in terms of q and q_0.

48 The radioactive isotope ^{210}Bi disintegrates according to the law $Q = k(2)^{-t/5}$ where t is in days. Use logarithms with base 2 to solve for t in terms of Q and k.

49 The number N of bacteria in a certain culture at time t is given by $N = 10^4(3)^t$. Use logarithms with base 3 to solve for t in terms of N.

50 When a linear dimension x (such as height) of an organism is related to volume or weight y, the relationship between $\log_a y$ and $\log_a x$ is often linear, in the sense that $\log_a y = k \log_a x + \log_a b$, for some constants k and $b > 0$. Rewrite this equation using laws of logarithms and show that y and x are related by an equation of the form $y = cx^d$ for some constants c and d.

5.4 # Graphs of Logarithmic Functions

Since the logarithmic function $\log_a$ is the inverse of the exponential function with base a, the graph of $y = \log_a x$ can be obtained by reflecting the graph of $y = a^x$ through the line $y = x$ that bisects quadrants I and III (see page 168). This is illustrated in Figure 5.13 for the case $a > 1$.

We can also sketch the graph by plotting points. Since

$$y = \log_a x \quad \text{if and only if} \quad x = a^y,$$

coordinates of points on the graph of $y = \log_a x$ may be found by using the equation $x = a^y$. This leads to the following table:

FIGURE 5.13

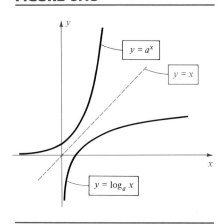

$y = a^x$

$y = x$

$y = \log_a x$

y	-3	-2	-1	0	1	2	3
x	$\dfrac{1}{a^3}$	$\dfrac{1}{a^2}$	$\dfrac{1}{a}$	1	a	a^2	a^3

If $a > 1$, we obtain the sketch in Figure 5.14(i). In this case f is an increasing function throughout its domain. If $0 < a < 1$, then the graph has the general shape shown in Figure 5.14(ii) and hence, f is a decreasing function.

FIGURE 5.14

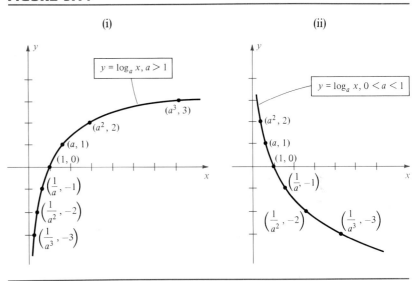

(i)

(ii)

Functions defined in terms of $\log_a p$, for some expression p that involves x, often occur in mathematics and applications. Functions of this type are classified as members of the logarithmic family; however, the graphs may differ from those sketched in Figures 5.13 and 5.14, as illustrated in the following examples.

FIGURE 5.15

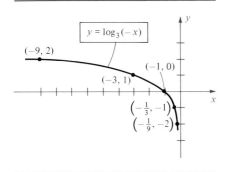

EXAMPLE 1 Sketch the graph of f if $f(x) = \log_3(-x)$ for $x < 0$.

SOLUTION If $x < 0$, then $-x > 0$, and hence, $\log_3(-x)$ exists. We wish to sketch the graph of the equation $y = \log_3(-x)$, which, by the definition of logarithm, is equivalent to $3^y = -x$. Thus, to find points on the graph of f, we may substitute for y in the equation $x = -(3^y)$. The following table displays the coordinates of several such points:

y	-2	-1	0	1	2
x	$-\frac{1}{9}$	$-\frac{1}{3}$	-1	-3	-9

Plotting these points leads to the sketch in Figure 5.15. ∎

FIGURE 5.16

$y = \log_3 |x|$

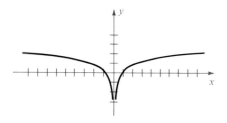

EXAMPLE 2   Sketch the graph of the equation $y = \log_3 |x|$ if $x \neq 0$.

SOLUTION   Since $|x| > 0$ for all $x \neq 0$, the graph includes points corresponding to negative values of x as well as to positive values. If $x > 0$, then $|x| = x$ and hence, to the right of the y-axis the graph coincides with the graph of $y = \log_3 x$, or equivalently, $x = 3^y$. If $x < 0$, then $|x| = -x$ and the graph is the same as that of $y = \log_3(-x)$ (see Example 1). The graph is sketched in Figure 5.16. Note that the graph is symmetric with respect to the y-axis.   ∎

EXAMPLE 3   Sketch the graph of

(a) $y = \log_3(x - 2)$ (b) $y = \log_3 x - 2$.

SOLUTION
(a) We can obtain the graph of $y = \log_3(x - 2)$ by shifting the graph of $y = \log_3 x$ two units to the right. (See the discussion of *horizontal shifts* in Section 3.4). Since the graph of $y = \log_3 x$ is the part of the graph in Figure 5.16 that lies to the right of the y-axis, this leads to the sketch in Figure 5.17(i).

(b) The graph of $y = \log_3 x - 2$ can be obtained by shifting the graph of $y = \log_3 x$ two units downward. (See *vertical shifts*, Section 3.4). This leads to the sketch in Figure 5.17(ii). Note that the x-intercept is given by $\log_3 x = 2$, or $x = 3^2 = 9$.

FIGURE 5.17

(i)

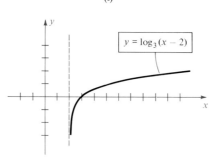

$y = \log_3(x - 2)$

(ii)

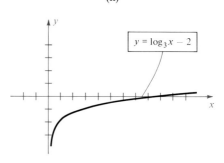

$y = \log_3 x - 2$

■

Exercises 5.4

Sketch the graph of f in Exercises 1–20.

1 $f(x) = \log_2 x$

2 $f(x) = \log_5 x$

3 $f(x) = \log_4 x$

4 $f(x) = \log_{10} x$

5 $f(x) = \log_2(x + 3)$

6 $f(x) = \log_2(x - 3)$

7 $f(x) = \log_2 x + 3$ **8** $f(x) = \log_2 x - 3$ **15** $f(x) = \log_3 (1/x)$ **16** $f(x) = \log_3 (2 - x)$

9 $f(x) = \log_3 (3x)$ **10** $f(x) = \log_2 (x^2)$ **17** $f(x) = 1/(\log_3 x)$ **18** $f(x) = |\log_2 x|$

11 $f(x) = 3 \log_3 x$ **12** $f(x) = \log_2 (x^3)$ **19** $f(x) = \log_2 |x - 5|$ **20** $f(x) = \log_3 |x + 1|$

13 $f(x) = \log_2 \sqrt{x}$ **14** $f(x) = \log_2 |x|$

5.5 Common and Natural Logarithms

Before electronic calculators were invented, logarithms with base 10 were used for complicated numerical computations involving products, quotients, and powers of real numbers. Base 10 was employed because it is well suited for numbers that are expressed in decimal form. Logarithms with base 10 are called **common logarithms.** The symbol log x is used as an abbreviation for $\log_{10} x$. Thus, we have the following definition.

DEFINITION OF COMMON LOGARITHMS

$$\log x = \log_{10} x \quad \text{for every } x > 0.$$

Since inexpensive calculators are now available, there is little need for common logarithms as a tool for computational work. However, base 10 does occur in applications, and hence many calculators have a $\boxed{\log}$ key that can be used to approximate common logarithms. Appendix II contains a table of common logarithms that may be used if a calculator either is not available or is inoperative. The use of the table of common logarithms is explained in Appendix I.

EXAMPLE 1 Using the *Richter scale*, the magnitude R of an earthquake of intensity I may be found by means of the formula

$$R = \log \frac{I}{I_0}$$

where I_0 is a certain minimum intensity.

(a) Find R assuming the intensity of an earthquake is $(1000)I_0$.

(b) Express I in terms of R and I_0.

SOLUTION

(a) If $I = (1000)I_0$, then

$$R = \log \frac{I}{I_0} = \log \frac{(1000)I_0}{I_0} = \log (1000).$$

We could use a calculator to find $\log (1000)$ or write

$$\log (1000) = \log 10^3 = 3 \log 10 = 3(1) = 3$$

where we have used the fact that $\log_a a = 1$. Hence $R = 3$.

(b) By the definition of logarithm with base 10,

$$\text{if} \quad R = \log \frac{I}{I_0}, \quad \text{then} \quad \frac{I}{I_0} = 10^R.$$

Thus, $$I = I_0(10^R).$$ ∎

In Section 5.2 we defined the natural exponential function f by means of the equation $f(x) = e^x$. The logarithmic function with base e is called the **natural logarithmic function.** We use **ln x** (read "*ell-en of x*") as an abbreviation for $\log_e x$ and refer to it as the **natural logarithm of x.** Thus *the natural logarithmic and natural exponential functions are inverse functions of one another.* Let us state the definition for reference.

DEFINITION OF NATURAL LOGARITHMS

$$\ln x = \log_e x \quad \text{for every} \quad x > 0.$$

Since $e \approx 3$, the graph of $y = \ln x$ is similar in appearance to the graph of $y = \log_3 x$. The Laws of Logarithms for natural logarithms are as follows.

LAWS OF NATURAL LOGARITHMS

(i) $\ln (uv) = \ln u + \ln v$

(ii) $\ln \dfrac{u}{v} = \ln u - \ln v$

(iii) $\ln u^c = c \ln u$

If we substitute e for a in the theorem on page 258, we obtain the following for $x > 0$:

$$e^{\ln x} = x, \qquad \ln e = 1, \qquad \ln 1 = 0.$$

Many calculators have a key labeled $\boxed{\ln x}$ that can be used to approximate natural logarithms. A short table of values of $\ln x$ is given in Table 3 of Appendix II.

EXAMPLE 2   Use a calculator to approximate log 436; ln 436; log 0.0436; and ln 0.0436.

SOLUTION   Entering the indicated numbers and pressing the appropriate keys, we obtain the following approximations:

$$\log 436 \approx 2.6394865$$

$$\ln 436 \approx 6.0776422$$

$$\log 0.0436 \approx -1.3605135$$

$$\ln 0.0436 \approx -3.1326981 \qquad \blacksquare$$

EXAMPLE 3   Newton's Law of Cooling states that the rate at which an object cools is directly proportional to the difference in temperature between the object and its surrounding medium. Newton's Law can be used to show that under certain conditions the temperature T of an object at time t is given by $T = 75e^{-2t}$. Express t as a function of T.

SOLUTION   The equation $T = 75e^{-2t}$ may be written

$$e^{-2t} = \frac{T}{75}.$$

Using natural logarithms gives us

$$-2t = \log_e \frac{T}{75} = \ln \frac{T}{75}.$$

Consequently,

$$t = -\tfrac{1}{2} \ln \frac{T}{75} \quad \text{or} \quad t = -\tfrac{1}{2}[\ln T - \ln 75]. \qquad \blacksquare$$

EXAMPLE 4   If a beam of light that has intensity k is projected vertically downward into water, then its intensity $I(x)$ at a depth of x meters is $I(x) = ke^{-1.4x}$. At what depth is the intensity one-half its value at the surface?

SOLUTION   At the surface $x = 0$, and the intensity is

$$I(0) = ke^0 = k.$$

We wish to find the value of x such that $I(x) = \frac{1}{2}k$; that is,

$$ke^{-1.4x} = \tfrac{1}{2}k \quad \text{or} \quad e^{-1.4x} = \tfrac{1}{2}.$$

Using natural logarithms,

$$-1.4x = \ln \tfrac{1}{2}$$

or $\qquad\qquad -1.4x = \ln 1 - \ln 2 = 0 - \ln 2 = -\ln 2.$

Hence, $\qquad\qquad x = \dfrac{-\ln 2}{-1.4} \approx \dfrac{-0.693}{-1.4} \approx 0.5 \text{ meter.}$ ∎

To solve certain problems, it is necessary to find x when given either $\log x$ or $\ln x$. One way to accomplish this is by using the inverse function key $\boxed{\text{INV}}$. If we first press $\boxed{\text{INV}}$ and then press $\boxed{\log}$, we obtain the *inverse logarithmic function* $\log^{-1}$. Recall that $\log^{-1}$ is the exponential function with base 10. Since $\log^{-1}(\log x) = x$, we can obtain x by entering $\log x$ and then pressing, successively, $\boxed{\text{INV}}$ and $\boxed{\log}$. Similarly, given $\ln x$, we can find x by entering $\ln x$ and pressing $\boxed{\text{INV}}\,\boxed{\ln x}$. This procedure is illustrated in the next example.

EXAMPLE 5 Approximate x to three decimal places if:

(a) $\log x = 1.7959$ (b) $\ln x = 4.7$

SOLUTION

(a) Given $\log x = 1.7959$,

$$\text{Enter: } \quad 1.7959$$
$$\text{Press } \boxed{\text{INV}}\,\boxed{\log}: \quad 62.502876$$

Thus, $x \approx 62.503$.

(b) Given $\ln x = 4.7$,

$$\text{Enter: } \quad 4.7$$
$$\text{Press } \boxed{\text{INV}}\,\boxed{\ln x}: \quad 109.94717$$

Hence, $x \approx 109.947$. ∎

Alternatively, Example 5(a) could be solved by noting that since common logarithms have base 10,

$$\text{if} \quad \log x = 1.7959, \quad \text{then} \quad x = 10^{1.7959}.$$

The number x can then be approximated using a $\boxed{y^x}$ key.

Similarly, part (b) could be solved by using the fact that natural logarithms have base e. Specifically,

$$\text{if} \quad \ln x = 4.7, \quad \text{then} \quad x = e^{4.7}.$$

Table 2 in Appendix II or a calculator could then be used to approximate x.

Finally, it is sometimes necessary to *change the base* of a logarithm by expressing $\log_b u$ in terms of $\log_a u$, for some positive real number $b \neq 1$. We begin with the equivalent equations

$$v = \log_b u \quad \text{and} \quad b^v = u.$$

Taking the logarithm, base a, of both sides of the second equation gives us

$$\log_a b^v = \log_a u,$$

or equivalently, $\quad\quad\quad\quad v \log_a b = \log_a u.$

Solving for v (that is, $\log_b u$), we obtain formula (i) in the next box.

<table>
<tr><td>**CHANGE OF
BASE FORMULAS**</td><td>(i) $\log_b u = \dfrac{\log_a u}{\log_a b}$ (ii) $\log_b a = \dfrac{1}{\log_a b}$</td></tr>
</table>

To obtain formula (ii) we let $u = a$ in (i) and use the fact that $\log_a a = 1$. If we let $a = e$ and $b = 10$ in the change of base formulas we obtain the following special cases:

$$\log u = \frac{\ln u}{\ln 10} \quad \text{and} \quad \log e = \frac{1}{\ln 10}.$$

Exercises 5.5

In Exercises 1–12 approximate x to three significant figures.

1 $\log x = 3.6274$ **2** $\log x = 1.8965$

3 $\log x = 0.9469$ **4** $\log x = 4.9680$

5 $\log x = -1.6253$ **6** $\log x = -2.2118$

7 $\ln x = 2.3$ **8** $\ln x = 3.7$

9 $\ln x = 0.05$ **10** $\ln x = 0.95$

11 $\ln x = -1.6$ **12** $\ln x = -5$

13 Using the Richter scale formula $R = \log (I/I_0)$, find the magnitude of an earthquake that has intensity

(a) 100 times that of I_0;
(b) 10,000 times that of I_0;
(c) 100,000 times that of I_0.

14 Refer to Exercise 13. The largest recorded magnitudes of earthquakes have been between 8 and 9 on the Richter scale. Find the corresponding intensities in terms of I_0.

15 The following formula, valid for earthquakes in the eastern U.S., relates the magnitude R of the earthquake to the surrounding area A (in square miles) that is affected by the quake:

$$R = 2.3 \log (A + 34{,}000) - 7.5.$$

Solve for A in terms of R.

16 For the western U.S., the area-magnitude formula (see Exercise 15) takes the form

$$R = 2.3 \log (A + 3000) - 5.1.$$

If A_1 is the area affected by an earthquake of magnitude R in the west, while A_2 is the area affected by a similar quake in the east, find a formula for A_1/A_2 in terms of R.

17 The loudness of a sound, as experienced by the human ear, is based upon intensity levels. A formula used for finding the intensity level α, in decibels, that corresponds to a sound intensity I is $\alpha = 10 \log (I/I_0)$ where I_0 is a special value of I agreed to be the weakest sound that can be detected by the ear under certain conditions. Find α if:

(a) I is 10 times as great as I_0;

(b) I is 1000 times as great as I_0;

(c) I is 10,000 times as great as I_0. (This is the intensity level of the average voice.)

18 A sound intensity level of 140 decibels produces pain in the average human ear. Approximately how many times greater than I_0 must I be in order for α to reach this level? (Refer to Exercise 17.)

19 Chemists use a number denoted by pH to describe quantitatively the acidity or basicity of solutions. By definition,

$$\text{pH} = -\log [H^+]$$

where $[H^+]$ is the hydrogen ion concentration in moles per liter. Approximate the pH of each substance:

(a) vinegar: $[H^+] \approx 6.3 \times 10^{-3}$;

(b) carrots: $[H^+] \approx 1.0 \times 10^{-5}$;

(c) sea water: $[H^+] \approx 5.0 \times 10^{-9}$.

20 Approximate the hydrogen ion concentration $[H^+]$ in each of the following substances. (Refer to Exercise 19).

(a) apples: pH ≈ 3.0 (b) beer: pH ≈ 4.2

(c) milk: pH ≈ 6.6

21 A solution is considered acidic if $[H^+] > 10^{-7}$ or basic if $[H^+] < 10^{-7}$. What are the corresponding inequalities involving pH?

22 Many solutions have a pH between 1 and 14. What is the corresponding range of the hydrogen ion content $[H^+]$?

23 The current I at time t in a certain electrical circuit is given by $I = 20e^{-Rt/L}$, where R and L denote the resistance and inductance, respectively. Use natural logarithms to solve for t in terms of the remaining variables.

24 An electrical condenser with initial charge Q_0 is allowed to discharge. After t seconds the charge Q is $Q = Q_0 e^{kt}$ where k is a constant. Use natural logarithms to solve for t in terms of Q_0, Q, and k.

25 Under certain conditions the atmospheric pressure p at altitude h is given by $p = 29e^{-0.000034h}$. Use natural logarithms to solve for h as a function of p.

26 If p denotes the selling price (in dollars) of a commodity, and x is the corresponding demand (in number sold per day), then frequently the relationship between p and x is $p = p_0 e^{-ax}$ where p_0 and a are positive constants. Express x as a function of p.

27 The *Ehrenberg relation* $\ln W = \ln 2.4 + (1.84)h$ is an empirically based formula relating the height h (in meters) to the weight W (in kg) for children aged 5 through 13 years old. The formula has been verified in many different countries. Express W as a function of h.

28 A rocket of mass m_1 is filled with fuel of initial mass m_2. Assuming frictional forces are neglected, the total mass m of the rocket at time t is related to its upward velocity by the formula $v = -a \ln m + b$ for some constants a and b. At time $t = 0$ we have $v = 0$ and $m = m_1 + m_2$. At burnout $m = m_1$. Using this information, find a formula for the velocity of the rocket at burnout. (Write the formula in terms of one logarithm.)

29 If n is the average number of earthquakes (worldwide) in a given year with magnitude (on the Richter scale) between R and $R + 1$, then $\log n = 7.7 - (0.9)R$.

(a) Solve for n in terms of R.

(b) Find n if $R = 4$, 5, and 6.

30 The energy E (in ergs) released during an earthquake of magnitude R is given by the formula $\log E = 1.4 + (1.5)R$.

(a) Solve for E in terms of R.

(b) Find the energy released during the famous Alaskan quake of 1964, which measured 8.4 on the Richter scale.

31 A certain radioactive substance decays according to the formula $q(t) = q_0 e^{-0.0063t}$ where q_0 is the initial amount of the substance and t is the time in days. Approximate its half-life, that is, the number of days it takes for half of the substance to decay.

32 The air pressure $p(h)$, in pounds per in², at an altitude of h feet above sea level may be approximated by the formula $p(h) = 14.7e^{-0.0000385h}$. At approximately what altitude h is the air pressure (a) 10 pounds per in²? (b) one-half its value at sea level?

33 Use the Compound Interest Formula to determine how long it will take a sum of money to double if it is invested at a rate of 6% per year compounded monthly.

34 If interest is compounded continuously at the rate of 10% per year, the compound interest formula takes the form $A = Pe^{0.1t}$. A man has determined that he will need to deposit about \$5600 to generate \$25,000 for his son's college education. To provide a cushion for unexpected expenses, he decides to deposit \$6000. After how many years will this initial deposit have grown to \$25,000?

35 The population $N(t)$ (in millions) of the United States t years after 1980 may be approximated by the formula $N(t) = 227e^{0.007t}$. When will the population be twice what it was in 1980?

36 The population $N(t)$ of India (in millions) t years after 1980 may be approximated by $N(t) = 651e^{0.02t}$. When will the population grow to one billion?

5.6 Exponential and Logarithmic Equations

If variables in equations appear as exponents or logarithms, the equations are often called **exponential** or **logarithmic equations,** respectively. The following examples illustrate techniques for solving such equations.

EXAMPLE 1 Solve the equation $3^x = 21$.

SOLUTION Taking the common logarithm of both sides and using (iii) of the Laws of Logarithms, we obtain

$$\log (3^x) = \log 21$$

$$x \log 3 = \log 21$$

$$x = \frac{\log 21}{\log 3}.$$

If an approximation is desired, we may use a calculator or Table 1 in Appendix II to obtain

$$x \approx \frac{1.3222}{0.4771} \approx 2.77$$

A partial check on the solution is to note that since $3^2 = 9$ and $3^3 = 27$, the number x such that $3^x = 21$ should lie between 2 and 3, somewhat closer to 3 than to 2.

We could also have solved for x by using natural logarithms. In this case, a calculator yields

$$x = \frac{\ln 21}{\ln 3} \approx \frac{3.0445224}{1.0986123}$$

$$\approx 2.7712437 \approx 2.77. \qquad \blacksquare$$

EXAMPLE 2   Solve the equation $5^{2x+1} = 6^{x-2}$.

SOLUTION   If we take the common logarithm of both sides and use (iii) of the Laws of Logarithms, we obtain

$$(2x + 1) \log 5 = (x - 2) \log 6.$$

We may now solve for x:

$$2x \log 5 + \log 5 = x \log 6 - 2 \log 6$$

$$2x \log 5 - x \log 6 = -\log 5 - 2 \log 6$$

$$x(2 \log 5 - \log 6) = -(\log 5 + \log 6^2)$$

$$x = \frac{-(\log 5 + \log 36)}{2 \log 5 - \log 6}$$

$$= \frac{-\log (5 \cdot 36)}{\log 5^2 - \log 6}.$$

Thus,

$$x = -\frac{\log 180}{\log \frac{25}{6}}.$$

If an approximation to the solution is desired, we could proceed as in Example 1:

$$x \approx -\frac{2.2553}{0.6198} \approx -3.64.$$

Natural logarithms could also have been used:

$$x = -\frac{\ln 180}{\ln \frac{25}{6}} \approx -\frac{5.1929569}{1.4271164} \approx -3.64.$$ ∎

EXAMPLE 3 Solve the equation $\log(5x - 1) - \log(x - 3) = 2$.

SOLUTION The equation may be written

$$\log\frac{5x - 1}{x - 3} = 2.$$

Using the definition of logarithm with $a = 10$ gives us

$$\frac{5x - 1}{x - 3} = 10^2.$$

Consequently,

$$5x - 1 = 10^2(x - 3) = 100x - 300 \quad \text{or} \quad 299 = 95x.$$

Hence, $$x = \tfrac{299}{95}.$$

We leave it to the reader to check that this is the solution of the equation. ∎

EXAMPLE 4 Solve the equation

$$\frac{5^x - 5^{-x}}{2} = 3.$$

SOLUTION Multiplying both sides of the equation by 2 gives us

$$5^x - 5^{-x} = 6.$$

If we now multiply both sides by 5^x, we obtain

$$5^{2x} - 1 = 6(5^x),$$

which may be written

$$(5^x)^2 - 6(5^x) - 1 = 0.$$

Letting $u = 5^x$ leads to the quadratic equation

$$u^2 - 6u - 1 = 0$$

in the variable u. Applying the Quadratic Formula,

$$u = \frac{6 \pm \sqrt{36 + 4}}{2} = 3 \pm \sqrt{10},$$

that is, $5^x = 3 \pm \sqrt{10}$. Since 5^x is never negative, the number $3 - \sqrt{10}$ must be discarded; therefore,

$$5^x = 3 + \sqrt{10}.$$

Taking the common logarithm of both sides and using (iii) of the Laws of Logarithms,

$$x \log 5 = \log (3 + \sqrt{10}) \quad \text{or} \quad x = \frac{\log (3 + \sqrt{10})}{\log 5}.$$

To obtain an approximate solution, we may write $3 + \sqrt{10} \approx 6.16$ and use Table 1 or a calculator. This gives us

$$x \approx \frac{\log 6.16}{\log 5} \approx \frac{0.7896}{0.6990} \approx 1.13.$$

Natural logarithms could also have been used to obtain

$$x = \frac{\ln (3 + \sqrt{10})}{\ln 5}.$$

EXAMPLE 5 The Beer-Lambert Law asserts that the amount of light I that penetrates to a depth of x meters in an ocean is given by $I = I_0 a^x$, where $0 < a < 1$ and I_0 is the amount of light at the surface.

(a) Solve for x in terms of I, I_0, and a by using (i) common logarithms; (ii) natural logarithms.

(b) If $a = \frac{1}{4}$, what is the depth at which $I = 0.01 I_0$? (This depth determines the zone where photosynthesis can take place.)

SOLUTION
(a) Taking the common logarithm of both sides of the equation $I = I_0 a^x$ and using laws of logarithms, we obtain

$$\log I = \log (I_0 a^x) = \log I_0 + \log a^x = \log I_0 + x \log a.$$

Hence, $\qquad x \log a = \log I - \log I_0 = \log (I/I_0)$

and
$$x = \frac{\log (I/I_0)}{\log a}.$$

If natural logarithms are used, then

$$x = \frac{\ln (I/I_0)}{\ln a}$$

(b) Letting $I = 0.01I_0$ and $a = \frac{1}{4}$ in the formula for x obtained in part (a),

$$x = \frac{\log (0.01I_0/I_0)}{\log \frac{1}{4}} = \frac{\log (0.01)}{\log 1 - \log 4} = \frac{\log 10^{-2}}{-\log 4}$$

Using a calculator or Table 1,

$$x \approx \frac{-2}{-0.6021} \approx 3.32 \text{ meters.} \qquad \blacksquare$$

Exercises 5.6

In Exercises 1–14 (a) find the solutions of the equations without using a calculator or a table; (b) find two-decimal-place approximations to the solutions.

1 $10^x = 7$

2 $5^x = 8$

3 $4^x = 3$

4 $10^x = 6$

5 $3^{4-x} = 5$

6 $(\frac{1}{3})^x = 100$

7 $3^{x+4} = 2^{1-3x}$

8 $4^{2x+3} = 5^{x-2}$

9 $2^{-x} = 8$

10 $2^{-x^2} = 5$

11 $\log x = 1 - \log (x - 3)$

12 $\log (5x + 1) = 2 + \log (2x - 3)$

13 $\log (x^2 + 4) - \log (x + 2) = 3 + \log (x - 2)$

14 $\log (x - 4) - \log (3x - 10) = \log (1/x)$

Solve the equations in Exercises 15–20 without using a calculator or a table.

15 $\log (x^2) = (\log x)^2$

16 $\log \sqrt{x} = \sqrt{\log x}$

17 $\log (\log x) = 2$

18 $\log \sqrt{x^3 - 9} = 2$

19 $x^{\sqrt{\log x}} = 10^8$

20 $\log (x^3) = (\log x)^3$

In Exercises 21–24 use common logarithms to solve for x in terms of y.

21 $y = \dfrac{10^x + 10^{-x}}{2}$

22 $y = \dfrac{10^x - 10^{-x}}{2}$

23 $y = \dfrac{10^x - 10^{-x}}{10^x + 10^{-x}}$

24 $y = \dfrac{1}{10^x - 10^{-x}}$

In Exercises 25–28 use natural logarithms to solve for x in terms of y.

25 $y = \dfrac{e^x - e^{-x}}{2}$

26 $y = \dfrac{e^x + e^{-x}}{2}$

27 $y = \dfrac{e^x - e^{-x}}{e^x + e^{-x}}$

28 $y = \dfrac{e^x + e^{-x}}{e^x - e^{-x}}$

29 The current I in a certain electrical circuit at time t is given by

$$I = \frac{E}{R}(1 - e^{-Rt/L})$$

where E, R, and L denote the electromotive force, the resistance, and the inductance, respectively (see figure).

Use natural logarithms to solve for t in terms of the remaining symbols.

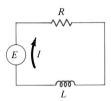

30 Solve the Compound Interest Formula

$$A = P\left(1 + \frac{r}{n}\right)^{nt}$$

for t in terms of the other symbols by using natural logarithms.

31 Refer to Example 5. The most important zone in the sea from the viewpoint of marine biology is the *photic zone*, the zone in which photosynthesis can take place. That zone must end at the depth where about 1% of the surface light penetrates. In very clear waters in the Caribbean, 50% of the light at the surface reaches a depth of about 13 meters. Estimate the depth of the photic zone.

32 In sharp contrast to the situation in Exercise 31, in parts of New York harbor, 50% of the surface light does not reach a depth of 10 cm! Estimate the depth of the photic zone.

33 A drug is eliminated from the body through urine. The initial dose is 10 mg and the amount $A(t)$ in the body t hours later is given by $A(t) = 10(0.8)^t$. In order for the drug to be effective, at least 2 mg must be in the body.

  (a) Determine when only 2 mg are left.

  (b) What is the half-life of the drug?

34 Radioactive Iodine, ^{131}I, is frequently used in tracer studies involving the thyroid gland. The substance decays according to the formula $A(t) = A_0 a^{-t}$ where A_0 is the initial dose and t is the time in days. Find a assuming the half-life of ^{131}I is eight days.

5.7 Review

Define or discuss each of the following.

1 The exponential function with base a

2 The natural exponential function

3 The logarithmic function with base a

4 Laws of Logarithms

5 Common logarithms

6 The natural logarithmic function

Exercises 5.7

Find the numbers in Exercises 1–10 without the aid of a calculator or table.

1 $\log_2 \frac{1}{16}$

2 $\log_5 \sqrt[3]{5}$

3 $6^{\log_6 4}$

4 $10^{3 \log 2}$

5 $\log 1{,}000{,}000$

6 $\ln e$

7 $\log_4 2$

8 $\log_\pi 1$

9 $e^{\ln 5}$

10 $\log \log 10^{10}$

In Exercises 11–24 sketch the graph of f.

11 $f(x) = 3^{x+2}$

12 $f(x) = \left(\frac{3}{5}\right)^x$

13 $f(x) = \left(\frac{3}{2}\right)^{-x}$

14 $f(x) = 3^{-2x}$

15 $f(x) = 3^{-x^2}$ **16** $f(x) = 1 - 3^{-x}$

17 $f(x) = \log_6 x$ **18** $f(x) = \log_3 (x^2)$

19 $f(x) = 2 \log_3 x$ **20** $f(x) = \log_2 (x + 4)$

21 $f(x) = e^{x/2}$ **22** $f(x) = \frac{1}{2}e^x$

23 $f(x) = e^{x-2}$ **24** $f(x) = e^{2-x}$

Find the solutions of the equations in Exercises 25–34 without using a calculator or table.

25 $\log_8 (x - 5) = \frac{2}{3}$

26 $\log_4 (x + 1) = 2 + \log_4 (3x - 2)$

27 $2 \log_3 (x + 3) - \log_3 (x + 1) = 3 \log_3 2$

28 $\log \sqrt[4]{x + 1} = \frac{1}{2}$ **29** $2^{5-x} = 6$

30 $3^{x^2} = 7$ **31** $2^{5x+3} = 3^{2x+1}$

32 $e^{\ln (x + 1)} = 3$

33 $x^2(-2xe^{-x^2}) + 2xe^{-x^2} = 0$

34 $\ln x = 1 + \ln (x + 1)$

35 Express $\log x^4 \sqrt[3]{y^2/z}$ in terms of logarithms of x, y, and z.

36 Express $\log (x^2/y^3) + 4 \log y - 6 \log \sqrt{xy}$ as one logarithm.

Solve the equations in Exercises 37 and 38 for x in terms of y.

37 $y = \dfrac{10^x + 10^{-x}}{10^x - 10^{-x}}$ **38** $y = \dfrac{1}{10^x + 10^{-x}}$

In Exercises 39–44 approximate x using (a) a calculator; (b) tables.

39 $\log x = 1.8938$ **40** $\log x = -2.4260$

41 $\ln x = 1.8$ **42** $\ln x = -0.75$

43 $x = \ln 6.6$ **44** $x = \log 8.4$

45 The number of bacteria in a certain culture at time t is given by $Q(t) = 2(3^t)$, where t is measured in hours and $Q(t)$ in thousands. What is the initial number of bacteria? What is the number after 10 minutes? after 30 minutes? after 1 hour?

46 If $1000 is invested at a rate of 12% compounded four times per year, what is the principal after one year?

47 Radioactive iodine, ^{131}I, is frequently used in tracer studies in the human body (and notably in tests of the thyroid gland). The substance decays according to the formula $N = N_0(0.5)^{t/8}$ where N_0 is the initial dose and t is the time in days.

(a) Sketch the graph of the equation if $N_0 = 64$.

(b) Show that the half-life of ^{131}I is eight days.

48 One thousand young trout are put in a fishing pond. Three months later, the owner estimates that there are about 600 left. Find an exponential formula $N = N_0 a^t$ that fits this information and use it to estimate the number of trout left after one year.

49 Ten thousand dollars is invested in a savings fund in which interest is compounded continuously at the rate of 11% per year. The amount A in the account t years later is given by $A = 10,000e^{0.11t}$.

(a) When will the account contain $35,000?

(b) How long does it take money to double in the account?

50 The current $I(t)$ at time t in a certain electrical circuit is given by $I(t) = I_0 e^{-Rt/L}$ where R and L denote the resistance and inductance, respectively, and I_0 is the current at the time $t = 0$. At what time is the current $\frac{1}{100}I_0$?

51 The sound intensity level formula is $\alpha = 10 \log (I/I_0)$.

(a) Solve for I in terms of α and I_0.

(b) Show that a one-decibel rise in the intensity level α corresponds to a 26% increase in the intensity I.

52 Stars are classified into categories of brightness called *magnitudes*. The faintest stars (with light flux L_0) are assigned a magnitude of 6. Brighter stars are assigned magnitudes according to the formula

$$m = 6 - (2.5) \log (L/L_0)$$

where L is the light flux from the star.

(a) Find m if $L = 10^{0.4}L_0$.

(b) Solve the formula for L in terms of m and L_0.

53 Solve the von-Bertalanffy growth formula

$$y = a(1 - be^{-kt})$$

for t in terms of y, a, b, and k. The resulting formula can be used to estimate the age of a fish from a length measurement.

54 For a population of female African elephants, the weight W (in kg) at age t (in years) is given by

$$W = 2600(1 - 0.51e^{-0.075t})^3.$$

(a) What is the weight of a newborn?

(b) Assuming an adult female weighs 1800 kg, estimate her age.

55 If the pollution of Lake Erie were suddenly stopped, it has been estimated that the level y of pollutants would decrease according to the formula $y = y_0 e^{-0.3821t}$ where t is in years and y_0 is the pollutant level at which further pollution ceased. How many years would it take to clear 50% of the pollutants?

56 Radioactive Strontium 90, ^{90}Sr, has been deposited into a large field by acid rain. If sufficient amounts make their way through the food chain to man, bone cancer can result. It has been determined that the radioactivity level in the field is 2.5 times the safe level S. For how many years will this field be contaminated? ^{90}Sr decays according to the formula $A(t) = A_0 e^{-0.0239t}$ where A_0 is the present amount in the field and t is the time in years.

CHAPTER 6

Systems of Equations and Inequalities

It is sometimes necessary to work simultaneously with more than one equation in several variables. ■ We then refer to the equations as a *system of equations.* ■ In this chapter we shall develop methods for finding solutions that are common to all equations in a a system. ■ Of particular importance are the matrix techniques introduced for systems of linear equations. ■ We shall also briefly discuss systems of inequalities and linear programming.

6.1 Systems of Equations

An ordered pair (a, b) is a **solution** of an equation in two variables x and y if a true statement is obtained when a and b are substituted for x and y, respectively. Two equations in x and y are **equivalent** if they have exactly the same solutions. For example, the following equations are equivalent:

$$x^2 - 4y = 3, \qquad x^2 - 3 = 4y$$

As we know, the **graph** of an equation in x and y consists of all points in a coordinate plane that correspond to the solutions of the equation.

A **system** of two equations in x and y is any two equations in those variables. An ordered pair (a, b) is called a **solution of the system** if (a, b) is a solution of both equations. The points that correspond to the solutions are the points at which the graphs of the two equations intersect.

As a concrete example, consider the system

$$\begin{cases} x^2 - y = 0 \\ y - 2x - 3 = 0. \end{cases}$$

The brace is used to indicate that the two equations are to be treated simultaneously. The following table lists some solutions of the equation $x^2 - y = 0$, or equivalently, $y = x^2$.

x	-2	-1	0	1	2	3	4
y	4	1	0	1	4	9	16

FIGURE 6.1

The graph is the parabola sketched in Figure 6.1. Several solutions for $y - 2x - 3 = 0$, or equivalently, $y = 2x + 3$, are listed in the following table.

x	-2	-1	0	1	2	3	4
y	-1	1	3	5	7	9	11

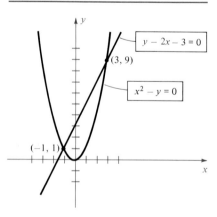

The graph is the line in Figure 6.1. The pairs $(3, 9)$ and $(-1, 1)$ are solutions of both equations and hence are solutions of the system. We shall see later that they are the *only* solutions. Note that the solutions of the system determine the points of intersection of the two graphs.

The solutions of the preceding system can be found algebraically, that is, without reference to graphs. Let us begin by solving the equation $x^2 - y = 0$ for y in terms of x, obtaining $y = x^2$. It follows that if (x, y) is a solution of the system, then it is of the form (x, x^2). In particular, (x, x^2)

is a solution of the equation $y - 2x - 3 = 0$; that is,

$$x^2 - 2x - 3 = 0.$$

Factoring gives us $\quad (x - 3)(x + 1) = 0$

and hence either $x = 3$ or $x = -1$. The corresponding values for y (obtained from $y = x^2$) are 9 and 1, respectively. Thus, the only possible solutions of the system are the ordered pairs $(3, 9)$ and $(-1, 1)$. That these are actually solutions can be seen by substitution.

When we found the solutions algebraically in the preceding discussion, we first solved one equation for y in terms of x, and then we substituted for y in the other equation, obtaining an equation in one variable, x. The solutions of the latter equation were the only possible x-values for the solutions of the system. The corresponding y-values were found by means of the equation that expressed y in terms of x. This technique is called the *method of substitution*. Sometimes it is convenient to begin by solving one equation for x in terms of y, and then to substitute for x in the other equation. In general, the steps used in this process may be listed as follows:

METHOD OF SUBSTITUTION

> (i) Solve one of the equations for one variable in terms of the other variable.
>
> (ii) Substitute the expression obtained in step (i) into the other equation, obtaining an equation in one variable.
>
> (iii) Find the solutions of the equation obtained in step (ii).
>
> (iv) Use the solutions from step (iii), together with the expression obtained in step (i), to find the solutions of the system.

In the next example we find the solutions of the system considered at the beginning of this section by first solving one equation for x in terms of y.

EXAMPLE 1 Find the solutions of the system

$$\begin{cases} x^2 - y = 0 \\ y - 2x - 3 = 0. \end{cases}$$

SOLUTION We may solve the second equation for x in terms of y as follows:

$$2x = y - 3 \quad \text{or} \quad x = \frac{y - 3}{2}.$$

Substituting for x in the first equation gives us

$$\left(\frac{y-3}{2}\right)^2 - y = 0$$

$$\frac{y^2 - 6y + 9}{4} - y = 0$$

$$y^2 - 6y + 9 - 4y = 0$$

$$y^2 - 10y + 9 = 0$$

$$(y - 9)(y - 1) = 0.$$

The last equation has solutions $y = 9$ and $y = 1$. If we substitute 9 and 1 for y in the equation $x = (y - 3)/2$, we obtain the values $x = 3$ and $x = -1$, respectively. Hence, as before, the solutions of the system are $(3, 9)$ and $(-1, 1)$. ∎

EXAMPLE 2   Find the solutions of the following system and then sketch the graph of each equation, showing the points of intersection:

$$\begin{cases} x^2 + y^2 = 25 \\ x^2 + y\ = 19. \end{cases}$$

SOLUTION   Solving the second equation for y, we obtain $y = 19 - x^2$. Substituting for y in the first equation leads to the following list of equivalent equations:

$$x^2 + (19 - x^2)^2 = 25$$

$$x^2 + (361 - 38x^2 + x^4) = 25$$

$$x^4 - 37x^2 + 336 = 0$$

$$(x^2 - 16)(x^2 - 21) = 0.$$

The solutions of the last equation are 4, -4, $\sqrt{21}$, and $-\sqrt{21}$. The corresponding values of y are found by substituting for x in $y = 19 - x^2$. Substitution of 4 or -4 for x gives us $y = 3$, whereas substitution of $\sqrt{21}$ or $-\sqrt{21}$ gives us $y = -2$. Hence, the only possible solutions of the system are

$$(4, 3), \quad (-4, 3), \quad (\sqrt{21}, -2), \quad \text{and} \quad (-\sqrt{21}, -2).$$

It can be seen by substitution in the given equations that all four pairs are solutions.

FIGURE 6.2

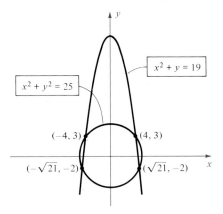

The graph of $x^2 + y^2 = 25$ is a circle of radius 5 with center at the origin, and the graph of $y = 19 - x^2$ is a parabola with a vertical axis. The graphs are illustrated in Figure 6.2. The points of intersection are determined by the solutions of the system. ∎

We can also consider equations in three variables x, y, and z, such as

$$x^2 y + xz + 3^y = 4z^3.$$

The equation has a **solution** (a, b, c) if substitution of a, b, and c, for x, y, and z, respectively, produces a true statement. We refer to (a, b, c) as an **ordered triple** of real numbers. Equivalent equations are defined as before. A system of equations in three variables and the solutions of such a system are defined as in the two-variable case. In like manner, we can consider systems of *any* number of equations in *any* number of variables.

The method of substitution can be extended to these more complicated systems. For example, given three equations in three variables, suppose that it is possible to solve one of the equations for one variable in terms of the remaining two variables. By substituting that expression in each of the other equations, we obtain a system of two equations in two variables. The solutions of the two-variable system can then be used to find the solutions of the original system, as illustrated in the following example.

EXAMPLE 3   Find the solutions of the system

$$\begin{cases} x - y + z = 2 \\ \qquad\quad xyz = 0 \\ \quad 2y + z = 1. \end{cases}$$

SOLUTION   Solving the third equation for z,

$$z = 1 - 2y.$$

Substituting for z in the first two equations of the system we obtain the following system of two equations in two variables:

$$\begin{cases} x - y + (1 - 2y) = 2 \\ \qquad xy(1 - 2y) = 0. \end{cases}$$

This system is equivalent to

$$\begin{cases} x - 3y - 1 = 0 \\ xy(1 - 2y) = 0. \end{cases}$$

We now find the solutions of the last system. Solving the first equation for x in terms of y gives us

$$x = 3y + 1.$$

Substituting $3y + 1$ for x in the second equation $xy(1 - 2y) = 0$, we obtain

$$(3y + 1)y(1 - 2y) = 0,$$

which has as solutions the numbers $-\frac{1}{3}$, 0, and $\frac{1}{2}$. These are the only possible y-values for the solutions of the system. To obtain the corresponding x-values, we use the equation $x = 3y + 1$, obtaining $x = 0$, 1, and $\frac{5}{2}$, respectively. Finally, substituting in $z = 1 - 2y$ gives us the z-values $\frac{5}{3}$, 1, and 0. It follows that the solutions of the original system consist of the ordered triples

$$(0, -\tfrac{1}{3}, \tfrac{5}{3}), \quad (1, 0, 1), \quad \text{and} \quad (\tfrac{5}{2}, \tfrac{1}{2}, 0). \qquad \blacksquare$$

EXAMPLE 4 Is it possible to construct an aquarium with a glass top and two square ends that holds 16 ft³ of water and requires 40 ft² of glass?

FIGURE 6.3

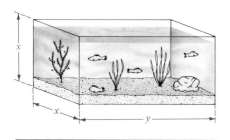

SOLUTION We begin by sketching a typical aquarium and labeling it as in Figure 6.3, in which x and y are in feet. Referring to the figure and using formulas for volume and area, we see that

$$\text{Volume of the aquarium} = x^2 y,$$

$$\text{Square feet of glass required} = 2x^2 + 4xy.$$

Since the volume and the glass required are 16 and 40, respectively, we obtain the following system of equations:

$$\begin{cases} x^2 y = 16 \\ 2x^2 + 4xy = 40 \end{cases}$$

Solving the first equation for y gives us $y = 16/x^2$. Substituting for y in the second equation leads to

$$2x^2 + 4x\left(\frac{16}{x^2}\right) = 40$$

$$2x^2 + \frac{64}{x} = 40$$

$$2x^3 + 64 = 40x$$

$$x^3 - 20x + 32 = 0$$

We next look for rational solutions of this equation. Dividing the polynomial $x^3 - 20x + 32$ synthetically by $x - 2$ gives us

$$\begin{array}{r|rrrr} 2 & 1 & 0 & -20 & 32 \\ & & 2 & 4 & -32 \\ \hline & 1 & 2 & -16 & 0 \end{array}$$

Thus, one solution of $x^3 - 20x + 32 = 0$ is 2, and the remaining two solutions satisfy

$$x^2 + 2x - 16 = 0.$$

Using the Quadratic Formula, we obtain

$$x = \frac{-2 \pm \sqrt{4 + 64}}{2} = \frac{-2 \pm 2\sqrt{17}}{2} = -1 \pm \sqrt{17}.$$

Since x is positive we may discard $-1 - \sqrt{17}$. Hence,

$$x = -1 + \sqrt{17} \approx 3.12.$$

The corresponding y-values can be determined from $y = 16/x^2$. On the one hand, letting $x = 2$ gives us $y = \frac{16}{4} = 4$. On the other hand, if $x = -1 + \sqrt{17}$, then it can be shown that $y = (9 + \sqrt{17})/8 \approx 1.64$. We have shown that there are two different ways to construct the aquarium. It is likely that most people (and fish) would prefer the dimensions 2 feet by 2 feet by 4 feet. ∎

Exercises 6.1

In Exercises 1–30 use the method of substitution to find the solutions of the system of equations.

1 $\begin{cases} y = x^2 - 4 \\ y = 2x - 1 \end{cases}$

2 $\begin{cases} y = x^2 + 1 \\ x + y = 3 \end{cases}$

3 $\begin{cases} y^2 = 1 - x \\ x + 2y = 1 \end{cases}$

4 $\begin{cases} y^2 = x \\ x + 2y + 3 = 0 \end{cases}$

5 $\begin{cases} 2y = x^2 \\ y = 4x^3 \end{cases}$

6 $\begin{cases} x - y^3 = 1 \\ 2x = 9y^2 + 2 \end{cases}$

7 $\begin{cases} x + 2y = -1 \\ 2x - 3y = 12 \end{cases}$

8 $\begin{cases} 3x - 4y + 20 = 0 \\ 3x + 2y + 8 = 0 \end{cases}$

9 $\begin{cases} 2x - 3y = 1 \\ -6x + 9y = 4 \end{cases}$

10 $\begin{cases} 4x - 5y = 2 \\ 8x - 10y = -5 \end{cases}$

11 $\begin{cases} x + 3y = 5 \\ x^2 + y^2 = 25 \end{cases}$

12 $\begin{cases} 3x - 4y = 25 \\ x^2 + y^2 = 25 \end{cases}$

13 $\begin{cases} x^2 + y^2 = 8 \\ y - x = 4 \end{cases}$

14 $\begin{cases} x^2 + y^2 = 25 \\ 3x + 4y = -25 \end{cases}$

15 $\begin{cases} x^2 + y^2 = 9 \\ y - 3x = 2 \end{cases}$

16 $\begin{cases} x^2 + y^2 = 16 \\ y + 2x = -1 \end{cases}$

17 $\begin{cases} x^2 + y^2 = 16 \\ 2y - x = 4 \end{cases}$

18 $\begin{cases} x^2 + y^2 = 1 \\ y + 2x = -3 \end{cases}$

19 $\begin{cases} (x - 1)^2 + (y + 2)^2 = 10 \\ x + y = 1 \end{cases}$

20 $\begin{cases} xy = 2 \\ 3x - y + 5 = 0 \end{cases}$

21 $\begin{cases} y = 20/x^2 \\ y = 9 - x^2 \end{cases}$

22 $\begin{cases} x = y^2 - 4y + 5 \\ x - y = 1 \end{cases}$

23 $\begin{cases} y^2 - 4x^2 = 4 \\ 9y^2 + 16x^2 = 140 \end{cases}$

24 $\begin{cases} 25y^2 - 16x^2 = 400 \\ 9y^2 - 4x^2 = 36 \end{cases}$

25 $\begin{cases} x^2 - y^2 = 4 \\ x^2 + y^2 = 12 \end{cases}$

26 $\begin{cases} 6x^3 - y^3 = 1 \\ 3x^3 + 4y^3 = 5 \end{cases}$

27 $\begin{cases} x + 2y - z = -1 \\ 2x - y + z = 9 \\ x + 3y + 3z = 6 \end{cases}$

28 $\begin{cases} 2x - 3y - z^2 = 0 \\ x - y - z^2 = -1 \\ x^2 - xy = 0 \end{cases}$

29 $\begin{cases} x^2 + z^2 = 5 \\ 2x + y = 1 \\ y + z = 1 \end{cases}$

30 $\begin{cases} x + 2z = 1 \\ 2y - z = 4 \\ xyz = 0 \end{cases}$

31 The perimeter of a rectangle is 40 inches and its area is 96 square inches. Find the length and width.

32 Find the values of b such that the system

$$\begin{cases} x^2 + y^2 = 4 \\ y = x + b \end{cases}$$

has solutions consisting of (a) one real number, (b) two real numbers, (c) no real numbers. Interpret the three cases geometrically.

33 Is there a real number x such that $x = 2^{-x}$? Decide by displaying graphically the system $y - x = 0$ and $y - 2^{-x} = 0$.

34 Is there a real number x such that $x = \log x$? Decide by displaying graphically the system $y - \log x = 0$ and $y - x = 0$.

35 Sections of tin cylindrical tubing are to be made from rectangular sheets that have an area of 200 in² (see figure). Is it possible to construct a tube that has a volume of 200 in³? If so, find the dimensions of the rectangular sheet.

FIGURE FOR EXERCISE 35

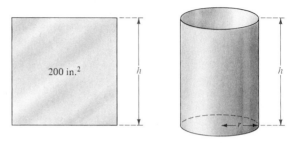

36 Shown in the figure is the graph of $y = x^2$ and a line of slope m that passes through the point (1, 1). Find a value of m such that the line intersects the graph only at (1, 1).

FIGURE FOR EXERCISE 36

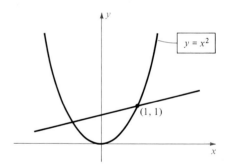

37 In fishery science, spawner-recruit functions are used to predict the number R in next year's breeding population from an estimate S of the number of fish presently spawning. (See Exercise 28 of Section 4.6.)

(a) If $R = aS/(S + b)$ estimate a and b from the data in the following table.

Year	1983	1984	1985
Number spawning	40,000	60,000	72,000

(b) Predict the breeding population for the year 1986.

38 Refer to Exercise 37. Ricker's spawner-recruit function is given by

$$R = aSe^{-bS}$$

where a and b are positive constants. This relationship predicts low recruitment from very high stocks and has been found to be appropriate for many species, such as arctic cod. Rework Exercise 37 using Ricker's spawner-recruit function.

39 A *competition model* is a collection of equations that specify how two or more species interact in competing for the food resources of an ecosystem. Let x and y denote the numbers (in hundreds) of two competing species

and suppose that the respective rates of growth R_1 and R_2 are given by

$$R_1 = 0.01x(50 - x - y),$$
$$R_2 = 0.02y(100 - y - 0.5x).$$

Determine the population levels (x, y) at which both rates of growth are zero. (Such population levels are called *stationary points*.)

40 A rancher has 2420 feet of fence to enclose pastureland that lies along a straight river. If no fence is used along the river, is it possible to enclose 10 acres of pastureland? Recall that 1 acre = 43,560 ft^2.

6.2 Systems of Linear Equations in Two Variables

An equation of the form $ax + by + c = 0$ (or, equivalently, $ax + by = -c$), where a and b are not both zero, is called a *linear equation* in x and y. Similarly, a linear equation in three variables x, y, and z is an equation of the form $ax + by + cz = d$, where the coefficients are real numbers. In general, if we let $x_1, x_2, \ldots, x_n$ denote variables where n is any positive integer, then an equation of the form

$$a_1x_1 + a_2x_2 + \cdots + a_nx_n = a$$

where $a_1, a_2, \ldots, a_n$ and a are real numbers and not every a_j is zero, is called a **linear equation in n variables** (with real coefficients).

The most common systems of equations are those in which all the equations are linear. In this section we shall only consider systems of two linear equations in two variables. Systems involving more than two variables are discussed in Section 6.3.

One method for solving a system of equations in several variables is to replace equations in the system by equivalent equations until we reach a system from which the solutions are easily obtained. Some rules for transforming systems of equations are stated in the next theorem. We shall use the notation $p = 0$ and $q = 0$ for typical equations in a system. Bear in mind that the symbols p and q represent expressions in the variables under consideration. For example, if we let $p = 3x - 2y + 5z - 1$ and $q = 6x + y - 7z - 4$, then the equations $p = 0$ and $q = 0$ have the forms

$$3x - 2y + 5z - 1 = 0, \qquad 6x + y - 7z - 4 = 0.$$

TRANSFORMATIONS THAT LEAD TO EQUIVALENT SYSTEMS OF EQUATIONS

> The following transformations do not change the solutions of a system of equations:
>
> (i) Interchanging the position of any two equations;
>
> (ii) Multiplying both sides of an equation in the system by a non-zero real number;
>
> (iii) Replacing an equation $q = 0$ of the system by $kp + q = 0$, where $p = 0$ is any other equation in the system and k is any real number.

PROOF   It is easy to show that (i) and (ii) do not change the solutions of the system, and therefore we shall omit the proofs.

To prove (iii), we first note that a solution of the original system is a solution of both equations $p = 0$ and $q = 0$. This means that each of the expressions p and q equals zero if the variables are replaced by appropriate real numbers, and hence the expression $kp + q$ will also equal zero. Since none of the other equations has been changed, this shows that any solution of the original system is also a solution of the transformed system, which is obtained by replacing the equation $q = 0$ by $kp + q = 0$.

Conversely, given a solution of the transformed system, both of the expressions $kp + q$ and p equal zero if the variables are replaced by appropriate numbers. This implies, however, that $(kp + q) - kp$ equals 0. Since $(kp + q) - kp = q$, we see that q must also equal zero when the substitution is made. Thus, a solution of the transformed system is also a solution of the original system. This completes the proof.   □

For convenience, we shall describe transformation (iii) by the phrase "add to one equation of the system k times any other equation of the system." Of course, to *add* two equations means to add corresponding sides. To *multiply* an equation by k means to multiply both sides of the equation by k. The process may also be applied if 0 is not on one side of each equation, as illustrated in the next example.

EXAMPLE 1   Find the solutions of the system

$$\begin{cases} x + 3y = -1 \\ 2x - y = 5. \end{cases}$$

SOLUTION By (ii) of the preceding theorem we may multiply the second equation by 3. This gives us the equivalent system

$$\begin{cases} x + 3y = -1 \\ 6x - 3y = 15. \end{cases}$$

Next, by (iii) of the theorem, we may add to the second equation 1 times the first equation, obtaining

$$\begin{cases} x + 3y = -1 \\ 7x = 14. \end{cases}$$

We see from the last equation that $x = 2$. The corresponding value for y may be found by substituting for x in the equation $x + 3y = -1$. This gives us

$$2 + 3y = -1, \qquad 3y = -3, \qquad y = -1.$$

Thus, the system has one solution, $(2, -1)$.

There are other methods of solution. For example, we could begin by multiplying the first equation by -2, obtaining

$$\begin{cases} -2x - 6y = 2 \\ 2x - y = 5. \end{cases}$$

If we next add the first equation to the second, we get

$$\begin{cases} -2x - 6y = 2 \\ - 7y = 7. \end{cases}$$

The last equation implies that $y = -1$. Substitution for y in the equation $-2x - 6y = 2$ gives us

$$-2x - 6(-1) = 2, \qquad -2x = -4, \qquad x = 2.$$

Again we see that the solution is $(2, -1)$.

The graphs of the two equations, showing the point of intersection $(2, -1)$, are sketched in Figure 6.4. ■

FIGURE 6.4

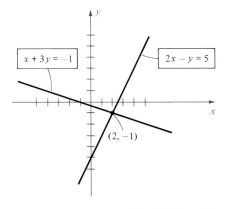

The technique used in Example 1 is called the **method of elimination,** since it involves eliminating a variable from one of the equations. The method of elimination usually leads to solutions in fewer steps than the method of substitution discussed in Section 6.1.

EXAMPLE 2 Find the solutions of the system

$$\begin{cases} 3x + y = 6 \\ 6x + 2y = 12. \end{cases}$$

SOLUTION Multiplying the second equation by $\frac{1}{2}$ gives us

$$\begin{cases} 3x + y = 6 \\ 3x + y = 6. \end{cases}$$

Thus, (a, b) is a solution if and only if $3a + b = 6$; that is, $b = 6 - 3a$. It follows that the solutions consist of all ordered pairs of the form $(a, 6 - 3a)$ where a is a real number. If we wish to find particular solutions we may substitute various values for a. A few solutions are $(0, 6)$, $(1, 3)$, $(3, -3)$, $(-2, 12)$ and $(\sqrt{2}, 6 - 3\sqrt{2})$. Note that the graph of each equation is the same line. ∎

EXAMPLE 3 Find the solutions of the system

$$\begin{cases} 3x + y = 6 \\ 6x + 2y = 20. \end{cases}$$

SOLUTION If we add to the second equation -2 times the first equation, we obtain the equivalent system

$$\begin{cases} 3x + y = 6 \\ 0 = 8. \end{cases}$$

The last equation can be written $0x + 0y = 8$, which is false for every ordered pair (x, y). Thus, the system has no solutions. This means that the graphs of the given equations do not intersect (see Figure 6.5). ∎

Since the graph of every linear equation $ax + by = c$ is a line, it follows that for every system of two such equations, precisely one of the following three possibilities occurs:

(i)   The lines intersect in exactly one point.

(ii) The lines are identical.

(iii) The lines are parallel.

Since the solutions of the system determine the points of intersection of the graphs of the two equations, we may interpret (i)–(iii) in the following way.

FIGURE 6.5

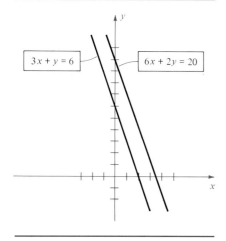

$3x + y = 6$ $6x + 2y = 20$

THEOREM

> For each system of two linear equations in two variables, one and only one of the following statements is true.
>
> (i)   The system has exactly one solution;
>
> (ii) The system has an infinite number of solutions;
>
> (iii) The system has no solution.

If (i) occurs the system is said to be **consistent**.

If (ii) occurs the equations are said to be **dependent**.

If (iii) occurs, the system is called **inconsistent**.

In practice, there should be little difficulty in determining which of the three cases occurs. The case of the unique solution (i) will become apparent when suitable transformations are applied to the system, as illustrated in Example 1. In case (ii) the solution is similar to that for Example 2 where one of the equations can be transformed into the other. In case (iii) the lack of a solution is indicated by an absurdity, such as the statement $0 = 8$, which appeared in Example 3.

Certain applied problems can be solved by introducing systems of two linear equations, as illustrated in the next example.

EXAMPLE 4   A motor boat, operating at full throttle, made a trip 4 miles upstream (against a constant current) in 15 minutes. The return trip (with the same current and at full throttle) took 12 minutes. Find the speed of the current and the equivalent speed of the boat in still water.

SOLUTION   We shall begin by introducing letters to denote the unknown quantities. Thus, let

$$x = \text{speed of boat (in mph)},$$

$$y = \text{speed of current (in mph)}.$$

We plan to use the formula $d = rt$, where d denotes the distance traveled, r the rate, and t the time. Since the current slows the boat as it travels upstream, but adds to its speed as it travels downstream, we obtain

$$\text{upstream rate} = x - y \quad \text{(in mph)}$$

$$\text{downstream rate} = x + y \quad \text{(in mph)}.$$

The time (in hours) traveled in each direction is

$$\text{upstream time} = \tfrac{15}{60} = \tfrac{1}{4} \text{ hr,}$$

$$\text{downstream time} = \tfrac{12}{60} = \tfrac{1}{5} \text{ hr.}$$

The distance is 4 miles for each trip. Substituting in $d = rt$ gives us the system

$$\begin{cases} 4 = (x - y)(\tfrac{1}{4}) \\ 4 = (x + y)(\tfrac{1}{5}) \end{cases}$$

or equivalently,

$$\begin{cases} x - y = 16 \\ x + y = 20. \end{cases}$$

Adding the last two equations, we see that $2x = 36$, or $x = 18$. Consequently, $y = 20 - x = 20 - 18 = 2$. Hence, the speed of the boat in still water is 18 miles per hour and the speed of the current is 2 miles per hour. ∎

Exercises 6.2

Find the solutions of the systems in Exercises 1–20.

1 $\begin{cases} 2x + 3y = 2 \\ x - 2y = 8 \end{cases}$

2 $\begin{cases} 4x + 5y = 13 \\ 3x + y = -4 \end{cases}$

3 $\begin{cases} 2x + 5y = 16 \\ 3x - 7y = 24 \end{cases}$

4 $\begin{cases} 7x - 8y = 9 \\ 4x + 3y = -10 \end{cases}$

5 $\begin{cases} 3r + 4s = 3 \\ r - 2s = -4 \end{cases}$

6 $\begin{cases} 9u + 2v = 0 \\ 3u - 5v = 17 \end{cases}$

7 $\begin{cases} 5x - 6y = 4 \\ 3x + 7y = 8 \end{cases}$

8 $\begin{cases} 2x + 8y = 7 \\ 3x - 5y = 4 \end{cases}$

9 $\begin{cases} \tfrac{1}{3}c + \tfrac{1}{2}d = 5 \\ c - \tfrac{2}{3}d = -1 \end{cases}$

10 $\begin{cases} \tfrac{1}{2}t - \tfrac{1}{5}v = \tfrac{3}{2} \\ \tfrac{2}{3}t + \tfrac{1}{4}v = \tfrac{5}{12} \end{cases}$

11 $\begin{cases} \sqrt{3}x - \sqrt{2}y = 2\sqrt{3} \\ 2\sqrt{2}x + \sqrt{3}y = \sqrt{2} \end{cases}$

12 $\begin{cases} 0.11x - 0.03y = 0.25 \\ 0.12x + 0.05y = 0.70 \end{cases}$

13 $\begin{cases} 2x - 3y = 5 \\ -6x + 9y = 12 \end{cases}$

14 $\begin{cases} 3p - q = 7 \\ -12p + 4q = 3 \end{cases}$

15 $\begin{cases} 3m - 4n = 2 \\ -6m + 8n = -4 \end{cases}$

16 $\begin{cases} x - 5y = 2 \\ 3x - 15y = 6 \end{cases}$

17 $\begin{cases} 2y - 5x = 0 \\ 3y + 4x = 0 \end{cases}$

18 $\begin{cases} 3x + 7y = 9 \\ y = 5 \end{cases}$

19 $\begin{cases} \dfrac{2}{x} + \dfrac{3}{y} = -2 \\ \dfrac{4}{x} - \dfrac{5}{y} = 1 \end{cases}$ (*Hint:* Let $x' = 1/x$ and $y' = 1/y$.)

20 $\begin{cases} \dfrac{3}{x - 1} + \dfrac{4}{y + 2} = 2 \\ \dfrac{6}{x - 1} - \dfrac{7}{y + 2} = -3 \end{cases}$

21 The price of admission for a high school play was $1.50 for students and $2.25 for nonstudents. If 450 tickets were sold for a total of $777.75, how many of each kind were purchased?

22 New West Airlines flies from Los Angeles to Albuquerque with a stopover in Phoenix. The airfare to Phoenix is $45 while the fare to Albuquerque is $60. A total of 185 passengers boarded the plane in Los Angeles and fares totaled $10,500. How many passengers got off the plane in Phoenix?

23 A crayon is to be 8 cm in length, 1 cm in diameter, and will be made from 5 cm³ of colored wax. The crayon is to have the shape of a cylinder surmounted by a small conical tip (see figure). Find the length of the cylinder and the height of the cone.

FIGURE FOR EXERCISE 23

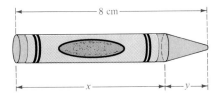

24 A man rows a boat 500 feet upstream against a constant current in 10 minutes. He then rows downstream (with the same current), covering 300 feet in 5 minutes. Find the speed of the current and the equivalent rate at which he can row in still water.

25 A large table for a conference room is to be constructed in the shape of a rectangle with two semicircles at the ends (see figure). Find the length and width of the rectangular portion assuming the table is to have a perimeter of 40 feet and the area of the rectangular portion is to be twice the sum of the areas of the two ends.

FIGURE FOR EXERCISE 25

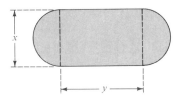

26 A woman has $10,000 invested in two funds that pay simple interest rates of 8% and 7%, respectively. If she receives yearly interest of $772, how much is invested in each fund?

27 A man receives income from two investments at simple interest rates of $6\frac{1}{2}\%$ and 8%, respectively. He has twice as much invested at $6\frac{1}{2}\%$ as at 8%. If his annual income from the two investments is $698.25, find how much is invested at each rate.

28 A 300 gallon water storage tank is filled by a single inlet pipe, and two identical outlet pipes can be used to supply water to the surrounding fields (see figure). It takes 5 hours to fill an empty tank when both outlet pipes are open. When one outlet pipe is closed, it takes 3 hours to fill the tank. Find the flow rates (in gallons per hour) in and out of the tank.

FIGURE FOR EXERCISE 28

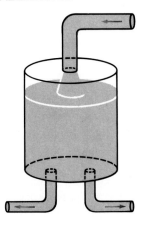

29 A silversmith has two alloys, the first containing 35% silver, and the second 60% silver. How much of each should be melted and combined to obtain 100 grams of an alloy containing 50% silver?

30 A merchant wishes to mix peanuts costing $2.00 per pound with cashews costing $3.50 per pound to obtain 60 pounds of a mixture costing $2.65 per pound. How many pounds of each variety should be mixed?

31 An airplane, flying with a tail wind, travels 1200 miles in 2 hours. The return trip, against the wind, takes $2\frac{1}{2}$ hours. Find the cruising speed of the plane and the speed of the wind (assume that both rates are constant).

32 A stationery company sells two types of notebooks to college bookstores, the first wholesaling for 50¢ and the second for 70¢. The company receives an order for 500 notebooks, together with a check for $286. If the order fails to specify the number of each type, how should the company fill the order?

33 As a ball rolls down an inclined plane, its velocity $v(t)$ (in cm/second) at time t (in seconds) is given by $v(t) = v_0 + at$, where v_0 is the initial velocity and a is

the acceleration (in cm/sec^2). If $v(2) = 16$ and $v(5) = 25$, find v_0 and a.

34 If an object is projected vertically upward from an altitude of s_0 feet with an initial velocity of v_0 feet/second, then its distance $s(t)$ above the ground after t seconds is

$$s(t) = -16t^2 + v_0 t + s_0.$$

If $s(1) = 84$ and $s(2) = 116$, what are v_0 and s_0?

35 A small furniture company manufactures sofas and recliners. Each sofa requires 8 hours of labor and $60 in materials, while a recliner can be built for $35 in 6 hours. The company has 340 hours of labor available each week and can afford to buy $2250 in materials. How many recliners and sofas can be produced if all labor hours and all materials will be used?

36 A rancher is preparing an oat–cornmeal mixture for livestock. Each ounce of oats contain 4 grams of protein and 18 grams of carbohydrates, while an ounce of corn meal has 3 grams of protein and 24 grams of carbohydrates. How many ounces of each can be used to meet the nutritional goals of 200 grams of protein and 1320 grams of carbohydrates per feeding?

37 A plumber and an electrician are each doing repairs on their places of business and agree to swap services. The number of hours spent on each of the projects is shown in the following table.

	Plumber's Business	Electrician's Business
Plumber's hours	6	4
Electrician's hours	5	6

Ordinarily they would call the matter even, but, due to tax laws, they must charge for all work performed. They agree to select hourly wage rates so that the total bill on each place of business will match the income that they would ordinarily receive on the two projects.

(a) If x and y denote the hourly wages of the plumber and electrician, respectively, show that $6x + 5y = 10x$ and $4x + 6y = 11y$. Describe the solutions to this system.

(b) If the plumber ordinarily makes $20 per hour, what should the electrician charge?

6.3 Systems of Linear Equations in More than Two Variables

For systems of linear equations containing more than two variables, we can use either the method of substitution explained in Section 6.1 or the method of elimination developed in Section 6.2. The method of elimination is the shorter and more straightforward technique for finding solutions. In addition, it leads to the matrix technique, discussed in this section.

EXAMPLE 1 Find the solutions of the system

$$\begin{cases} x - 2y + 3z = 4 \\ 2x + y - 4z = 3 \\ -3x + 4y - z = -2. \end{cases}$$

SOLUTION We shall begin by eliminating x from the second and third equations. If we add, to the second equation, -2 times the first equation, we get the equivalent system

$$\begin{cases} x - 2y + 3z = 4 \\ 5y - 10z = -5 \\ -3x + 4y - z = -2. \end{cases}$$

Next we add, to the third equation, 3 times the first equation. This gives us

$$\begin{cases} x - 2y + 3z = 4 \\ 5y - 10z = -5 \\ -2y + 8z = 10. \end{cases}$$

To simplify computations let us multiply the second equation by $\frac{1}{5}$, obtaining

$$\begin{cases} x - 2y + 3z = 4 \\ y - 2z = -1 \\ -2y + 8z = 10. \end{cases}$$

We now eliminate y from the third equation by adding to it 2 times the second equation. This gives us

$$\begin{cases} x - 2y + 3z = 4 \\ y - 2z = -1 \\ 4z = 8. \end{cases}$$

Finally, multiplying the third equation by $\frac{1}{4}$, we obtain

$$\begin{cases} x - 2y + 3z = 4 \\ y - 2z = -1 \\ z = 2. \end{cases}$$

The solutions of the last system are easy to find by a procedure called **back substitution.** Thus, from the third equation we obtain $z = 2$. Substituting 2 for z in the second equation, $y - 2z = -1$, we get

$$y - 2(2) = -1 \quad \text{or} \quad y = 3.$$

Finally, the x-value is found by substituting for y and z in the first equation.

This gives us

$$x - 2(3) + 3(2) = 4 \quad \text{and hence} \quad x = 4.$$

Thus, there is one solution, (4, 3, 2). ■

It can be shown that any system of three linear equations in three variables has either a *unique solution,* an *infinite number of solutions,* or *no solutions.* As with the case of two equations in two variables, the terminology used to describe these cases is *consistent, dependent,* or *inconsistent,* respectively.

If we analyze the method of solution in Example 1, we see that the symbols used for the variables are immaterial. The *coefficients* of the variables are the important things to consider. Thus, if different symbols such as r, s, and t are used for the variables, we obtain the system

$$\begin{cases} r - 2s + 3t = 4 \\ 2r + s - 4t = 3 \\ -3r + 4s - t = -2. \end{cases}$$

The method of elimination could then proceed exactly as before. Since this is true, it is possible to simplify the process. Specifically, we introduce a scheme for keeping track of the coefficients in such a way that the variables do not have to be written down. Referring to the preceding system and checking that the variables appear in the same order in each equation, and that terms not involving variables are to the right of the equal signs, we list the numbers that are involved in the equations in the following manner:

$$\begin{bmatrix} 1 & -2 & 3 & 4 \\ 2 & 1 & -4 & 3 \\ -3 & 4 & -1 & -2 \end{bmatrix}$$

If some variable had not appeared in one of the equations, we would have used a zero in the appropriate position. An array of numbers of this type is called a **matrix.** The **rows** of the matrix are the numbers that appear next to one another *horizontally.* Thus, the first row R_1 is 1 -2 3 4, the second row R_2 is 2 1 -4 3, and the third row R_3 is -3 4 -1 -2. The **columns** of the matrix are the numbers that appear *vertically.* For example, the second column consists of the numbers -2, 1, 4 (in that order); the fourth column consists of 4, 3, -2; and so on.

Before discussing a matrix method of solving a system of linear equations, let us state a general definition of matrix. We shall use a **double subscript notation.** Specifically, a_{ij} will denote the number that appears in row i and column j. We call i the **row subscript** and j the **column subscript** of a_{ij}.

DEFINITION

Let m and n be positive integers. An $m \times n$ **matrix** is an array of the form

$$\begin{bmatrix} a_{11} & a_{12} & a_{13} & \cdots & a_{1n} \\ a_{21} & a_{22} & a_{23} & \cdots & a_{2n} \\ a_{31} & a_{32} & a_{33} & \cdots & a_{3n} \\ \vdots & \vdots & \vdots & & \vdots \\ a_{m1} & a_{m2} & a_{m3} & \cdots & a_{mn} \end{bmatrix}$$

where each a_{ij} is a real number.

The notation $m \times n$ in the definition is read "m by n." It is possible to consider matrices in which the symbols a_{ij} represent complex numbers, polynomials, or other mathematical objects; however, we shall not do so in this text. The rows and columns of a matrix are defined as before. Thus, the matrix in the definition has m rows and n columns. Note that a_{23} is in row 2 and column 3, whereas a_{32} is in row 3 and column 2. Each a_{ij} is called an **element of the matrix.** The elements $a_{11}, a_{22}, a_{33}, \ldots$ are called the **main diagonal elements.** If $m \neq n$, the matrix is a **rectangular matrix.** If $m = n$, we refer to the matrix as a **square matrix of order n.**

Let us return to the system of equations considered in Example 1. The 3×4 matrix on page 298 is called the **matrix of the system.** The matrix of the system is also called the **augmented matrix.** In certain cases we may wish to consider only the *coefficients* of the variables. The corresponding matrix is called the **coefficient matrix.** The coefficient matrix in Example 1 is

$$\begin{bmatrix} 1 & -2 & 3 \\ 2 & 1 & -4 \\ -3 & 4 & -1 \end{bmatrix}.$$

After forming the matrix of the system, we work with the rows of the matrix *just as though they were equations.* The only items missing are the symbols for the variables, the addition signs used between terms, and the equal signs. We simply keep in mind that the numbers in the first column are the coefficients of the first variable, the numbers in the second column are the coefficients of the second variable, and so on. The rules for transforming a matrix are formulated so that they always produce a matrix of an equivalent system of equations. The following theorem follows from the theorem on transformations of systems, stated on page 290.

MATRIX ROW TRANSFORMATION THEOREM

Given a matrix of a system of linear equations, each of the following transformations results in a matrix of an equivalent system of linear equations:

(i) Interchanging any two rows;

(ii) Multiplying all of the elements in a row by the same nonzero real number k;

(iii) Adding to the elements in one row, k times the corresponding elements of any other row, for any real number k.

We shall refer to (ii) as "multiplying a row by the number k" and (iii) will be described by "add k times any other row to a row." Rules (i)—(iii) of the preceding theorem are called the **elementary row transformations** of a matrix.

It is convenient to use the following symbols to specify elementary row transformations of a matrix.

Symbol	*Meaning*
R_{ij}	Interchange rows i and j
kR_i	Multiply the ith row by k
$kR_i + R_j$	Add to the jth row, k times the ith row

We shall next rework Example 1 using matrices. It may be enlightening to compare the two solutions, since analogous steps are employed in each case.

EXAMPLE 2 Find the solutions of the system

$$\begin{cases} x - 2y + 3z = 4 \\ 2x + y - 4z = 3 \\ -3x + 4y - z = -2. \end{cases}$$

SOLUTION We begin with the matrix of the system and then change its form by means of elementary row transformations. An arrow, together with the symbols we have introduced, will specify each transformation. Thus, in the first transformation that follows, $-2R_1 + R_2$ indicates that we have added to the second row, -2 times the first row. In the second transformation, $3R_1 + R_3$ indicates that we have added to the third row, 3 times the first row. In the third transformation, $\frac{1}{5}R_2$ denotes the fact that the second row has been multiplied by $\frac{1}{5}$, and so on.

$$
\begin{bmatrix} 1 & -2 & 3 & 4 \\ 2 & 1 & -4 & 3 \\ -3 & 4 & -1 & -2 \end{bmatrix} \xrightarrow{\ -2R_1 + R_2\ } \begin{bmatrix} 1 & -2 & 3 & 4 \\ 0 & 5 & -10 & -5 \\ -3 & 4 & -1 & -2 \end{bmatrix}
$$

$$
\xrightarrow{\ 3R_1 + R_3\ } \begin{bmatrix} 1 & -2 & 3 & 4 \\ 0 & 5 & -10 & -5 \\ 0 & -2 & 8 & 10 \end{bmatrix}
$$

$$
\xrightarrow{\ \frac{1}{5}R_2\ } \begin{bmatrix} 1 & -2 & 3 & 4 \\ 0 & 1 & -2 & -1 \\ 0 & -2 & 8 & 10 \end{bmatrix}
$$

$$
\xrightarrow{\ 2R_2 + R_3\ } \begin{bmatrix} 1 & -2 & 3 & 4 \\ 0 & 1 & -2 & -1 \\ 0 & 0 & 4 & 8 \end{bmatrix}
$$

$$
\xrightarrow{\ \frac{1}{4}R_3\ } \begin{bmatrix} 1 & -2 & 3 & 4 \\ 0 & 1 & -2 & -1 \\ 0 & 0 & 1 & 2 \end{bmatrix}
$$

We use the final matrix to return to the system of equations

$$
\begin{cases} x - 2y + 3z = 4 \\ y - 2z = -1 \\ z = 2, \end{cases}
$$

which is equivalent to the original system. The solutions may now be found by back substitution as in Example 1. ∎

The final matrix in the solution of Example 2 is said to be in **echelon form.** In general, a matrix is in echelon form if it satisfies the following conditions.

ECHELON FORM OF A MATRIX

(i) The first nonzero number in each row, reading from left to right, is 1.

(ii) The column containing the first nonzero number in any row is to the left of the column containing the first nonzero number in the next row.

(iii) Rows consisting entirely of zeros appear at the bottom of the matrix.

We can use elementary row operations to transform the matrix of any system of linear equations to echelon form. The echelon form can then be used to produce a system of equations that is equivalent to the original system. The solutions of the latter system may be found by back substitution. The next example illustrates this technique for a system of four linear equations.

EXAMPLE 3 Find the solutions of the system

$$\begin{cases} x & -2z + 2w = & 1 \\ -2x + 3y + 4z & = -1 \\ & y + z - w = & 0 \\ 3x + & y - 2z - w = & 3. \end{cases}$$

SOLUTION We have arranged the equations so that the same variables appear in vertical columns. We begin with the matrix of the system and proceed as follows:

$$\begin{bmatrix} 1 & 0 & -2 & 2 & 1 \\ -2 & 3 & 4 & 0 & -1 \\ 0 & 1 & 1 & -1 & 0 \\ 3 & 1 & -2 & -1 & 3 \end{bmatrix} \xrightarrow{2R_1 + R_2} \begin{bmatrix} 1 & 0 & -2 & 2 & 1 \\ 0 & 3 & 0 & 4 & 1 \\ 0 & 1 & 1 & -1 & 0 \\ 3 & 1 & -2 & -1 & 3 \end{bmatrix}$$

$$\xrightarrow{-3R_1 + R_4} \begin{bmatrix} 1 & 0 & -2 & 2 & 1 \\ 0 & 3 & 0 & 4 & 1 \\ 0 & 1 & 1 & -1 & 0 \\ 0 & 1 & 4 & -7 & 0 \end{bmatrix}$$

$$\xrightarrow{R_{23}} \begin{bmatrix} 1 & 0 & -2 & 2 & 1 \\ 0 & 1 & 1 & -1 & 0 \\ 0 & 3 & 0 & 4 & 1 \\ 0 & 1 & 4 & -7 & 0 \end{bmatrix}$$

$$\xrightarrow{-3R_2 + R_3} \begin{bmatrix} 1 & 0 & -2 & 2 & 1 \\ 0 & 1 & 1 & -1 & 0 \\ 0 & 0 & -3 & 7 & 1 \\ 0 & 1 & 4 & -7 & 0 \end{bmatrix}$$

$$\xrightarrow{(-1)R_2 + R_4} \begin{bmatrix} 1 & 0 & -2 & 2 & 1 \\ 0 & 1 & 1 & -1 & 0 \\ 0 & 0 & -3 & 7 & 1 \\ 0 & 0 & 3 & -6 & 0 \end{bmatrix}$$

$$\xrightarrow{R_3 + R_4} \begin{bmatrix} 1 & 0 & -2 & 2 & 1 \\ 0 & 1 & 1 & -1 & 0 \\ 0 & 0 & -3 & 7 & 1 \\ 0 & 0 & 0 & 1 & 1 \end{bmatrix}$$

$$\xrightarrow{-\frac{1}{3}R_3} \begin{bmatrix} 1 & 0 & -2 & 2 & 1 \\ 0 & 1 & 1 & -1 & 0 \\ 0 & 0 & 1 & -\frac{7}{3} & -\frac{1}{3} \\ 0 & 0 & 0 & 1 & 1 \end{bmatrix}$$

The final matrix is in echelon form and corresponds to the following system of equations

$$\begin{cases} x & -2z + 2w = & 1 \\ y + & z - & w = & 0 \\ & z - \frac{7}{3}w = & -\frac{1}{3} \\ & w = & 1. \end{cases}$$

We now use back substitution to find the solution. From the last equation we see that $w = 1$. Substituting in the third equation $z - \frac{7}{3}w = -\frac{1}{3}$, we get

$$z - \frac{7}{3}(1) = -\frac{1}{3} \quad \text{or} \quad z = \frac{6}{3} = 2.$$

Substituting $w = 1$ and $z = 2$ in the second equation $y + z - w = 0$, we obtain

$$y + 2 - 1 = 0 \quad \text{or} \quad y = -1.$$

Finally, from the first equation $x - 2z + 2w = 1$, we have

$$x - 2(2) + 2(1) = 1 \quad \text{or} \quad x = 3.$$

Hence, the system has one solution, $x = 3$, $y = -1$, $z = 2$, and $w = 1$. ∎

Sometimes it is necessary to consider systems in which the number of equations is not the same as the number of variables. The same matrix techniques are applicable, as illustrated in the next example.

EXAMPLE 4   Find the solutions of the system

$$\begin{cases} 2x + 3y + 4z = 1 \\ 3x + 4y + 5z = 3. \end{cases}$$

SOLUTION When the number of equations is less than the number of unknowns we shall go beyond the echelon form, applying elementary row transformations until there are zeros both below *and* above the first non-zero entry in each row. Thus,

$$\begin{bmatrix} 2 & 3 & 4 & 1 \\ 3 & 4 & 5 & 3 \end{bmatrix} \xrightarrow{-\frac{3}{2}R_1 + R_2} \begin{bmatrix} 2 & 3 & 4 & 1 \\ 0 & -\frac{1}{2} & -1 & \frac{3}{2} \end{bmatrix}$$

$$\xrightarrow{\frac{1}{2}R_1} \begin{bmatrix} 1 & \frac{3}{2} & 2 & \frac{1}{2} \\ 0 & -\frac{1}{2} & -1 & \frac{3}{2} \end{bmatrix}$$

$$\xrightarrow{-2R_2} \begin{bmatrix} 1 & \frac{3}{2} & 2 & \frac{1}{2} \\ 0 & 1 & 2 & -3 \end{bmatrix}$$

$$\xrightarrow{-\frac{3}{2}R_2 + R_1} \begin{bmatrix} 1 & 0 & -1 & 5 \\ 0 & 1 & 2 & -3 \end{bmatrix}$$

The last matrix is the matrix of the system

$$\begin{cases} x & - z = & 5 \\ & y + 2z = & -3 \end{cases}$$

or equivalently,

$$\begin{cases} x = & z + 5 \\ y = & -2z - 3. \end{cases}$$

There are an infinite number of solutions to this system; they can be found by assigning z any value c and then using the last two equations to express x and y in terms of c. This gives us

$$x = c + 5, \qquad y = -2c - 3, \qquad z = c.$$

Thus, the solutions of the system consist of all ordered triples of the form

$$(c + 5, -2c - 3, c)$$

where c is any real number. This may be checked by substituting $c + 5$ for x, $-2c - 3$ for y, and c for z in the two given equations.

We can obtain any number of solutions for the system by substituting specific real numbers for c. For example, if $c = 0$, we obtain $(5, -3, 0)$; if $c = 2$, we have $(7, -7, 2)$; and so on.

There are other ways to specify the general solution. For example, starting with $x = z + 5$ and $y = -2z - 3$, we could let $z = d - 5$, where d is any real number. In this event,

$$x = z + 5 = (d - 5) + 5 = d$$

$$y = -2z - 3 = -2(d - 5) - 3 = -2d + 7$$

and the solutions of the system have the form

$$(d, -2d + 7, d - 5).$$

These triples produce the same solutions as $(c + 5, -2c - 3, c)$. For example, if $d = 5$ we get $(5, -3, 0)$, if $d = 7$ we obtain $(7, -7, 2)$, and so on. ∎

A system of linear equations is said to be **homogeneous** if all the terms that do not contain variables are zero. A system of homogeneous equations always has the **trivial solution** obtained by substituting zero for each variable. Nontrivial solutions sometimes exist. The procedure for finding solutions is the same as that used for nonhomogeneous systems.

EXAMPLE 5   Find the solutions of the homogeneous system

$$\begin{cases} x - y + 4z = 0 \\ 2x + y - z = 0 \\ -x - y + 2z = 0. \end{cases}$$

SOLUTION   We proceed as follows:

$$\begin{bmatrix} 1 & -1 & 4 & 0 \\ 2 & 1 & -1 & 0 \\ -1 & -1 & 2 & 0 \end{bmatrix} \xrightarrow{-2R_1 + R_2} \begin{bmatrix} 1 & -1 & 4 & 0 \\ 0 & 3 & -9 & 0 \\ -1 & -1 & 2 & 0 \end{bmatrix}$$

$$\xrightarrow{R_1 + R_3} \begin{bmatrix} 1 & -1 & 4 & 0 \\ 0 & 3 & -9 & 0 \\ 0 & -2 & 6 & 0 \end{bmatrix}$$

$$\xrightarrow{\frac{1}{3}R_2} \begin{bmatrix} 1 & -1 & 4 & 0 \\ 0 & 1 & -3 & 0 \\ 0 & -2 & 6 & 0 \end{bmatrix}$$

$$\xrightarrow{2R_2 + R_3} \begin{bmatrix} 1 & -1 & 4 & 0 \\ 0 & 1 & -3 & 0 \\ 0 & 0 & 0 & 0 \end{bmatrix}$$

$$\xrightarrow{R_2 + R_1} \begin{bmatrix} 1 & 0 & 1 & 0 \\ 0 & 1 & -3 & 0 \\ 0 & 0 & 0 & 0 \end{bmatrix}$$

Note that the second from the last matrix is in echelon form. As in Example 4, we have gone one step further, obtaining a zero *above* the first nonzero entry in the second row. The final matrix corresponds to the system

$$\begin{cases} x & + & z = 0 \\ & y - 3z = 0 \end{cases} \quad \text{or} \quad \begin{cases} x = -z \\ y = 3z \end{cases}$$

Assigning any value c to z, we obtain $x = -c$ and $y = 3c$. Thus, the solutions consist of all ordered triples of the form $(-c, 3c, c)$ where c is any real number. ∎

EXAMPLE 6   Find the solutions of the system

$$\begin{cases} x + y + z = 0 \\ x - y + z = 0 \\ x - y - z = 0. \end{cases}$$

SOLUTION

$$\begin{bmatrix} 1 & 1 & 1 & 0 \\ 1 & -1 & 1 & 0 \\ 1 & -1 & -1 & 0 \end{bmatrix} \xrightarrow{(-1)R_1 + R_2} \begin{bmatrix} 1 & 1 & 1 & 0 \\ 0 & -2 & 0 & 0 \\ 1 & -1 & -1 & 0 \end{bmatrix}$$

$$\xrightarrow{(-1)R_1 + R_3} \begin{bmatrix} 1 & 1 & 1 & 0 \\ 0 & -2 & 0 & 0 \\ 0 & -2 & -2 & 0 \end{bmatrix}$$

$$\xrightarrow{-\frac{1}{2}R_2} \begin{bmatrix} 1 & 1 & 1 & 0 \\ 0 & 1 & 0 & 0 \\ 0 & -2 & -2 & 0 \end{bmatrix}$$

$$\xrightarrow{2R_2 + R_3} \begin{bmatrix} 1 & 1 & 1 & 0 \\ 0 & 1 & 0 & 0 \\ 0 & 0 & -2 & 0 \end{bmatrix}$$

$$\xrightarrow{-\frac{1}{2}R_3} \begin{bmatrix} 1 & 1 & 1 & 0 \\ 0 & 1 & 0 & 0 \\ 0 & 0 & 1 & 0 \end{bmatrix}$$

The last matrix is in echelon form and is the matrix of the system

$$x + y + z = 0, \qquad y = 0, \qquad z = 0.$$

It follows that the only solution for the given system is the trivial one, $(0, 0, 0)$. ∎

The next example is an illustration of an applied problem that can be solved by means of a system of three linear equations.

EXAMPLE 7 A merchant wishes to mix two grades of peanuts costing $1.50 and $2.50 per pound, respectively, with cashews costing $4.00 per pound, to obtain 130 pounds of a mixture costing $3.00 per pound. If the merchant also wants the amount of cheaper grade peanuts to be twice that of the better grade, how many pounds of each variety should be mixed?

SOLUTION Let us introduce three variables as follows:

$$x = \text{pounds of peanuts at \$1.50 per pound},$$

$$y = \text{pounds of peanuts at \$2.50 per pound},$$

$$z = \text{pounds of cashews at \$4.00 per pound}.$$

We refer to the statement of the problem and obtain the following system of equations

$$\begin{cases} x + y + z = 130 \\ 1.50x + 2.50y + 4.00z = 3.00(130) \\ x = 2y. \end{cases}$$

We leave it to the reader to verify that the solution to this system is $x = 40$, $y = 20$, $z = 70$. Thus, the merchant should use 40 pounds of the $1.50 peanuts, 20 pounds of the $2.50 peanuts, and 70 pounds of cashews. ∎

Exercises 6.3

Use matrices to solve the systems in Exercises 1–26.

1 $\begin{cases} x - 2y - 3z = -1 \\ 2x + y + z = 6 \\ x + 3y - 2z = 13 \end{cases}$

2 $\begin{cases} x + 3y - z = -3 \\ 3x - y + 2z = 1 \\ 2x - y + z = -1 \end{cases}$

3 $\begin{cases} 5x + 2y - z = -7 \\ x - 2y + 2z = 0 \\ 3y + z = 17 \end{cases}$

4 $\begin{cases} 4x - y + 3z = 6 \\ -8x + 3y - 5z = -6 \\ 5x - 4y = -9 \end{cases}$

5 $\begin{cases} 2x + 6y - 4z = 1 \\ x + 3y - 2z = 4 \\ 2x + y - 3z = -7 \end{cases}$

6 $\begin{cases} 2x - y = 5 \\ 5y + 3z = -2 \\ x - 7z = 3 \end{cases}$

7 $\begin{cases} 2x - 3y + 2z = -3 \\ -3x + 2y + z = 1 \\ 4x + y - 3z = 4 \end{cases}$

8 $\begin{cases} 2x - 3y + z = 2 \\ 3x + 2y - z = -5 \\ 5x - 2y + z = 0 \end{cases}$

9 $\begin{cases} x + 3y + z = 0 \\ x + y - z = 0 \\ x - 2y - 4z = 0 \end{cases}$

10 $\begin{cases} 2x - y + z = 0 \\ x - y - 2z = 0 \\ 2x - 3y - z = 0 \end{cases}$

11 $\begin{cases} 2x + y + z = 0 \\ x - 2y - 2z = 0 \\ x + y + z = 0 \end{cases}$

12 $\begin{cases} x + y - 2z = 0 \\ x - y - 4z = 0 \\ y + z = 0 \end{cases}$

13 $\begin{cases} 3x - 2y + 5z = 7 \\ x + 4y - z = -2 \end{cases}$

14 $\begin{cases} 2x - y + 4z = 8 \\ -3x + y - 2z = 5 \end{cases}$

15 $\begin{cases} 4x - 2y + z = 5 \\ 3x + y - 4z = 0 \end{cases}$

16 $\begin{cases} 5x + 2y - z = 10 \\ y + z = -3 \end{cases}$

17 $\begin{cases} x + 2y - z - 3w = 2 \\ 3x + y - 2z - w = 6 \\ x + y + 3z - 2w = -3 \\ 4x - 3y - z - 2w = -8 \end{cases}$

18 $\begin{cases} x - 2y - 5z + w = -1 \\ 2x - y + z + w = 1 \\ 3x - 2y - 4z - 2w = 1 \\ x + y + 3z - 2w = 2 \end{cases}$

19 $\begin{cases} 2x - y - 2z + 2s - 5t = 2 \\ x + 3y - 2z + s - 2t = -5 \\ -x + 4y + 2z - 3s + 8t = -4 \\ 3x - 2y - 4z + s - 3t = -3 \\ 4x - 6y + z - 2s + t = 10 \end{cases}$

20 $\begin{cases} 3x + 2y + z + 3u + v + w = 1 \\ 2x + y - 2z + 3u - v + 4w = 6 \\ 6x + 3y + 4z - u + 2v + w = -6 \\ x + y + z + u - v - w = 8 \\ -2x - 2y + z - 3u + 2v - 3w = -10 \\ x - 3y + 2z + u + 3v + w = -1 \end{cases}$

21 $\begin{cases} 5x + 2z = 1 \\ y - 3z = 2 \\ 2x + y = 3 \end{cases}$

22 $\begin{cases} 2x - 5y = 4 \\ 3y + 2z = -3 \\ 7x - 3z = 1 \end{cases}$

23 $\begin{cases} 4x - 3y = 1 \\ 2x + y = -7 \\ -x + y = -1 \end{cases}$

24 $\begin{cases} 2x + 3y = -2 \\ x + y = 1 \\ x - 2y = 13 \end{cases}$

25 $\begin{cases} 2x + 3y = 5 \\ x - 3y = 4 \\ x + y = -2 \end{cases}$

26 $\begin{cases} 4x - y = 2 \\ 2x + 2y = 1 \\ 4x - 5y = 3 \end{cases}$

27 A chemist has three solutions containing a certain acid. The first contains 10% acid, the second 30%, and the third 50%. He wishes to use all three solutions to obtain a mixture of 50 liters containing 32% acid, using twice as much of the 50% solution as the 30% solution. How many liters of each solution should be used?

28 A swimming pool can be filled by three pipes A, B, and C. Pipe A alone can fill the pool in 8 hours. If pipes A and C are used together, the pool can be filled in 6 hours. If B and C are used together, it takes 10 hours. How long does it take to fill the pool if all three pipes are used?

29 A company has three machines A, B, and C that are each capable of producing a certain item. However, because of a lack of skilled operators, only two of the machines can be used simultaneously. The following table indicates production over a three-day period using various combinations of the machines.

Machines used	Hours used	Items produced
A and B	6	4500
A and C	8	3600
B and C	7	4900

How long would it take each machine, if used alone, to produce 1000 items?

30 In electrical circuits, the formula $1/R = (1/R_1) + (1/R_2)$ is used to find the total resistance R if two resistors R_1 and R_2 are connected in parallel. Given three resis-

tors A, B, and C, suppose that the total resistance is 48 ohms if A and B are connected in parallel; 80 ohms if B and C are connected in parallel, and 60 ohms if A and C are connected in parallel. Find A, B, and C.

31 A supplier of lawn products has three types of grass fertilizer, G_1, G_2, and G_3, having nitrogen contents of 30%, 20%, and 15%, respectively. The supplier plans to mix them, obtaining 600 pounds of fertilizer with a 25% nitrogen content. In addition, the mixture is to contain 100 pounds more of type G_3 than of type G_2. How much of each type should be used?

32 If a particle moves along a coordinate line with a constant acceleration a (cm/sec^2), then at time t (sec) its distance $s(t)$ (cm) from the origin is

$$s(t) = \tfrac{1}{2}at^2 + v_0t + s_0$$

where v_0 and s_0 are the velocity and distance from the origin, respectively, at $t = 0$. If the distances of the particle from the origin at $t = \tfrac{1}{2}$, $t = 1$ and $t = \tfrac{3}{2}$ are 7, 11, and 17, respectively, find a, v_0, and s_0.

33 Shown in the figure is an electrical circuit containing three resistors, a 6-volt battery, and a 12-volt battery. It can be shown, using Kirchoff's Laws, that the three currents I_1, I_2, and I_3 satisfy the following system of equations:

$$\begin{cases} I_1 - I_2 + I_3 = 0 \\ R_1I_1 + R_2I_2 = 6 \\ R_2I_2 + R_3I_3 = 12. \end{cases}$$

Find the three currents if

(a) $R_1 = R_2 = R_3 = 3$ ohms.

(b) $R_1 = 4$ ohms, $R_2 = 1$ ohm, and $R_3 = 4$ ohms.

FIGURE FOR EXERCISE 33

34 A stable population of 35,000 birds lives on three islands. Each year 10% of the population on island A migrates to island B, 20% of the population on island B migrates to island C, and 5% of the population on island C migrates to A. Find the number of birds on each island if the populations on each island do not vary from year to year.

35 A shop specializes in preparing blends of gourmet coffees. The owner wishes to prepare 1-pound bags that will sell for $3.50 from Columbian, Brazilian, and Kenyan coffees. The cost per pound of these coffees is $4, $2, and $3, respectively.

(a) Show that at least $\tfrac{1}{2}$ pound of Columbian coffee and no more than $\tfrac{1}{4}$ pound of Brazilian coffee can be used.

(b) Find the amount of each type of coffee assuming the proprietor decides to use $\tfrac{1}{8}$ pound of Brazilian coffee.

36 A rancher has 750 head of cattle consisting of 400 adults (aged 2 or more years), 150 yearlings, and 200 calves. The following information is known about this particular species. Each spring an adult female gives birth to a single calf and 75% of these calves will survive the year. The yearly survival percentages for yearlings and adults are 80% and 90% respectively. Finally, the male–female ratio is 1 in all age classes.

(a) Estimate the population of each age class next spring.

(b) Estimate the population of each age class last spring.

37 Find a quadratic function f such that $f(1) = 3$, $f(2) = \tfrac{5}{2}$, and $f(-1) = 1$.

38 If $f(x) = ax^3 + bx + c$, determine a, b, and c such that the graph of f passes through the points $(-3, -12)$, $(-1, 22)$, and $(2, 13)$.

39 Find an equation of the circle that passes through the three points $P_1(2, 1)$, $P_2(-1, -4)$, and $P_3(3, 0)$. (*Hint:* An equation of the circle has the form $x^2 + y^2 + ax + by + c = 0$.)

40 Determine a, b, and c such that the graph of the equation $y = ax^2 + bx + c$ passes through the points $P_1(3, -1)$, $P_2(1, -7)$, and $P_3(-2, 14)$.

6.4 Partial Fractions

In this section we show how systems of equations can be used to help decompose rational expressions into sums of simpler expressions. This technique is useful in certain parts of advanced mathematics courses, such as calculus.

It is easy to verify that

$$\frac{2}{x^2 - 1} = \frac{1}{x - 1} + \frac{-1}{x + 1}.$$

The expression on the right side of this equation is called *the partial fraction decomposition* of $2/(x^2 - 1)$.

It is theoretically possible to write *any* rational expression as a sum of rational expressions whose denominators involve powers of polynomials of degree not greater than two. Specifically, if $f(x)$ and $g(x)$ are polynomials *and the degree of $f(x)$ is less than the degree of $g(x)$,* it can be proved that

$$\frac{f(x)}{g(x)} = F_1 + F_2 + \cdots + F_r,$$

where each F_k has one of the forms

$$\frac{A}{(px + q)^m} \quad \text{or} \quad \frac{Cx + D}{(ax^2 + bx + c)^n}$$

for some nonnegative integers m and n, and where $ax^2 + bx + c$ is **irreducible** in the sense that this quadratic polynomial has no real zeros, that is, $b^2 - 4ac < 0$. The sum $F_1 + F_2 + \cdots + F_r$ is called the **partial fraction decomposition** of $f(x)/g(x)$ and each F_k is called a **partial fraction.** We shall not prove this result but will, instead, give rules for obtaining the decomposition.

To find the partial fraction decomposition of $f(x)/g(x)$ *it is essential that $f(x)$ have lower degree than $g(x)$.* If this is not the case, then long division should be employed to arrive at such an expression. For example, given

$$\frac{x^3 - 6x^2 + 5x - 3}{x^2 - 1}$$

we obtain
$$\frac{x^3 - 6x^2 + 5x - 3}{x^2 - 1} = x - 6 + \frac{6x - 9}{x^2 - 1}.$$

The partial fraction decomposition is then found for $(6x - 9)/(x^2 - 1)$.

GUIDELINES: **Finding Partial Fraction Decompositions of** $\dfrac{f(x)}{g(x)}$

A If the degree of $f(x)$ is not lower than the degree of $g(x)$, use long division to obtain the proper form.

B Express $g(x)$ as a product of linear factors $px + q$ or irreducible quadratic factors $ax^2 + bx + c$, and collect repeated factors so that $g(x)$ is a product of *different* factors of the form $(px + q)^m$ or $(ax^2 + bx + c)^n$, where m and n are nonnegative integers.

C Apply the following rules.

Rule 1. For each factor of the form $(px + q)^m$ where $m \geq 1$, the partial fraction decomposition contains a sum of m partial fractions of the form

$$\frac{A_1}{px + q} + \frac{A_2}{(px + q)^2} + \cdots + \frac{A_m}{(px + q)^m}$$

where each A_k is a real number.

Rule 2. For each factor of the form $(ax^2 + bx + c)^n$ where $n \geq 1$ and $ax^2 + bx + c$ is irreducible, the partial fraction decomposition contains a sum of n partial fractions of the form

$$\frac{A_1 x + B_1}{ax^2 + bx + c} + \frac{A_2 x + B_2}{(ax^2 + bx + c)^2} + \cdots + \frac{A_n x + B_n}{(ax^2 + bx + c)^n}$$

where each A_k and B_k is a real number. ■ ■ ■

EXAMPLE 1 Find the partial fraction decomposition of

$$\frac{4x^2 + 13x - 9}{x^3 + 2x^2 - 3x}.$$

SOLUTION   The denominator has the factored form $x(x + 3)(x - 1)$. Each factor has the form stated in Rule 1, with $m = 1$. Thus, for the factor x there corresponds a partial fraction of the form A/x. Similarly, for the factors $x + 3$ and $x - 1$ there correspond partial fractions $B/(x + 3)$ and $C/(x - 1)$, respectively. The partial fraction decomposition has the form

$$\frac{4x^2 + 13x - 9}{x(x + 3)(x - 1)} = \frac{A}{x} + \frac{B}{x + 3} + \frac{C}{x - 1}.$$

Multiplying by the lowest common denominator, $x(x + 3)(x - 1)$, gives us

$$4x^2 + 13x - 9 = A(x + 3)(x - 1) + Bx(x - 1) + Cx(x + 3)$$
$$= A(x^2 + 2x - 3) + B(x^2 - x) + C(x^2 + 3x).$$

This may also be written

$$4x^2 + 13x - 9 = (A + B + C)x^2 + (2A - B + 3C)x - 3A.$$

If we equate the coefficients of like powers of x on each side of the last equation, we obtain the system of equations

$$\begin{cases} A + B + C = 4 \\ 2A - B + 3C = 13 \\ -3A = -9. \end{cases}$$

It can be shown that the solution is $A = 3$, $B = -1$, $C = 2$. Thus, the partial fraction decomposition is

$$\frac{4x^2 + 13x - 9}{x(x + 3)(x - 1)} = \frac{3}{x} + \frac{-1}{x + 3} + \frac{2}{x - 1}. \qquad \blacksquare$$

EXAMPLE 2 Find the partial fraction decomposition of

$$\frac{x^2 + 10x - 36}{x(x - 3)^2}.$$

SOLUTION By Rule 1 with $m = 1$, there is a partial fraction A/x corresponding to the factor x. Next, applying Rule 1 with $m = 2$, the factor $(x - 3)^2$ determines a sum of two partial fractions $B/(x - 3) + C/(x - 3)^2$. Thus, the partial fraction decomposition has the form

$$\frac{x^2 + 10x - 36}{x(x - 3)^2} = \frac{A}{x} + \frac{B}{x - 3} + \frac{C}{(x - 3)^2}.$$

Multiplying both sides by $x(x - 3)^2$ gives us

$$\begin{aligned} x^2 + 10x - 36 &= A(x - 3)^2 + Bx(x - 3) + Cx \\ &= A(x^2 - 6x + 9) + B(x^2 - 3x) + Cx \end{aligned}$$

or equivalently,

$$x^2 + 10x - 36 = (A + B)x^2 + (-6A - 3B + C)x + 9A.$$

As in Example 1 we equate the coefficients of like powers of x, obtaining

$$\begin{cases} A + B = 1 \\ -6A - 3B + C = 10 \\ 9A = -36. \end{cases}$$

This system of equations has the solution $A = -4$, $B = 5$, $C = 1$. The partial fraction decomposition is therefore

$$\frac{x^2 + 10x - 36}{x(x - 3)^2} = \frac{-4}{x} + \frac{5}{x - 3} + \frac{1}{(x - 3)^2}.$$

■

EXAMPLE 3 Find the partial fraction decomposition of

$$\frac{x^2 - x - 21}{2x^3 - x^2 + 8x - 4}.$$

SOLUTION The denominator may be factored by grouping, as follows:

$$2x^3 - x^2 + 8x - 4 = x^2(2x - 1) + 4(2x - 1) = (x^2 + 4)(2x - 1).$$

Applying Rule 2 to the irreducible quadratic factor $x^2 + 4$, we see that one of the partial fractions has the form $(Ax + B)/(x^2 + 4)$. By Rule 1, there is also a partial fraction $C/(2x - 1)$ corresponding to factor $2x - 1$. Consequently,

$$\frac{x^2 - x - 21}{2x^3 - x^2 + 8x - 4} = \frac{Ax + B}{x^2 + 4} + \frac{C}{2x - 1}.$$

As in previous examples, this leads to

$$x^2 - x - 21 = (Ax + B)(2x - 1) + C(x^2 + 4)$$
$$= 2Ax^2 - Ax + 2Bx - B + Cx^2 + 4C$$

or $\quad x^2 - x - 21 = (2A + C)x^2 + (-A + 2B)x - B + 4C.$

This gives us the system

$$\begin{cases} 2A + C = 1 \\ -A + 2B = -1 \\ - B + 4C = -21, \end{cases}$$

which has the solution $A = 3$, $B = 1$, $C = -5$. Thus, the partial fraction decomposition is

$$\frac{x^2 - x - 21}{2x^3 - x^2 + 8x - 4} = \frac{3x + 1}{x^2 + 4} + \frac{-5}{2x - 1}.$$

■

EXAMPLE 4 Find the partial fraction decomposition of

$$\frac{5x^3 - 3x^2 + 7x - 3}{(x^2 + 1)^2}.$$

SOLUTION Applying Rule 2 with $n = 2$,

$$\frac{5x^3 - 3x^2 + 7x - 3}{(x^2 + 1)^2} = \frac{Ax + B}{x^2 + 1} + \frac{Cx + D}{(x^2 + 1)^2}.$$

Multiplying both sides by $(x^2 + 1)^2$ gives us

$$5x^3 - 3x^2 + 7x - 3 = (Ax + B)(x^2 + 1) + Cx + D$$

or $\quad 5x^3 - 3x^2 + 7x - 3 = Ax^3 + Bx^2 + (A + C)x + (B + D).$

Comparing the coefficients of x^3 and x^2, we obtain $A = 5$ and $B = -3$. From the coefficients of x we see that $A + C = 7$, or equivalently, $C = 7 - A = 7 - 5 = 2$. Finally, the constant terms give us $B + D = -3$, or $D = -3 - B = -3 - (-3) = 0$. Therefore,

$$\frac{5x^3 - 3x^2 + 7x - 3}{(x^2 + 1)^2} = \frac{5x - 3}{x^2 + 1} + \frac{2x}{(x^2 + 1)^2}. \qquad \blacksquare$$

Exercises 6.4

Find the partial fraction decompositions in Exercises 1–26.

1 $\dfrac{8x - 1}{(x - 2)(x + 3)}$

2 $\dfrac{x - 29}{(x - 4)(x + 1)}$

3 $\dfrac{x + 34}{x^2 - 4x - 12}$

4 $\dfrac{5x - 12}{x^2 - 4x}$

5 $\dfrac{4x^2 - 15x - 1}{(x - 1)(x + 2)(x - 3)}$

6 $\dfrac{x^2 + 19x + 20}{x(x + 2)(x - 5)}$

7 $\dfrac{4x^2 - 5x - 15}{x^3 - 4x^2 - 5x}$

8 $\dfrac{37 - 11x}{(x + 1)(x^2 - 5x + 6)}$

9 $\dfrac{2x + 3}{(x - 1)^2}$

10 $\dfrac{5x^2 - 4}{x^2(x + 2)}$

11 $\dfrac{19x^2 + 50x - 25}{3x^3 - 5x^2}$

12 $\dfrac{10 - x}{x^2 + 10x + 25}$

13 $\dfrac{x^2 - 6}{(x + 2)^2(2x - 1)}$

14 $\dfrac{2x^2 + x}{(x - 1)^2(x + 1)^2}$

15 $\dfrac{3x^3 + 11x^2 + 16x + 5}{x(x + 1)^3}$

16 $\dfrac{4x^3 + 3x^2 + 5x - 2}{x^3(x + 2)}$

17 $\dfrac{x^2 + x - 6}{(x^2 + 1)(x - 1)}$

18 $\dfrac{x^2 - x - 21}{(x^2 + 4)(2x - 1)}$

19 $\dfrac{9x^2 - 3x + 8}{x^3 + 2x}$

20 $\dfrac{2x^3 + 2x^2 + 4x - 3}{x^4 + x^2}$

21 $\dfrac{4x^3 - x^2 + 4x + 2}{(x^2 + 1)^2}$

22 $\dfrac{3x^3 + 13x - 1}{(x^2 + 4)^2}$

23 $\dfrac{2x^4 - 2x^3 + 6x^2 - 5x + 1}{x^3 - x^2 + x - 1}$

24 $\dfrac{x^3}{x^3 - 3x^2 + 9x - 27}$

25 $\dfrac{4x^3 + 4x^2 - 4x + 2}{2x^2 - x - 1}$

26 $\dfrac{x^5 - 5x^4 + 7x^3 - x^2 - 4x + 12}{x^3 - 3x^2}$

6.5 The Algebra of Matrices

It is possible to develop a comprehensive theory for matrices that has many mathematical and scientific applications. In this section we discuss algebraic properties of matrices that serve as the starting point for that theory.

To conserve space it is sometimes convenient to use the symbol (a_{ij}) to denote an $m \times n$ matrix A of the type displayed in the definition given in Section 6.3. If (b_{ij}) denotes another $m \times n$ matrix B, then we say that A and B are **equal,** and we write

$$A = B \quad \text{if and only if} \quad a_{ij} = b_{ij}$$

for every i and j. For example,

$$\begin{bmatrix} 1 & 0 & 5 \\ \sqrt[3]{8} & 3^2 & -2 \end{bmatrix} = \begin{bmatrix} (-1)^2 & 0 & \sqrt{25} \\ 2 & 9 & -2 \end{bmatrix}.$$

If $A = (a_{ij})$ and $B = (b_{ij})$ are $m \times n$ matrices, then their **sum $A + B$** is defined as the $m \times n$ matrix $C = (c_{ij})$ such that $c_{ij} = a_{ij} + b_{ij}$ for all i and j. Thus, to add two matrices we add the elements that appear in corresponding positions in each matrix. Two matrices can be added only if they have the same number of rows and the same number of columns. Using the parentheses notation, the definition of addition may be expressed as follows:

$$(a_{ij}) + (b_{ij}) = (a_{ij} + b_{ij}).$$

Although we have used the symbol $+$ in two different ways, there is little chance for confusion, since whenever $+$ appears between symbols for matrices it refers to matrix addition, and when $+$ is used between real numbers it denotes their sum. An example of the sum of two 3×2 matrices is

$$\begin{bmatrix} 4 & -5 \\ 0 & 4 \\ -6 & 1 \end{bmatrix} + \begin{bmatrix} 3 & 2 \\ 7 & -4 \\ -2 & 1 \end{bmatrix} = \begin{bmatrix} 7 & -3 \\ 7 & 0 \\ -8 & 2 \end{bmatrix}.$$

It is not difficult to prove that addition of matrices is both commutative and associative, that is,

$$A + B = B + A, \qquad A + (B + C) = (A + B) + C$$

for $m \times n$ matrices A, B, and C.

The **$m \times n$ zero matrix,** denoted by O, is the matrix with m rows and n columns in which every element is 0. It is an identity element relative to addition, since

$$A + O = A$$

for every $m \times n$ matrix A. For example,

$$\begin{bmatrix} a_{11} & a_{12} \\ a_{21} & a_{22} \\ a_{31} & a_{32} \end{bmatrix} + \begin{bmatrix} 0 & 0 \\ 0 & 0 \\ 0 & 0 \end{bmatrix} = \begin{bmatrix} a_{11} & a_{12} \\ a_{21} & a_{22} \\ a_{31} & a_{32} \end{bmatrix}.$$

The **additive inverse** $-A$ of the matrix $A = (a_{ij})$ is, by definition, the matrix $(-a_{ij})$ obtained by changing the sign of each element of A. For example,

$$-\begin{bmatrix} 2 & -3 & 4 \\ -1 & 0 & 5 \end{bmatrix} = \begin{bmatrix} -2 & 3 & -4 \\ 1 & 0 & -5 \end{bmatrix}.$$

It follows that for every $m \times n$ matrix A,

$$A + (-A) = O.$$

Subtraction of two $m \times n$ matrices is defined by

$$A - B = A + (-B).$$

Using the parentheses notation for matrices, this implies that

$$(a_{ij}) - (b_{ij}) = (a_{ij}) + (-b_{ij}) = (a_{ij} - b_{ij}).$$

Thus, to subtract two matrices, we subtract the elements that are in corresponding positions.

The **product** of a real number c and an $m \times n$ matrix $A = (a_{ij})$ is defined by

$$cA = (ca_{ij}).$$

Thus, to find cA, we multiply each element of A by c. For example,

$$3\begin{bmatrix} 4 & -1 \\ 2 & 3 \end{bmatrix} = \begin{bmatrix} 12 & -3 \\ 6 & 9 \end{bmatrix}.$$

The following results may be established, where A and B are $m \times n$ matrices and c and d are any real numbers.

$$c(A + B) = cA + cB$$

$$(c + d)A = cA + dA$$

$$(cd)A = c(dA)$$

The following definition of the product of two matrices may appear unusual to the beginning student; however, there are many applications that justify the form of the definition. To define the product AB of two matrices A and B, *the number of columns of A must be the same as the number of rows of B.* Suppose that $A = (a_{ij})$ is $m \times n$ and $B = (b_{ij})$ is $n \times p$. To determine the element c_{ij} of the product, we single out row i of A and column j of B as follows:

$$\begin{bmatrix} a_{11} & a_{12} & \cdots & a_{1n} \\ \vdots & \vdots & & \vdots \\ a_{i1} & a_{i2} & \cdots & a_{in} \\ \vdots & \vdots & & \vdots \\ a_{m1} & a_{m2} & \cdots & a_{mn} \end{bmatrix} \begin{bmatrix} b_{11} & \cdots & b_{1j} & \cdots & b_{1p} \\ b_{21} & \cdots & b_{2j} & \cdots & b_{2p} \\ \vdots & & \vdots & & \vdots \\ b_{n1} & \cdots & b_{nj} & \cdots & b_{np} \end{bmatrix}$$

Next we multiply pairs of elements and then add them, using the formula

$$c_{ij} = a_{i1}b_{1j} + a_{i2}b_{2j} + \cdots + a_{in}b_{nj}.$$

For example, the element c_{11} in the first row and the first column of AB is

$$c_{11} = a_{11}b_{11} + a_{12}b_{21} + \cdots + a_{1n}b_{n1}.$$

The element c_{12} in the first row and second column of AB is

$$c_{12} = a_{11}b_{12} + a_{12}b_{22} + \cdots + a_{1n}b_{n2}.$$

By definition, the product AB has the same number of rows as A and the same number of columns as B. In particular, if A is $m \times n$ and B is $n \times p$, then AB is $m \times p$. This is illustrated by the following product of a 2×3 matrix and a 3×4 matrix.

$$\begin{bmatrix} 1 & 2 & -3 \\ 4 & 0 & -2 \end{bmatrix} \begin{bmatrix} 5 & -4 & 2 & 0 \\ -1 & 6 & 3 & 1 \\ 7 & 0 & 4 & 8 \end{bmatrix} = \begin{bmatrix} -18 & 8 & -4 & -22 \\ 6 & -16 & 0 & -16 \end{bmatrix}.$$

Here are some typical computations of the elements c_{ij} in the product:

$$c_{11} = (1)(5) + (2)(-1) + (-3)(7) = 5 - 2 - 21 = -18$$

$$c_{13} = (1)(2) + (2)(3) + (-3)(4) = 2 + 6 - 12 = -4$$

$$c_{23} = (4)(2) + (0)(3) + (-2)(4) = 8 + 0 - 8 = 0$$

$$c_{24} = (4)(0) + (0)(1) + (-2)(8) = 0 + 0 - 16 = -16.$$

The reader should calculate the remaining elements.

The product operation for matrices is not commutative. Indeed, if A is 2×3 and B is 3×4, then AB may be found, but BA is undefined, since the number of columns of B is different from the number of rows of A. Even if AB and BA are both defined, it is often true that these products are different. This is illustrated in the next example, along with the fact that the product of two nonzero matrices may equal a zero matrix.

EXAMPLE 1

If $A = \begin{bmatrix} 2 & 2 \\ -1 & -1 \end{bmatrix}$ and $B = \begin{bmatrix} 1 & 2 \\ 1 & 2 \end{bmatrix}$, show that $AB \neq BA$.

SOLUTION Using the definition of product we obtain

$$AB = \begin{bmatrix} 4 & 8 \\ -2 & -4 \end{bmatrix} \quad \text{and} \quad BA = \begin{bmatrix} 0 & 0 \\ 0 & 0 \end{bmatrix}.$$

Hence, $AB \neq BA$. Note that BA is a zero matrix. ∎

It can be shown that matrix multiplication is associative. Thus

$$A(BC) = (AB)C$$

provided that the indicated products are defined, which will be the case if A is $m \times n$, B is $n \times p$, and C is $p \times q$.

The Distributive Properties also hold if the matrices involved have the proper number of rows and columns. If A_1 and A_2 are $m \times n$ matrices, and if B_1 and B_2 are $n \times p$ matrices, then

$$A_1(B_1 + B_2) = A_1 B_1 + A_1 B_2$$

and $$(A_1 + A_2)B_1 = A_1 B_1 + A_2 B_1.$$

As a special case, if all matrices are square, of order n, then the Associative and Distributive Properties are true. We shall not prove these properties.

Throughout the remainder of this section we shall concentrate on square matrices. The symbol I_n will denote the square matrix of order n that has 1 in each position on the main diagonal and 0 elsewhere. For example,

$$I_2 = \begin{bmatrix} 1 & 0 \\ 0 & 1 \end{bmatrix}, \qquad I_3 = \begin{bmatrix} 1 & 0 & 0 \\ 0 & 1 & 0 \\ 0 & 0 & 1 \end{bmatrix},$$

and so on. It can be shown that if A is any square matrix of order n, then

$$AI_n = A = I_n A.$$

For that reason I_n is called the **identity matrix of order n.** To illustrate, if $A = (a_{ij})$ is of order 2, then a direct calculation shows that

$$\begin{bmatrix} a_{11} & a_{12} \\ a_{21} & a_{22} \end{bmatrix}\begin{bmatrix} 1 & 0 \\ 0 & 1 \end{bmatrix} = \begin{bmatrix} a_{11} & a_{12} \\ a_{21} & a_{22} \end{bmatrix} = \begin{bmatrix} 1 & 0 \\ 0 & 1 \end{bmatrix}\begin{bmatrix} a_{11} & a_{12} \\ a_{21} & a_{22} \end{bmatrix}.$$

Some, but not all, $n \times n$ matrices A have an **inverse** in the sense that there is a matrix B such that $AB = I_n = BA$. If A has an inverse we denote it by A^{-1} and write

$$AA^{-1} = I_n = A^{-1}A.$$

The symbol A^{-1} is read "A inverse." In matrix theory it is *not* acceptable to use the symbol $1/A$ in place of A^{-1}.

There is an interesting technique for finding the inverse of a square matrix A, whenever it exists. We shall not attempt to justify the following procedure, since that would require advanced concepts. Given the $n \times n$ matrix $A = (a_{ij})$, we begin by forming the $n \times 2n$ matrix:

$$\begin{bmatrix} a_{11} & a_{12} & \cdots & a_{1n} & 1 & 0 & \cdots & 0 \\ a_{21} & a_{22} & \cdots & a_{2n} & 0 & 1 & \cdots & 0 \\ \vdots & \vdots & & \vdots & \vdots & \vdots & & \vdots \\ a_{n1} & a_{n2} & \cdots & a_{nn} & 0 & 0 & \cdots & 1 \end{bmatrix}$$

where the $n \times n$ identity matrix I_n appears "to the right" of the matrix A, as indicated. We next apply a succession of elementary row transformations until we arrive at a matrix of the form

$$\begin{bmatrix} 1 & 0 & \cdots & 0 & b_{11} & b_{12} & \cdots & b_{1n} \\ 0 & 1 & \cdots & 0 & b_{21} & b_{22} & \cdots & b_{2n} \\ \vdots & \vdots & & \vdots & \vdots & \vdots & & \vdots \\ 0 & 0 & \cdots & 1 & b_{n1} & b_{n2} & \cdots & b_{nn} \end{bmatrix}$$

where the identity matrix I_n appears "to the left" of the $n \times n$ matrix (b_{ij}). It can be shown that (b_{ij}) is the desired inverse A^{-1}.

EXAMPLE 2 Find A^{-1} if

$$A = \begin{bmatrix} 3 & 5 \\ 1 & 4 \end{bmatrix}.$$

SOLUTION We begin with the matrix $\begin{bmatrix} 3 & 5 & 1 & 0 \\ 1 & 4 & 0 & 1 \end{bmatrix}$ and then perform elementary row transformations until the identity matrix I_2 appears on the left, as follows:

$$\begin{bmatrix} 3 & 5 & 1 & 0 \\ 1 & 4 & 0 & 1 \end{bmatrix} \xrightarrow{R_{12}} \begin{bmatrix} 1 & 4 & 0 & 1 \\ 3 & 5 & 1 & 0 \end{bmatrix}$$

$$\xrightarrow{-3R_1 + R_2} \begin{bmatrix} 1 & 4 & 0 & 1 \\ 0 & -7 & 1 & -3 \end{bmatrix}$$

$$\xrightarrow{-\frac{1}{7}R_2} \begin{bmatrix} 1 & 4 & 0 & 1 \\ 0 & 1 & -\frac{1}{7} & \frac{3}{7} \end{bmatrix}$$

$$\xrightarrow{-4R_2 + R_1} \begin{bmatrix} 1 & 0 & \frac{4}{7} & -\frac{5}{7} \\ 0 & 1 & -\frac{1}{7} & \frac{3}{7} \end{bmatrix}$$

According to the previous discussion,

$$A^{-1} = \begin{bmatrix} \frac{4}{7} & -\frac{5}{7} \\ -\frac{1}{7} & \frac{3}{7} \end{bmatrix}.$$

Checking this fact, we see that

$$\begin{bmatrix} 3 & 5 \\ 1 & 4 \end{bmatrix}\begin{bmatrix} \frac{4}{7} & -\frac{5}{7} \\ -\frac{1}{7} & \frac{3}{7} \end{bmatrix} = \begin{bmatrix} 1 & 0 \\ 0 & 1 \end{bmatrix} = \begin{bmatrix} \frac{4}{7} & -\frac{5}{7} \\ -\frac{1}{7} & \frac{3}{7} \end{bmatrix}\begin{bmatrix} 3 & 5 \\ 1 & 4 \end{bmatrix}. \qquad \blacksquare$$

EXAMPLE 3 Find A^{-1} if

$$A = \begin{bmatrix} -1 & 3 & 1 \\ 2 & 5 & 0 \\ 3 & 1 & -2 \end{bmatrix}.$$

SOLUTION   We proceed as follows:

$$
\begin{bmatrix} -1 & 3 & 1 & 1 & 0 & 0 \\ 2 & 5 & 0 & 0 & 1 & 0 \\ 3 & 1 & -2 & 0 & 0 & 1 \end{bmatrix} \xrightarrow{(-1)R_1} \begin{bmatrix} 1 & -3 & -1 & -1 & 0 & 0 \\ 2 & 5 & 0 & 0 & 1 & 0 \\ 3 & 1 & -2 & 0 & 0 & 1 \end{bmatrix}
$$

$$
\xrightarrow{-2R_1 + R_2} \begin{bmatrix} 1 & -3 & -1 & -1 & 0 & 0 \\ 0 & 11 & 2 & 2 & 1 & 0 \\ 3 & 1 & -2 & 0 & 0 & 1 \end{bmatrix}
$$

$$
\xrightarrow{-3R_1 + R_3} \begin{bmatrix} 1 & -3 & -1 & -1 & 0 & 0 \\ 0 & 11 & 2 & 2 & 1 & 0 \\ 0 & 10 & 1 & 3 & 0 & 1 \end{bmatrix}
$$

$$
\xrightarrow{(-1)R_3 + R_2} \begin{bmatrix} 1 & -3 & -1 & -1 & 0 & 0 \\ 0 & 1 & 1 & -1 & 1 & -1 \\ 0 & 10 & 1 & 3 & 0 & 1 \end{bmatrix}
$$

$$
\xrightarrow{3R_2 + R_1} \begin{bmatrix} 1 & 0 & 2 & -4 & 3 & -3 \\ 0 & 1 & 1 & -1 & 1 & -1 \\ 0 & 10 & 1 & 3 & 0 & 1 \end{bmatrix}
$$

$$
\xrightarrow{-10R_2 + R_3} \begin{bmatrix} 1 & 0 & 2 & -4 & 3 & -3 \\ 0 & 1 & 1 & -1 & 1 & -1 \\ 0 & 0 & -9 & 13 & -10 & 11 \end{bmatrix}
$$

$$
\xrightarrow{-\frac{1}{9}R_3} \begin{bmatrix} 1 & 0 & 2 & -4 & 3 & -3 \\ 0 & 1 & 1 & -1 & 1 & -1 \\ 0 & 0 & 1 & -\frac{13}{9} & \frac{10}{9} & -\frac{11}{9} \end{bmatrix}
$$

$$
\xrightarrow{-2R_3 + R_1} \begin{bmatrix} 1 & 0 & 0 & -\frac{10}{9} & \frac{7}{9} & -\frac{5}{9} \\ 0 & 1 & 1 & -1 & 1 & -1 \\ 0 & 0 & 1 & -\frac{13}{9} & \frac{10}{9} & -\frac{11}{9} \end{bmatrix}
$$

$$
\xrightarrow{(-1)R_3 + R_2} \begin{bmatrix} 1 & 0 & 0 & -\frac{10}{9} & \frac{7}{9} & -\frac{5}{9} \\ 0 & 1 & 0 & \frac{4}{9} & -\frac{1}{9} & \frac{2}{9} \\ 0 & 0 & 1 & -\frac{13}{9} & \frac{10}{9} & -\frac{11}{9} \end{bmatrix}
$$

Consequently,

$$
A^{-1} = \begin{bmatrix} -\frac{10}{9} & \frac{7}{9} & -\frac{5}{9} \\ \frac{4}{9} & -\frac{1}{9} & \frac{2}{9} \\ -\frac{13}{9} & \frac{10}{9} & -\frac{11}{9} \end{bmatrix} = \tfrac{1}{9} \begin{bmatrix} -10 & 7 & -5 \\ 4 & -1 & 2 \\ -13 & 10 & -11 \end{bmatrix}.
$$

It can be shown that

$$AA^{-1} = I_3 = A^{-1}A.$$

∎

There are many uses for inverses of matrices. One application concerns solutions of systems of linear equations. To illustrate, let us consider the case of two linear equations in two unknowns:

$$\begin{cases} a_{11}x + a_{12}y = k_1 \\ a_{21}x + a_{22}y = k_2. \end{cases}$$

We may express this system in terms of matrices as follows:

$$\begin{bmatrix} a_{11}x + a_{12}y \\ a_{21}x + a_{22}y \end{bmatrix} = \begin{bmatrix} k_1 \\ k_2 \end{bmatrix}.$$

If we let

$$A = \begin{bmatrix} a_{11} & a_{12} \\ a_{21} & a_{22} \end{bmatrix}, \quad X = \begin{bmatrix} x \\ y \end{bmatrix}, \quad \text{and} \quad B = \begin{bmatrix} k_1 \\ k_2 \end{bmatrix},$$

we have $AX = B.$

If A^{-1} exists, then multiplying both sides of the last equation by A^{-1} gives us $A^{-1}AX = A^{-1}B$. Since $A^{-1}A = I_2$ and $I_2X = X$, this leads to

$$X = A^{-1}B$$

from which the solution (x, y) may be found. This technique may be extended to systems of n linear equations in n unknowns.

EXAMPLE 4   Solve the following system of equations:

$$\begin{cases} -x + 3y + z = 1 \\ 2x + 5y = 3 \\ 3x + y - 2z = -2. \end{cases}$$

SOLUTION   If we let

$$A = \begin{bmatrix} -1 & 3 & 1 \\ 2 & 5 & 0 \\ 3 & 1 & -2 \end{bmatrix}, \quad X = \begin{bmatrix} x \\ y \\ z \end{bmatrix}, \quad \text{and} \quad B = \begin{bmatrix} 1 \\ 3 \\ -2 \end{bmatrix},$$

then, as in the preceding discussion, the given system may be written in terms of matrices as $AX = B$. This implies that $X = A^{-1}B$. The matrix A^{-1} was found in Example 3. Substituting for X, A^{-1}, and B in the last equation gives us

$$
\begin{bmatrix} x \\ y \\ z \end{bmatrix} = \tfrac{1}{9} \begin{bmatrix} -10 & 7 & -5 \\ 4 & -1 & 2 \\ -13 & 10 & -11 \end{bmatrix} \begin{bmatrix} 1 \\ 3 \\ -2 \end{bmatrix} = \tfrac{1}{9} \begin{bmatrix} 21 \\ -3 \\ 39 \end{bmatrix} = \begin{bmatrix} \tfrac{7}{3} \\ -\tfrac{1}{3} \\ \tfrac{13}{3} \end{bmatrix}.
$$

It follows that $x = \tfrac{7}{3}$, $y = -\tfrac{1}{3}$, and $z = \tfrac{13}{3}$. Hence, the ordered triple $(\tfrac{7}{3}, -\tfrac{1}{3}, \tfrac{13}{3})$ is the solution of the given system. ∎

The method of solution employed in Example 4 is beneficial only if A^{-1} is known, or if many systems with the same coefficient matrix are to be considered. The preferred technique for solving an arbitrary system of linear equations is the matrix method discussed in Section 6.3.

Exercises 6.5

In Exercises 1–8 find $A + B$, $A - B$, $2A$, and $-3B$.

1 $A = \begin{bmatrix} 5 & -2 \\ 1 & 3 \end{bmatrix}$, $B = \begin{bmatrix} 4 & 1 \\ -3 & 2 \end{bmatrix}$

2 $A = \begin{bmatrix} 3 & 0 \\ -1 & 2 \end{bmatrix}$, $B = \begin{bmatrix} 3 & -4 \\ 1 & 1 \end{bmatrix}$

3 $A = \begin{bmatrix} 6 & -1 \\ 2 & 0 \\ -3 & 4 \end{bmatrix}$, $B = \begin{bmatrix} 3 & 1 \\ -1 & 5 \\ 6 & 0 \end{bmatrix}$

4 $A = \begin{bmatrix} 0 & -2 & 7 \\ 5 & 4 & -3 \end{bmatrix}$, $B = \begin{bmatrix} 8 & 4 & 0 \\ 0 & 1 & 4 \end{bmatrix}$

5 $A = \begin{bmatrix} 4 & -3 & 2 \end{bmatrix}$, $B = \begin{bmatrix} 7 & 0 & -5 \end{bmatrix}$

6 $A = \begin{bmatrix} 7 \\ -16 \end{bmatrix}$, $B = \begin{bmatrix} -11 \\ 9 \end{bmatrix}$

7 $A = \begin{bmatrix} 0 & 4 & 0 & 3 \\ 1 & 2 & 0 & -5 \end{bmatrix}$,

$B = \begin{bmatrix} -3 & 0 & 1 & 3 \\ 2 & 0 & 7 & -2 \end{bmatrix}$

8 $A = \begin{bmatrix} -7 \end{bmatrix}$, $B = \begin{bmatrix} 9 \end{bmatrix}$

In Exercises 9–18 find AB and BA.

9 $A = \begin{bmatrix} 2 & 6 \\ 3 & -4 \end{bmatrix}$, $B = \begin{bmatrix} 5 & -2 \\ 1 & 7 \end{bmatrix}$

10 $A = \begin{bmatrix} 4 & -2 \\ -2 & 1 \end{bmatrix}$, $B = \begin{bmatrix} 2 & 1 \\ 4 & 2 \end{bmatrix}$

11 $A = \begin{bmatrix} 3 & 0 & -1 \\ 0 & 4 & 2 \\ 5 & -3 & 1 \end{bmatrix}$, $B = \begin{bmatrix} 1 & -5 & 0 \\ 4 & 1 & -2 \\ 0 & -1 & 3 \end{bmatrix}$

12 $A = \begin{bmatrix} 5 & 0 & 0 \\ 0 & -3 & 0 \\ 0 & 0 & 2 \end{bmatrix}$, $B = \begin{bmatrix} 3 & 0 & 0 \\ 0 & 4 & 0 \\ 0 & 0 & -2 \end{bmatrix}$

13 $A = \begin{bmatrix} 4 & -3 & 1 \\ -5 & 2 & 2 \end{bmatrix}$, $B = \begin{bmatrix} 2 & 1 \\ 0 & 1 \\ -4 & 7 \end{bmatrix}$

14 $A = \begin{bmatrix} 2 & 1 & -1 & 0 \\ 3 & -2 & 0 & 5 \\ -2 & 1 & 4 & 2 \end{bmatrix}$, $B = \begin{bmatrix} 5 & -3 & 1 \\ 1 & 2 & 0 \\ -1 & 0 & 4 \\ 0 & -2 & 3 \end{bmatrix}$

15 $A = \begin{bmatrix} 1 & 2 & 3 \\ 4 & 5 & 6 \\ 7 & 8 & 9 \end{bmatrix}$, $B = \begin{bmatrix} 1 & 0 & 0 \\ 0 & 1 & 0 \\ 0 & 0 & 1 \end{bmatrix}$

16 $A = \begin{bmatrix} 1 & 2 & 3 \\ 2 & 3 & 1 \\ 3 & 1 & 2 \end{bmatrix}$, $B = \begin{bmatrix} 2 & 0 & 0 \\ 0 & 2 & 0 \\ 0 & 0 & 2 \end{bmatrix}$

17 $A = [-3 \quad 7 \quad 2]$, $B = \begin{bmatrix} 1 \\ 4 \\ -5 \end{bmatrix}$

18 $A = [4 \quad 8]$, $B = \begin{bmatrix} -3 \\ 2 \end{bmatrix}$

In Exercises 19–22 find AB.

19 $A = \begin{bmatrix} 4 & -2 \\ 0 & 3 \\ -7 & 5 \end{bmatrix}$, $B = \begin{bmatrix} 3 \\ 4 \end{bmatrix}$

20 $A = \begin{bmatrix} 4 \\ -3 \\ 2 \end{bmatrix}$, $B = [5 \quad 1]$

21 $A = \begin{bmatrix} 2 & 1 & 0 & -3 \\ -7 & 0 & -2 & 4 \end{bmatrix}$, $B = \begin{bmatrix} 4 & -2 & 0 \\ 1 & 1 & -2 \\ 0 & 0 & 5 \\ -3 & -1 & 0 \end{bmatrix}$

22 $A = \begin{bmatrix} 1 & 2 & -3 \\ 4 & -5 & 6 \end{bmatrix}$, $B = \begin{bmatrix} 1 & -1 & 0 & 2 \\ -2 & 3 & 1 & 0 \\ 0 & 4 & 0 & -3 \end{bmatrix}$

23 If $A = \begin{bmatrix} 1 & 2 \\ 0 & -3 \end{bmatrix}$, and $B = \begin{bmatrix} 2 & -1 \\ 3 & 1 \end{bmatrix}$, show that $(A + B)(A - B) \neq A^2 - B^2$, where $A^2 = AA$ and $B^2 = BB$.

24 If A and B are the matrices of Exercise 23, show that $(A + B)(A + B) \neq A^2 + 2AB + B^2$.

25 If A and B are the matrices of Exercise 23 and $C = \begin{bmatrix} 3 & 1 \\ -2 & 0 \end{bmatrix}$, show that $A(B + C) = AB + AC$.

26 If A, B, and C are the matrices of Exercise 25, show that $A(BC) = (AB)C$.

Prove the identities stated in Exercises 27–30, where

$A = \begin{bmatrix} a_{11} & a_{12} \\ a_{21} & a_{22} \end{bmatrix}$, $B = \begin{bmatrix} b_{11} & b_{12} \\ b_{21} & b_{22} \end{bmatrix}$, $C = \begin{bmatrix} c_{11} & c_{22} \\ c_{21} & c_{22} \end{bmatrix}$

and c, d are real numbers.

27 $c(A + B) = cA + cB$ **28** $(c + d)A = cA + dA$

29 $A(B + C) = AB + AC$ **30** $A(BC) = (AB)C$

In Exercises 31–42 find the inverse of the matrix, if it exists.

31 $\begin{bmatrix} 2 & -4 \\ 1 & 3 \end{bmatrix}$ **32** $\begin{bmatrix} 3 & 2 \\ 4 & 5 \end{bmatrix}$

33 $\begin{bmatrix} 2 & 4 \\ 4 & 8 \end{bmatrix}$ **34** $\begin{bmatrix} 3 & -1 \\ 6 & -2 \end{bmatrix}$

35 $\begin{bmatrix} 3 & -1 & 0 \\ 2 & 2 & 0 \\ 0 & 0 & 4 \end{bmatrix}$ **36** $\begin{bmatrix} 3 & 0 & 2 \\ 0 & 1 & 0 \\ -4 & 0 & 2 \end{bmatrix}$

37 $\begin{bmatrix} -2 & 2 & 3 \\ 1 & -1 & 0 \\ 0 & 1 & 4 \end{bmatrix}$ **38** $\begin{bmatrix} 1 & 2 & 3 \\ -2 & 1 & 0 \\ 3 & -1 & 1 \end{bmatrix}$

39 $\begin{bmatrix} 2 & 0 & 0 \\ 0 & 4 & 0 \\ 0 & 0 & 6 \end{bmatrix}$ **40** $\begin{bmatrix} 1 & 1 & 1 \\ 2 & 2 & 2 \\ 3 & 3 & 3 \end{bmatrix}$

41 $\begin{bmatrix} 1 & -1 & 0 & 1 \\ 0 & 1 & -2 & 0 \\ -1 & 2 & 1 & 2 \\ -2 & 1 & 2 & 0 \end{bmatrix}$ **42** $\begin{bmatrix} 1 & 2 & 0 & 1 \\ 0 & -1 & 1 & -2 \\ 0 & 0 & 2 & 0 \\ 0 & 0 & 0 & 1 \end{bmatrix}$

43 State conditions on a and b that guarantee that the matrix $\begin{bmatrix} a & 0 \\ 0 & b \end{bmatrix}$ has an inverse, and find a formula for the inverse, if it exists.

44 If $abc \neq 0$, find the inverse of $\begin{bmatrix} a & 0 & 0 \\ 0 & b & 0 \\ 0 & 0 & c \end{bmatrix}$.

45 If $A = \begin{bmatrix} a_{11} & a_{12} & a_{13} \\ a_{21} & a_{22} & a_{23} \\ a_{31} & a_{32} & a_{33} \end{bmatrix}$, prove that $AI_3 = A = I_3A$.

46 Prove that $AI_4 = A = I_4A$ for every square matrix A of order 4.

Solve the systems in Exercises 47–50 by the method of Example 4. (Refer to inverses of matrices found in Exercises 31, 32, 37, and 38.)

47 $\begin{cases} 2x - 4y = 3 \\ x + 3y = 1 \end{cases}$ **48** $\begin{cases} 3x + 2y = -1 \\ 4x + 5y = 1 \end{cases}$

49 $\begin{cases} -2x + 2y + 3z = 1 \\ x - y = 3 \\ y + 4z = -2 \end{cases}$ **50** $\begin{cases} x + 2y + 3z = -1 \\ -2x + y = 4 \\ 3x - y + z = 2 \end{cases}$

6.6 **Determinants**

Throughout this section and the next we will assume that all matrices under discussion are *square* matrices. Associated with each square matrix A is a number called the **determinant of** A, denoted by $|A|$. This notation should not be confused with the symbol for the absolute value of a real number. To avoid any misunderstanding, the expression det A is sometimes used instead of $|A|$. We shall define $|A|$ by beginning with the case in which A has order 1 and then by increasing the order a step at a time.

If A is a square matrix of order 1, then A has only one element. Thus, $A = [a_{11}]$ and we define $|A| = a_{11}$. If A is a square matrix of order 2, then

$$A = \begin{bmatrix} a_{11} & a_{12} \\ a_{21} & a_{22} \end{bmatrix}$$

and the determinant of A is defined by

$$|A| = a_{11}a_{22} - a_{21}a_{12}.$$

Another notation for $|A|$ is obtained by replacing the brackets used for A with vertical bars as follows.

DEFINITION

$$|A| = \begin{vmatrix} a_{11} & a_{12} \\ a_{21} & a_{22} \end{vmatrix} = a_{11}a_{22} - a_{21}a_{12}$$

EXAMPLE 1

Find $|A|$ if $A = \begin{bmatrix} 2 & -1 \\ 4 & -3 \end{bmatrix}$.

SOLUTION By definition,

$$|A| = \begin{vmatrix} 2 & -1 \\ 4 & -3 \end{vmatrix} = (2)(-3) - (4)(-1) = -6 + 4 = -2. \qquad \blacksquare$$

For square matrices of order 3 it is convenient to introduce the following terminology.

DEFINITION

> Let A be a square matrix of order 3. The **minor** M_{ij} of the element a_{ij} is the determinant of the matrix of order 2 obtained by deleting row i and column j.

To determine the minor of an element, we discard the row and column in which the element appears and then find the determinant of the resulting matrix. To illustrate, given the 3×3 matrix

$$A = \begin{bmatrix} a_{11} & a_{12} & a_{13} \\ a_{21} & a_{22} & a_{23} \\ a_{31} & a_{32} & a_{33} \end{bmatrix}$$

we obtain

$$M_{11} = \begin{vmatrix} a_{22} & a_{23} \\ a_{32} & a_{33} \end{vmatrix} = a_{22}a_{33} - a_{32}a_{23}$$

$$M_{12} = \begin{vmatrix} a_{21} & a_{23} \\ a_{31} & a_{33} \end{vmatrix} = a_{21}a_{33} - a_{31}a_{23}$$

$$M_{13} = \begin{vmatrix} a_{21} & a_{22} \\ a_{31} & a_{32} \end{vmatrix} = a_{21}a_{32} - a_{31}a_{22}$$

$$M_{23} = \begin{vmatrix} a_{11} & a_{12} \\ a_{31} & a_{32} \end{vmatrix} = a_{11}a_{32} - a_{31}a_{12}$$

and likewise for the other minors M_{21}, M_{22}, M_{31}, M_{32}, and M_{33}.

We shall also make use of the following concept.

DEFINITION

> The **cofactor** A_{ij} of the element a_{ij} is
>
> $$A_{ij} = (-1)^{i+j} M_{ij}.$$

To obtain the cofactor of a_{ij}, we find the minor and multiply it by 1 or -1, depending on whether the sum of i and j is even or odd, respectively. An easy way to remember the sign $(-1)^{i+j}$ associated with the cofactor A_{ij} is to consider the following "checkerboard" scheme:

$$\begin{bmatrix} + & - & + \\ - & + & - \\ + & - & + \end{bmatrix}$$

EXAMPLE 2 If

$$A = \begin{bmatrix} 1 & -3 & 3 \\ 4 & 2 & 0 \\ -2 & -7 & 5 \end{bmatrix}$$

find M_{11}, M_{21}, M_{22}, A_{11}, A_{21}, and A_{22}.

SOLUTION By definition,

$$M_{11} = \begin{vmatrix} 2 & 0 \\ -7 & 5 \end{vmatrix} = (2)(5) - (-7)(0) = 10$$

$$M_{21} = \begin{vmatrix} -3 & 3 \\ -7 & 5 \end{vmatrix} = (-3)(5) - (-7)(3) = 6$$

$$M_{22} = \begin{vmatrix} 1 & 3 \\ -2 & 5 \end{vmatrix} = (1)(5) - (-2)(3) = 11.$$

To obtain the cofactors, we prefix the corresponding minors with the proper signs. Thus, using the definition of cofactor,

$$A_{11} = (-1)^{1+1} M_{11} = (1)(10) = 10$$
$$A_{21} = (-1)^{2+1} M_{21} = (-1)(6) = -6$$
$$A_{22} = (-1)^{2+2} M_{22} = (1)(11) = 11.$$

The checkerboard scheme can also be used to determine the proper signs. ∎

The determinant $|A|$ of a square matrix of order 3 is defined as follows.

DEFINITION

$$|A| = \begin{vmatrix} a_{11} & a_{12} & a_{13} \\ a_{21} & a_{22} & a_{23} \\ a_{31} & a_{32} & a_{33} \end{vmatrix} = a_{11}A_{11} + a_{12}A_{12} + a_{13}A_{13}$$

Since $A_{11} = (-1)^{1+1} M_{11} = M_{11}$, $A_{12} = (-1)^{1+2} M_{12} = -M_{12}$, and $A_{13} = (-1)^{1+3} M_{13} = M_{13}$, the preceding definition may also be written

$$|A| = a_{11}M_{11} - a_{12}M_{12} + a_{13}M_{13}.$$

If we express M_{11}, M_{12}, and M_{13} in terms of elements of A, we obtain the following formula for $|A|$:

$$|A| = a_{11}a_{22}a_{33} - a_{11}a_{32}a_{23} - a_{12}a_{21}a_{33} + a_{12}a_{31}a_{23}$$
$$+ a_{13}a_{21}a_{32} - a_{13}a_{31}a_{22}.$$

The definition of $|A|$ for a square matrix A of order 3 displays a pattern of multiplying each element in row 1 by its cofactor, and then adding to find $|A|$. This is referred to as *expanding* $|A|$ *by the first row*. By actually carrying out the computations, it is not difficult to show that $|A|$ *can be expanded in similar fashion by using any row or column*. As an illustration, the expansion by the second column is

$$|A| = a_{12}A_{12} + a_{22}A_{22} + a_{32}A_{32}$$
$$= a_{12}\left(-\begin{vmatrix} a_{21} & a_{23} \\ a_{31} & a_{32} \end{vmatrix}\right) + a_{22}\left(+\begin{vmatrix} a_{11} & a_{13} \\ a_{31} & a_{33} \end{vmatrix}\right) + a_{32}\left(-\begin{vmatrix} a_{11} & a_{13} \\ a_{21} & a_{23} \end{vmatrix}\right).$$

Applying the definition to the determinants in parentheses, multiplying as indicated, and rearranging the terms in the sum, we could arrive at the formula for $|A|$ in terms of the elements of A. Similarly, the expansion by the third row is

$$|A| = a_{31}A_{31} + a_{32}A_{32} + a_{33}A_{33}.$$

Once again it can be shown that this agrees with previous expansions.

EXAMPLE 3

Find $|A|$ if $A = \begin{bmatrix} -1 & 3 & 1 \\ 2 & 5 & 0 \\ 3 & 1 & -2 \end{bmatrix}$.

SOLUTION   Since there is a zero in the second row we shall expand $|A|$ by that row, because then we need only evaluate two cofactors. Thus,

$$|A| = (2)A_{21} + (5)A_{22} + (0)A_{23}.$$

Using the definition of cofactor,

$$A_{21} = (-1)^3 M_{21} = -\begin{vmatrix} 3 & 1 \\ 1 & -2 \end{vmatrix} = -[(3)(-2) - (1)(1)] = 7$$

$$A_{22} = (-1)^4 M_{22} = \begin{vmatrix} -1 & 1 \\ 3 & -2 \end{vmatrix} = [(-1)(-2) - (3)(1)] = -1.$$

Consequently,

$$|A| = (2)(7) + (5)(-1) + (0)A_{23} = 14 - 5 + 0 = 9.$$ ∎

The definition of the determinant of a matrix of arbitrary order n may be patterned after that used for order 3. Specifically, the **minor** M_{ij} is defined as the determinant of the matrix of order $n-1$ obtained by deleting row i and column j. The **cofactor** A_{ij} is defined as $(-1)^{i+j}M_{ij}$. The sign $(-1)^{i+j}$ associated with A_{ij} can be remembered by using a checkerboard similar to that used for order 3, extending the rows and columns as far as necessary. We then define the determinant $|A|$ of a matrix A of order n as the expansion by the first row, that is,

$$|A| = a_{11}A_{11} + a_{12}A_{12} + \cdots + a_{1n}A_{1n}$$

or, in terms of minors,

$$|A| = a_{11}M_{11} - a_{12}M_{12} + \cdots + a_{1n}(-1)^{1+n}M_{1n}.$$

The number $|A|$ may be found by using *any* row or column, as stated in the following theorem.

EXPANSION THEOREM FOR DETERMINANTS

If A is a square matrix of order $n > 1$, then the determinant $|A|$ may be found by multiplying the elements of any row (or column) by their respective cofactors, and adding the resulting products.

The proof of this theorem is difficult and may be found in texts on matrix theory. The theorem is quite useful if many zeros appear in a row or column, as illustrated in the following example.

EXAMPLE 4

Find $|A|$ if $A = \begin{bmatrix} 1 & 0 & 2 & 5 \\ -2 & 1 & 5 & 0 \\ 0 & 0 & -3 & 0 \\ 0 & -1 & 0 & 3 \end{bmatrix}$.

SOLUTION Note that all but one of the elements in the third row is zero. Hence if we expand $|A|$ by the third row, there will be at most one nonzero term. Specifically,

$$|A| = (0)A_{31} + (0)A_{32} + (-3)A_{33} + (0)A_{34} = -3A_{33}$$

where

$$A_{33} = \begin{vmatrix} 1 & 0 & 5 \\ -2 & 1 & 0 \\ 0 & -1 & 3 \end{vmatrix}.$$

Expanding A_{33} by column 1, we obtain

$$A_{33} = (1)\begin{vmatrix} 1 & 0 \\ -1 & 3 \end{vmatrix} + (-2)\left(-\begin{vmatrix} 0 & 5 \\ -1 & 3 \end{vmatrix}\right) + 0\begin{vmatrix} 0 & 5 \\ 1 & 0 \end{vmatrix} = 3 + 10 + 0 = 13.$$

Thus, $$|A| = -3A_{33} = (-3)(13) = -39. \qquad \blacksquare$$

In general, if all but one element a in some row (or column) of A is zero, and if the determinant $|A|$ is expanded by that row (or column), then all terms drop out except the product of the element a with its cofactor. We will make important use of this fact in the next section.

If *every* element in a row (or column) of a matrix A is zero, then upon expanding $|A|$ by that row (or column), we obtain the number 0. This gives us the following result.

THEOREM

> If every element of a row (or column) of a square matrix A is zero, then $|A| = 0$.

Exercises 6.6

In Exercises 1–4 find all the minors and cofactors of the elements in the matrix.

1 $\begin{bmatrix} 2 & 4 & -1 \\ 0 & 3 & 2 \\ -5 & 7 & 0 \end{bmatrix}$ **2** $\begin{bmatrix} 5 & -2 & 1 \\ 4 & 7 & 0 \\ -3 & 4 & -1 \end{bmatrix}$

3 $\begin{bmatrix} 7 & -1 \\ 5 & 0 \end{bmatrix}$ **4** $\begin{bmatrix} -6 & 4 \\ 3 & 2 \end{bmatrix}$

5–8 Find the determinants of the matrices in Exercises 1–4.

Find the determinants of the matrices in Exercises 9–20.

9 $\begin{bmatrix} -5 & 4 \\ -3 & 2 \end{bmatrix}$ **10** $\begin{bmatrix} 6 & 4 \\ -3 & 2 \end{bmatrix}$

11 $\begin{bmatrix} a & -a \\ b & -b \end{bmatrix}$ **12** $\begin{bmatrix} c & d \\ -d & c \end{bmatrix}$

13 $\begin{bmatrix} 3 & 1 & -2 \\ 4 & 2 & 5 \\ -6 & 3 & -1 \end{bmatrix}$ **14** $\begin{bmatrix} 2 & -5 & 1 \\ -3 & 1 & 6 \\ 4 & -2 & 3 \end{bmatrix}$

15 $\begin{bmatrix} -5 & 4 & 1 \\ 3 & -2 & 7 \\ 2 & 0 & 6 \end{bmatrix}$ **16** $\begin{bmatrix} 2 & 7 & -3 \\ 1 & 0 & 4 \\ 4 & -1 & -2 \end{bmatrix}$

17 $\begin{bmatrix} 3 & -1 & 2 & 0 \\ 4 & 0 & -3 & 5 \\ 0 & 6 & 0 & 0 \\ 1 & 3 & -4 & 2 \end{bmatrix}$

18 $\begin{bmatrix} 2 & 5 & 1 & 0 \\ -4 & 0 & -3 & 0 \\ 3 & -2 & 1 & 6 \\ -1 & 4 & 2 & 0 \end{bmatrix}$

19 $\begin{bmatrix} 0 & b & 0 & 0 \\ 0 & 0 & c & 0 \\ a & 0 & 0 & 0 \\ 0 & 0 & 0 & d \end{bmatrix}$ **20** $\begin{bmatrix} a & u & v & w \\ 0 & b & x & y \\ 0 & 0 & c & z \\ 0 & 0 & 0 & d \end{bmatrix}$

33 $A = \begin{bmatrix} -3 & -2 \\ 2 & 2 \end{bmatrix}$ **34** $A = \begin{bmatrix} 2 & -4 \\ -3 & 5 \end{bmatrix}$

In Exercises 35–38 let $I = I_3$ and let $f(x) = |A - xI|$. Find (a) the polynomial $f(x)$, and (b) the zeros of $f(x)$.

Verify the identities in Exercises 21–28 by expanding each determinant.

21 $\begin{vmatrix} a & b \\ c & d \end{vmatrix} = - \begin{vmatrix} c & d \\ a & b \end{vmatrix}$ **22** $\begin{vmatrix} a & b \\ c & d \end{vmatrix} = - \begin{vmatrix} b & a \\ d & c \end{vmatrix}$

23 $\begin{vmatrix} a & kb \\ c & kd \end{vmatrix} = k \begin{vmatrix} a & b \\ c & d \end{vmatrix}$ **24** $\begin{vmatrix} a & b \\ kc & kd \end{vmatrix} = k \begin{vmatrix} a & b \\ c & d \end{vmatrix}$

25 $\begin{vmatrix} a & b \\ c & d \end{vmatrix} = \begin{vmatrix} a & b \\ ka + c & kb + d \end{vmatrix}$

26 $\begin{vmatrix} a & b \\ c & d \end{vmatrix} = \begin{vmatrix} a & ka + b \\ c & kc + d \end{vmatrix}$

27 $\begin{vmatrix} a & b \\ c & d \end{vmatrix} + \begin{vmatrix} a & e \\ c & f \end{vmatrix} = \begin{vmatrix} a & b + e \\ c & d + f \end{vmatrix}$

28 $\begin{vmatrix} a & b \\ c & d \end{vmatrix} + \begin{vmatrix} a & b \\ e & f \end{vmatrix} = \begin{vmatrix} a & b \\ c + e & d + f \end{vmatrix}$

29 Prove that if a square matrix A of order 2 has two identical rows or columns, then $|A| = 0$.

30 Repeat Exercise 29 for a matrix of order 3.

In Exercises 31–34 let $I = I_2$ be the identity matrix of order 2, and let $f(x) = |A - xI|$. Find (a) the polynomial $f(x)$, and (b) the zeros of $f(x)$. (In the study of matrices, $f(x)$ is called the **characteristic polynomial** of A, and the zeros of $f(x)$ are called the **characteristic values,** or **eigenvalues,** of A.

31 $A = \begin{bmatrix} 1 & 2 \\ 3 & 2 \end{bmatrix}$ **32** $A = \begin{bmatrix} 3 & 1 \\ 2 & 2 \end{bmatrix}$

35 $A = \begin{bmatrix} 1 & 0 & 0 \\ 1 & 0 & -2 \\ -1 & 1 & -3 \end{bmatrix}$

36 $A = \begin{bmatrix} 2 & 1 & 0 \\ -1 & 0 & 0 \\ 1 & 3 & 2 \end{bmatrix}$

37 $A = \begin{bmatrix} 0 & 2 & -2 \\ -1 & 3 & 1 \\ -3 & 3 & 1 \end{bmatrix}$

38 $A = \begin{bmatrix} 3 & 2 & 2 \\ 1 & 0 & 2 \\ -1 & -1 & 0 \end{bmatrix}$

In Exercises 39–42 express the determinants in the form $ai + bj + ck$, where a, b, and c are real numbers.

39 $\begin{vmatrix} i & j & k \\ 2 & -1 & 6 \\ -3 & 5 & 1 \end{vmatrix}$ **40** $\begin{vmatrix} i & j & k \\ 1 & -2 & 3 \\ 2 & 1 & -4 \end{vmatrix}$

41 $\begin{vmatrix} i & j & k \\ 5 & -6 & -1 \\ 3 & 0 & 1 \end{vmatrix}$ **42** $\begin{vmatrix} i & j & k \\ 4 & -6 & 2 \\ -2 & 3 & -1 \end{vmatrix}$

6.7 Properties of Determinants

Evaluating a determinant by using the Expansion Theorem stated in Section 6.6 is inefficient for matrices of high order. For example, if a determinant of a matrix of order 10 is expanded by any row, a sum of 10 terms is obtained, and each term contains the determinant of a matrix of order 9, which is a cofactor of the original matrix. If any of the latter determinants are expanded by a row (or column), a sum of 9 terms is obtained, each containing the determinant of a matrix of order 8. Hence, at this stage there are 90 determinants of matrices of order 8 to evaluate! The process could be continued until only determinants of matrices of order 2 remain. Unless many elements of the original matrix are zero, it is an enormous task to carry out all of the computations.

In this section we discuss rules that simplify the process of evaluating determinants. The main use for these rules is to introduce zeros into the determinant. They may also be used to change the determinant to **echelon form,** that is, a form in which the elements below the main diagonal elements are all zero. The transformations on rows stated in the next theorem are the same as the elementary row transformations of a matrix introduced in Section 6.3. However, for determinants we may also employ similar transformations on columns.

THEOREM ON ROW AND COLUMN TRANSFORMATIONS OF A DETERMINANT

Let A be a square matrix of order n.

(i) If a matrix B is obtained from A by interchanging two rows (or columns), then $|B| = -|A|$.

(ii) If B is obtained from A by multiplying every element of one row (or column) of A by a real number k, then $|B| = k|A|$.

(iii) If B is obtained from A by adding to any row (or column) of A, k times another row (or column), where k is a real number, then $|B| = |A|$.

We shall not prove this theorem. When using the theorem, we refer to the rows (or columns) of the *determinant* in the obvious way. For example, property (iii) may be phrased: "Adding the product of k times another row (or column) to any row (or column) of a determinant does not affect the value of the determinant."

Row transformations of determinants will be specified by means of the symbols R_{ij}, kR_i, and $kR_i + R_j$, which were introduced in Section 6.3. Analogous symbols are used for column transformations. For example, $kC_i + C_j$ means "add to the jth column, k times the ith column." The following are illustrations of the preceding theorem, with the reason for each equality stated at the right.

$$\begin{vmatrix} 2 & 0 & 1 \\ 6 & 4 & 3 \\ 0 & 3 & 5 \end{vmatrix} = - \begin{vmatrix} 6 & 4 & 3 \\ 2 & 0 & 1 \\ 0 & 3 & 5 \end{vmatrix} \qquad R_{12}$$

$$\begin{vmatrix} 2 & 0 & 1 \\ 6 & 4 & 3 \\ 0 & 3 & 5 \end{vmatrix} = - \begin{vmatrix} 1 & 0 & 2 \\ 3 & 4 & 6 \\ 5 & 3 & 0 \end{vmatrix} \qquad C_{13}$$

$$\begin{vmatrix} 1 & -3 & 4 \\ 2 & -1 & 0 \\ 3 & 1 & 6 \end{vmatrix} = \begin{vmatrix} 1 & -3 & 4 \\ 0 & 5 & -8 \\ 3 & 1 & 6 \end{vmatrix} \qquad -2R_1 + R_2$$

$$\begin{vmatrix} 1 & -3 & 4 \\ 2 & -1 & 0 \\ 3 & 1 & 6 \end{vmatrix} = \begin{vmatrix} -5 & -3 & 4 \\ 0 & -1 & 0 \\ 5 & 1 & 6 \end{vmatrix} \qquad 2C_2 + C_1$$

THEOREM

> If two rows (or columns) of a square matrix A are identical, then $|A| = 0$.

PROOF If B is the matrix obtained from A by interchanging the two identical rows (or columns), then B and A are the same and, consequently, $|B| = |A|$. However, by (i) of the Theorem on Row and Column Transformations of a Determinant, $|B| = -|A|$ and hence $-|A| = |A|$, which implies that $|A| = 0$. □

EXAMPLE 1

Find $|A|$ if $A = \begin{bmatrix} 2 & 3 & 0 & 4 \\ 0 & 5 & -1 & 6 \\ 1 & 0 & -2 & 3 \\ -3 & 2 & 0 & -5 \end{bmatrix}$.

SOLUTION We plan to use (iii) of the Theorem on Row and Column Transformations to introduce many zeros in some row or column. To do this, it is convenient to work with an element of the matrix that equals 1, since this enables us to avoid the use of fractions. If 1 is not an element of the original matrix, it is always possible to introduce the number 1 by using (iii) or (ii) of the theorem. In this example 1 appears in row 3, and we proceed as follows, where the reason for each equality is stated at the right.

$$
\begin{vmatrix} 2 & 3 & 0 & 4 \\ 0 & 5 & -1 & 6 \\ 1 & 0 & -2 & 3 \\ -3 & 2 & 0 & -5 \end{vmatrix} = \begin{vmatrix} 0 & 3 & 4 & -2 \\ 0 & 5 & -1 & 6 \\ 1 & 0 & -2 & 3 \\ -3 & 2 & 0 & -5 \end{vmatrix}
$$
$-2R_3 + R_1$

$$
= \begin{vmatrix} 0 & 3 & 4 & -2 \\ 0 & 5 & -1 & 6 \\ 1 & 0 & -2 & 3 \\ 0 & 2 & -6 & 4 \end{vmatrix}
$$
$3R_3 + R_4$

$$
= (1) \begin{vmatrix} 3 & 4 & -2 \\ 5 & -1 & 6 \\ 2 & -6 & 4 \end{vmatrix}
$$
Expand by the first column.

$$
= \begin{vmatrix} 23 & 4 & -2 \\ 0 & -1 & 6 \\ -28 & -6 & 4 \end{vmatrix}
$$
$5C_2 + C_1$

$$
= \begin{vmatrix} 23 & 4 & 22 \\ 0 & -1 & 0 \\ -28 & -6 & -32 \end{vmatrix}
$$
$6C_2 + C_3$

$$
= (-1) \begin{vmatrix} 23 & 22 \\ -28 & -32 \end{vmatrix}
$$
Expand by the second row.

$$
= (-1)[(23)(-32) - (-28)(22)]
$$
Definition of determinant

$$
= 120
$$

∎

Part (ii) of the Theorem on Row and Column Transformations is useful for finding factors of determinants. To illustrate, for a determinant of a matrix of order 3 we have the following:

$$
\begin{vmatrix} a_{11} & a_{12} & a_{13} \\ ka_{21} & ka_{22} & ka_{23} \\ a_{31} & a_{32} & a_{33} \end{vmatrix} = k \begin{vmatrix} a_{11} & a_{12} & a_{13} \\ a_{21} & a_{22} & a_{23} \\ a_{31} & a_{32} & a_{33} \end{vmatrix}.
$$

Similar formulas hold if k is a common factor of the elements of some other row or column. When referring to this manipulation, we often use the phrase "k is a common factor in the row (or column)."

EXAMPLE 2

Find $|A|$ if $A = \begin{bmatrix} 14 & -6 & 4 \\ 4 & -5 & 12 \\ -21 & 9 & -6 \end{bmatrix}$.

SOLUTION

$$|A| = 2 \begin{vmatrix} 7 & -3 & 2 \\ 4 & -5 & 12 \\ -21 & 9 & -6 \end{vmatrix} \qquad \text{2 is a common factor in row 1.}$$

$$= (2)(-3) \begin{vmatrix} 7 & -3 & 2 \\ 4 & -5 & 12 \\ 7 & -3 & 2 \end{vmatrix} \qquad \text{-3 is a common factor in row 3.}$$

$$= 0 \qquad \text{Two rows are identical.} \qquad \blacksquare$$

EXAMPLE 3   Without expanding, show that $a - b$ is a factor of

$$\begin{vmatrix} 1 & 1 & 1 \\ a & b & c \\ a^2 & b^2 & c^2 \end{vmatrix}.$$

SOLUTION

$$\begin{vmatrix} 1 & 1 & 1 \\ a & b & c \\ a^2 & b^2 & c^2 \end{vmatrix} = \begin{vmatrix} 0 & 1 & 1 \\ a-b & b & c \\ a^2-b^2 & b^2 & c^2 \end{vmatrix} \qquad (-1)C_2 + C_1$$

$$= (a-b) \begin{vmatrix} 0 & 1 & 1 \\ 1 & b & c \\ a+b & b^2 & c^2 \end{vmatrix} \qquad \begin{array}{l} a-b \text{ is a common factor} \\ \text{of column 1.} \end{array} \qquad \blacksquare$$

Exercises 6.7

Without expanding, explain why the statements in Exercises 1–14 are true.

1 $\begin{vmatrix} 1 & 0 & 1 \\ 0 & 1 & 1 \\ 1 & 1 & 0 \end{vmatrix} = - \begin{vmatrix} 1 & 0 & 1 \\ 1 & 1 & 0 \\ 0 & 1 & 1 \end{vmatrix}$

2 $\begin{vmatrix} 1 & 0 & 1 \\ 0 & 1 & 1 \\ 1 & 1 & 0 \end{vmatrix} = - \begin{vmatrix} 1 & 1 & 0 \\ 0 & 1 & 1 \\ 1 & 0 & 1 \end{vmatrix}$

3 $\begin{vmatrix} 1 & 0 & 1 \\ 2 & 1 & 0 \\ 1 & 1 & 2 \end{vmatrix} = \begin{vmatrix} 1 & 0 & 1 \\ 2 & 1 & 0 \\ 0 & 1 & 1 \end{vmatrix}$

4 $\begin{vmatrix} 1 & 1 & 2 \\ 1 & 0 & 1 \\ 2 & 1 & 1 \end{vmatrix} = \begin{vmatrix} 0 & 1 & 1 \\ 1 & 0 & 1 \\ 2 & 1 & 1 \end{vmatrix}$

5 $\begin{vmatrix} 2 & 4 & 2 \\ 1 & 2 & 4 \\ 2 & 6 & 4 \end{vmatrix} = 4\begin{vmatrix} 1 & 2 & 1 \\ 1 & 2 & 4 \\ 1 & 3 & 2 \end{vmatrix}$

6 $\begin{vmatrix} 2 & 1 & 6 \\ 4 & 3 & 3 \\ 2 & 1 & 3 \end{vmatrix} = 6\begin{vmatrix} 1 & 1 & 2 \\ 2 & 3 & 1 \\ 1 & 1 & 1 \end{vmatrix}$

7 $\begin{vmatrix} 1 & -1 & 2 \\ 1 & 2 & -1 \\ 1 & -1 & 2 \end{vmatrix} = 0$

8 $\begin{vmatrix} 1 & -1 & 1 \\ 0 & 1 & 0 \\ -1 & 0 & -1 \end{vmatrix} = 0$

9 $\begin{vmatrix} 1 & 5 \\ -3 & 2 \end{vmatrix} = -\begin{vmatrix} 1 & 5 \\ 3 & -2 \end{vmatrix}$

10 $\begin{vmatrix} 2 & -2 \\ 1 & 1 \end{vmatrix} = -\begin{vmatrix} -2 & 2 \\ 1 & 1 \end{vmatrix}$

11 $\begin{vmatrix} 0 & 0 & 1 \\ 1 & 0 & 0 \\ 0 & 0 & 2 \end{vmatrix} = 0$

12 $\begin{vmatrix} 1 & 0 & 1 \\ 0 & 0 & 0 \\ 1 & 1 & 0 \end{vmatrix} = 0$

13 $\begin{vmatrix} 1 & -1 & -2 \\ -1 & 2 & 1 \\ 0 & 1 & 1 \end{vmatrix} = \begin{vmatrix} 1 & -1 & 0 \\ -1 & 2 & -1 \\ 0 & 1 & 1 \end{vmatrix}$

14 $\begin{vmatrix} a & 0 & 0 \\ 0 & b & 0 \\ 0 & 0 & c \end{vmatrix} = -\begin{vmatrix} 0 & 0 & a \\ 0 & b & 0 \\ c & 0 & 0 \end{vmatrix}$

In Exercises 15–24 find the determinant of the matrix after introducing zeros, as in Example 1.

15 $\begin{bmatrix} 3 & 1 & 0 \\ -2 & 0 & 1 \\ 1 & 3 & -1 \end{bmatrix}$

16 $\begin{bmatrix} -3 & 0 & 4 \\ 1 & 2 & 0 \\ 4 & 1 & -1 \end{bmatrix}$

17 $\begin{bmatrix} 5 & 4 & 3 \\ -3 & 2 & 1 \\ 0 & 7 & -2 \end{bmatrix}$

18 $\begin{bmatrix} 0 & 2 & -6 \\ 5 & 1 & -3 \\ 6 & -2 & 5 \end{bmatrix}$

19 $\begin{bmatrix} 2 & 2 & -3 \\ 3 & 6 & 9 \\ -2 & 5 & 4 \end{bmatrix}$

20 $\begin{bmatrix} 3 & 8 & 5 \\ 5 & 3 & -6 \\ 2 & 4 & -2 \end{bmatrix}$

21 $\begin{bmatrix} 3 & 1 & -2 & 2 \\ 2 & 0 & 1 & 4 \\ 0 & 1 & 3 & 5 \\ -1 & 2 & 0 & -3 \end{bmatrix}$

22 $\begin{bmatrix} 3 & 2 & 0 & 4 \\ -2 & 0 & 5 & 0 \\ 4 & -3 & 1 & 6 \\ 2 & -1 & 2 & 0 \end{bmatrix}$

23 $\begin{bmatrix} 2 & -2 & 0 & 0 & -3 \\ 3 & 0 & 3 & 2 & -1 \\ 0 & 1 & -2 & 0 & 2 \\ -1 & 2 & 0 & 3 & 0 \\ 0 & 4 & 1 & 0 & 0 \end{bmatrix}$

24 $\begin{bmatrix} 2 & 0 & -1 & 0 & 2 \\ 1 & 3 & 0 & 0 & 1 \\ 0 & 4 & 3 & 0 & -1 \\ -1 & 2 & 0 & -2 & 0 \\ 0 & 1 & 5 & 0 & -4 \end{bmatrix}$

25 Prove that

$$\begin{vmatrix} 1 & 1 & 1 \\ a & b & c \\ a^2 & b^2 & c^2 \end{vmatrix} = (a - b)(b - c)(c - a).$$

(*Hint:* See Example 3.)

26 Prove that

$$\begin{vmatrix} 1 & 1 & 1 \\ a & b & c \\ a^3 & b^3 & c^3 \end{vmatrix} = (a - b)(b - c)(c - a)(a + b + c).$$

27 If A is a matrix of order 4 of the form

$$A = \begin{bmatrix} a_{11} & a_{12} & a_{13} & a_{14} \\ 0 & a_{22} & a_{23} & a_{24} \\ 0 & 0 & a_{33} & a_{34} \\ 0 & 0 & 0 & a_{44} \end{bmatrix}$$

show that $|A| = a_{11}a_{22}a_{33}a_{44}$.

28 If

$$A = \begin{bmatrix} a & b & 0 & 0 \\ c & d & 0 & 0 \\ 0 & 0 & e & f \\ 0 & 0 & g & h \end{bmatrix}$$

prove that

$$|A| = \begin{vmatrix} a & b \\ c & d \end{vmatrix} \begin{vmatrix} e & f \\ g & h \end{vmatrix}.$$

29 If $A = (a_{ij})$ and $B = (b_{ij})$ are arbitrary square matrices of order 2, prove that $|AB| = |A||B|$.

30 If $A = (a_{ij})$ is a square matrix of order n and k is any real number, prove that $|kA| = k^n|A|$. (*Hint:* Use (ii) of the Theorem on Row and Column Transformations of a Determinant.)

31 Use properties of determinants to show that

$$\begin{vmatrix} x & y & 1 \\ x_1 & y_1 & 1 \\ x_2 & y_2 & 1 \end{vmatrix} = 0$$

is an equation of a line through the points (x_1, y_1) and (x_2, y_2).

32 Use properties of determinants to show that

$$\begin{vmatrix} x^2 + y^2 & x & y & 1 \\ x_1^2 + y_1^2 & x_1 & y_1 & 1 \\ x_2^2 + y_2^2 & x_2 & y_2 & 1 \\ x_3^2 + y_3^2 & x_3 & y_3 & 1 \end{vmatrix} = 0$$

is an equation of a circle through the three points (x_1, y_1), (x_2, y_2), and (x_3, y_3).

6.8 Cramer's Rule

Determinants arise in the study of solutions of systems of linear equations. To illustrate, let us consider the following case of two linear equations in two variables x and y:

$$\begin{cases} a_{11}x + a_{12}y = k_1 \\ a_{21}x + a_{22}y = k_2 \end{cases}$$

where at least one nonzero coefficient appears in each equation. We may as well assume that $a_{11} \neq 0$, for otherwise $a_{12} \neq 0$, and we could regard y as the "first" variable instead of x. We shall use elementary row transformations to obtain the matrix of an equivalent system as follows:

$$\begin{bmatrix} a_{11} & a_{12} & k_1 \\ a_{21} & a_{22} & k_2 \end{bmatrix} \xrightarrow{-\frac{a_{21}}{a_{11}}R_1 + R_2} \begin{bmatrix} a_{11} & a_{12} & k_1 \\ 0 & a_{22} - \left(\frac{a_{12}a_{21}}{a_{11}}\right) & k_2 - \left(\frac{a_{21}k_1}{a_{11}}\right) \end{bmatrix}$$

$$\xrightarrow{a_{11}R_2} \begin{bmatrix} a_{11} & a_{12} & k_1 \\ 0 & (a_{11}a_{22} - a_{12}a_{21}) & (a_{11}k_2 - a_{21}k_1) \end{bmatrix}$$

Thus, the given system is equivalent to

$$\begin{cases} a_{11}x + a_{12}y = k_1 \\ (a_{11}a_{22} - a_{12}a_{21})y = a_{11}k_2 - a_{21}k_1 \end{cases}$$

which may also be written

$$\begin{cases} a_{11}x + a_{12}y = k_1 \\ \begin{vmatrix} a_{11} & a_{12} \\ a_{21} & a_{22} \end{vmatrix} y = \begin{vmatrix} a_{11} & k_1 \\ a_{21} & k_2 \end{vmatrix}. \end{cases}$$

If $\begin{vmatrix} a_{11} & a_{12} \\ a_{21} & a_{22} \end{vmatrix} \neq 0$, we can solve the second equation for y, obtaining

$$y = \frac{\begin{vmatrix} a_{11} & k_1 \\ a_{21} & k_2 \end{vmatrix}}{\begin{vmatrix} a_{11} & a_{12} \\ a_{21} & a_{22} \end{vmatrix}}$$

The corresponding value for x may be found by substituting for y in the first equation. It can be shown that this leads to

$$x = \frac{\begin{vmatrix} k_1 & a_{12} \\ k_2 & a_{22} \end{vmatrix}}{\begin{vmatrix} a_{11} & a_{12} \\ a_{21} & a_{22} \end{vmatrix}}$$

This proves that *if the determinant of the coefficient matrix of a system of two linear equations in two variables is not zero, then the system has a unique solution.* The last two formulas for x and y as quotients of certain determinants constitute what is known as **Cramer's Rule.**

There is an easy way to remember Cramer's Rule. Let

$$D = \begin{bmatrix} a_{11} & a_{12} \\ a_{21} & a_{22} \end{bmatrix}$$

be the coefficient matrix of the system and let D_x denote the matrix obtained from D by replacing the coefficients a_{11}, a_{21} of x by the numbers k_1, k_2, respectively. Similarly, let D_y denote the matrix obtained from D by replacing the coefficients a_{12}, a_{22} of y by the numbers k_1, k_2, respectively. Thus,

$$D_x = \begin{bmatrix} k_1 & a_{12} \\ k_2 & a_{22} \end{bmatrix}, \qquad D_y = \begin{bmatrix} a_{11} & k_1 \\ a_{21} & k_2 \end{bmatrix}.$$

If $|D| \neq 0$, the solution (x, y) is given by

CRAMER'S RULE

$$x = \frac{|D_x|}{|D|}, \qquad y = \frac{|D_y|}{|D|}.$$

EXAMPLE 1 Use Cramer's Rule to solve the system

$$\begin{cases} 2x - 3y = -4 \\ 5x + 7y = 1. \end{cases}$$

SOLUTION The determinant of the coefficient matrix is

$$|D| = \begin{vmatrix} 2 & -3 \\ 5 & 7 \end{vmatrix} = 29.$$

Using the notation introduced previously,

$$|D_x| = \begin{vmatrix} -4 & -3 \\ 1 & 7 \end{vmatrix} = -25, \qquad |D_y| = \begin{vmatrix} 2 & -4 \\ 5 & 1 \end{vmatrix} = 22.$$

Hence,

$$x = \frac{|D_x|}{|D|} = \frac{-25}{29}, \qquad y = \frac{|D_y|}{|D|} = \frac{22}{29}.$$

Thus, the system has the unique solution $(-\frac{25}{29}, \frac{22}{29})$. ∎

Cramer's Rule can be extended to systems of n linear equations in n variables $x_1, x_2, \ldots x_n$, where each equation has the form

$$a_1 x_1 + a_2 x_2 + \cdots + a_n x_n = a.$$

To solve such a system, let D denote the coefficient matrix and let D_{x_i} denote the matrix obtained by replacing the coefficients of x_i in D by the column of numbers $k_1, \ldots, k_n$ that appears to the right of the equal signs in the system. It can be shown that if $|D| \neq 0$, then the system has the following unique solution.

CRAMER'S RULE (GENERAL FORM)

$$x_1 = \frac{|D_{x_1}|}{|D|}, \qquad x_2 = \frac{|D_{x_2}|}{|D|}, \qquad \ldots, \qquad x_n = \frac{|D_{x_n}|}{|d|}.$$

EXAMPLE 2 Use Cramer's Rule to solve the system

$$\begin{cases} x - 2z = 3 \\ - y + 3z = 1 \\ 2x + 5z = 0. \end{cases}$$

SOLUTION We shall merely list the various determinants, leaving the reader to check the answers:

$$|D| = \begin{vmatrix} 1 & 0 & -2 \\ 0 & -1 & 3 \\ 2 & 0 & 5 \end{vmatrix} = -9, \qquad |D_x| = \begin{vmatrix} 3 & 0 & -2 \\ 1 & -1 & 3 \\ 0 & 0 & 5 \end{vmatrix} = -15,$$

$$|D_y| = \begin{vmatrix} 1 & 3 & -2 \\ 0 & 1 & 3 \\ 2 & 0 & 5 \end{vmatrix} = 27, \qquad |D_z| = \begin{vmatrix} 1 & 0 & 3 \\ 0 & -1 & 1 \\ 2 & 0 & 0 \end{vmatrix} = 6.$$

By Cramer's Rule, the solution is

$$x = \frac{|D_x|}{|D|} = \frac{-15}{-9} = \frac{5}{3}, \quad y = \frac{|D_y|}{|D|} = \frac{27}{-9} = -3, \quad z = \frac{|D_z|}{|D|} = \frac{6}{-9} = -\frac{2}{3}.$$

■

Cramer's Rule is an inefficient method to apply if there are a large number of equations, since many determinants of matrices of high order must be evaluated. Note also that Cramer's Rule cannot be used directly if $|D| = 0$ or if the number of equations is not the same as the number of variables. In general, the matrix method is far superior to Cramer's Rule.

Exercises 6.8

1–18 Use Cramer's Rule to solve the systems in Exercises 1–18 of Section 6.2.

19–26 Use Cramer's Rule to solve the systems in Exercises 1–8 of Section 6.3.

27–30 Use Cramer's Rule to solve the systems in Exercises 17, 18, 21, and 22 of Section 6.3.

6.9 Systems of Inequalities

The discussion of inequalities in Chapter 2 was restricted to inequalities in one variable. Inequalities in several variables can also be considered. For example, expressions of the form

$$3x + y < 5y^2 + 1,$$

$$2x^2 \geq 4 - 3y,$$

and so on are called **inequalities in x and y.** A **solution** of an inequality in x and y is defined as an ordered pair (a, b) that produces a true statement if a and b are substituted for x and y, respectively. The **graph of an inequality** is the graph of all the solutions. Two inequalities are **equivalent** if they have exactly the same solutions. An inequality in x and y can often be simplified by adding an expression in x and y to both sides or by multiplying both sides by some expression (provided care is taken with regard to signs). Similar remarks apply to inequalities in more than two variables. We shall restrict our discussion, however, to the case of inequalities in two variables.

EXAMPLE 1 Find the solutions and sketch the graph of the inequality

$$3x - 3 < 5x - y.$$

SOLUTION The inequalities in the following list are equivalent:

$$3x - 3 < 5x - y$$
$$y + 3x - 3 < 5x$$
$$y < 5x - (3x - 3)$$
$$y < 2x + 3.$$

Hence, the solutions consist of all ordered pairs (x, y) such that $y < 2x + 3$. It is convenient to denote the solutions as follows:

$$\{(x, y) : y < 2x + 3\}.$$

There is a close relationship between the graph of the inequality $y < 2x + 3$ and the graph of the equation $y = 2x + 3$. The graph of the equation is the line sketched in Figure 6.6. For each real number a, the point on the line with x-coordinate a has coordinates $(a, 2a + 3)$. A point $P(a, b)$ belongs to the graph of the *inequality* if and only if $b < 2a + 3$; that is, if and only if the point $P(a, b)$ lies directly below the point with coordinates $(a, 2a + 3)$ as shown in Figure 6.6. It follows that the graph of the inequality $y < 2x + 3$ consists of all points that lie below the line $y = 2x + 3$. In Figure 6.7 we have shaded a portion of the graph of the inequality. Dashes used for the line indicate that it is not part of the graph. ∎

A region of the type shown in Figure 6.7 is called a **half-plane.** More precisely, if the line is *not* included, we refer to the region as an **open half-plane.** If the line *is* included, as would be the case for the graph of the inequality $y \leq 2x + 3$, then the region is called a **closed half-plane.**

By an argument similar to that used in Example 1, it can be shown that the graph of the inequality $y > 2x + 3$ is the open half-plane that lies *above* the line $y = 2x + 3$.

FIGURE 6.6

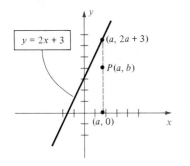

FIGURE 6.7

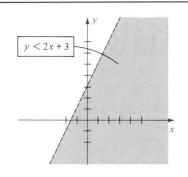

If an inequality involves only polynomials of the first degree in x and y, as was the case in Example 1, it is called a **linear inequality.**

The procedure used in Example 1 can be generalized to inequalities of the form $y < f(x)$ for any function f. Specifically, the following theorem is true.

THEOREM

> Let f be a function.
>
> (i) The graph of the inequality $y < f(x)$ is the set of points that lie *below* the graph of the equation $y = f(x)$.
>
> (ii) The graph of $y > f(x)$ is the set of points that lie *above* the graph of $y = f(x)$.

EXAMPLE 2   Find the solutions and sketch the graph of the inequality

$$x(x + 1) - 2y > 3(x - y).$$

FIGURE 6.8

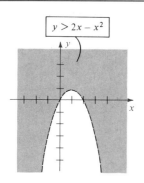

$y > 2x - x^2$

SOLUTION   The inequality is equivalent to

$$x^2 + x - 2y > 3x - 3y.$$

Adding $3y - x^2 - x$ to both sides, we obtain

$$y > 2x - x^2.$$

Hence, the solutions are $\{(x, y) : y > 2x - x^2\}$.

To find the graph of $y > 2x - x^2$, we begin by sketching the graph of $y = 2x - x^2$ (a parabola) with dashes as illustrated in Figure 6.8. By (ii) of the preceding theorem, the graph is the region above the parabola, as indicated by the shaded portion of the figure.   ∎

The following can also be proved.

THEOREM

> Let g be a function.
>
> (i) The graph of the inequality $x < g(y)$ is the set of points to the *left* of the graph of the equation $x = g(y)$.
>
> (ii) The graph of $x > g(y)$ is the set of points to the *right* of the graph of $x = g(y)$.

FIGURE 6.9

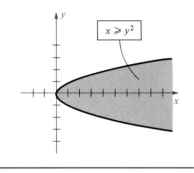

$x \geqslant y^2$

EXAMPLE 3   Sketch the graph of $x \geq y^2$.

SOLUTION   The graph of the equation $x = y^2$ is a parabola that is symmetric to the x-axis. By the preceding theorem, the graph of the inequality consists of all points on the parabola, together with the points in the region to the right of the parabola (see Figure 6.9).   ■

It is sometimes necessary to work simultaneously with several inequalities in two variables. In this case we refer to the inequalities as a **system of inequalities.** The **solutions of a system** of inequalities are, by definition, the solutions that are common to all inequalities in the system. It should be clear how to define **equivalent systems** and the **graph of a system** of inequalities. The following examples illustrate a method for solving systems of inequalities.

EXAMPLE 4   Find the solutions and sketch the graph of the system

$$\begin{cases} x + y \leq 4 \\ 2x - y \leq 4. \end{cases}$$

FIGURE 6.10

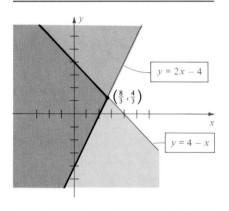

$y = 2x - 4$

$\left(\frac{8}{3}, \frac{4}{3}\right)$

$y = 4 - x$

SOLUTION   The system is equivalent to

$$\begin{cases} y \leq 4 - x \\ y \geq 2x - 4. \end{cases}$$

We begin by sketching the graphs of the lines $y = 4 - x$ and $y = 2x - 4$. The lines intersect at the point $\left(\frac{8}{3}, \frac{4}{3}\right)$ shown in Figure 6.10. The graph of $y \leq 4 - x$ includes the points on the graph of $y = 4 - x$ together with the points that lie below this line. The graph of $y \geq 2x - 4$ includes the points on the graph of $y = 2x - 4$ together with the points that lie above this line. A portion of each of these regions is shown in Figure 6.10. The graph of the system consists of the points that are in *both* regions as indicated by the double-shaded portion of the figure.   ■

EXAMPLE 5   Sketch the graph of the system

$$\begin{cases} x + y \leq 4 \\ 2x - y \leq 4 \\ \quad\quad x \geq 0 \\ \quad\quad y \geq 0. \end{cases}$$

FIGURE 6.11

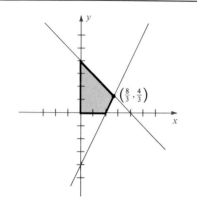

SOLUTION   The first two inequalities are the same as those considered in Example 4, and hence the points on the graph of the system must lie within the double-shaded region shown in Figure 6.10. In addition, the third and fourth inequalities in the system tell us that the points must lie in the first quadrant or on its boundaries. This gives us the region shown in Figure 6.11.   ■

EXAMPLE 6   Sketch the graph of the system

$$\begin{cases} x^2 + y^2 \le 1 \\ (x - 1)^2 + y^2 \le 1. \end{cases}$$

SOLUTION   The graph of the equation $x^2 + y^2 = 1$ is a unit circle with center at the origin, and the graph of $(x - 1)^2 + y^2 = 1$ is a unit circle with center at the point $C(1, 0)$. To find the points of intersection of the two circles, let us solve the equation $x^2 + y^2 = 1$ for y^2, obtaining $y^2 = 1 - x^2$. Substituting for y^2 in $(x - 1)^2 + y^2 = 1$ leads to the following equations:

FIGURE 6.12

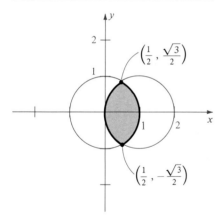

$$(x - 1)^2 + (1 - x^2) = 1$$
$$x^2 - 2x + 1 + 1 - x^2 = 1$$
$$-2x = -1$$
$$x = \tfrac{1}{2}.$$

The corresponding values for y are given by

$$y^2 = 1 - x^2 = 1 - (\tfrac{1}{2})^2 = \tfrac{3}{4}$$

and hence $y = \pm\sqrt{3}/2$. Thus, the points of intersection are $(\tfrac{1}{2}, \sqrt{3}/2)$ and $(\tfrac{1}{2}, -\sqrt{3}/2)$ as shown in Figure 6.12. By the Distance Formula, the graphs of the inequalities are the regions within and on the two circles. The graph of the system consists of the points common to both regions, as indicated by the shaded portion of the figure.   ■

EXAMPLE 7   The manager of a baseball team wishes to buy bats and balls costing $12 and $3 each, respectively. At least five bats and ten balls are required, and the total cost is not to exceed $180. Find a system of inequalities that describes all possibilities and sketch the graph.

FIGURE 6.13

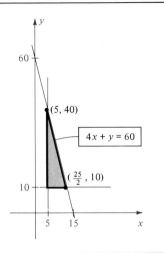

SOLUTION   We begin by introducing the following variables:

$$x = \text{number of bats},$$
$$y = \text{number of balls}.$$

Thus, the bats will cost $12x$ dollars and the balls will cost $3y$ dollars. Since the total cost is not to exceed $180, we must have

$$12x + 3y \le 180 \quad \text{or} \quad 4x + y \le 60.$$

The other restrictions are

$$x \ge 5 \quad \text{and} \quad y \ge 10.$$

The graph of the system is sketched in Figure 6.13. ■

Exercises 6.9

In Exercises 1–10 find the solutions and sketch the graph of the inequality.

1 $3x - 2y < 6$ **2** $4x + 3y < 12$

3 $2x + 3y \ge 2y + 1$ **4** $2x - y > 3$

5 $y + 2 < x^2$ **6** $y^2 - x \le 0$

7 $x^2 + 1 \le y$ **8** $y - x^3 < 1$

9 $yx^2 \ge 1$ **10** $x^2 + 4 \ge y$

In Exercises 11–24 sketch the graph of the system of inequalities.

11 $\begin{cases} 3x + y < 3 \\ 4 - y < 2x \end{cases}$ **12** $\begin{cases} y + 2 < 2x \\ y - x > 4 \end{cases}$

13 $\begin{cases} y - x < 0 \\ 2x + 5y < 10 \end{cases}$ **14** $\begin{cases} 2y - x \le 4 \\ 3y + 2x < 6 \end{cases}$

15 $\begin{cases} 3x + y \le 6 \\ y - 2x \ge 1 \\ x \ge -2 \\ y \le 4 \end{cases}$ **16** $\begin{cases} 3x - 4y \ge 12 \\ x - 2y \le 2 \\ x \ge 9 \\ y \le 5 \end{cases}$

17 $\begin{cases} x^2 + y^2 \le 4 \\ x + y \ge 1 \end{cases}$ **18** $\begin{cases} x^2 + y^2 > 1 \\ x^2 + y^2 < 4 \end{cases}$

19 $\begin{cases} x^2 \le 1 - y \\ x \ge 1 + y \end{cases}$ **20** $\begin{cases} x - y^2 < 0 \\ x + y^2 > 0 \end{cases}$

21 $\begin{cases} y < 3^x \\ y > 2^x \\ x \ge 0 \end{cases}$ **22** $\begin{cases} y \ge \log x \\ y - x \le 1 \\ x \ge 1 \end{cases}$

23 $\begin{cases} y \le \log x \\ y + x \ge 1 \\ x \le 10 \end{cases}$ **24** $\begin{cases} y \le 3^{-x} \\ y \ge 2^{-x} \\ y < 9 \end{cases}$

25 A store sells two brands of television sets. Customer demand indicates that it is necessary to stock at least twice as many sets of brand A as of brand B. It is also necessary to have on hand at least 20 of brand A and 10 of brand B. If there is room for not more than 100 sets in the store, find a system of inequalities that describes all possibilities, and sketch the graph.

26 An auditorium contains 600 seats. For a certain event it is planned to charge $8.00 for some seats and $5.00 for others. At least 225 tickets are to be sold for $5.00, and total sales of more than $3000 are desired. Find a system of inequalities that describes all possibilities, and sketch the graph.

27 A woman wishes to invest $15,000 in two different savings accounts. She also wants to have at least $2000 in each account, with the amount in one account being at least three times that in the other. Find a system of inequalities that describes all possibilities, and sketch the graph.

28 The manager of a college bookstore stocks two types of notebooks, the first wholesaling for 55 cents and the second for 85 cents. If the maximum amount to be spent is $600 and if an inventory of at least 300 of the 85-cent variety and 400 of the 55-cent variety is desired, find a system of inequalities that describes all possibilities and sketch the graph.

29 An aerosol can is to be constructed in the shape of a circular cylinder with a small cone on the top. The total height of the can is to be no more than 9 inches and the cylinder must contain at least 75% of the total volume. In addition, the height of the conical top must be at least 1 inch. Find and graph the system of inequalities that describes the possible relationships between the height of the cylinder and the height of the cone.

30 A stained-glass window is to be constructed in the form of a rectangle surmounted by a semicircle. The total height of the window can be no more than 6 feet, and the area of the rectangular part must be at least twice the area of the semicircle. In addition, the diameter of the semicircle must be at least 2 feet. Find and graph the system of inequalities that describes the possibilities for the length and width of the rectangle.

31 A nuclear power plant will be constructed to serve the power needs of cities A and B. City B is 100 miles due east of A. The state has promised that the plant will be at least 60 miles from each city. It is not possible, however, to locate the plant south of either city because of rough terrain, and the plant must be within 100 miles of both A and B. Assuming A is at the origin, find and graph the system of inequalities that describes all possible locations for the plant.

32 A man has a rectangular back yard that is 50 feet wide and 60 feet deep. He plans to construct a pool area and a patio area as shown in the figure, and can spend at most $12,000 on the project. The patio area must be at least as large as the pool area. The pool area will cost $5 per square foot and the patio will cost $3 per square foot. Find and graph the system of inequalities that describes the possibilities for the width of the patio and pool areas.

FIGURE FOR EXERCISE 32

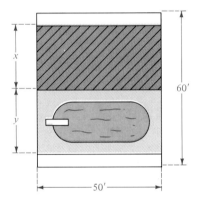

6.10 Linear Programming

Certain applications require finding particular solutions of systems of inequalities. A typical problem consists of finding maximum and minimum values of expressions involving variables that are subject to various constraints. If all the expressions and inequalities are linear in the variables, then a technique called **linear programming** may be used to help solve such problems. This technique has become very important in businesses that require decisions to be made concerning the best use of stock, parts, or manufacturing processes. Usually the objective of management is to maximize profit or to minimize cost. Since there are often many choices, it may be extremely difficult to arrive at a correct decision. A mathematical theory such as that afforded by linear programming can simplify the task considerably. The logical development of the theorems and methods that are needed would take us beyond the objectives of this text. We shall, therefore, limit ourselves to several examples.

EXAMPLE 1   A manufacturer of a certain product has two warehouses, W_1 and W_2. There are 80 units of the product stored at W_1 and 70 units at W_2. Two customers, A and B, order 35 units and 60 units, respectively. The shipping cost from each warehouse to A and B is determined according to the following table. How should the order be filled to minimize the total shipping cost?

Warehouse	Customer	Shipping cost per unit
W_1	A	$ 8
W_1	B	12
W_2	A	10
W_2	B	13

SOLUTION   Let

$$x = \text{number of units sent to } A \text{ from } W_1,$$

$$y = \text{number of units sent to } B \text{ from } W_1.$$

Thus, to fill the orders we must have

$$35 - x = \text{number sent to } A \text{ from } W_2,$$

$$60 - y = \text{number sent to } B \text{ from } W_2.$$

We wish to determine values for x and y that make the total shipping costs minimal. Since x and y are between 35 and 60, respectively, the pair (x, y) must be a solution of the following system of inequalities:

$$\begin{cases} 0 \leq x \leq 35 \\ 0 \leq y \leq 60. \end{cases}$$

The graph of this system is the rectangular region shown in Figure 6.14.

FIGURE 6.14

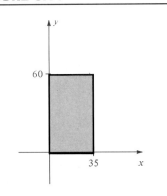

FIGURE 6.15

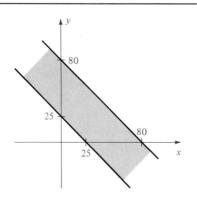

There are further constraints on x and y that make it possible to reduce the size of the region in Figure 6.14. Since the total number of units shipped from W_1 cannot exceed 80 and the total shipped from W_2 cannot exceed 70, the pair (x, y) must also be a solution of the system

$$\begin{cases} x + y \leq 80 \\ (35 - x) + (60 - y) \leq 70. \end{cases}$$

This system is equivalent to

$$\begin{cases} x + y \leq 80 \\ x + y \geq 25. \end{cases}$$

FIGURE 6.16

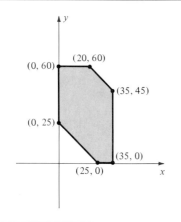

The graph of this system is the region between the parallel lines $x + y = 80$ and $x + y = 25$ (see Figure 6.15). Since the pair (x, y) that we seek must be a solution of this system, and also of the system $0 \leq x \leq 35$, $0 \leq y \leq 60$, the corresponding point must lie in the region shown in Figure 6.16.

Let C denote the total cost (in dollars) of shipping the merchandise to A and B. We see from the table listing shipping costs that the following are true:

Cost of shipping 35 units to $A = 8x + 10(35 - x)$,

Cost of shipping 60 units to $B = 12y + 13(60 - y)$.

Hence, the total cost is

$$C = 8x + 10(35 - x) + 12y + 13(60 - y)$$
$$= (8x + 350 - 10x) + (12y + 780 - 13y)$$

or $\qquad C = 1130 - 2x - y.$

For each point (x, y) of the region shown in Figure 6.16 there corresponds a value for C. For example, at $(20, 40)$

$$C = 1130 - 40 - 40 = 1050,$$

and at $(10, 50)$, $\qquad C = 1130 - 20 - 50 = 1060.$

Since x and y are integers, there is only a finite number of possible values for C. By checking each possibility, we could find the pair (x, y) that produces the smallest cost. However, since there is a very large number of pairs, the task of checking each one would be very tedious. This is where the theory developed in linear programming is helpful. It can be shown that if we are interested in the value C of a linear expression $ax + by + c$, and if each pair (x, y) is a solution of a system of linear inequalities, and hence corresponds to a point that is common to several half-planes, then C takes on its maximum or minimum value at a point of intersection of the lines that determine the half-planes. This means that to determine the minimum (or maximum) value of C we need only check the points $(0, 25)$, $(0, 60)$, $(20, 60)$, $(35, 45)$, $(35, 0)$, and $(25, 0)$ shown in Figure 6.16. The values are displayed in the following table.

Point	$1130 - 2x - y = C$
$(0. 25)$	$1130 - 2(0) - 25 = 1105$
$(0, 60)$	$1130 - 2(0) - 60 = 1070$
$(20, 60)$	$1130 - 2(20) - 60 = 1030$
$(35, 45)$	$1130 - 2(35) - 45 = 1015$
$(35, 0)$	$1130 - 2(35) - 0 = 1060$
$(25, 0)$	$1130 - 2(25) - 0 = 1080$

According to our remarks, the minimal shipping cost $1015 occurs if $x = 35$ and $y = 45$. This means that the manufacturer should ship all of the units to A from W_1 and none from W_2. In addition, the manufacturer should ship 45 units to B from W_1 and 15 units to B from W_2. Note that the *maximum* shipping cost will occur if $x = 0$ and $y = 25$, that is, if all 35 units are shipped to A from W_2 and if B receives 25 units from W_1 and 35 units from W_2. ■

The preceding example demonstrates how linear programming can be used to minimize the cost in a certain situation. The next example illustrates maximization of profit.

EXAMPLE 2 A firm manufactures two products X and Y. For each product it is necessary to use three different machines, A, B, and C. To manufacture one unit of product X, machine A must be used for 3 hours, machine B for 1 hour, and machine C for 1 hour. To manufacture one unit of product Y requires 2 hours on A, 2 hours on B, and 1 hour on C. The profit on product X is $500 per unit and the profit on product Y is $350 per unit. Machine A is available for a total of 24 hours per day; however, B can only be used for 16 hours and C for 9 hours. Assuming the machines are available when needed (subject to the noted total hour restrictions), determine the number of units of each product that should be manufactured each day in order to maximize the profit.

SOLUTION The following table summarizes the data given in the statement of the problem.

Machine	Hours required for 1 unit of X	Hours required for one unit of Y	Hours available
A	3	2	24
B	1	2	16
C	1	1	9

Let x and y denote the number of units of products X and Y, respectively, to be produced per day. Since each unit of product X requires 3 hours on machine A, x units require $3x$ hours. Similarly, since each unit of product Y requires 2 hours on A, y units require $2y$ hours. Hence, the total number of hours per day that machine A must be used is $3x + 2y$. Since A can be used for at most 24 hours per day,

$$3x + 2y \leq 24.$$

Using the same type of reasoning on rows two and three of the table, we see that

$$x + 2y \leq 16$$
$$x + \ y \leq \ 9.$$

This system of three linear inequalities together with the obvious inequalities

$$x \geq 0, \qquad y \geq 0$$

states, in mathematical form, the constraints that occur in the manufacturing process. The graph of the preceding system of five linear inequalities is sketched in Figure 6.17.

The points shown in the figure are found by solving systems of linear equations. Specifically, (6, 3) is the solution of the system $3x + 2y = 24$, $x + y = 9$, and (2, 7) is the solution of the system $x + 2y = 16$, $x + y = 9$.

Since the production of each unit of product X yields a profit of $500, and each unit of product Y yields a profit of $350, the profit P obtained by producing x units of X together with y units of Y is

$$P = 500x + 350y.$$

The maximum value of P must occur at one of the points shown in Figure 6.17. The values of P at all the points are shown in the following table.

FIGURE 6.17

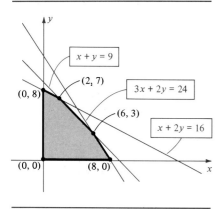

(x, y)	$500x + 350y = P$
$(0, 8)$	$500(0) + 350(8) = 2800$
$(2, 7)$	$500(2) + 350(7) = 3450$
$(6, 3)$	$500(6) + 350(3) = 4050$
$(8, 0)$	$500(8) + 350(0) = 4000$
$(0, 0)$	$500(0) + 350(0) = 0$

We see from the table that a maximum profit of $4050 occurs for a daily production of 6 units of product X and 3 units of product Y. ∎

The two illustrations given in this section are elementary problems in linear programming that can be solved by rather crude methods. The much more complicated problems that occur in practice are usually solved by employing matrix techniques that are adapted for solutions by computers.

Exercises 6.10

1 A manufacturer of tennis rackets makes a profit of $15 on each Set Point racket and $8 on each Double Fault racket. To meet dealer demand, daily production of Double Faults should be between 30 and 80, whereas the number of Set Points produced should be between 10 and 30. To maintain high quality, the total number of rackets produced should not exceed 80 per day. How many of each type should be manufactured daily to maximize the profit?

2 A manufacturer of CB radios makes a profit of $25 on a deluxe model and $30 on a standard model. The company wishes to produce at least 80 deluxe models and at least 100 standard models per day. To maintain high quality, the daily production should not exceed 200 radios. How many of each type should be produced daily in order to maximize the profit?

3 Two substances S and T each contain two types of ingredients I and G. One pound of S contains 2 ounces of I and 4 ounces of G. One pound of T contains 2 ounces of I and 6 ounces of G. A manufacturer plans to combine quantities of the two substances to obtain a mixture that contains at least 9 ounces of I and 20 ounces of G. If the cost of S is $3.00 per pound and the cost of T is $4.00

per pound, how much of each substance should be used to keep the cost to a minimum?

4 A stationery company makes two types of notebooks: a deluxe notebook, with subject dividers, that sells for $1.25, and a regular notebook that sells for $0.90. It costs the company $1.00 to produce each deluxe notebook and $0.75 to produce each regular notebook. The company has the facilities to manufacture between 2000 and 3000 deluxe and between 3000 and 6000 regular, but not more than 7000 altogether. How many notebooks of each type should be manufactured to maximize the difference between the selling prices and the costs of production?

5 Refer to Example 1 of this section. If the shipping costs are $12 per unit from W_1 to A, $10 per unit from W_2 to A, $16 per unit from W_1 to B, and $12 per unit from W_2 to B, determine how the order should be filled to minimize shipping costs.

6 A coffee company purchases mixed lots of coffee beans and then grades them into premium, regular, and unusable beans. The company needs at least 280 tons of premium-grade and 200 tons of regular-grade coffee beans. The company can purchase ungraded coffee from

two suppliers in any amount desired. Samples from the two suppliers contain the following percentages of premium, regular, and unusable beans:

Supplier	Premium	Regular	Unusable
A	20%	50%	30%
B	40%	20%	40%

If supplier A charges $125 per ton and B charges $200 per ton, how much should the company purchase from each supplier to fulfill its needs at minimum cost?

7 A farmer, in the business of growing fodder for livestock, has 100 acres available for planting alfalfa and corn. The cost of seed per acre is $4 for alfalfa and $6 for corn. The total cost of labor will amount to $20 per acre for alfalfa and $10 per acre for corn. The expected income from alfalfa is $110 per acre, and from corn, $150 per acre. If the farmer does not wish to spend more than $480 for seed and $1400 for labor, how many acres of each should be planted to obtain the maximum profit?

8 A small firm manufactures bookshelves and desks for microcomputers. For each product it is necessary to use a table saw and a power router. To manufacture each bookshelf, the saw must be used for $\frac{1}{2}$ hour and the router for 1 hour. A desk requires the use of each machine for 2 hours. The profits are $20 per bookshelf and $50 per desk. If the saw can be used 8 hours per day, and the router 12 hours per day, how many bookshelves and desks should be manufactured each day to maximize profits?

9 Three substances X, Y, and Z each contain four ingredients A, B, C, and D. The percentage of each ingredient and the cost in cents per ounce of each substance are given in the following table.

Substance	A	B	C	D	Cost/ ounce
X	20%	10%	25%	45%	25¢
Y	20%	40%	15%	25%	35¢
Z	10%	20%	25%	45%	50¢

If the cost is to be minimum, how many ounces of each substance should be combined to obtain a mixture of

20 ounces containing at least 14% A, 16% B, and 20% C? What combination would make the cost greatest?

10 A man plans to operate a stand at a one-day fair at which he will sell bags of peanuts and bags of candy. He has $100 available to purchase his stock, which will cost 10¢ per bag of peanuts and 20¢ per bag of candy. He intends to sell the peanuts at 25¢ and the candy at 40¢ per bag. His stand can accommodate up to 500 bags of peanuts and 400 bags of candy. From past experience he knows that he will sell no more than a total of 700 bags. Find the number of bags of each that he should have available in order to maximize his profit. What is the maximum profit?

11 A small community wishes to purchase vans and small buses for its public transportation system. The community can spend no more than $100,000 for the vehicles and no more than $500 per month for maintenance. The vans sell for $10,000 each and average $100 per month in maintenance costs. The corresponding cost estimates for each bus are $20,000 and $75 per month. If each van can carry 10 passengers and each bus can accomodate 15 riders, determine the number of vans and buses that should be purchased to maximize the passenger capacity of the system.

12 Refer to Exercise 11. The monthly fuel costs (based on 5000 miles of service) for each van is $550 while each bus consumes $850 in fuel. Find the number of vans and buses that should be purchased to minimize the monthly fuel costs if the passenger capacity of the system must be at least 75.

13 A fish farmer will purchase no more than 5000 young trout and bass from the hatchery and feed them a special diet over the next year. Each trout will consume $0.50 in food while $0.75 will be spent per bass. The total amount spent on the special diet is not to exceed $3000. At the end of the year, a typical trout will weigh 2 pounds and a bass will weigh 3 pounds. How many fish of each type should be stocked in the pond in order to maximize the total number of pounds of fish at the end of the year?

14 A hospital dietician wishes to prepare a corn–squash vegetable dish that will provide at least 3 grams of protein and cost no more than 36 cents per serving. An ounce of creamed corn provides $\frac{1}{2}$ gram of protein and costs 4 cents. An ounce of squash supplies $\frac{1}{4}$ gram of protein and

costs 3 cents. For taste, there must be at least 2 ounces of corn and at least as much squash as corn. It is important to keep the total number of ounces in a serving as small as possible. Find the combination of corn and squash that minimizes the size of the dish.

15 A contractor has a large building that he wishes to convert into a series of self-storage spaces. He will construct basic 8 × 10 foot units and deluxe 12 × 10 foot units that contain extra shelves and a clothes closet. Market

considerations dictate that there be at least twice as many smaller units as larger units and that he rent the smaller units for $40 per month and the deluxe units for $75 per month. There is at most 7200 ft^2 for the storage spaces and no more than $30,000 can be spent on construction. If each small unit will cost $300 to make while a deluxe unit will cost $600, how many units of each type should be constructed to maximize monthly revenues?

6.11 Review

Define or discuss each of the following.

1 System of equations
2 Solution of a system of equations
3 Equivalent systems of equations
4 System of linear equations
5 Partial fraction decompositions
6 An $m \times n$ matrix
7 A square matrix of order n
8 The coefficient matrix of a system of linear equations; the augmented matrix
9 Elementary row transformations
10 Homogeneous system of linear equations

11 The sum and product of two matrices
12 Zero matrix
13 Identity matrix
14 Inverse of a matrix
15 Determinant
16 Minor
17 Cofactor
18 Properties of determinants
19 Cramer's Rule
20 System of inequalities
21 Linear programming

Exercises 6.11

Find the solutions of the systems of equations in Exercises 1–16.

1 $\begin{cases} 2x - 3y = 4 \\ 5x + 4y = 1 \end{cases}$

2 $\begin{cases} x - 3y = 4 \\ -2x + 6y = 2 \end{cases}$

3 $\begin{cases} y + 4 = x^2 \\ 2x + y = -1 \end{cases}$

4 $\begin{cases} x^2 + y^2 = 25 \\ x - y = 7 \end{cases}$

5 $\begin{cases} 9x^2 + 16y^2 = 140 \\ x^2 - 4y^2 = 4 \end{cases}$

6 $\begin{cases} 2x = y^2 + 3z \\ x = y^2 + z - 1 \\ x^2 = xz \end{cases}$

7 $\begin{cases} \dfrac{1}{x} + \dfrac{3}{y} = 7 \\ \dfrac{4}{x} - \dfrac{2}{y} = 1 \end{cases}$

8 $\begin{cases} 2^x + 3^{y+1} = 10 \\ 2^{x+1} - 3^y = 5 \end{cases}$

9 $\begin{cases} 3x + y - 2z = -1 \\ 2x - 3y + z = 4 \\ 4x + 5y - z = -2 \end{cases}$

10 $\begin{cases} x + 3y = 0 \\ y - 5z = 3 \\ 2x + z = -1 \end{cases}$

11 $\begin{cases} 4x - 3y - z = 0 \\ x - y - z = 0 \\ 3x - y + 3z = 0 \end{cases}$

12 $\begin{cases} 2x + y - z = 0 \\ x - 2y + z = 0 \\ 3x + 3y + 2z = 0 \end{cases}$

13 $\begin{cases} 4x + 2y - z = 1 \\ 3x + 2y + 4z = 2 \end{cases}$

14 $\begin{cases} 2x + y = 6 \\ x - 3y = 17 \\ 3x + 2y = 7 \end{cases}$

15 $\begin{cases} \dfrac{4}{x} + \dfrac{1}{y} + \dfrac{2}{z} = 4 \\ \dfrac{2}{x} + \dfrac{3}{y} - \dfrac{1}{z} = 1 \\ \dfrac{1}{x} + \dfrac{1}{y} + \dfrac{1}{z} = 4 \end{cases}$

16 $\begin{cases} 2x - y + 3z - w = -3 \\ 3x + 2y - z + w = 13 \\ x - 3y + z - 2w = -4 \\ -x + y + 4z + 3w = 0 \end{cases}$

Find the solutions and sketch the graphs of the systems in Exercises 17–20.

17 $\begin{cases} x^2 + y^2 < 16 \\ y - x^2 > 0 \end{cases}$

18 $\begin{cases} y - x \le 0 \\ y + x \ge 2 \\ x \le 5 \end{cases}$

19 $\begin{cases} x - 2y \le 2 \\ y - 3x \le 4 \\ 2x + y \le 4 \end{cases}$

20 $\begin{cases} x^2 - y < 0 \\ y - 2x < 5 \\ xy < 0 \end{cases}$

Find the determinants of the matrices in Exercises 21–30.

21 $[-6]$

22 $\begin{bmatrix} 3 & 4 \\ -6 & -5 \end{bmatrix}$

23 $\begin{bmatrix} 3 & -4 \\ 6 & 8 \end{bmatrix}$

24 $\begin{bmatrix} 0 & 4 & -3 \\ 2 & 0 & 4 \\ -5 & 1 & 0 \end{bmatrix}$

25 $\begin{bmatrix} 2 & -3 & 5 \\ -4 & 1 & 3 \\ 3 & 2 & -1 \end{bmatrix}$

26 $\begin{bmatrix} 3 & 1 & -2 \\ -5 & 2 & -4 \\ 7 & 3 & -6 \end{bmatrix}$

27 $\begin{bmatrix} 5 & 0 & 0 & 0 \\ 6 & -3 & 0 & 0 \\ 1 & 4 & -4 & 0 \\ 7 & 2 & 3 & 2 \end{bmatrix}$

28 $\begin{bmatrix} 1 & 2 & 0 & 3 & 1 \\ -2 & -1 & 4 & 1 & 2 \\ 3 & 0 & -1 & 0 & -1 \\ 2 & -3 & 2 & -4 & 2 \\ -1 & 1 & 0 & 1 & 3 \end{bmatrix}$

29 $\begin{bmatrix} 2 & 0 & 1 & 0 & -1 \\ 0 & 1 & 0 & 1 & 2 \\ 2 & -2 & 1 & -2 & 0 \\ 0 & 0 & -2 & 0 & 1 \\ 1 & -1 & 0 & -1 & 0 \end{bmatrix}$

30 $\begin{bmatrix} 1 & 2 & 0 & 0 & 0 \\ 3 & 4 & 0 & 0 & 0 \\ 0 & 0 & 1 & 2 & 3 \\ 0 & 0 & 2 & -1 & 1 \\ 0 & 0 & 1 & 3 & -1 \end{bmatrix}$

31 Find the determinant of the $n \times n$ matrix (a_{ij}) where $a_{ij} = 0$ if $i \ne j$.

32 Without expanding, show that

$$\begin{vmatrix} 1 & a & b + c \\ 1 & b & a + c \\ 1 & c & a + b \end{vmatrix} = 0.$$

Find the inverses of the matrices in Exercises 33–36.

33 $\begin{bmatrix} 5 & -4 \\ -3 & 2 \end{bmatrix}$

34 $\begin{bmatrix} 2 & -1 & 0 \\ 1 & 4 & 2 \\ 3 & -2 & 1 \end{bmatrix}$

35 $\begin{bmatrix} 3 & -1 & 0 & 0 \\ 1 & 2 & 0 & 0 \\ 0 & 0 & -1 & -2 \\ 0 & 0 & 5 & 3 \end{bmatrix}$

36 $\begin{bmatrix} 2 & 0 & 0 & 0 \\ 0 & 3 & 0 & 0 \\ 0 & 0 & 4 & 0 \\ 0 & 0 & 0 & 5 \end{bmatrix}$

Express each product or sum in Exercises 37–46 as a single matrix.

37 $\begin{bmatrix} 2 & -1 & 0 \\ 3 & 0 & -2 \end{bmatrix} \begin{bmatrix} 2 & -1 & 3 \\ 0 & 3 & 0 \\ 1 & 4 & 2 \end{bmatrix}$

38 $\begin{bmatrix} 4 & 2 \\ 5 & -3 \end{bmatrix}\begin{bmatrix} 3 \\ 7 \end{bmatrix}$

39 $\begin{bmatrix} 2 & 0 \\ 1 & 4 \\ -2 & 3 \end{bmatrix}\begin{bmatrix} 0 & 2 & -3 \\ 4 & 5 & 1 \end{bmatrix}$

40 $\begin{bmatrix} 0 & -2 & 3 \\ 4 & 1 & 2 \end{bmatrix}\begin{bmatrix} 2 & 0 \\ 3 & 8 \\ 2 & -7 \end{bmatrix}$

41 $2\begin{bmatrix} 0 & -1 & -4 \\ 3 & 2 & 1 \end{bmatrix} - 3\begin{bmatrix} 4 & -2 & 1 \\ 0 & 5 & -1 \end{bmatrix}$

42 $\begin{bmatrix} 1 & 3 \\ 2 & 4 \end{bmatrix}\begin{bmatrix} a & 0 \\ 0 & a \end{bmatrix}$ **43** $\begin{bmatrix} a & 0 \\ 0 & b \end{bmatrix}\begin{bmatrix} 1 & 3 \\ 2 & 4 \end{bmatrix}$

44 $\begin{bmatrix} 3 & 2 \\ 0 & 0 \end{bmatrix}\begin{bmatrix} -2 & 0 \\ 3 & 0 \end{bmatrix}$

45 $\begin{bmatrix} 1 & 2 \\ 3 & 4 \end{bmatrix}\left\{\begin{bmatrix} 2 & -4 \\ 3 & 7 \end{bmatrix} + \begin{bmatrix} 1 & 5 \\ -2 & -3 \end{bmatrix}\right\}$

46 $\begin{bmatrix} 3 & 2 & 5 \\ -3 & 4 & 7 \\ 6 & 5 & 1 \end{bmatrix}\begin{bmatrix} 3 & 2 & 5 \\ -3 & 4 & 7 \\ 6 & 5 & 1 \end{bmatrix}^{-1}$

Verify Exercises 47 and 48 without expanding the determinants.

47 $\begin{vmatrix} 2 & 4 & -6 \\ 1 & 4 & 3 \\ 2 & 2 & 0 \end{vmatrix} = 12\begin{vmatrix} 1 & 1 & -1 \\ 1 & 2 & 1 \\ 2 & 1 & 0 \end{vmatrix}$

48 $\begin{vmatrix} a & b & c \\ d & e & f \\ g & h & k \end{vmatrix} = \begin{vmatrix} d & e & f \\ g & h & k \\ a & b & c \end{vmatrix}$

In Exercises 49 and 50, find the solutions of the equation $|A - xI| = 0$, where x denotes a real number.

49 $A = \begin{bmatrix} 2 & 3 \\ 1 & -4 \end{bmatrix}$, $I = I_2$

50 $A = \begin{bmatrix} 2 & -1 & 3 \\ 0 & 4 & 0 \\ 1 & 0 & -2 \end{bmatrix}$, $I = I_3$

51 Suppose that $A = (a_{ij})$ is a square matrix of order n such that $a_{ij} = 0$ if $i < j$. Prove that
$$|A| = a_{11}a_{22}\cdots a_{nn}.$$

52 If $A = (a_{ij})$ is any 2×2 matrix such that $|A| \neq 0$, prove that A has an inverse and find a general formula for A^{-1}.

In Exercises 53 and 54 find the partial fraction decompositions.

53 $\dfrac{4x^2 + 54x + 134}{(x + 3)(x^2 + 4x - 5)}$ **54** $\dfrac{x^2 + 14x - 13}{x^3 + 5x^2 + 4x + 20}$

55 A rotating sprinkler head with a range of 50 feet is to be placed in the center of a rectangular field (see figure). Find the dimensions of the field assuming it is to contain 4000 ft² and the water is to just reach the corners of the field.

FIGURE FOR EXERCISE 55

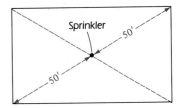

56 Find equations of the two lines that are tangent to the circle $x^2 + y^2 = 1$ and pass through the point $(0, 3)$. (*Hint:* Let $y = mx + 3$ and determine conditions on m that will ensure that the system has only one solution.)

57 An accountant must pay taxes and payroll bonuses to employees from the company's profits of $50,000. The total tax is 40% of the amount left after bonuses are paid, and the total paid in bonuses is 10% of the amount left after taxes. Find the total tax and the total bonus amount.

58 A circular track is to have a 10-foot wide running lane around the outside, and the inside distance around the track is to be 90% of the outside distance. Find the dimensions of the track.

59 Three inlet pipes A, B, and C can be used to fill a 1000-ft³ water storage tank. When all three pipes are in operation, the tank can be filled in 10 hours. When only A and B are used, the time increases to 20 hours. Pipes A and C can fill the tank in 12.5 hours. Find the individual flow rates (in cubic feet per hour) for each of the three pipes.

60 A deer spends the day in three basic activities: rest, searching for food, and grazing. At least 6 hours each day must be spent resting, and the number of hours spent searching for food will be at least two times the number of hours spent grazing. Using x as the number of hours spent searching for food and y as the number of hours spent grazing, find and graph the system of inequalities that describes the possible divisions of the day.

61 The Mower-for-the-Money Company manufactures a power lawn mower and a power edger. These two products are of such high quality that the company can sell all the products it makes, but production capability is limited in the areas of machining, welding, and assembly. Each week, the company has available 600 hours for machining, 300 hours for welding, and 550 hours for assembly. The number of hours required for the production of a single item is shown in the following table:

Product	Machining	Welding	Assembly
Lawn mower	6	2	5
Edger	4	3	5

The profits from the sale of a mower and an edger are $100 and $80, respectively. How many mowers and edgers should be produced each week in order to maximize profits?

CHAPTER 7

Sequences, Series, and Probability

The method of *mathematical induction*, considered in the first section of this chapter, is needed to prove that certain statements are true for every positive integer. ■ This is followed by material on *sequences* and *summation notation*. ■ Of special interest are the *arithmetic* and *geometric sequences* discussed in Sections 7.3 and 7.4. ■ The last part of the chapter deals with counting processes that arise frequently in mathematics and everyday life. ■ These include the concepts of *permutations*, *combinations*, and *probability*.

7.1 Mathematical Induction

If n is a positive integer, let P_n denote the statement

$$(xy)^n = x^n y^n$$

where x and y are real numbers. Thus, P_1 represents $(xy)^1 = x^1 y^1$, P_2 denotes $(xy)^2 = x^2 y^2$, P_3 is $(xy)^3 = x^3 y^3$, and so on. It is easy to show that P_1, P_2, and P_3 are *true* statements. However, since the set of positive integers is infinite, it is impossible to check the validity of P_n for every positive integer n. To give a proof, the method of mathematical induction is required. This method is based on the following fundamental axiom.

AXIOM OF MATHEMATICAL INDUCTION

> Suppose a set S of positive integers has the following two properties:
>
> (i) S contains the integer 1.
>
> (ii) Whenever S contains a positive integer k, S also contains $k + 1$.
>
> Then S contains every positive integer.

If S is a set of positive integers that satisfies property (ii), then whenever S contains an arbitrary positive integer k, it must also contain the next positive integer, $k + 1$. If S also satisfies property (i), then S contains 1 and hence by (ii), S contains $1 + 1$, or 2. Applying (ii) again, we see that S contains $2 + 1$, or 3. Once again, S must contain $3 + 1$, or 4. If we continue in this manner, it can be argued that if n is any *specific* positive integer, then n is in S, since we can proceed a step at a time, eventually reaching n. Although this argument does not *prove* the axiom, it certainly makes it plausible.

We shall use the preceding axiom to establish the following fundamental principle.

PRINCIPLE OF MATHEMATICAL INDUCTION

> If with each positive integer n there is associated a statement P_n, then all the statements P_n are true, provided the following two conditions are satisfied:
>
> (i) P_1 is true.
>
> (ii) Whenever k is a positive integer such that P_k is true, then P_{k+1} is also true.

PROOF Assume that conditions (i) and (ii) of the Principle hold, and let S denote the set of all positive integers n such that P_n is true. By assumption, P_1 is true, and consequently, 1 is in S. Thus, S satisfies property (i) of the Axiom of Mathematical Induction. Whenever S contains a positive integer k, then by the definition of S, P_k is true and hence from condition (ii) of the Principle, P_{k+1} is also true. This means that S contains $k + 1$. We have shown that whenever S contains a positive integer k, then S also contains $k + 1$. Consequently, property (ii) of the Axiom of Mathematical Induction is true, and hence S contains every positive integer; that is, P_n is true for every positive integer n. □

When applying the Principle of Mathematical Induction always follow these two steps:

Step (i) Prove that P_1 is true.

Step (ii) Assume that P_k is true and prove that P_{k+1} is true.

Step (ii) is usually the most confusing for the beginning student. We do not *prove* that P_k is true (except for $k = 1$). Instead, we show that *if* P_k is true, then the statement P_{k+1} is true. That is all that is necessary according to the Principle of Mathematical Induction. The assumption that P_k is true is referred to as the **induction hypothesis.**

EXAMPLE 1 Prove that for every positive integer n, the sum of the first n positive integers is

$$\frac{n(n + 1)}{2}.$$

SOLUTION If n is any positive integer, let P_n denote the statement

$$1 + 2 + 3 + \cdots + n = \frac{n(n + 1)}{2}.$$

The following are some special cases of P_n:
If $n = 2$, then P_2 is

$$1 + 2 = \frac{2(2 + 1)}{2} \quad \text{or} \quad 3 = 3.$$

If $n = 3$, then P_3 is

$$1 + 2 + 3 = \frac{3(3 + 1)}{2} \quad \text{or} \quad 6 = 6.$$

If $n = 5$, then P_5 is

$$1 + 2 + 3 + 4 + 5 = \frac{5(5 + 1)}{2} \quad \text{or} \quad 15 = 15.$$

Although it is instructive to check the validity of P_n for several values of n as we have done, it is unnecessary to do so. We need only apply the two-step process outlined prior to this example. Thus, we proceed as follows:

Step (i): If we substitute $n = 1$ in P_n, then the left side contains only the number 1 and the right side is $\frac{1(1 + 1)}{2}$, which also equals 1. This proves that P_1 is true.

Step (ii): Assume that P_k is true. Thus, the induction hypothesis is

$$1 + 2 + 3 + \cdots + k = \frac{k(k + 1)}{2}.$$

Our goal is to prove that P_{k+1} is true; that is,

$$1 + 2 + 3 + \cdots + (k + 1) = \frac{(k + 1)[(k + 1) + 1]}{2}.$$

By the induction hypothesis we already have a formula for the sum of the first k positive integers. Hence, a formula for the sum of the first $k + 1$ positive integers may be found simply by adding $(k + 1)$ to both sides. Doing so and simplifying, we obtain

$$1 + 2 + 3 + \cdots + k + (k + 1) = \frac{k(k + 1)}{2} + (k + 1)$$

$$= \frac{k(k + 1) + 2(k + 1)}{2}$$

$$= \frac{k^2 + 3k + 2}{2}$$

$$= \frac{(k + 1)(k + 2)}{2}$$

$$= \frac{(k + 1)[(k + 1) + 1]}{2}.$$

We have shown that P_{k+1} is true and, therefore, the proof by mathematical induction is complete. ∎

EXAMPLE 2 Prove that for each positive integer n,

$$1^2 + 3^2 + \cdots + (2n - 1)^2 = \frac{n(2n - 1)(2n + 1)}{3}.$$

SOLUTION For each positive integer n, let P_n denote the given statement. Note that this is a formula for the sum of the squares of the first n odd positive integers. We again follow the two-step procedure.

Step (i): Substituting 1 for n in P_n, we obtain

$$1^2 = \frac{(1)(2 - 1)(2 + 1)}{3} = \frac{3}{3} = 1.$$

This shows that P_1 is true.

Step (ii): Assume that P_k is true. Thus, the induction hypothesis is

$$1^2 + 3^2 + \cdots + (2k - 1)^2 = \frac{k(2k - 1)(2k + 1)}{3}.$$

We wish to prove that P_{k+1} is true, that is,

$$1^2 + 3^2 + \cdots + [2(k + 1) - 1]^2 = \frac{(k + 1)[2(k + 1) - 1][2(k + 1) + 1]}{3}.$$

This equation for P_{k+1} simplifies to

$$1^2 + 3^2 + \cdots + (2k + 1)^2 = \frac{(k + 1)(2k + 1)(2k + 3)}{3}.$$

Observe that the second from the last term on the left-hand side is $(2k - 1)^2$. In a manner similar to the solution of Example 1, we may obtain the left side of P_{k+1} by adding $(2k + 1)^2$ to both sides of the induction hypothesis P_k. This gives us

$$1^2 + 3^2 + \cdots + (2k - 1)^2 + (2k + 1)^2 = \frac{k(2k - 1)(2k + 1)}{3} + (2k + 1)^2.$$

It is left to the reader to show that the right side of the preceding equation may be written in the form of the right side of P_{k+1}. This proves that P_{k+1} is true, and hence P_n is true for every n. ■

Let j be a positive integer and suppose that with each integer $n \geq j$ there is associated a statement P_n. For example, if $j = 6$, then the statements are numbered P_6, P_7, P_8, ... The principle of mathematical induction may be extended to cover this situation. Just as before, two steps are used. Specifically, to prove that the statements P_n are true for $n \geq j$, we use the following steps.

EXTENDED PRINCIPLE OF MATHEMATICAL INDUCTION FOR P_k, $k \geq j$

(i′) Prove that P_j is true.

(ii′) Assume that P_k is true for $k \geq j$ and prove that P_{k+1} is true.

EXAMPLE 3 Let a be a nonzero real number such that $a > -1$. Prove that $(1 + a)^n > 1 + na$ for every integer $n \geq 2$.

SOLUTION For each positive integer n, let P_n denote the inequality $(1 + a)^n > 1 + na$. Note that P_1 is *false*, since $(1 + a)^1 = 1 + (1)(a)$. However, we can show that P_n is true for $n \geq 2$ by using the Extended Principle with $j = 2$.

Step (i′): We first note that $(1 + a)^2 = 1 + 2a + a^2$. Since $a \neq 0$, we have $a^2 > 0$ and therefore $1 + 2a + a^2 > 1 + 2a$. This gives us $(1 + a)^2 > 1 + 2a$, and hence P_2 is true.

Step (ii′): Assume that P_k is true. Thus, the induction hypothesis is

$$(1 + a)^k > 1 + ka.$$

We wish to show that P_{k+1} is true, that is,

$$(1 + a)^{k+1} > 1 + (k + 1)a.$$

Since $a > -1$, we have $1 + a > 0$, and hence multiplying both sides of the induction hypothesis by $1 + a$ will not change the inequality sign. Consequently,

$$(1 + a)^k(1 + a) > (1 + ka)(1 + a),$$

which may be rewritten as

$$(1 + a)^{k+1} > 1 + ka + a + ka^2$$

or as $\qquad\qquad (1 + a)^{k+1} > 1 + (k + 1)a + ka^2.$

Since $ka^2 > 0$,

$$1 + (k + 1)a + ka^2 > 1 + (k + 1)a$$

and therefore, $\qquad\qquad (1 + a)^{k+1} > 1 + (k + 1)a.$

Thus, P_{k+1} is true and the proof is complete. ∎

The Binomial Theorem was stated without proof in Section 1.5. Let us conclude this section by restating this important result and giving a proof using mathematical induction.

THE BINOMIAL THEOREM

$$(a + b)^n = a^n + na^{n-1}b + \frac{n(n - 1)}{2!} a^{n-2}b^2 + \cdots$$

$$+ \frac{n(n - 1)(n - 2) \cdots (n - r + 1)}{r!} a^{n-r}b^r$$

$$+ \cdots + nab^{n-1} + b^n$$

PROOF For each positive integer n, let P_n denote the statement given in the Binomial Theorem.

Step (i): If $n = 1$, the statement reduces to $(a + b)^1 = a^1 + b^1$. Consequently, P_1 is true.

Step (ii): Assume that P_k is true. Thus, the induction hypothesis is

$$(a + b)^k = a^k + ka^{k-1}b + \frac{k(k - 1)}{2!} a^{k-2}b^2 + \cdots$$

$$+ \frac{k(k - 1)(k - 2) \cdots (k - r + 2)}{(r - 1)!} a^{k-r+1}b^{r-1}$$

$$+ \frac{k(k - 1)(k - 2) \cdots (k - r + 1)}{r!} a^{k-r}b^r$$

$$+ \cdots + kab^{k-1} + b^k.$$

We have shown both the rth and the $(r + 1)$st terms in the expansion. Multiplying both sides of the last equation by $(a + b)$, we obtain

$$(a + b)^{k+1} = \left[a^{k+1} + ka^kb + \frac{k(k - 1)}{2!} a^{k-1}b^2 + \cdots \right.$$

$$\left. + \frac{k(k - 1) \cdots (k - r + 1)}{r!} a^{k-r+1}b^r + \cdots + ab^k \right]$$

$$+ \left[a^kb + ka^{k-1}b^2 + \cdots + \frac{k(k - 1) \cdots (k - r + 2)}{(r - 1)!} a^{k-r+1}b^r \right.$$

$$\left. + \cdots + kab^k + b^{k+1} \right]$$

where the terms in the first pair of brackets result from multiplying the right side of the induction hypothesis by a, and the terms in the second pair of brackets result from multiplying by b. Rearranging and combining terms,

$$(a + b)^{k+1} = a^{k+1} + (k + 1)a^k b + \left[\frac{k(k - 1)}{2!} + k \right] a^{k-1} b^2 + \cdots$$

$$+ \left[\frac{k(k - 1) \cdots (k - r + 1)}{r!} + \frac{k(k - 1) \cdots (k - r + 2)}{(r - 1)!} \right] a^{k-r+1} b^r$$

$$+ \cdots + (1 + k)ab^k + b^{k+1}.$$

It can be shown that if the coefficients are simplifed we obtain statement P_n with $k + 1$ substituted for n. Thus, P_{k+1} is true and therefore P_n holds for every positive integer n. This completes the proof. □

Exercises 7.1

In Exercises 1–18 prove that the formula is true for every positive integer n.

1 $2 + 4 + 6 + \cdots + 2n = n(n + 1)$

2 $1 + 4 + 7 + \cdots + (3n - 2) = \dfrac{n(3n - 1)}{2}$

3 $1 + 3 + 5 + \cdots + (2n - 1) = n^2$

4 $3 + 9 + 15 + \cdots + (6n - 3) = 3n^2$

5 $2 + 7 + 12 + \cdots + (5n - 3) = \frac{1}{2}n(5n - 1)$

6 $2 + 6 + 18 + \cdots + 2 \cdot 3^{n-1} = 3^n - 1$

7 $1 + 2 \cdot 2 + 3 \cdot 2^2 + \cdots + n \cdot 2^{n-1} = 1 + (n - 1) \cdot 2^n$

8 $(-1)^1 + (-1)^2 + (-1)^3 + \cdots + (-1)^n = \dfrac{(-1)^n - 1}{2}$

9 $1^2 + 2^2 + 3^2 + \cdots + n^2 = \dfrac{n(n + 1)(2n + 1)}{6}$

10 $1^3 + 2^3 + 3^3 + \cdots + n^3 = \left[\dfrac{n(n + 1)}{2} \right]^2$

11 $\dfrac{1}{1 \cdot 2} + \dfrac{1}{2 \cdot 3} + \dfrac{1}{3 \cdot 4} + \cdots + \dfrac{1}{n(n + 1)} = \dfrac{n}{n + 1}$

12 $\dfrac{1}{1 \cdot 2 \cdot 3} + \dfrac{1}{2 \cdot 3 \cdot 4} + \dfrac{1}{3 \cdot 4 \cdot 5} + \cdots + \dfrac{1}{n(n + 1)(n + 2)}$

$$= \dfrac{n(n + 3)}{4(n + 1)(n + 2)}$$

13 $3 + 3^2 + 3^3 + \cdots + 3^n = \frac{3}{2}(3^n - 1)$

14 $1^3 + 3^3 + 5^3 + \cdots + (2n - 1)^3 = n^2(2n^2 - 1)$

15 $n < 2^n$

16 $1 + 2n \le 3^n$

17 $1 + 2 + 3 + \cdots + n < \frac{1}{8}(2n + 1)^2$

18 If $0 < a < b$, then $\left(\dfrac{a}{b} \right)^{n+1} < \left(\dfrac{a}{b} \right)^n$.

Prove that the statements in Exercises 19–22 are true for every positive integer n.

19 3 is a factor of $n^3 - n + 3$.

20 2 is a factor of $n^2 + n$.

21 4 is a factor of $5^n - 1$.

22 9 is a factor of $10^{n+1} + 3 \cdot 10^n + 5$.

23 Use mathematical induction to prove that if a is any real number greater than 1, then $a^n > 1$ for every positive integer n.

24 Prove that

$$a + ar + ar^2 + \cdots + ar^{n-1} = \frac{a(1 - r^n)}{1 - r}$$

where n is any positive integer and a and r are real numbers with $r \neq 1$.

25 Use mathematical induction to prove that $a - b$ is a factor of $a^n - b^n$ for every positive integer n. (*Hint:* $a^{k+1} - b^{k+1} = a^k(a - b) + (a^k - b^k)b$.)

26 Prove that $a + b$ is a factor of $a^{2n-1} + b^{2n-1}$ for every positive integer n.

7.2 Infinite Sequences and Summation Notation

Recall that a function f from a set D to a set E is a correspondence that associates with each element x of D a unique element $f(x)$ of E. Up to now the domain D has usually been a set of real numbers. In this section we shall consider a different class of functions.

DEFINITION

> An **infinite sequence** is a function whose domain is the set of positive integers.

For convenience we sometimes refer to an infinite sequence as a **sequence.** In this book the range of an infinite sequence will be a set of real numbers.

If f is an infinite sequence, then to each positive integer n there corresponds a real number $f(n)$. These numbers in the range of f may be represented by writing

$$f(1), f(2), f(3), \ldots, f(n), \ldots$$

where the dots at the end indicate that the sequence does not terminate. The number $f(1)$ is called the **first term** of the sequence, $f(2)$ the **second**

term and, in general, $f(n)$ the **nth term** of the sequence. It is customary to use a subscript notation instead of the function notation and to write these numbers as

$$a_1, a_2, a_3, \ldots, a_n, \ldots$$

where it is understood that for each positive integer n, the symbol a_n denotes the real number $f(n)$. In this way we obtain an infinite collection of real numbers that is *ordered* in the sense that there is a first number, a second number, a forty-fifth number, and so on. Although sequences are functions, an ordered collection of the type displayed above will also be referred to as an infinite sequence. If we wish to convert the collection to a function f, we let $f(n) = a_n$ for every positive integer n.

From the definition of equality of functions we see that a sequence

$$a_1, a_2, a_3, \ldots, a_n, \ldots$$

is **equal** to a sequence $\qquad b_1, b_2, b_3, \ldots, b_n, \ldots$

if and only if $a_k = b_k$ for every positive integer k. Infinite sequences are often defined by stating a formula for the nth term, as in the following example.

EXAMPLE 1 List the first four terms and the tenth term of the sequence whose nth term is as follows:

(a) $a_n = \dfrac{n}{n+1}$ (b) $a_n = 2 + (0.1)^n$

(c) $a_n = (-1)^{n+1} \dfrac{n^2}{3n-1}$ (d) $a_n = 4$

SOLUTION To find the first four terms we substitute, successively, $n = 1, 2, 3,$ and 4 in the formula for a_n. The tenth term is found by substituting 10 for n. Doing this and simplifying gives us the following:

	nth term	First four terms	Tenth term
(a)	$\dfrac{n}{n+1}$	$\dfrac{1}{2}, \dfrac{2}{3}, \dfrac{3}{4}, \dfrac{4}{5}$	$\dfrac{10}{11}$
(b)	$2 + (0.1)^n$	2.1, 2.01, 2.001, 2.0001	2.0000000001
(c)	$(-1)^{n+1} \dfrac{n^2}{3n-1}$	$\dfrac{1}{2}, -\dfrac{4}{5}, \dfrac{9}{8}, -\dfrac{16}{11}$	$-\dfrac{100}{29}$
(d)	4	4, 4, 4, 4	4

■

Some infinite sequences are described by stating the first term a_1, together with a rule that shows how to obtain any term a_{k+1} from the preceding term a_k whenever $k \geq 1$. A description of this type is called a **recursive definition,** and the sequence is said to be defined **recursively.**

EXAMPLE 2   Find the first four terms and the nth term of the infinite sequence defined as follows:

$$a_1 = 3 \quad \text{and} \quad a_{k+1} = 2a_k \quad \text{for } k \geq 1.$$

SOLUTION   The sequence is defined recursively since the first term is given and, moreover, whenever a term a_k is known, then the next term a_{k+1} can be found. Thus,

$$a_1 = 3$$
$$a_2 = 2a_1 = 2 \cdot 3 = 6$$
$$a_3 = 2a_2 = 2 \cdot 2 \cdot 3 = 2^2 \cdot 3 = 12$$
$$a_4 = 2a_3 = 2 \cdot 2 \cdot 2 \cdot 3 = 2^3 \cdot 3 = 24.$$

We have written the terms as products in order to gain some insight into the nature of the nth term. Continuing, we obtain $a_5 = 2^4 \cdot 3$ and $a_6 = 2^5 \cdot 3$; and it appears that

$$a_n = 2^{n-1} \cdot 3$$

for every positive integer n. We shall prove that this guess is correct by mathematical induction. If we let P_n denote the statement $a_n = 2^{n-1} \cdot 3$, then P_1 is true since $a_1 = 2^0 \cdot 3 = 3$. Next, *assume* that P_k is true, that is, $a_k = 2^{k-1} \cdot 3$. We then have

$$a_{k+1} = 2a_k \qquad \text{(definition of } a_{k+1})$$
$$= 2 \cdot 2^{k-1} \cdot 3 \qquad \text{(induction hypothesis)}$$
$$= 2^k \cdot 3 \qquad \text{(a law of exponents)}$$
$$= 2^{(k+1)-1} \cdot 3, \qquad \text{(rewriting } 2^k \cdot 3)$$

which shows that P_{k+1} is true. Hence, $a_n = 2^{n-1} \cdot 3$ for every positive integer n.   ∎

It is important to observe that if only the first few terms of an infinite sequence are known, then it is impossible to predict additional terms. For example, if we were given $3, 6, 9, \ldots$ and asked to find the fourth term, we could not proceed without further information. The infinite sequence with nth term

$$a_n = 3n + (1 - n)^3(2 - n)^2(3 - n)$$

has for its first four terms 3, 6, 9, and 120. It is possible to describe sequences in which the first three terms are 3, 6, and 9, and the fourth term is *any* given number. This shows that when we work with an infinite sequence it is essential to have specific information about the *n*th term or to know a general scheme for obtaining each term from the preceding one.

It is sometimes necessary to find the sum of many terms of an infinite sequence. For ease in expressing such sums we use the following **summation notation.** Given an infinite sequence

$$a_1, a_2, a_3, \ldots, a_n, \ldots$$

the symbol $\sum_{k=1}^m a_k$ represents the sum of the first *m* terms, that is,

SUMMATION NOTATION

$$\sum_{k=1}^m a_k = a_1 + a_2 + a_3 + \cdots + a_m.$$

The Greek capital letter $\sum$ (sigma) indicates a sum and the symbol a_k represents the *k*th term. The letter *k* is called the **index of summation** or the **summation variable,** and the numbers 1 and *m* indicate the smallest and largest values of the summation variable.

EXAMPLE 3 Find the sum

$$\sum_{k=1}^4 k^2(k-3).$$

SOLUTION   In this case, $a_k = k^2(k-3)$. To find the sum we merely substitute, in succession, the integers 1, 2, 3, and 4 for *k* and add the resulting terms.

$$\sum_{k=1}^4 k^2(k-3) = 1^2(1-3) + 2^2(2-3) + 3^2(3-3) + 4^2(4-3)$$

$$= (-2) + (-4) + 0 + 16 = 10. \qquad \blacksquare$$

The letter used for the summation variable is arbitrary. To illustrate, if *j* is the summation variable, then

$$\sum_{j=1}^m a_j = a_1 + a_2 + a_3 + \cdots + a_m,$$

which is the same as $\sum_{k=1}^{m} a_k$. Other symbols can also be used. As a numerical example, the sum in Example 3 can be written

$$\sum_{j=1}^{4} j^2(j-3).$$

If n is a positive integer, then the sum of the first n terms of an infinite sequence will be denoted by S_n. For example, given $a_1, a_2, a_3, \ldots, a_n, \ldots,$

$$S_1 = a_1$$
$$S_2 = a_1 + a_2$$
$$S_3 = a_1 + a_2 + a_3$$
$$S_4 = a_1 + a_2 + a_3 + a_4$$

and, in general,

$$S_n = \sum_{k=1}^{n} a_k = a_1 + a_2 + \cdots + a_n.$$

The real number S_n is called the **nth partial sum** of the infinite sequence $a_1, a_2, a_3, \ldots, a_n, \ldots,$ and the sequence

$$S_1, S_2, S_3, \ldots, S_n, \ldots$$

is called a **sequence of partial sums.** Sequences of partial sums are very important in calculus, where the concept of *infinite series* is studied. We shall discuss some special types of infinite series in Section 7.4.

EXAMPLE 4 Find the first four terms and the nth term of the sequence of partial sums associated with the sequence $1, 2, 3, \ldots, n, \ldots$ of positive integers.

SOLUTION The first four terms of the sequence of partial sums are

$$S_1 = 1$$
$$S_2 = 1 + 2 = 3$$
$$S_3 = 1 + 2 + 3 = 6$$
$$S_4 = 1 + 2 + 3 + 4 = 10.$$

From Example 1 of Section 7.1 we see that

$$S_n = 1 + 2 + 3 + \cdots + n = \frac{n(n+1)}{2}.$$

■

If a_k is the same for every positive integer k, say $a_k = c$, where c is a real number, then

$$\sum_{k=1}^{n} a_k = a_1 + a_2 + a_3 + \cdots + a_n$$

$$= c + c + c + \cdots + c = nc.$$

This gives us the following result.

THEOREM

$$\sum_{k=1}^{n} c = nc$$

The domain of the summation variable does not have to begin at 1. For example, the following is self-explanatory:

$$\sum_{k=4}^{8} a_k = a_4 + a_5 + a_6 + a_7 + a_8.$$

As another variation, if the first term of an infinite sequence is a_0, as in

$$a_0, a_1, a_2, \ldots, a_n, \ldots,$$

then sums of the form

$$\sum_{k=0}^{n} a_i = a_0 + a_1 + a_2 + \cdots + a_n$$

may be considered. Note that this is the sum of the first $n + 1$ terms of the sequence.

EXAMPLE 5 Find the sum

$$\sum_{k=0}^{3} \frac{2^k}{(k+1)}.$$

SOLUTION

$$\sum_{k=0}^{3} \frac{2^k}{(k+1)} = \frac{2^0}{(0+1)} + \frac{2^1}{(1+1)} + \frac{2^2}{(2+1)} + \frac{2^3}{(3+1)}$$

$$= 1 + 1 + \frac{4}{3} + 2 = \frac{16}{3}.$$ ∎

Summation notation can be used to denote polynomials compactly. For example, in place of

$$f(x) = a_0 + a_1 x + a_2 x^2 + \cdots + a_n x^n$$

we may write

$$f(x) = \sum_{k=0}^{n} a_k x^k.$$

As another illustration, the rather cumbersome formula for the Binomial Theorem (see page 40) can be written

$$(a + b)^n = \sum_{k=0}^{n} \binom{n}{k} a^{n-k} b^k.$$

The following theorem concerning sums has many uses in advanced courses in mathematics.

THEOREM ON SUMS

If $a_1, a_2, \ldots, a_n, \ldots$ and $b_1, b_2, \ldots, b_n, \ldots$ are infinite sequences, then for every positive integer n,

(i) $\displaystyle\sum_{k=1}^{n} (a_k + b_k) = \sum_{k=1}^{n} a_k + \sum_{k=1}^{n} b_k$

(ii) $\displaystyle\sum_{k=1}^{n} (a_k - b_k) = \sum_{k=1}^{n} a_k - \sum_{k=1}^{n} b_k$

(iii) $\displaystyle\sum_{k=1}^{n} c a_k = c \left(\sum_{k=1}^{n} a_k \right)$ for every number c.

PROOF Although the theorem can be proved by mathematical induction, we shall use an argument that makes the truth of the formulas transparent. We begin as follows:

$$\sum_{k=1}^{n} (a_k + b_k) = (a_1 + b_1) + (a_2 + b_2) + (a_3 + b_3) + \cdots + (a_n + b_n).$$

Using commutative and associative properties many times, we may rearrange the terms on the right to produce

$$\sum_{k=1}^{n} (a_k + b_k) = (a_1 + a_2 + a_3 + \cdots + a_n) + (b_1 + b_2 + b_3 + \cdots + b_n).$$

Expressing the right side in summation notation gives us formula (i).

For formula (iii) we have

$$\sum_{k=1}^{n} (ca_k) = ca_1 + ca_2 + ca_3 + \cdots + ca_n$$

$$= c(a_1 + a_2 + a_3 + \cdots + a_n)$$

$$= c\left(\sum_{k=1}^{n} a_k\right).$$

The proof of (ii) is left as an exercise. □

Exercises 7.2

In Exercises 1–16 find the first five terms and the eighth term of the sequence that has the given nth term.

1 $a_n = 12 - 3n$

2 $a_n = \dfrac{3}{5n - 2}$

3 $a_n = \dfrac{3n - 2}{n^2 + 1}$

4 $a_n = 10 + \dfrac{1}{n}$

5 $a_n = 9$

6 $a_n = (n - 1)(n - 2)(n - 3)$

7 $a_n = 2 + (-0.1)^n$

8 $a_n = 4 + (0.1)^n$

9 $a_n = (-1)^{n-1} \dfrac{n + 7}{2n}$

10 $a_n = (-1)^n \dfrac{6 - 2n}{\sqrt{n + 1}}$

11 $a_n = 1 + (-1)^{n+1}$

12 $a_n = (-1)^{n+1} + (0.1)^{n-1}$

13 $a_n = \dfrac{2^n}{n^2 + 2}$

14 $a_n = \sqrt{2}$

15 a_n is the number of decimal places in $(0.1)^n$.

16 a_n is the number of positive integers less than n^3.

Find the first five terms of the infinite sequences defined recursively in Exercises 17–24.

17 $a_1 = 2,\ a_{k+1} = 3a_k - 5$

18 $a_1 = 5,\ a_{k+1} = 7 - 2a_k$

19 $a_1 = -3,\ a_{k+1} = a_k^2$

20 $a_1 = 128,\ a_{k+1} = a_k/4$

21 $a_1 = 5,\ a_{k+1} = ka_k$

22 $a_1 = 3,\ a_{k+1} = 1/a_k$

23 $a_1 = 2,\ a_{k+1} = (a_k)^k$

24 $a_1 = 2,\ a_{k+1} = (a_k)^{1/k}$

25 A test question lists the first four terms of a sequence as 2, 4, 6, and 8 and asks for the fifth term. Show that the fifth term can be any real number a by finding the nth term of a sequence that has for its first five terms, 2, 4, 6, 8, and a.

26 The number of bacteria in a certain culture doubles every day. If the initial number of bacteria is 500, how many are present after one day? Two days? Three days? Find a formula for the number of bacteria present after n days.

Find the number given in each of Exercises 27–42.

27 $\displaystyle\sum_{k=1}^{5} (2k - 7)$

28 $\displaystyle\sum_{k=1}^{6} (10 - 3k)$

29 $\displaystyle\sum_{k=1}^{4} (k^2 - 5)$

30 $\displaystyle\sum_{k=1}^{10} [1 + (-1)^k]$

31 $\displaystyle\sum_{k=0}^{5} k(k - 2)$

32 $\displaystyle\sum_{k=0}^{4} (k - 1)(k - 3)$

33 $\displaystyle\sum_{k=3}^{6} \dfrac{k - 5}{k - 1}$

34 $\displaystyle\sum_{k=1}^{6} \dfrac{3}{k + 1}$

35 $\displaystyle\sum_{k=1}^{5} (-3)^{k-1}$

36 $\displaystyle\sum_{k=0}^{4} 3(2^k)$

37 $\displaystyle\sum_{k=1}^{100} 100$

38 $\displaystyle\sum_{k=1}^{1000} 5$

39 $\sum_{k=1}^{n} (k^2 + 3k + 5)$ (*Hint:* Use the Theorem on Sums to write the sum as $\sum_{k=1}^{n} k^2 + 3 \sum_{k=1}^{n} k + \sum_{k=1}^{n} 5$. Next employ Exercise 9 and Example 1 of Section 7.1, together with the formula for $\sum_{k=1}^{n} c$.)

40 $\sum_{k=1}^{n} (3k^2 - 2k + 1)$ **41** $\sum_{k=1}^{n} (2k - 3)^2$

42 $\sum_{k=1}^{n} (k^3 + 2k^2 - k + 4)$

(*Hint:* See Exercise 10 of Section 7.1).

Express the sums in Exercises 43–52 in terms of summation notation.

43 $1 + 5 + 9 + 13 + 17$ **44** $2 + 5 + 8 + 11 + 14$

45 $\frac{1}{2} + \frac{2}{5} + \frac{3}{8} + \frac{4}{11}$ **46** $\frac{1}{4} + \frac{2}{9} + \frac{3}{14} + \frac{4}{19}$

47 $1 - \frac{x^2}{2} + \frac{x^4}{4} - \frac{x^6}{6} + \cdots + (-1)^n \frac{x^{2n}}{2n}$

48 $2 - 4 + 8 - 16 + 32 - 64$

49 $1 - \frac{1}{2} + \frac{1}{3} - \frac{1}{4} + \frac{1}{5} - \frac{1}{6} + \frac{1}{7}$

50 $1 + x + \frac{x^2}{2} + \frac{x^3}{3} + \cdots + \frac{x^n}{n}$

51 $\frac{1}{1 \cdot 2} + \frac{1}{2 \cdot 3} + \frac{1}{3 \cdot 4} + \cdots + \frac{1}{99 \cdot 100}$

52 $\frac{1}{1 \cdot 2 \cdot 3} + \frac{1}{2 \cdot 3 \cdot 4} + \frac{1}{3 \cdot 4 \cdot 5} + \cdots + \frac{1}{98 \cdot 99 \cdot 100}$

53 Prove (ii) of the Theorem on Sums.

54 Extend (i) of the Theorem on Sums to $\sum_{k=1}^{n} (a_k + b_k + c_k)$.

55 Prove the Theorem on Sums by mathematical induction.

56 Prove $\sum_{k=1}^{n} c = nc$ by mathematical induction.

Calculator Exercises 7.2

1 Terms of the sequence defined recursively by $a_1 = 5$, $a_{k+1} = \sqrt{a_k}$ may be generated by entering 5 and pressing the square root key repeatedly. Describe what happens to the terms of the sequence as k increases.

2 Approximations to $\sqrt{N}$ may be generated from the sequence

$$x_1 = \frac{N}{2}, \quad x_{k+1} = \frac{1}{2}\left(x_k + \frac{N}{x_k}\right).$$

Approximate x_2, x_3, x_4, x_5, x_6 if $N = 10$.

3 *Bode's Sequence*, defined by the formulas $a_1 = \frac{4}{10}$ and $a_k = (3 \cdot 2^{k-2} + 4)/10$ for $k \geq 2$, can be used to approximate distances of the planets from the sun. The third term of the sequence is $a_3 = 1$ *astronomical unit* (or 92,900,000 miles) and corresponds to earth. Approximate the first five terms of the sequence. (The fifth term corresponds to the minor planet Ceres.)

4 The famous *Fibonacci sequence* is defined recursively by $a_{k+1} = a_k + a_{k-1}$ with $a_1 = a_2 = 1$.

(a) Find the first ten terms of the sequence.

(b) The terms of the sequence $r_k = a_{k+1}/a_k$ give progressively better approximations to τ, the *golden ratio*. Approximate the first ten terms of this sequence.

5 The *discrete logistic sequence* is the sequence defined recursively by

$$y_{k+1} = y_k + \frac{r}{K} y_k(K - y_k)$$

where r and K are positive constants. This sequence is used to model the growth of a seasonally breeding animal population in an environment with limited resources.

(a) Find the terms in the sequence if $y_1 = K$.

(b) Approximate the first ten terms in the sequence if $y_1 = 400$, $r = 2$, and $K = 500$. Describe the behavior of this sequence, which predicts the number y_k in the population after k years.

7.3 Arithmetic Sequences

In this section and the next we shall concentrate on two special types of sequences. The first may be defined as follows:

DEFINITION

> An **arithmetic sequence** is a sequence such that successive terms differ by the same real number.

Arithmetic sequences are also called **arithmetic progressions.** By definition, a sequence

$$a_1, a_2, a_3, \ldots, a_n, \ldots$$

is arithmetic if and only if there is a real number d such that

$$a_{k+1} - a_k = d$$

for every positive integer k. The number d is called the **common difference** associated with the arithmetic sequence.

EXAMPLE 1   Show that the sequence

$$1, 4, 7, 10, \ldots, 3n - 2, \ldots$$

is arithmetic and find the common difference.

SOLUTION   If $a_n = 3n - 2$, then for every positive integer k,

$$
\begin{aligned}
a_{k+1} - a_k &= [3(k + 1) - 2] - (3k - 2) \\
&= 3k + 3 - 2 - 3k + 2 = 3.
\end{aligned}
$$

Hence, by definition, the given sequence is arithmetic with common difference 3. ∎

Given an arithmetic sequence, we know that

$$a_{k+1} = a_k + d$$

for every positive integer k. This provides a recursive formula for obtaining successive terms. Beginning with any real number a_1, we can obtain an arithmetic sequence with common difference d simply by adding d to a_1, then to $a_1 + d$, and so on, obtaining

$$a_1, a_1 + d, a_1 + 2d, a_1 + 3d, a_1 + 4d, \ldots$$

It is evident that the nth term a_n of this sequence is given by the next formula.

nth TERM OF AN ARITHMETIC SEQUENCE

$$a_n = a_1 + (n - 1)d$$

EXAMPLE 2   Find the fifteenth term of the arithmetic sequence whose first three terms are 20, 16.5, and 13.

SOLUTION   The common difference is -3.5. Substituting $a_1 = 20$, $d = -3.5$, and $n = 15$ in the formula $a_n = a_1 + (n - 1)d$,

$$a_{15} = 20 + (15 - 1)(-3.5) = 20 - 49 = -29. \qquad \blacksquare$$

EXAMPLE 3   If the fourth term of an arithmetic sequence is 5 and the ninth term is 20, find the sixth term.

SOLUTION   Substituting $n = 4$ and $n = 9$ in $a_n = a_1 + (n - 1)d$, and using the fact that $a_4 = 5$ and $a_9 = 20$, we obtain the following system of linear equations in the variables a_1 and d:

$$\begin{cases} 5 = a_1 + (4 - 1)d \\ 20 = a_1 + (9 - 1)d \end{cases} \quad \text{or} \quad \begin{cases} 5 = a_1 + 3d \\ 20 = a_1 + 8d \end{cases}$$

This system has the solution $d = 3$ and $a_1 = -4$. (Verify this fact.) Substitution in the formula $a_n = a_1 + (n - 1)d$ gives us

$$a_6 = (-4) + (6 - 1)(3) = 11. \qquad \blacksquare$$

THEOREM

If $a_1, a_2, \ldots, a_n, \ldots$ is an arithmetic sequence with common difference d, then the nth partial sum S_n is given by both

$$S_n = \frac{n}{2}[2a_1 + (n - 1)d] \quad \text{and} \quad S_n = \frac{n}{2}(a_1 + a_n).$$

PROOF We may write

$$S_n = a_1 + a_2 + a_3 + \cdots + a_n$$
$$= a_1 + (a_1 + d) + (a_1 + 2d) + \cdots + [a_1 + (n-1)d].$$

Employing commutative and associative properties many times, we obtain

$$S_n = (a_1 + a_1 + a_1 + \cdots + a_1) + [d + 2d + \cdots + (n-1)d]$$

where a_1 appears n times within the first parentheses. It follows that

$$S_n = na_1 + d[1 + 2 + \cdots + (n-1)].$$

The expression within brackets is the sum of the first $n - 1$ positive integers. From Example 1 of Section 7.1 (with $n - 1$ in place of n),

$$1 + 2 + \cdots + (n-1) = \frac{(n-1)n}{2}.$$

Substituting in the last equation for S_n and factoring,

$$S_n = na_1 + d\frac{n(n-1)}{2} = \frac{n}{2}[2a_1 + (n-1)d].$$

Since $a_n = a_1 + (n-1)d$, this is the same as

$$S_n = \frac{n}{2}(a_1 + a_n). \qquad \square$$

EXAMPLE 4 Find the sum of all the even integers from 2 through 100.

SOLUTION This problem is equivalent to finding the sum of the first 50 terms of the arithmetic sequence $2, 4, 6, \ldots, 2n, \ldots$ Substituting $n = 50$, $a_1 = 2$, and $a_{50} = 100$ in the second formula of the preceding theorem,

$$S_{50} = \tfrac{50}{2}(2 + 100) = 2550.$$

As a check on our work we may use the first formula of the theorem. Thus,

$$S_{50} = \tfrac{50}{2}[2 \cdot 2 + (50-1)2] = 25[4 + 98] = 2550. \qquad \blacksquare$$

The **arithmetic mean** of two numbers a and b is defined as $(a + b)/2$. This is also called the **average** of a and b. Note that

$$a, \frac{a + b}{2}, b$$

is an arithmetic sequence. This concept may be generalized as follows: If $c_1, c_2, \ldots, c_k$ are real numbers such that

$$a, c_1, c_2, \ldots, c_k, b$$

is an arithmetic sequence, then $c_1, c_2, \ldots, c_k$ are called the **k arithmetic means** of the numbers a and b. The process of determining these numbers is referred to as *inserting k arithmetic means between a and b.*

EXAMPLE 5 Insert three arithmetic means between 2 and 9.

SOLUTION We wish to find three real numbers $c_1, c_2,$ and c_3 such that $2, c_1, c_2, c_3, 9$ is an arithmetic sequence. The common difference d may be found by using the formula $a_n = a_1 + (n - 1)d$ with $n = 5$, $a_5 = 9$, and $a_1 = 2$. This gives us

$$9 = 2 + (5 - 1)d \quad \text{or} \quad d = \tfrac{7}{4}.$$

The three arithmetic means are

$$c_1 = a_1 + d = 2 + \tfrac{7}{4} = \tfrac{15}{4}$$
$$c_2 = c_1 + d = \tfrac{15}{4} + \tfrac{7}{4} = \tfrac{22}{4} = \tfrac{11}{2}$$
$$c_3 = c_2 + d = \tfrac{11}{2} + \tfrac{7}{4} = \tfrac{29}{4}. \qquad \blacksquare$$

Exercises 7.3

In Exercises 1–8 find the fifth term, the tenth term, and the nth term of the arithmetic sequence.

1 2, 6, 10, 14, . . .

2 16, 13, 10, 7, . . .

3 3, 2.7, 2.4, 2.1, . . .

4 $-6, -4.5, -3, -1.5, \ldots$

5 $-7, -3.9, -0.8, 2.3, \ldots$

6 $x - 8, x - 3, x + 2, x + 7, \ldots$

7 $\ln 3, \ln 9, \ln 27, \ln 81, \ldots$

8 $\log 1000, \log 100, \log 10, 0, \ldots$

9 Find the twelfth term of the arithmetic sequence whose first two terms are 9.1 and 7.5.

10 Find the eleventh term of the arithmetic sequence whose first two terms are $2 + \sqrt{2}$ and 3.

11 The sixth and seventh terms of an arithmetic sequence are 2.7 and 5.2. Find the first term.

12 Given an arithmetic sequence with $a_3 = 7$ and $a_{20} = 43$, find a_{15}.

In Exercises 13–16 find the sum S_n of the arithmetic sequence that satisfies the stated condition.

13 $a_1 = 40$, $d = -3$, $n = 30$

14 $a_1 = 5$, $d = 0.1$, $n = 40$

15 $a_1 = -9$, $a_{10} = 15$, $n = 10$

16 $a_7 = \frac{7}{3}$, $d = -\frac{2}{3}$, $n = 15$

Find the sums in Exercises 17–20.

17 $\displaystyle\sum_{k=1}^{20} (3k - 5)$ **18** $\displaystyle\sum_{k=1}^{12} (7 - 4k)$

19 $\displaystyle\sum_{k=1}^{18} \left(\frac{1}{2}k + 7\right)$ **20** $\displaystyle\sum_{k=1}^{10} \left(\frac{1}{4}k + 3\right)$

21 How many integers between 32 and 395 are divisible by 6? Find their sum.

22 How many negative integers greater than -500 are divisible by 33? Find their sum.

23 How many terms are in an arithmetic sequence with first term -2, common difference $\frac{1}{4}$, and sum 21?

24 How many terms are in an arithmetic sequence with sixth term -3, common difference 0.2, and sum -33?

25 Insert five arithmetic means between 2 and 10.

26 Insert three arithmetic means between 3 and -5.

27 A pile of logs has 24 logs in the first layer, 23 in the second, 22 in the third, and so on. The top layer contains 10 logs. Find the total number of logs in the pile.

28 A seating section in a certain athletic stadium has 30 seats in the first row, 32 seats in the second, 34 in the third, and so on, until the tenth row is reached, after which there are 10 more rows, each containing 50 seats. Find the total number of seats in the section.

29 A man wishes to construct a ladder with nine rungs that diminish uniformly from 24 inches at the base to 18 inches at the top. Determine the lengths of the seven intermediate rungs.

30 A boy on a bicycle coasts down a hill, covering 4 feet the first second and in each succeeding second 5 feet more than in the preceding second. If he reaches the bottom of the hill in 11 seconds, find the total distance traveled.

31 A contest will have five cash prizes totaling $5000, and there will be a $100 difference between successive prizes. Find the first prize.

32 A company is to distribute $46,000 in bonuses to its top ten salespeople. The tenth salesperson on the list will receive $1000, and the difference in bonus money between successively ranked salespeople is to be constant. Find the bonuses for each salesperson.

33 Certain polygons with n sides have the property that the measures of the interior angles can be arranged in an arithmetic sequence. If, in a polygon of this type, the smallest angle is $20°$ and the largest angle is $160°$, determine the number of sides of the polygon. (*Hint:* Find an expression for the sum of the interior angles if the polygon has n sides.)

34 If f is a linear function, show that the sequence with nth term $a_n = f(n)$ is an arithmetic sequence.

35 The sequence defined recursively by $x_{k+1} = x_k/(1 + x_k)$ arises in genetics in the study of the elimination of a deficient gene from a population. Show that the sequence with nth term $1/x_n$ is arithmetic.

36 Find the total length of the curve in the figure if the width of the maze formed by the curve is 16 inches and all halls in the maze have width 1 inch. What is the length if the width of the maze is 32 inches?

FIGURE FOR EXERCISE 36

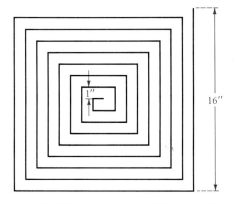

7.4 Geometric Sequences

Another important type of infinite sequence is defined as follows.

DEFINITION

> A sequence $a_1, a_2, \ldots, a_n, \ldots$ is a **geometric sequence** if there is a real number $r \neq 0$ such that
>
> $$\frac{a_{k+1}}{a_k} = r$$
>
> for every positive integer k.

Geometric sequences are also called **geometric progressions.** The number r in the definition is called the **common ratio** associated with the geometric sequence. Thus $a_1, a_2, \ldots, a_n, \ldots$ is geometric (with common ratio r) if

$$a_{k+1} = a_k r$$

for every positive integer k. This provides a recursive method for obtaining terms. Beginning with any nonzero real number a_1, we multiply by the number r successively, obtaining

$$a_1, a_1 r, a_1 r^2, a_1 r^3, \ldots$$

The nth term a_n of this sequence is given by the next formula.

nth TERM OF A GEOMETRIC SEQUENCE

$$a_n = a_1 r^{n-1}$$

EXAMPLE 1   Find the first five terms and the tenth term of the geometric sequence having first term 3 and common ratio $-\frac{1}{2}$.

SOLUTION   If we let $a_1 = 3$ and $r = -\frac{1}{2}$, then the first five terms are

$$3, -\tfrac{3}{2}, \tfrac{3}{4}, -\tfrac{3}{8}, \tfrac{3}{16}.$$

Using the formula $a_n = a_1 r^{n-1}$ with $n = 10$,

$$a_{10} = 3(-\tfrac{1}{2})^9 = -\tfrac{3}{512}.$$

■

EXAMPLE 2 If the third term of a geometric sequence is 5 and the sixth term is -40, find the eighth term.

SOLUTION We are given $a_3 = 5$ and $a_6 = -40$. Substituting $n = 3$ and $n = 6$ in the formula $a_n = a_1 r^{n-1}$ leads to the following system of equations:

$$\begin{cases} 5 = a_1 r^2 \\ -40 = a_1 r^5. \end{cases}$$

Since $r \neq 0$, the first equation is equivalent to $a_1 = 5/r^2$. Substituting for a_1 in the second equation,

$$-40 = \left(\frac{5}{r^2}\right) \cdot r^5 = 5r^3.$$

Hence, $r^3 = -8$ and $r = -2$. If we now substitute -2 for r in the equation $5 = a_1 r^2$, we obtain $a_1 = \frac{5}{4}$. Finally, using $a_n = a_1 r^{n-1}$ with $n = 8$,

$$a_8 = (\tfrac{5}{4})(-2)^7 = -160. \qquad \blacksquare$$

Let us find a formula for S_n, the nth partial sum of a geometric sequence. If the first term is a_1 and the common ratio is r,

$$S_n = a_1 + a_1 r + a_1 r^2 + \cdots + a_1 r^{n-2} + a_1 r^{n-1}.$$

If $r = 1$, then $S_n = na_1$. Next suppose that $r \neq 1$. Multiplying both sides of the general formula for S_n by r,

$$rS_n = a_1 r + a_1 r^2 + a_1 r^3 + \cdots + a_1 r^{n-1} + a_1 r^n.$$

If we subtract the preceding equation from that for S_n, then many terms on the right side drop out, leaving

$$S_n - rS_n = a_1 - a_1 r^n \quad \text{or} \quad (1 - r)S_n = a_1(1 - r^n).$$

Since $r \neq 1$, we have $1 - r \neq 0$, and dividing both sides by $1 - r$ gives us

$$S_n = \frac{a_1(1 - r^n)}{1 - r}.$$

We have proved the following theorem.

THEOREM The nth partial sum of a geometric sequence with first term a_1 and common ratio $r \neq 1$ is

$$S_n = a_1 \frac{(1 - r^n)}{1 - r}.$$

EXAMPLE 3 Find the sum of the first five terms of the geometric sequence that begins as follows: 1, 0.3, 0.09, 0.027, . . .

SOLUTION If we let $a_1 = 1$, $r = 0.3$, and $n = 5$ in the formula for S_n, then

$$S_5 = 1\left(\frac{1 - (0.3)^5}{1 - 0.3}\right).$$

This reduces to $S_5 = 1.4251$. ∎

EXAMPLE 4 A man wishes to save money by setting aside 1 cent the first day, 2 cents the second day, 4 cents the third day and so on, doubling the amount each day. If this is continued, how much must be set aside on the fifteenth day? Assuming he does not run out of money, what is the total amount saved at the end of 30 days?

SOLUTION The amount (in cents) set aside on successive days forms a geometric sequence

$$1, 2, 4, 8, \ldots$$

with first term 1 and common ratio 2. The amount needed for the fifteenth day is found by using $a_n = a_1 r^{n-1}$ with $a_1 = 1$ and $n = 15$. This gives us $1 \cdot 2^{14}$, or \$163.84. To find the total amount set aside after 30 days, we use the formula for S_n with $n = 30$. Thus,

$$S_{30} = 1\frac{(1 - 2^{30})}{1 - 2},$$

which simplifies to \$10,737,418.23. ∎

Given the geometric series with first term a_1 and common ratio $r \neq 1$, we may write the formula for S_n of the last theorem in the form

$$S_n = \frac{a_1}{1 - r} - \frac{a_1}{1 - r}r^n.$$

Although we shall not do it here, it can be shown that if $|r| < 1$, then r^n *approaches* 0 *as n increases* without bound; that is, we can make r^n as close as we wish to 0 by taking n sufficiently large. It follows that S_n approaches $a_1/(1 - r)$ as n increases without bound. Using the notation introduced in our work with rational functions in Chapter 4, this may be expressed symbolically as

$$S_n \to \frac{a_1}{1 - r} \quad \text{as} \quad n \to \infty.$$

The number $a_1/(1 - r)$ is called the *sum* of the **infinite geometric series**

$$a_1 + a_1 r + a_1 r^2 + \cdots + a_1 r^{n-1} + \cdots$$

This gives us the next result.

THEOREM

> If $|r| < 1$, then the infinite geometric series
>
> $$a_1 + a_1 r + a_1 r^2 + \cdots + a_1 r^{n-1} + \cdots$$
>
> has the sum $a_1/(1 - r)$.

The preceding theorem implies that if we add more and more terms of the indicated infinite geometric series, the sums get closer and closer to $a_1/(1 - r)$. The next example illustrates how the theorem can be used to show that every real number represented by a repeating decimal is rational.

EXAMPLE 5　Find the rational number that corresponds to the infinite repeating decimal $5.4\overline{27}$, where the bar means that the block of digits underneath is repeated indefinitely.

SOLUTION　From the decimal expression $5.4272727\ldots$, we obtain the infinite series

$$5.4 + 0.027 + 0.00027 + 0.0000027 + \cdots$$

The part of the expression after the first term is

$$0.027 + 0.00027 + 0.0000027 + \cdots,$$

which has the form given in the last theorem with $a_1 = 0.027$ and $r = 0.01$. Hence, the sum of this infinite geometric series is

$$\frac{0.027}{1 - 0.01} = \frac{0.027}{0.99} = \frac{27}{990} = \frac{3}{110}.$$

Thus, it appears that the desired number is $5.4 + \frac{3}{110}$, or $\frac{597}{110}$. A check by division shows that $\frac{597}{110}$ does equal the given repeating decimal.　∎

In general, given any infinite sequence $a_1, a_2, \ldots, a_n, \ldots$, the expression

$$a_1 + a_2 + \cdots + a_n + \cdots$$

is called an **infinite series,** or simply a **series.** In summation notation, this series is denoted by

$$\sum_{n=1}^{\infty} a_n.$$

Each number a_k is called a **term** of the series and a_n is called the **nth term.** Since only finite sums may be added algebraically, it is necessary to *define* what is meant by an infinite sum. This is done by considering the sequence of partial sums

$$S_1, S_2, \ldots, S_n, \ldots$$

If there is a number S such that $S_n \to S$ as $n \to \infty$, then, as in our discussion of infinite geometric series, we call S the **sum** of the series and write

$$S = a_1 + a_2 + \cdots + a_n + \cdots$$

Thus, in Example 5 we may write

$$\tfrac{597}{110} = 5.4 + 0.027 + 0.00027 + 0.0000027 + \cdots$$

The next example provides an illustration of a nongeometric infinite series. Since a complete discussion is beyond the scope of this text, the solution is presented in an intuitive manner.

EXAMPLE 6   Show that the following infinite series has a sum:

$$\frac{1}{1 \cdot 2} + \frac{1}{2 \cdot 3} + \frac{1}{3 \cdot 4} + \cdots + \frac{1}{n(n+1)} + \cdots$$

SOLUTION   The nth term $a_n = 1/n(n+1)$ has the partial fraction decomposition

$$a_n = \frac{1}{n(n+1)} = \frac{1}{n} - \frac{1}{n+1}.$$

Consequently, the nth partial sum of the series may be written

$$S_n = a_1 + a_2 + a_3 + \cdots + a_n$$

$$= \left(1 - \frac{1}{2}\right) + \left(\frac{1}{2} - \frac{1}{3}\right) + \left(\frac{1}{3} - \frac{1}{4}\right) + \cdots + \left(\frac{1}{n} - \frac{1}{n+1}\right)$$

$$= 1 - \frac{1}{n+1}.$$

Since $1/(n + 1) \to 0$ as $n \to \infty$, it follows that $S_n \to 1$, and we may write

$$1 = \frac{1}{1 \cdot 2} + \frac{1}{2 \cdot 3} + \frac{1}{3 \cdot 4} + \cdots + \frac{1}{n(n + 1)} + \cdots$$

This means that, as we add more and more terms of the series, the sums get closer and closer to 1. ■

If the terms of an infinite sequence are alternately positive and negative, and if we consider the expression

$$a_1 + (-a_2) + a_3 + (-a_4) + \cdots + [(-1)^{n+1}a_n] + \cdots$$

where all the a_k are positive real numbers, then this expression is referred to as an **alternating infinite series,** and we write it in the form

$$a_1 - a_2 + a_3 - a_4 + \cdots + (-1)^{n+1}a_n + \cdots$$

Illustrations of alternating infinite series can be obtained by using infinite geometric series with negative common ratio.

Infinite series have many applications in mathematics and the sciences. Texts on calculus contain careful treatments of this important concept.

Exercises 7.4

In Exercises 1–12 find the fifth term, the eighth term, and the nth term of the geometric sequence.

1 $8, 4, 2, 1, \ldots$

2 $4, 1.2, 0.36, 0.108, \ldots$

3 $300, -30, 3, -0.3, \ldots$

4 $1, -\sqrt{3}, 3, -\sqrt{27}, \ldots$

5 $5, 25, 125, 625, \ldots$

6 $2, 6, 18, 54, \ldots$

7 $4, -6, 9, -13.5, \ldots$

8 $162, -54, 18, -6, \ldots$

9 $1, -x^2, x^4, -x^6, \ldots$

10 $1, -\dfrac{x}{3}, \dfrac{x^2}{9}, -\dfrac{x^3}{27}, \ldots$

11 $2, 2^{x+1}, 2^{2x+1}, 2^{3x+1}, \ldots$

12 $10, 10^{2x-1}, 10^{4x-3}, 10^{6x-5}, \ldots$

13 Find the sixth term of the geometric sequence whose first two terms are 4 and 6.

14 Find the seventh term of the geometric sequence that has 2 and $-\sqrt{2}$ for its second and third terms, respectively.

15 In a certain geometric sequence $a_5 = \frac{1}{16}$ and $r = \frac{3}{2}$. Find a_1 and S_5.

16 Given a geometric sequence such that $a_4 = 4$ and $a_7 = 12$, find r and a_{10}.

Find the sums in Exercises 17–20.

17 $\displaystyle\sum_{k=1}^{10} 3^k$

18 $\displaystyle\sum_{k=1}^{9} (-\sqrt{5})^k$

19 $\displaystyle\sum_{k=0}^{9} (-\tfrac{1}{2})^{k+1}$

20 $\displaystyle\sum_{k=1}^{7} (3^{-k})$

21 A vacuum pump removes one-half of the air in a container at each stroke. After 10 strokes, what percentage of the original amount of air remains in the container?

22 The yearly depreciation of a certain machine is 25% of its value at the beginning of the year. If the original cost of the machine is $20,000, what is its value after 6 years?

23 A culture of bacteria increases 20% every hour. For the original culture containing 10,000 bacteria find a formula for the number of bacteria present after t hours. How many bacteria are in the culture at the end of 10 hours?

24 If an amount of money P is deposited in a savings account that pays interest at a rate of r percent per year compounded quarterly, and if the principal and accumulated interest are left in the account, find a formula for the total amount in the account after n years.

Find the sums of the infinite geometric series in Exercises 25–30, whenever they exist.

25 $1 - \dfrac{1}{2} + \dfrac{1}{4} - \dfrac{1}{8} + \cdots$

26 $2 + \dfrac{2}{3} + \dfrac{2}{9} + \dfrac{2}{27} + \cdots$

27 $1.5 + 0.015 + 0.00015 + \cdots$

28 $1 - 0.1 + 0.01 - 0.001 + \cdots$

29 $\sqrt{2} - 2 + \sqrt{8} - 4 + \cdots$

30 $250 - 100 + 40 - 16 + \cdots$

In Exercises 31–38 find the rational number represented by the repeating decimal.

31 $0.\overline{23}$ **32** $0.0\overline{71}$

33 $2.4\overline{17}$ **34** $10.\overline{55}$

35 $5.\overline{146}$ **36** $3.2\overline{394}$

37 $1.\overline{6124}$ **38** $123.6\overline{183}$

39 A rubber ball is dropped from a height of 10 meters. If it rebounds approximately one-half the distance after each fall, use an infinite geometric series to approximate the total distance the ball travels before coming to rest.

40 The bob of a pendulum swings through an arc 24 cm long on its first swing. If each successive swing is approximately five-sixths the length of the preceding swing, use an infinite geometric series to approximate the total distance it travels before coming to rest.

41 A branch of a clothing manufacturing company has just located in a small community and will pay two million dollars per year in salaries. It has been estimated that 60% of these salaries will be spent in the local area, and

of this money spent another 60% will again change hands within the community. This process will be repeated ad infinitum. This is called the Multiplier Effect. Find the total amount of spending that will be generated by company salaries.

42 In a pest eradication program, N sterilized male flies are released into the general population each day, and 90% of these flies will survive a given day.

(a) Show that the number of sterilized flies in the population n days after the program has begun is

$$N + (0.9)N + (0.9)^2 N + \cdots + (0.9)^{n-1} N.$$

(b) If the *long-range* goal of the program is to keep 20,000 sterilized males in the population, how many flies should be released each day?

43 A certain drug has a half-life of about 2 hours in the bloodstream. Doses of D mg will be administered every 4 hours, where D is still to be determined.

(a) Show that the number of milligrams of drug in the bloodstream after the nth dose has been administered is

$$D + \tfrac{1}{4}D + \cdots + (\tfrac{1}{4})^{n-1} D$$

and that this sum is approximately $\tfrac{4}{3}D$ for large values of n.

(b) If it is considered dangerous to have more than 500 mg of the drug in the bloodstream, find the largest possible dose that can be given repeatedly over a long period of time.

44 Shown in the figure is a family tree displaying 3 prior generations and a total of 12 grandparents. If you were

---

FIGURE FOR EXERCISE 44

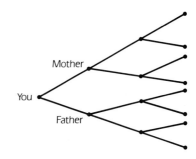

to trace your family history back 10 generations, how many grandparents would you find?

45 Shown in the first figure is a nested sequence of squares $S_1, S_2, \ldots, S_k, \ldots$. Let a_k, A_k, and P_k denote the side, area, and perimeter, respectively, of the square S_k. The square S_{k+1} is constructed from S_k by selecting four points on S_k that are at a distance of $\frac{1}{4}a_k$ from the vertices and connecting them (see second figure).

(a) Find the relationship between a_{k+1} and a_k.

(b) Find a_n, A_n, and P_n.

(c) Calculate $\sum\limits_{n=1}^{\infty} P_n$.

FIGURES FOR EXERCISE 45

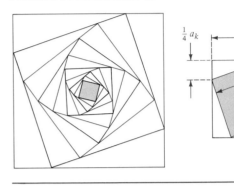

7.5 Permutations

FIGURE 7.1

First Place	Second Place	Final Standings	
a	b	a	b
	c	a	c
	d	a	d
b	a	b	a
	c	b	c
	d	b	d
c	a	c	a
	b	c	b
	d	c	d
d	a	d	a
	b	d	b
	c	d	c

Start

Suppose that four teams are involved in a tournament in which first, second, third, and fourth places will be determined. For identification purposes, we label the teams a, b, c, and d. Let us find the number of different ways that first and second place can be decided. It is convenient to use a **tree diagram** as in Figure 7.1. Beginning at the word "start," the four possibilities for first place are listed. From each of these an arrow points to a possible second-place finisher. The final standings list the possible outcomes, from left to right. They are found by following the different paths (*branches* of the tree) that lead from the word "start" to the second-place team. The total number of outcomes is 12, which is the product of the number of choices (4) for first place and the number of choices (3) for second place (after the first has been determined).

Let us now find the total number of ways that first, second, third, and fourth positions can be filled. To sketch a tree diagram we may begin by drawing arrows from the word "start" to each possible first-place finisher a, b, c, or d. Then we draw arrows from those to possible second-place finishers, as was done in Figure 7.1. Next, from each second-place position we draw arrows indicating the possible third-place positions. Finally, we draw arrows to the fourth-place team. If we consider only the case in which team a finishes in first place, we have the diagram shown in Figure 7.2. Note that there are six possible final standings in which team a occupies first place. In a complete tree diagram there would also be three other branches of this type corresponding to first place for b, c, and d, respec-

FIGURE 7.2

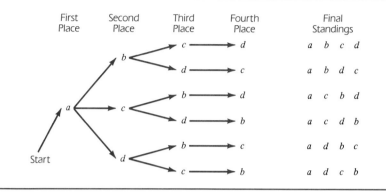

| | First Place | Second Place | Third Place | Fourth Place | Final Standings |

tively. A complete diagram would display the following 24 possibilities for the final standings:

$$abcd, \quad abdc, \quad acbd, \quad acdb, \quad adbc, \quad adcb,$$
$$bacd, \quad badc, \quad bcad, \quad bcda, \quad bdac, \quad bdca,$$
$$cabd, \quad cadb, \quad cbad, \quad cbda, \quad cdab, \quad cdba,$$
$$dabc, \quad dacb, \quad dbac, \quad dbca, \quad dcab, \quad dcba.$$

Note that the number 24 is the product of the number of ways (4) that first place may occur, the number of ways (3) that second place may occur (after first place has been determined), the number of possible outcomes (2) for third place (after second place has been decided), and the number of ways (1) that fourth place can occur (after the first three places have been taken).

The preceding discussion illustrates the following general rule, which we accept as a basic axiom of counting.

FUNDAMENTAL COUNTING PRINCIPLE

Let $E_1, E_2, \ldots, E_k$ be a sequence of k events. If for each i, E_i can occur in m_i ways, then the total number of ways the events may take place is the product $m_1 m_2 \cdots m_k$.

Returning to our first illustration, we let E_1 represent the determination of the first-place team, so that $m_1 = 4$. If E_2 denotes the determination of the second place team, then $m_2 = 3$. Hence, the number of outcomes for the sequence E_1, E_2 is $4 \cdot 3 = 12$, which is the same as that found by means of the tree diagram. If we proceed to E_3, the determination of the third-place team, then $m_3 = 2$, and hence, $m_1 m_2 m_3 = 24$. Finally, if E_1, E_2, and E_3 have occurred, there is only one possible outcome for E_4. Thus, $m_4 = 1$ and $m_1 m_2 m_3 m_4 = 24$.

Instead of teams, let us now regard a, b, c, and d merely as symbols and consider the various *orderings* or *arrangements* that may be assigned to these symbols, taking them either two at a time, three at a time, or four at a time. By abstracting in this way we may apply our methods to other similar situations. For example, the problem of determining the number of two-digit numbers that can be formed from the digits 1, 2, 3, and 4 so that no digit occurs twice in any number is essentially the same as our illustration of listing first and second place in the tournament. Incidentally, the arrangements we have discussed are called **arrangements without repetitions,** since a symbol may not be used twice in the same arrangement. In Example 1 we shall consider arrangements in which repetitions *are* allowed.

Previously we defined ordered pairs and ordered triples. Similarly, an **ordered 4-tuple** is a set containing four elements x_1, x_2, x_3, x_4 in which an ordering has been specified, so that one of the elements may be referred to as the *first element*, another as the *second element*, and so on. The symbol (x_1, x_2, x_3, x_4) is used for the ordered 4-tuple having first element x_1, second element x_2, third element x_3, and fourth element x_4. In general, for any positive integer r, we speak of the **ordered r-tuple**

$$(x_1, x_2, \ldots, x_r)$$

as a set of elements in which x_1 is-designated as the first element, x_2 as the second element, and so on.

EXAMPLE 1   How many ordered triples can be obtained if we use the letters a, b, c, and d? How many ordered 4-tuples can be obtained? How many r-tuples?

SOLUTION   To answer the first question we must determine the number of symbols of the form (x_1, x_2, x_3) that can be obtained by using only the letters a, b, c, and d. This is not the same as listing first, second, and third place as in our previous illustration, since we have not ruled out the possibility of repetitions. For example, (a, b, a), (a, a, b), and (b, a, a) are different triples of the desired type. If for $i = 1, 2, 3$, we let E_i represent the determination of x_i in the triple (x_1, x_2, x_3), then, since repetitions are allowed, there are four possibilities for each of E_1, E_2, and E_3. Hence, by the Fundamental Counting Principle, the total number of ordered triples is $4 \cdot 4 \cdot 4$, or 64. Similarly, the number of possible 4-tuples (x_1, x_2, x_3, x_4)

is $4 \cdot 4 \cdot 4 \cdot 4 = 256$. Evidently, the number of r-tuples is the product $4 \cdot 4 \cdot \cdots \cdot 4$, where 4 appears as a factor r times. That product equals 4^r.

∎

EXAMPLE 2 A class consists of 60 girls and 40 boys. In how many ways can a president, vice-president, treasurer, and secretary be chosen if the treasurer must be a girl, the secretary must be a boy, and a student may not hold more than one office?

SOLUTION If an event is specialized in some way (for example, the treasurer *must* be a girl), then that event should be performed before any nonspecialized events. Thus, we let E_1 represent the choice of treasurer and E_2 the choice of secretary. Next we let E_3 and E_4 denote the choices for president and vice-president, respectively. As in the Fundamental Counting Principle, let m_i denote the number of different ways E_i can occur, for $i = 1, 2, 3,$ and 4. It follows that $m_1 = 60$, $m_2 = 40$, $m_3 = 98$, and $m_4 = 97$. By the Fundamental Counting Principle, the total number of possibilities is

$$60 \cdot 40 \cdot 98 \cdot 97 = 22,814,400.$$

∎

When working with sets, we are usually not concerned about the order or arrangement of the elements. In the remainder of this section, however, the arrangement of the elements will be our main concern. In the following definition, S denotes a set consisting of n distinct elements, and r is a positive integer such that $r \leq n$.

DEFINITION

> A **permutation** of r elements of a set S is an arrangement, without repetitions, of r elements of S.

We also use the phrase **permutation of n elements taken r at a time.** The symbol $_nP_r$ will denote the number of different permutations of r elements that can be obtained from a set containing n elements. As a special case, $_nP_n$ denotes the number of arrangements of n elements of S; that is, $_nP_n$ is the number of ways of arranging *all* the elements of S.

In our first illustration involving the four teams $a, b, c,$ and d, we had $_4P_2 = 12$, since there were 12 different ways of arranging the four teams in groups of two. It was also shown that the number of ways to arrange all the elements $a, b, c,$ and d is 24. In the present notation this would be written $_4P_4 = 24$.

Let us find a general formula for $_nP_r$. If S is a set containing n elements, then the problem of determining $_nP_r$ is equivalent to determining the number of different r-tuples $(x_1, x_2, \ldots, x_r)$, where each x_i is an element of S and no element of S appears twice in the same r-tuple. We may find this number by means of the Fundamental Counting Principle. For each $i = 1, 2, \ldots, r$, let E_i represent the determination of the element x_i and let m_i be the number of different ways of choosing x_i. We wish to apply the sequence $E_1, E_2, \ldots, E_r$. There are n possible choices for x_1 and, consequently, we have $m_1 = n$. Since repetitions are not allowed, there are $n - 1$ choices for x_2, so that $m_2 = n - 1$. Continuing in this manner, we successively obtain $m_3 = n - 2$, $m_4 = n - 3$, and ultimately $m_r = n - (r - 1)$, or $m_r = n - r + 1$. Hence, by the Fundamental Counting Principle, we have

$$_nP_r = n(n - 1)(n - 2) \cdots (n - r + 1)$$

where there are r factors on the right side of the equation. As illustrations, if we consider the special cases $r = 1, 2, 3,$ and 4, then $n - r + 1$ equals $n, n - 1, n - 2,$ and $n - 3$, respectively, and using the formula for $_nP_r$ gives us

$$_nP_1 = n$$
$$_nP_2 = n(n - 1)$$
$$_nP_3 = n(n - 1)(n - 2)$$
$$_nP_4 = n(n - 1)(n - 2)(n - 3).$$

EXAMPLE 3   Find (a) $_5P_2$; (b) $_6P_4$; (c) $_5P_5$.

SOLUTION   Using the formula for $_nP_r$,

(a) $_5P_2 = 5 \cdot 4 = 20$

(b) $_6P_4 = 6 \cdot 5 \cdot 4 \cdot 3 = 360$

(c) $_5P_5 = 5 \cdot 4 \cdot 3 \cdot 2 \cdot 1 = 120.$ ∎

EXAMPLE 4   A baseball team consists of nine players. Find the number of ways of arranging the first four positions in the batting order, if the pitcher is excluded.

SOLUTION   We wish to find the number of permutations of 8 objects taken 4 at a time. Using the formula for $_nP_r$ with $n = 8$ and $r = 4$,

$$_8P_4 = 8 \cdot 7 \cdot 6 \cdot 5 = 1680.$$ ∎

If we let $r = n$ in the formula for $_nP_r$ we obtain the number of different arrangements of *all* the elements of S. In this case, $n - r + 1 = n - n + 1 = 1$ and hence,

$$_nP_n = n(n - 1)(n - 2) \cdots 3 \cdot 2 \cdot 1 = n!$$

Consequently, $_nP_n$ is the product of the first n positive integers. We may also obtain a form for $_nP_r$ that involves the factorial notation. If r and n are positive integers with $r \leq n$, then

$$\frac{n!}{(n - r)!} = \frac{n(n - 1)(n - 2) \cdots (n - r + 1) \cdot [(n - r)!]}{(n - r)!}$$

$$= n(n - 1)(n - 2) \cdots (n - r + 1).$$

Comparison with the formula for $_nP_r$ gives us

$$_nP_r = \frac{n!}{(n - r)!}.$$

Exercises 7.5

Solve Exercises 1–14 by using tree diagrams or the Fundamental Counting Principle.

1 How many three-digit numbers can be formed from the digits 1, 2, 3, 4, and 5 if repetitions (a) are not allowed? (b) are allowed?

2 Repeat Exercise 1 for four-digit numbers.

3 How many numbers can be formed from the digits 1, 2, 3, and 4 if repetitions are not allowed? (*Note:* 42 and 231 are examples of such numbers.)

4 Work Exercise 3 assuming repetitions are allowed.

5 If eight basketball teams take part in a tournament, find the number of different ways that first, second, and third place can be decided, assuming ties are not allowed.

6 Repeat Exercise 5 for 12 teams.

7 A girl has four skirts and six blouses. How many different skirt–blouse combinations can she wear?

8 If the girl in Exercise 7 also has three sweaters, in how many different ways can she combine the three articles of clothing?

9 In a certain state, automobile license plates start with one letter of the alphabet followed by five numerals, using the digits 0, 1, 2, ... , 9. Find how many different license plates are possible if:

(a) the first digit following the letter cannot be 0;

(b) the first letter cannot be *O* or *I* and the first digit cannot be 0.

10 Two dice are tossed, one after the other. In how many different ways can they fall? List the number of different ways the sum of the dots can equal:

(a) three (b) five

(c) seven (d) nine

(e) eleven

11 A row of six seats in a classroom is to be filled by selecting individuals from a group of ten students. In how many different ways can the seats be occupied? If there are six boys and four girls in the group and if boys and girls are to be alternated, find the number of different seating arrangements.

12 A student in a certain school may take mathematics at 8, 10, 11, or 2 o'clock; English at 9, 10, 1, or 2; and History at 8, 11, 2, or 3. Find the number of different ways in which the student can schedule the three courses.

13 In how many different ways can a test consisting of ten true-or-false questions be answered?

14 A test consists of six multiple-choice questions, and the number of choices for each question is five. In how many different ways can the test be answered?

Find the numbers in Exercises 15–22.

15 $_7P_3$

16 $_8P_5$

17 $_9P_6$

18 $_5P_3$

19 $_5P_5$

20 $_4P_4$

21 $_6P_1$

22 $_5P_1$

Use permutations to solve Exercises 23–30.

23 In how many different ways can eight people be seated in a row?

24 In how many different ways can ten books be arranged on a shelf?

25 A signal man has six different flags. How many different signals can be sent by placing three flags, one above the other, on a flag pole?

26 In how many different ways can five books be selected from a set of twelve different books?

27 How many three-digit numbers can be formed from the digits 2, 4, 6, 8, and 9 if

(a) repetitions are not allowed?

(b) repetitions are allowed?

28 There are 24 letters in the Greek alphabet. How many fraternities may be specified by choosing three Greek letters if repetitions (a) are not allowed? (b) are allowed?

29 How many seven-digit phone numbers can be formed from the digits 0, 1, 2, 3, . . . , 9 if the first digit may not be 0?

30 After selecting nine players for a baseball game, the manager of the team arranges the batting order so that the the pitcher bats last and the best hitter bats fourth. In how many different ways can the remainder of the batting order be arranged?

7.6 Distinguishable Permutations and Combinations

Certain problems involve finding different arrangements of objects, some of which are indistinguishable. For example, suppose we are given five discs of the same size, where three are black, one is white, and one is red. Let us find the number of ways they can be arranged in a row so that different color arrangements are obtained. If the discs were all different, then the number of arrangements would be 5!, or 120. However, since some of the discs have the same appearance, we cannot obtain 120 different arrangements. To clarify this point, let us write

B B B W R

for the arrangement having black discs in the first three positions in the row, the white disc in the fourth position, and the red disc in the fifth position. Now the first three discs can be arranged in 3!, or 6, different ways, but these arrangements cannot be distinguished from one another because the first three discs look alike. We say that those 3! permutations are **nondistinguishable.** Similarly, given any other arrangement, say

$$B \quad R \quad B \quad W \quad B,$$

there are 3! different ways of arranging the three black discs, but again each such arrangement is nondistinguishable from the others. Let us call two arrangements of the five objects **distinguishable permutations** if one arrangement cannot be obtained from the other by rearranging the like objects. Thus, B B B W R and B R B W B are distinguishable permutations. Let n denote the number of distinguishable permutations. Since with each such arrangement there correspond 3! *nondistinguishable* permutations, we must have $3!n = 5!$, the number of permutations of five *different* objects. Hence, $n = 5!/3! = 5 \cdot 4 = 20$. By the same type of reasoning we can obtain the following extension of this discussion.

THEOREM

> If r objects in a collection of n objects are alike, and if the remaining objects are different from each other and also from the r objects, then the number of distinguishable permutations of the n objects is $n!/r!$.

This theorem can be generalized to the case in which there are several subcollections of indistinguishable objects. For example, consider eight discs, four of which are black, three white, and one red. In this case, with each arrangement, such as

$$B \quad W \quad B \quad W \quad B \quad W \quad B \quad R,$$

there are 4! arrangements of the black discs and 3! arrangements of the white discs that have no effect on the color arrangement. Hence there are 4!3! arrangements of discs that can be made that will not produce distinguishable permutations. If we let n denote the number of *distinguishable* permutations of the objects, then it follows that $4!3!n = 8!$, since 8! is the number of permutations we would obtain if the objects were all different. This gives us

$$n = \frac{8!}{4!3!} = \frac{8 \cdot 7 \cdot 6 \cdot 5}{3!} \cdot \frac{4!}{4!} = 280.$$

The following general result can be proved.

THEOREM ON DISTINGUISHABLE PERMUTATIONS

Consider a collection of n objects in which n_1 are alike, n_2 are alike of another kind, . . . , n_k are alike of a further kind, and let

$$n = n_1 + n_2 + \cdots + n_k.$$

Then the number of distinguishable permutations of the n objects is

$$\frac{n!}{n_1! n_2! \cdots n_k!}.$$

EXAMPLE 1   Find the number of distinguishable permutations of the letters in the word "MISSISSIPPI."

SOLUTION   In this example we are given a collection of eleven objects in which four are of one kind (the letter S), four are of another kind (I), two are of a third kind (P), and one is of a fourth kind (M). Hence, by the preceding theorem, the number of distinguishable permutations is

$$\frac{11!}{4!\,4!\,2!\,1!}$$

The reader should show that this equals 34,650. ∎

When we work with permutations our concern is with the orderings or arrangements of elements. Let us now ignore the order or arrangement of elements and consider the following question: Given a set containing n distinct elements, in how many ways can a subset of r elements be chosen, where $r \le n$? Before answering, let us state a definition.

DEFINITION

A **combination** of r elements of a set S is a subset of S that contains r distinct elements.

If S contains n elements we also use the phrase **combination of n elements taken r at a time.** The symbol $_nC_r$ will denote the number of combinations of r elements that can be obtained from a set containing n elements.

If S contains n elements, then, to find $_nC_r$, we must find the total number of subsets of the form

$$\{x_1, x_2, \ldots, x_r\}$$

where the x_i are *different* elements of S. Since the elements $x_1, x_2, \ldots, x_r$ can be arranged in $r!$ different ways, each such subset produces $r!$ different r-tuples. Hence, the total number of different r-tuples is $r! \, ({}_nC_r)$. However, in the previous section we found that the number of r-tuples is

$$_nP_r = \frac{n!}{(n-r)!}.$$

This implies that

$$r!({}_nC_r) = \frac{n!}{(n-r)!}$$

and hence,

$$_nC_r = \frac{n!}{(n-r)!\,r!}.$$

Note that ${}_nC_r$ *is identical with the binomial coefficient* $\dbinom{n}{r}$ defined in Section 1.5.

EXAMPLE 2   A little league baseball squad has six outfielders, seven infielders, five pitchers, and two catchers. Assuming that each outfielder can play any outfield position and each infielder can play any infield position, in how many ways can a team of nine players be chosen?

SOLUTION   The number of ways of choosing three outfielders from the six candidates is

$$_6C_3 = \frac{6!}{3!(6-3)!} = \frac{6!}{3!3!} = \frac{6 \cdot 5 \cdot 4 \cdot 3 \cdot 2 \cdot 1}{(3 \cdot 2 \cdot 1)(3 \cdot 2 \cdot 1)} = 20.$$

The number of ways of choosing the four infielders is

$$_7C_4 = \frac{7!}{4!(7-4)!} = \frac{(7 \cdot 6 \cdot 5)4!}{4!3!} = \frac{7 \cdot 6 \cdot 5}{3 \cdot 2 \cdot 1} = 35.$$

There are five ways of choosing a pitcher and two choices for the catcher. It follows from the Fundamental Counting Principle that the total number of ways to choose a team is

$$20 \cdot 35 \cdot 5 \cdot 2 = 7000. \qquad \blacksquare$$

It should be noted that if $r = n$, then the formula for $_nC_r$ becomes

$$_nC_n = \frac{n!}{(n - n)!n!} = \frac{n!}{0!n!} = 1.$$

Moreover, $_nC_n$ is the number of subsets consisting of n elements that can be obtained from a set of n elements. This is also equal to 1. Hence, the formula for $_nC_r$ is true for $r = n$.

It is convenient to assign a meaning to $_nC_r$ if $r = 0$. If the formula is to be true in this case, then we must have

$$_nC_0 = \frac{n!}{n!0!} = 1.$$

Hence, we *define* $_nC_0 = 1$, which is the same as $_nC_n$. Finally, for consistency we also *define* $_0C_0 = 1$. Thus, $_nC_r$ has meaning for all nonnegative integers n and r with $r \le n$.

EXAMPLE 3　　If a set S contains n elements, find the number of distinct subsets of S.

SOLUTION　　Let r be any nonnegative integer such that $r \le n$. From our previous work the number of subsets of S that contain r elements is $_nC_r$, or $\binom{n}{r}$. Hence, to find the total number of subsets, we find the sum

$$\binom{n}{0} + \binom{n}{1} + \binom{n}{2} + \binom{n}{3} + \cdots + \binom{n}{n}.$$

This is precisely the binomial expansion of $(1 + 1)^n$. Thus, there are 2^n subsets of a set of n elements. In particular, a set of 3 elements has 2^3, or 8, different subsets. A set of 4 elements has 2^4, or 16, subsets. A set of 10 elements has 2^{10}, or 1024, subsets.　　■

Exercises 7.6

Find the numbers in Exercises 1–8.

1　$_7C_3$

2　$_8C_4$

3　$_9C_8$

4　$_6C_2$

5　$_nC_{n-1}$

6　$_nC_1$

7　$_7C_0$

8　$_5C_5$

9　If five black, three red, two white, and two green discs

are to be arranged in a row, find the number of possible color arrangements for the discs.

10 Work Exercise 9 assuming there are three black, three red, three white, and three green discs.

11 Find the number of distinguishable permutations of the letters in the word "BOOKKEEPER."

12 Find the number of distinguishable permutations of the letters in the word "MOON." List all of the permutations.

13 Ten boys wish to play a basketball game. In how many different ways can two teams consisting of five players each be chosen?

14 A student may answer any six of ten questions on an examination. In how many ways can a choice be made? How many choices are possible if the first two questions must be answered?

15 How many lines are determined by eight points if no three of the points are collinear? How many triangles are determined?

16 A committee of five persons is to be chosen from a group of twelve men and eight women. If the committee is to

consist of three men and two women, determine the number of ways of selecting the committee.

17 A student has five mathematics books, four history books, and eight fiction books. In how many ways can they be arranged on a shelf if books in the same category are kept next to one another?

18 A basketball squad consists of twelve individuals. Disregarding positions, in how many ways can a team of five be selected? If the center of a team must be selected from two specific individuals on the squad and the other four members of the team from the remaining ten players, find the number of different teams possible.

19 A football squad consists of three centers, ten linemen who can play either guard or tackle, three quarterbacks, six halfbacks, four ends, and four fullbacks. In how many ways can a team consisting of one center, two guards, two tackles, two ends, two halfbacks, a quarterback, and a fullback be selected?

20 In how many different ways can seven keys be arranged on a key ring if the keys can slide completely around the ring?

7.7 Probability

If two dice are tossed, what are the chances of rolling a 7? If a person is dealt five cards from a standard deck of 52 playing cards, what is the likelihood of obtaining three aces? In the 17th century, similar questions about games of chance led to the study of what is now called *probability*. Since that time, the theory of probability has grown extensively. It is now used to predict outcomes of a large variety of situations that arise in the natural and social sciences.

Any chance process, such as flipping a coin, rolling a die, being dealt a card from a deck, determining whether or not a manufactured item is defective, or finding the blood pressure of an arbitrary individual, will be called an **experiment**. A result of an experiment is called an **outcome**. We shall restrict our discussion to experiments for which outcomes are **equally likely.** This means, for example, that if a coin is flipped, it is assumed that the possibility of obtaining a head is the same as that of obtaining a tail. Similarly, if a die is tossed, we assume that the die is *fair*, in the sense that there is an equal chance of obtaining either a 1, 2, 3, 4, 5, or 6. The set S of all possible outcomes of an experiment is called the **sample space** of the experiment. Thus, if the experiment consists of flipping a coin, and we let H or T denote the outcome of obtaining a head or tail, respectively, then the sample space S may be denoted by

$$S = \{H, T\}.$$

If a fair die is tossed, then the set S of all possible outcomes (the sample space) is

$$S = \{1, 2, 3, 4, 5, 6\}.$$

The following definition expresses, in mathematical terms, the notion of obtaining *particular* outcomes of an experiment.

DEFINITION

> Let S be the sample space of an experiment. Any subset E of S is called an **event** associated with the experiment.

Let us consider the experiment of tossing a single die, so that the sample space is $S = \{1, 2, 3, 4, 5, 6\}$. If $E = \{4\}$, then the event E associated with the experiment consists of the outcome of obtaining a 4 on the toss. Different events may be associated with the same experiment. For example, if we let $E = \{1, 3, 5\}$, then this event consists of obtaining an odd number on a toss of the die.

As another illustration, suppose the experiment consists of flipping two coins, one after the other. If we let HH denote the outcome of two heads appearing, HT that of a head appearing on the first coin and a tail on the second, and so on, then the sample space S of the experiment may be denoted by

$$S = \{HH, HT, TH, TT\}.$$

If we let $\qquad\qquad\qquad E = \{HT, TH\}$

then the event E consists of a head appearing on one of the coins and a tail on the other.

Next we shall define what is meant by the *probability* of an event. Throughout our discussion it is assumed that the sample space S of an experiment contains only a finite number of elements. If E is an event, the symbols $n(E)$ and $n(S)$ will denote the number of elements in E and S, respectively.

DEFINITION

> Let S be the sample space of an experiment and E an event. The **probability** $P(E)$ of E is given by
>
> $$P(E) = \frac{n(E)}{n(S)}.$$

Since E is a subset of S, it follows that $0 \leq n(E) \leq n(S)$ and hence,

$$0 \leq P(E) \leq 1.$$

Note that $P(E) = 1$ if $E = S$, whereas $P(E) = 0$ if E contains no elements.

The next example provides several illustrations of the preceding definition for the case where E contains exactly one element.

EXAMPLE 1

(a) If a coin is flipped, find the probability that a head will turn up.

(b) If a fair die is tossed, find the probability of obtaining a 4.

(c) If two coins are flipped, find the probability that both coins turn up heads.

SOLUTION For each experiment we shall list sets S and E, and then use the definition to find $P(E)$.

(a) $\quad S = \{H, T\}, \quad E = \{H\}, \quad P(E) = \frac{n(E)}{n(S)} = \frac{1}{2}$

(b) $\quad S = \{1, 2, 3, 4, 5, 6\}, \quad E = \{4\}, \quad P(E) = \frac{n(E)}{n(S)} = \frac{1}{6}$

(c) $\quad S = \{HH, HT, TH, TT\}, \quad E = \{HH\}, \quad P(E) = \frac{n(E)}{n(S)} = \frac{1}{4}$ ■

In (a) of Example 1 we found that the probability of obtaining a head on a flip of a coin is $\frac{1}{2}$. We take this to mean that if a coin is flipped many times, the number of heads that turn up should be approximately one-half the total number of flips. Thus, for 100 flips, a head should turn up approximately 50 times. Of course, it is unlikely that this number will be *exactly* 50. A probability of $\frac{1}{2}$ implies that if we let the number of flips increase, then the number of heads that turn up *approaches* $\frac{1}{2}$ the total number of flips. Similar remarks can be made for (b) and (c) of Example 1.

In the next two examples we consider experiments in which an event contains more than one element.

EXAMPLE 2 If two dice are tossed, what is the probability of rolling a sum of (a) 7? (b) 9?

SOLUTION
(a) Let us refer to one die as "the first die" and the other as "the second die." We shall use ordered pairs to represent outcomes as follows: (2, 4) will denote the outcome of obtaining a 2 on the first die and a 4 on the second; (5, 3) represents a 5 on the first die and a 3 on the second, and so on. Since there are six different possibilities for the first number of the ordered pair and, with each of these, six possibilities for the second number, the total number of ordered pairs is 36. Hence, if S is the sample space, then $n(S) = 36$. The event E corresponding to rolling a sum of 7 is given by

$$E = \{(1, 6), (2, 5), (3, 4), (4, 3), (5, 2), (6, 1)\}$$

and consequently, $\qquad P(E) = \dfrac{n(E)}{n(S)} = \dfrac{6}{36} = \dfrac{1}{6}.$

(b) If E is the event corresponding to rolling a sum of 9, then

$$E = \{(3, 6), (4, 5), (5, 4), (6, 3)\}$$

and $\qquad\qquad\qquad P(E) = \dfrac{n(E)}{n(S)} = \dfrac{4}{36} = \dfrac{1}{9}.$ ∎

In the next example (and in the exercises), when it is stated that several cards are drawn from a deck, we mean that each card is removed from the deck and is *not* replaced before the next card is drawn.

EXAMPLE 3 Suppose five cards are drawn from a standard deck of 52 playing cards. Find the probability that all five cards are hearts.

SOLUTION The sample space S of the experiment is the set of all possible five-card hands that can be formed from the 52 cards in the deck. It follows from our work in Section 7.6 that $n(S) = {}_{52}C_5$.

Since there are 13 cards in the heart suit, the number of different ways of obtaining a hand that contains five hearts is ${}_{13}C_5$. Hence, if E represents this event, then

$$P(E) = \frac{n(E)}{n(S)} = \frac{{}_{13}C_5}{{}_{52}C_5}$$

$$= \frac{13!}{5!8!} \div \frac{52!}{5!47!} = \frac{13!47!}{8!52!}.$$

It is left to the reader to check that

$$P(E) = \frac{1287}{2,598,960} \approx 0.0005$$

or $$P(E) \approx \frac{5}{10,000} \approx \frac{1}{2000}.$$

This implies that if the experiment is performed many times, then a five-card heart hand should be drawn approximately once every 2000 times.

∎

Suppose S is the sample space of an experiment, and E_1 and E_2 are two events associated with the experiment. Suppose further that E_1 and E_2 have no elements in common, that is, E_1 and E_2 are *disjoint* sets. If $E = E_1 \cup E_2$, then

$$n(E) = n(E_1 \cup E_2) = n(E_1) + n(E_2)$$

and hence $$P(E) = \frac{n(E_1) + n(E_2)}{n(S)} = \frac{n(E_1)}{n(S)} + \frac{n(E_2)}{n(S)},$$

that is, $$P(E) = P(E_1) + P(E_2).$$

Thus, the probability of E is the sum of the probabilities of E_1 and E_2.

The next theorem states that this result can be extended to any number of events $E_1, E_2, \ldots, E_k$ that are *mutually disjoint*, in the sense that, if $i \neq j$, then E_i and E_j have no elements in common.

THEOREM

Let $E_1, E_2, \ldots, E_k$ be mutually disjoint events associated with the same experiment. If

$$E = E_1 \cup E_2 \cup \cdots \cup E_k,$$

then $$P(E) = P(E_1) + P(E_2) + \cdots + P(E_k).$$

The theorem may be proved by mathematical induction.

EXAMPLE 4   If two dice are tossed, find the probability of rolling a sum of either 7 or 9.

SOLUTION   Let E_1 denote the event of rolling 7, and E_2 that of rolling 9. We wish to find the probability of the event $E = E_1 \cup E_2$. From Example 3, $P(E_1) = \frac{1}{6}$ and $P(E_2) = \frac{1}{9}$. Hence, by the last theorem,

$$P(E) = P(E_1) + P(E_2)$$

$$= \frac{1}{6} + \frac{1}{9} = \frac{3+2}{18} = \frac{5}{18}.$$ ∎

The results of this section provide merely an introduction to the theory of probability. There are many other types of problems that may be considered. For example, given two events E_1 and E_2, we have not discussed how to find $P(E_1 \cup E_2)$ if E_1 and E_2 have elements in common, or how to find $P(E_2)$ *after* E_1 has occurred. Such questions are left for independent investigation or for future courses.

Exercises 7.7

1 A single card is drawn from a standard 52-card deck. Find the probability that the card is (a) a king, (b) a king or a queen, (c) a king, a queen, or a jack.

2 A single card is drawn from a 52-card deck. Find the probability that the card is (a) a heart, (b) a heart or a diamond, (c) a heart, a diamond, or a club.

3 If a single die is tossed, find the probability of obtaining (a) a 4, (b) a 6, (c) either a 4 or a 6.

4 An urn contains five red balls, six green balls, and four white balls. If a single ball is drawn, find the probability that it is (a) red, (b) green, (c) either red or white.

5 If two dice are tossed, find the probability of rolling a sum of (a) 11, (b) 8, (c) either 11 or 8.

6 If two dice are tossed, find the probability that the sum of the dots is greater than 5.

7 If three dice are tossed, find the probability that the sum of the dots is 5.

8 If three dice are tossed, find the probability that a 6 turns up on exactly one die.

9 If three coins are flipped, find the probability that exactly two heads turn up.

10 If four coins are flipped, find the probability of obtaining two heads and two tails.

In Exercises 11–16, suppose five cards are drawn from a 52-card deck. Find the probability of obtaining the indicated cards.

11 Four of a kind (such as four aces or four kings)

12 Three aces and two kings

13 Four diamonds and one spade

14 Five face cards

15 A flush (five cards, all of the same suit)

16 A royal flush (an ace, king, queen, jack, and 10 of the same suit)

17 A true-or-false test contains eight questions. If a student guesses the answer for each question, find the probability that:

(a) 8 answers are correct;

(b) 7 answers are correct and 1 is incorrect;

(c) 6 answers are correct and 2 are incorrect;

(d) at least 6 answers are correct.

18 A six-member committee is to be chosen by drawing names of individuals from a hat. If the hat contains the names of eight men and fourteen women, find the probability that the committee will consist of three men and three women.

19 (a) If S is the sample space of an experiment and E is an event, let E' denote the elements of S that are not in E. Prove that $P(E') = 1 - P(E)$.

(b) If five cards are drawn from a 52-card deck, use (a) to find the probability of obtaining at least one ace.

20 If five cards are drawn from a 52-card deck, use (a) of Exercise 19 to find the probability of obtaining at least one heart.

21 In the popular dice game of craps, the shooter rolls two dice and wins on the first toss if a sum of 7 or 11 is obtained. The shooter automatically loses if the first roll results in a sum of 2, 3, or 12. Find the probability of (a) winning on the first roll. (b) losing on the first roll.

22 A standard slot machine contains three reels and each reel contains 20 symbols. If the first reel has five bells, the middle reel four bells, and the last reel two bells, find the probability of obtaining three bells in a row.

23 Assuming that girl-boy births are equiprobable, find the probability that a family with five children has (a) all boys; (b) at least one girl. (See Exercise 19(b).)

24 Three cards are placed in a hat. One card is red on both sides, the second card is black on both sides, and the final card is red on one side and black on the other. A card is selected from the hat and placed on a table. If the card shows red, what is the probability that the other side is also red?

25 In a simple experiment designed to test ESP, four cards (jack, queen, king, and ace) are shuffled and then placed face down on a table. The subject then attempts to identify each of the four cards, giving a different name to each

of the cards. If the individual is guessing, find the probability of correctly identifying (a) all four cards; (b) exactly two of the four cards.

26 If three dice are tossed, what is the probability that:

(a) All dice show the same number?

(b) The numbers on the dice are all different?

(c) Answer (a) and (b) for n dice.

27 When a die is viewed from a certain angle, only three of its six faces are visible. This fact has been used to construct trick dice known as 'tops and bottoms'. Shown in the figure are two such dice. The same number of dots are on each pair of opposite faces. Find the probability of (a) rolling a sum of 7; (b) rolling a sum of 8.

FIGURE FOR EXERCISE 27

28 In a common carnival game, three balls are rolled down an incline into slots numbered 1 through 9 (see figure). Players have no control over where the balls will collect, because the slots are so narrow. A large prize is given if the sum of the resulting three numbers is less than 7. Find the probability of this event.

FIGURE FOR EXERCISE 28

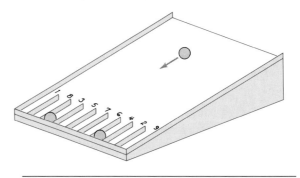

29 Shown in the figure on the next page is a small version of a probability demonstration device. A small ball is dropped into the top of the maze and tumbles to the bottom. Each time the ball strikes an obstacle, there is a 50% chance that the ball will move to the left. Find the probability that the ball ends up in the slot (a) on the far left; (b) in the middle.

FIGURE FOR EXERCISE 29

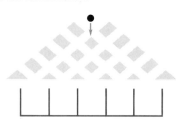

30 In the American version of roulette, a ball is spun around a wheel and has an equal chance of landing in any one of 38 slots numbered 0, 00, 1, 2, ... , 36. Shown in the

figure is a standard betting layout for roulette. Find the probability that the ball lands (a) in a black slot; (b) in a black slot two times in succession.

FIGURE FOR EXERCISE 30

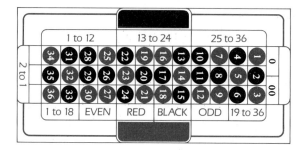

7.8 Review

Define or discuss each of the following.

1 Axiom of Mathematical Induction
2 Principle of Mathematical Induction
3 Infinite sequence
4 Summation notation
5 The nth partial sum of an infinite sequence
6 Arithmetic sequence
7 Arithmetic mean of two numbers
8 Geometric sequence
9 Infinite geometric series
10 Infinite series
11 Tree diagram

12 Fundamental Counting Principle
13 Ordered r-tuple
14 Permutation
15 $_nP_r$
16 Nondistinguishable permutations
17 Distinguishable permutations
18 Combination
19 $_nC_r$
20 Sample space
21 Event
22 Probability of an event

Exercises 7.8

Prove that the statements in Exercises 1–5 are true for every positive integer n.

1 $2 + 5 + 8 + \cdots + (3n - 1) = \dfrac{n(3n + 1)}{2}$

2 $2^2 + 4^2 + 6^2 + \cdots + (2n)^2 = \dfrac{2n(2n + 1)(n + 1)}{3}$

3 $\dfrac{1}{1 \cdot 3} + \dfrac{1}{3 \cdot 5} + \dfrac{1}{5 \cdot 7} + \cdots + \dfrac{1}{(2n - 1)(2n + 1)} = \dfrac{n}{2n + 1}$

4 $1 \cdot 2 + 2 \cdot 3 + 3 \cdot 4 + \cdots + n(n + 1) = \dfrac{n(n + 1)(n + 2)}{3}$

5 3 is a factor of $n^3 + 2n$.

6 Prove that $2^n > n^2$ for every positive integer $n \geq 5$.

In Exercises 7–10 find the first four terms and the seventh term of the sequence that has the given nth term.

7 $a_n = \dfrac{5n}{3 - 2n^2}$ **8** $a_n = (-1)^{n+1} - (0.1)^n$

9 $a_n = 1 + (-\tfrac{1}{2})^{n-1}$ **10** $a_n = \dfrac{2^n}{(n + 1)(n + 2)(n + 3)}$

Find the first five terms of the infinite sequences defined recursively in Exercises 11–14.

11 $a_1 = 10, \; a_{k+1} = 1 + (1/a_k)$

12 $a_1 = 2, \; a_{k+1} = a_k!$

13 $a_1 = 9, \; a_{k+1} = \sqrt{a_k}$

14 $a_1 = 1, \; a_{k+1} = (1 + a_k)^{-1}$

Find the numbers represented by the sums in Exercises 15–18.

15 $\displaystyle\sum_{k=1}^{5} (k^2 + 4)$ **16** $\displaystyle\sum_{k=2}^{6} \dfrac{2k - 8}{k - 1}$

17 $\displaystyle\sum_{k=1}^{100} 10$ **18** $\displaystyle\sum_{k=1}^{4} (2^k - 10)$

In Exercises 19–22 use summation notation to represent the sums.

19 $3 + 6 + 9 + 12 + 15$

20 $2 + 4 + 8 + 16 + 32 + 64 + 128$

21 $100 - 95 + 90 - 85 + 80$

22 $a_0 + a_4 x^4 + a_8 x^8 + \cdots + a_{100} x^{100}$

23 Find the tenth term and the sum of the first ten terms of the arithmetic sequence whose first two terms are $4 + \sqrt{3}$ and 3.

24 Find the sum of the first eight terms of an arithmetic sequence in which the fourth term is 9 and the common difference is -5.

25 The fifth and thirteenth terms of an arithmetic sequence are 5 and 77, respectively. Find the first term and the tenth term.

26 Insert four arithmetic means between 20 and -10.

27 Find the tenth term of the geometric sequence whose first two terms are $\tfrac{1}{8}$ and $\tfrac{1}{4}$.

28 If a geometric sequence has 3 and -0.3 as its third and fourth terms, find the eighth term.

29 Find a positive number c such that 4, c, 8 are successive terms of a geometric sequence.

30 In a certain geometric sequence the eighth term is 100 and the common ratio is $-\tfrac{3}{2}$. Find the first term.

Find the sums in Exercises 31–34.

31 $\displaystyle\sum_{k=1}^{15} (5k - 2)$ **32** $\displaystyle\sum_{k=1}^{10} (6 - \tfrac{1}{2}k)$

33 $\displaystyle\sum_{k=1}^{10} (2^k - \tfrac{1}{2})$ **34** $\displaystyle\sum_{k=1}^{8} (\tfrac{1}{2} - 2^k)$

35 Find the sum of the infinite geometric series

$$1 - \frac{2}{5} + \frac{4}{25} - \frac{8}{125} + \cdots .$$

36 Find the rational number whose decimal representation is $6.\overline{274}$.

37 Ten-foot lengths of 2×2 lumber are to be cut into five pieces to form children's building blocks, and the lengths of the five blocks are to form an arithmetic sequence.

(a) Show that the difference d in lengths must be less than 1 foot.

(b) If the smallest block is to have a length of 6 inches, find the lengths of the other four pieces.

38 When a ball is dropped from a height of h feet, it takes $\sqrt{h}/4$ seconds to reach the ground. It then takes $\sqrt{d}/4$ seconds for the ball to rebound to a height of d feet. If a rubber ball is dropped from a height of 10 feet and rebounds to three-fourths of its height after each fall, how many seconds elapse before the ball comes to rest?

39 Two boys each toss a coin. Find the probability that the coins will match. If each of three boys toss a coin, what is the probability that the coins will match?

40 If four cards are dealt from a 52-card deck, find the probability that (a) all four cards will be the same color; (b) the cards dealt will alternate red-black-red-black.

41 In how many ways can thirteen cards be selected from a deck of 52 cards? In how many ways can thirteen cards be selected if you wish to obtain five spades, three hearts, three clubs, and two diamonds?

42 How many four-digit numbers can be formed from the digits 1, 2, 3, 4, 5, and 6 if repetitions (a) are not allowed? (b) are allowed?

43 If a student must answer eight of twelve questions on an examination, in how many ways can a choice be made? How many choices are possible if the first three questions must be answered?

44 If six black, five red, four white, and two green discs are to be arranged in a row, what is the number of possible color arrangements?

45 If 1000 tickets are sold for a raffle, find the probability of winning if an individual purchases (a) 1 ticket, (b) 10 tickets, (c) 50 tickets.

46 If four coins are flipped, find the probability of obtaining 1 head and 3 tails.

47 A quiz consists of six true-or-false questions, and at least four correct answers are required for a passing grade. If a student guesses at each answer, what is the probability of (a) passing? (b) failing?

48 If a single die is tossed and then a card is drawn from a 52-card deck, what is the probability of obtaining:

(a) a 6 on the die and the king of hearts?

(b) *either* a 6 on the die or the king of hearts?

Using Logarithmic Tables

If x is any positive real number and we write

$$x = c \cdot 10^k$$

where $1 \le c < 10$ and k is an integer, then applying (iii) of the Laws of Logarithms,

$$\log x = \log c + \log 10^k.$$

Since $\log 10^k = k$, we see that

$$\log x = \log c + k.$$

The last equation tells us that to find $\log x$ for any positive real number x it is sufficient to know the logarithms of numbers between 1 and 10. The number $\log c$, for $1 \le c < 10$, is called the **mantissa,** and the integer k is called the **characteristic** of $\log x$.

If $1 \le c < 10$, then, since $\log x$ increases as x increases,

$$\log 1 \le \log c < \log 10,$$

or equivalently, $\qquad\qquad 0 \le \log c < 1.$

Hence, the mantissa of a logarithm is a number between 0 and 1. In numerical problems it is usually necessary to approximate logarithms. For example, it can be shown that

$$\log 2 = 0.3010299957 \ldots$$

where the decimal is nonrepeating and nonterminating. We often round off such logarithms to four decimal places and write

$$\log 2 \approx 0.3010.$$

If a number between 0 and 1 is written as a finite decimal, it is sometimes referred to as a **decimal fraction.** Thus, the equation $\log x = \log c + k$ implies that if x is any positive real number, then $\log x$ *may be approximated by the sum of a positive decimal fraction (the mantissa) and an integer k (the characteristic).* We shall refer to this representation as the **standard form** for $\log x$.

Common logarithms of many of the numbers between 1 and 10 have been calculated. Table 1 of Appendix II contains four-decimal-place approximations for logarithms of numbers between 1.00 and 9.99 at intervals of 0.01. This table can be used to find the common logarithm of any three-digit number to four-decimal-place accuracy. The use of Table 1 is illustrated in the following examples.

EXAMPLE 1   Approximate each of the following:

(a) log 43.6   (b) log 43,600   (c) log 0.0436

SOLUTION

(a) Since $43.6 = (4.36)10^1$, the characteristic of log 43.6 is 1. Referring to Table 1, we find that the mantissa of log 4.36 may be approximated by 0.6395. Hence, as in the preceding discussion,

$$\log 43.6 \approx 0.6395 + 1 = 1.6395.$$

(b) Since $43,600 = (4.36)10^4$, the mantissa is the same as in part (a); however, the characteristic is 4. Consequently,

$$\log 43,600 \approx 0.6395 + 4 = 4.6395.$$

(c) If we write $0.0436 = (4.36)10^{-2}$, then

$$\log 0.0436 = \log 4.36 + (-2).$$

Hence, $\log 0.0436 \approx 0.6395 + (-2).$

We could subtract 2 from 0.6395 and obtain

$$\log 0.0436 \approx -1.3605$$

but this is not standard form, since $-1.3605 = -0.3605 + (-1)$, a number in which the decimal part is *negative*. A common error is to write $0.6395 + (-2)$ as -2.6395. This is incorrect since

$$-2.6395 = -0.6395 + (-2),$$

which is not the same as $0.6395 + (-2)$. ∎

If a logarithm has a negative characteristic, we usually either leave it in standard form or rewrite the logarithm, keeping the decimal part positive. To illustrate the latter technique, let us add and subtract 8 on the right side of the equation

$$\log 0.0436 \approx 0.6395 + (-2).$$

This gives us $\quad \log 0.0436 \approx 0.6395 + (8 - 8) + (-2)$

or $\qquad\qquad\qquad \log 0.0436 \approx 8.6395 - 10.$

We could also write

$$\log 0.0436 \approx 18.6395 - 20 = 43.6395 - 45$$

and so on, as long as the *integral part* of the logarithm is -2.

EXAMPLE 2　Approximate each of the following:

(a)　$\log (0.00652)^2$　　(b)　$\log (0.00652)^{-2}$　　(c)　$\log (0.00652)^{1/2}$

SOLUTION

(a)　By (iii) of the Laws of Logarithms,

$$\log (0.00652)^2 = 2 \log 0.00652.$$

Since $0.00652 = (6.52)10^{-3}$,

$$\log 0.00652 = \log 6.52 + (-3).$$

Referring to Table 1, we see that $\log 6.52$ is approximately 0.8142 and, therefore,

$$\log 0.00652 \approx 0.8142 + (-3).$$

Hence, $\qquad\qquad \log (0.00652)^2 = 2 \log 0.00652$
$$\approx 2[0.8142 + (-3)]$$
$$= 1.6284 + (-6).$$

The standard form is $0.6284 + (-5)$.

(b) Again using Law (iii) and the value for log 0.00652 found in part (a),

$$\log (0.00652)^{-2} = -2 \log 0.00652$$
$$\approx -2[0.8142 + (-3)]$$
$$= -1.6284 + 6.$$

It is important to note that -1.6284 means $-0.6284 + (-1)$ and, consequently, the decimal part is negative. To obtain the standard form, we may write

$$-1.6284 + 6 = 6.0000 - 1.6284$$
$$= 4.3716.$$

This shows that the mantissa is 0.3716 and the characteristic is 4.

(c) By Law (iii),

$$\log (0.00652)^{1/2} = \tfrac{1}{2} \log 0.00652$$
$$\approx \tfrac{1}{2}[0.8142 + (-3)].$$

If we multiply by $\tfrac{1}{2}$, the standard form is not obtained, since neither number in the resulting sum is the characteristic. In order to avoid this, we may adjust the expression within brackets by adding and subtracting a suitable number. If we use 1 in this way, we obtain

$$\log (0.00652)^{1/2} \approx \tfrac{1}{2}[1.8142 + (-4)]$$
$$= 0.9071 + (-2),$$

which is in standard form. We could also have added and subtracted a number other than 1. For example,

$$\tfrac{1}{2}[0.8142 + (-3)] = \tfrac{1}{2}[17.8142 + (-20)]$$
$$= 8.9071 + (-10). \qquad \blacksquare$$

Table 1 can be used to find an approximation to x if log x is given, as illustrated in the following example.

EXAMPLE 3   Find a decimal approximation to x for each log x:

(a) log $x = 1.7959$ (b) log $x = -3.5918$

SOLUTION

(a) The mantissa 0.7959 determines the sequence of digits in x and the characteristic determines the position of the decimal point. Referring to the *body* of Table 1, we see that the mantissa 0.7959 is the logarithm of 6.25.

Since the characteristic is 1, x lies between 10 and 100. Consequently, $x \approx 62.5$.

(b) To find x from Table 1, $\log x$ must be written in standard form. To change $\log x = -3.5918$ to standard form, we may add and subtract 4, obtaining

$$\log x = (4 - 3.5918) - 4$$
$$= 0.4082 - 4.$$

Referring to Table 1, we see that the mantissa 0.4082 is the logarithm of 2.56. Since the characteristic of $\log x$ is -4, it follows that $x \approx 0.000256$.
∎

If a calculator with a $\boxed{\log}$ key is used to determine common logarithms, then the standard form for $\log x$ is obtained only if $x \geq 1$. For example, to find $\log 43.6$ on a typical calculator, we enter 43.6 and press $\boxed{\log}$, obtaining the standard form

$$1.6394865$$

If we find $\log 0.0436$ in similar fashion, then the following number appears on the display panel:

$$-1.3605135$$

This is not the standard form for the logarithm, but is similar to that which occurred in the solution to Example 1(c). To find the standard form we could add 2 to the logarithm (using a calculator) and then subtract 2 as follows:

$$\log 0.0436 \approx -1.3605135$$
$$= (-1.3605135 + 2) - 2$$
$$= 0.6394865 - 2$$
$$= 0.6394865 + (-2)$$

The only common logarithms that can be found *directly* from Table 1 are logarithms of numbers that contain at most three nonzero digits. If *four* nonzero digits are involved, then it is possible to obtain an approximation by using the method of linear interpolation described next. The terminology **linear interpolation** is used because, as we shall see, the method is based upon approximating portions of the graph of $y = \log x$ by line segments.

FIGURE A1.1

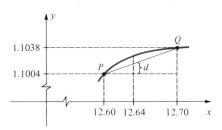

FIGURE A1.2

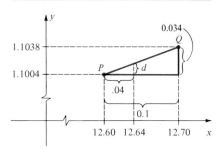

To illustrate the process of linear interpolation, and at the same time give some justification for it, let us consider the specific example log 12.64. Since the logarithmic function with base 10 is increasing, this number lies between log 12.60 ≈ 1.1004 and log 12.70 ≈ 1.1038. Examining the graph of $y = \log x$, we have the situation shown in Figure A1.1, where we have distorted the units on the x- and y-axes and also the portion of the graph shown. A more accurate drawing would indicate that the graph of $y = \log x$ is much closer to the line segment joining $P(12.60, 1.1004)$ to $Q(12.70, 1.1038)$ than is shown in the figure. Since log 12.64 is the y-coordinate of the point on the graph having x-coordinate 12.64, it can be approximated by the y-coordinate of the point with x-coordinate 12.64 on the *line segment PQ*. Referring to Figure A1.1, we see that the latter y-coordinate is $1.1004 + d$. The number d can be approximated by using similar triangles. Referring to Figure A1.2, where the graph of $y = \log x$ has been deleted, we may form the following proportion:

$$\frac{d}{0.0034} = \frac{0.04}{0.1}.$$

Hence

$$d = \frac{(0.04)(0.0034)}{0.1} = 0.00136.$$

When using this technique, we always round off decimals to the same number of places as appear in the body of the table. Consequently, $d \approx 0.0014$ and

$$\log 12.64 \approx 1.1004 + 0.0014 = 1.1018.$$

Hereafter we shall not sketch a graph when interpolating. Instead we shall use the scheme illustrated in the next example.

EXAMPLE 4 Approximate log 572.6.

SOLUTION It is convenient to arrange our work as follows:

$$1.0 \left\{ 0.6 \begin{cases} \log 572.0 \approx 2.7574 \\ \log 572.6 = ? \end{cases} d \atop \log 573.0 \approx 2.7582 \right\} 0.0008$$

where differences are indicated by appropriate symbols alongside of the braces. This leads to the proportion

$$\frac{d}{0.0008} = \frac{0.6}{1.0} = \frac{6}{10} \quad \text{or} \quad d = \left(\frac{6}{10}\right)(0.0008) = 0.00048 \approx 0.0005.$$

Hence, $\qquad$ $\log 572.6 \approx 2.7574 + 0.0005 = 2.7579$.

Another way of working this type of problem is to reason that since 572.6 is $\frac{6}{10}$ of the way from 572.0 to 573.0, then log 572.6 is (approximately) $\frac{6}{10}$ of the way from 2.7574 to 2.7582. Hence,

$$\log 572.6 \approx 2.7574 + (\tfrac{6}{10})(0.0008) \approx 2.7574 + 0.0005 = 2.7579. \qquad \blacksquare$$

EXAMPLE 5   Approximate log 0.003678.

SOLUTION   We begin by arranging our work as in the solution of Example 1. Thus,

$$10\left\{8\left\{\begin{matrix}\log 0.003670 \approx 0.5647 + (-3)\\ \log 0.003678 = ?\\ \log 0.003680 \approx 0.5658 + (-3)\end{matrix}\right\}d\right\}0.0011$$

Since we are only interested in ratios, we have used the numbers 8 and 10 on the left side because their ratio is the same as the ratio of 0.000008 to 0.000010. This leads to the proportion

$$\frac{d}{0.0011} = \frac{8}{10} = 0.8 \quad \text{or} \quad d = (0.0011)(0.8) = 0.00088 \approx 0.0009.$$

Hence, $\qquad$ $\log 0.003678 \approx [0.5647 + (-3)] + 0.0009$
$$= 0.5656 + (-3). \qquad \blacksquare$$

If a number x is written in the form $x = c \cdot 10^k$, where $1 \le c < 10$, then before using Table 1 to find log x by interpolation, c should be rounded off to three decimal places. Another way of saying this is that x should be rounded off to four **significant figures.** Some examples will help to clarify the procedure. If $x = 36.4635$, we round off to 36.46 before approximating log x. The number 684,279 should be rounded off to 684,300. For a decimal such as 0.096202 we use 0.09620. The reason for doing this is that Table 1 does not guarantee more than four-digit accuracy, since the mantissas that appear in it are approximations. This means that if *more* than four-digit accuracy is required in a problem, then Table 1 cannot be used. If, in more extensive tables, the logarithm of a number containing n digits can be found directly, then interpolation is allowed for numbers involving $n + 1$ digits, and numbers should be rounded off accordingly.

The method of interpolation can also be used to find x when we are given log x. If we use Table 1, then x may be found to four significant figures. In this case we are given the *y-coordinate* of a point on the graph of $y = \log x$ and are asked to find the *x-coordinate*. A geometric argument similar to the one given earlier can be used to justify the procedure illustrated in the next example.

EXAMPLE 6 Find x to four significant figures if $\log x = 1.7949$.

SOLUTION The mantissa 0.7949 does not appear in Table 1, but it can be isolated between adjacent entries, namely the mantissas corresponding to 6.230 and 6.240. We shall arrange our work as follows:

$$0.1 \left\{ r \begin{cases} \log 62.30 \approx 1.7945 \\ \log x \quad\ = 1.7949 \end{cases} 0.0004 \\ \log 62.40 \approx 1.7952 \right\} 0.0007 .$$

This leads to the proportion

$$\frac{r}{0.1} = \frac{0.0004}{0.0007} = \frac{4}{7} \quad \text{or} \quad r = (0.1)\left(\frac{4}{7}\right) \approx 0.06 .$$

Hence, $x \approx 62.30 + 0.06 = 62.36 .$ ■

Exercises A.1

In Exercises 1–16 use Table 1 and the Laws of Logarithms to approximate the common logarithms of the numbers.

1 347; 0.00347; 3.47

2 86.2; 8620; 0.862

3 0.54; 540; 540,000

4 208; 2.08; 20,800

5 60.2; 0.0000602; 602

6 5; 0.5; 0.0005

7 $(44.9)^2$; $(44.9)^{1/2}$; $(44.9)^{-2}$

8 $(1810)^4$; $(1810)^{40}$; $(1810)^{1/4}$

9 $(0.943)^3$; $(0.943)^{-3}$; $(0.943)^{1/3}$

10 $(0.017)^{10}$; $10^{0.017}$; $10^{1.43}$

11 $(638)(17.3)$

12 $\dfrac{(2.73)(78.5)}{621}$

13 $\dfrac{(47.4)^3}{(29.5)^2}$

14 $\dfrac{(897)^4}{\sqrt{17.8}}$

15 $\sqrt[3]{20.6}(371)^3$

16 $\dfrac{(0.0048)^{10}}{\sqrt{0.29}}$

In Exercises 17–30 use Table 1 to find a decimal approximation to x.

17 $\log x = 3.6274$

18 $\log x = 1.8965$

19 $\log x = 0.9469$

20 $\log x = 0.5729$

21 $\log x = 5.2095$

22 $\log x = 6.7300 - 10$

23 $\log x = 9.7348 - 10$

24 $\log x = 7.6739 - 10$

25 $\log x = 8.8306 - 10$

26 $\log x = 4.9680$

27 $\log x = 2.2765$

28 $\log x = 3.0043$

29 $\log x = -1.6253$

30 $\log x = -2.2118$

Use interpolation in Table 1 to approximate the common logarithms of the numbers in Exercises 31–50.

31 25.48

32 421.6

33 5363

34 0.3817

35 0.001259

36 69,450

37 123,400

38 0.0212

39 0.7786

40 1.203   **41** 384.7   **42** 54.44   **55** $\log x = 9.1664 - 10$ **56** $\log x = 8.3902 - 10$

43 0.9462   **44** 7259   **45** 66,590   **57** $\log x = 3.8153 - 6$ **58** $\log x = 5.9306 - 9$

46 0.001428   **47** 0.04321   **48** 300,100   **59** $\log x = 2.3705$ **60** $\log x = 4.2867$

49 3.003   **50** 1.236   **61** $\log x = 0.1358$ **62** $\log x = 0.0194$

63 $\log x = 8.9752 - 10$ **64** $\log x = 2.4979 - 5$

In Exercises 51–70 use interpolation in Table 1 to approximate x.   **65** $\log x = 5.0409$ **66** $\log x = 1.3796$

51 $\log x = 1.4437$ **52** $\log x = 3.7455$ **67** $\log x = -2.8712$ **68** $\log x = -1.8164$

53 $\log x = 4.6931$ **54** $\log x = 0.5883$ **69** $\log x = -0.6123$ **70** $\log x = -3.1426$

TABLE 1 Common Logarithms

N	0	1	2	3	4	5	6	7	8	9
1.0	.0000	.0043	.0086	.0128	.0170	.0212	.0253	.0294	.0334	.0374
1.1	.0414	.0453	.0492	.0531	.0569	.0607	.0645	.0682	.0719	.0755
1.2	.0792	.0828	.0864	.0899	.0934	.0969	.1004	.1038	.1072	.1106
1.3	.1139	.1173	.1206	.1239	.1271	.1303	.1335	.1367	.1399	.1430
1.4	.1461	.1492	.1523	.1553	.1584	.1614	.1644	.1673	.1703	.1732
1.5	.1761	.1790	.1818	.1847	.1875	.1903	.1931	.1959	.1987	.2014
1.6	.2041	.2068	.2095	.2122	.2148	.2175	.2201	.2227	.2253	.2279
1.7	.2304	.2330	.2355	.2380	.2405	.2430	.2455	.2480	.2504	.2529
1.8	.2553	.2577	.2601	.2625	.2648	.2672	.2695	.2718	.2742	.2765
1.9	.2788	.2810	.2833	.2856	.2878	.2900	.2923	.2945	.2967	.2989
2.0	.3010	.3032	.3054	.3075	.3096	.3118	.3139	.3160	.3181	.3201
2.1	.3222	.3243	.3263	.3284	.3304	.3324	.3345	.3365	.3385	.3404
2.2	.3424	.3444	.3464	.3483	.3502	.3522	.3541	.3560	.3579	.3598
2.3	.3617	.3636	.3655	.3674	.3692	.3711	.3729	.3747	.3766	.3784
2.4	.3802	.3820	.3838	.3856	.3874	.3892	.3909	.3927	.3945	.3962
2.5	.3979	.3997	.4014	.4031	.4048	.4065	.4082	.4099	.4116	.4133
2.6	.4150	.4166	.4183	.4200	.4216	.4232	.4249	.4265	.4281	.4298
2.7	.4314	.4330	.4346	.4362	.4378	.4393	.4409	.4425	.4440	.4456
2.8	.4472	.4487	.4502	.4518	.4533	.4548	.4564	.4579	.4594	.4609
2.9	.4624	.4639	.4654	.4669	.4683	.4698	.4713	.4728	.4742	.4757
3.0	.4771	.4786	.4800	.4814	.4829	.4843	.4857	.4871	.4886	.4900
3.1	.4914	.4928	.4942	.4955	.4969	.4983	.4997	.5011	.5024	.5038
3.2	.5051	.5065	.5079	.5092	.5105	.5119	.5132	.5145	.5159	.5172
3.3	.5185	.5198	.5211	.5224	.5237	.5250	.5263	.5276	.5289	.5302
3.4	.5315	.5328	.5340	.5353	.5366	.5378	.5391	.5403	.5416	.5428
3.5	.5441	.5453	.5465	.5478	.5490	.5502	.5514	.5527	.5539	.5551
3.6	.5563	.5575	.5587	.5599	.5611	.5623	.5635	.5647	.5658	.5670
3.7	.5682	.5694	.5705	.5717	.5729	.5740	.5752	.5763	.5775	.5786
3.8	.5798	.5809	.5821	.5832	.5843	.5855	.5866	.5877	.5888	.5899
3.9	.5911	.5922	.5933	.5944	.5955	.5966	.5977	.5988	.5999	.6010
4.0	.6021	.6031	.6042	.6053	.6064	.6075	.6085	.6096	.6107	.6117
4.1	.6128	.6138	.6149	.6160	.6170	.6180	.6191	.6201	.6212	.6222
4.2	.6232	.6243	.6253	.6263	.6274	.6284	.6294	.6304	.6314	.6325
4.3	.6335	.6345	.6355	.6365	.6375	.6385	.6395	.6405	.6415	.6425
4.4	.6435	.6444	.6454	.6464	.6474	.6484	.6493	.6503	.6513	.6522
4.5	.6532	.6542	.6551	.6561	.6571	.6580	.6590	.6599	.6609	.6618
4.6	.6628	.6637	.6646	.6656	.6665	.6675	.6684	.6693	.6702	.6712
4.7	.6721	.6730	.6739	.6749	.6758	.6767	.6776	.6785	.6794	.6803
4.8	.6812	.6821	.6830	.6839	.6848	.6857	.6866	.6875	.6884	.6893
4.9	.6902	.6911	.6920	.6928	.6937	.6946	.6955	.6964	.6972	.6981
5.0	.6990	.6998	.7007	.7016	.7024	.7033	.7042	.7050	.7059	.7067
5.1	.7076	.7084	.7093	.7101	.7110	.7118	.7126	.7135	.7143	.7152
5.2	.7160	.7168	.7177	.7185	.7193	.7202	.7210	.7218	.7226	.7235
5.3	.7243	.7251	.7259	.7267	.7275	.7284	.7292	.7300	.7308	.7316
5.4	.7324	.7332	.7340	.7348	.7356	.7364	.7372	.7380	.7388	.7396
5.5	.7404	.7412	.7419	.7427	.7435	.7443	.7451	.7459	.7466	.7474
5.6	.7482	.7490	.7497	.7505	.7513	.7520	.7528	.7536	.7543	.7551
5.7	.7559	.7566	.7574	.7582	.7589	.7597	.7604	.7612	.7619	.7627
5.8	.7634	.7642	.7649	.7657	.7664	.7672	.7679	.7686	.7694	.7701
5.9	.7709	.7716	.7723	.7731	.7738	.7745	.7752	.7760	.7767	.7774
6.0	.7782	.7789	.7796	.7803	.7810	.7818	.7825	.7832	.7839	.7846
6.1	.7853	.7860	.7868	.7875	.7882	.7889	.7896	.7903	.7910	.7917
6.2	.7924	.7931	.7938	.7945	.7952	.7959	.7966	.7973	.7980	.7987
6.3	.7993	.8000	.8007	.8014	.8021	.8028	.8035	.8041	.8048	.8055
6.4	.8062	.8069	.8075	.8082	.8089	.8096	.8102	.8109	.8116	.8122
6.5	.8129	.8136	.8142	.8149	.8156	.8162	.8169	.8176	.8182	.8189
6.6	.8195	.8202	.8209	.8215	.8222	.8228	.8235	.8241	.8248	.8254
6.7	.8261	.8267	.8274	.8280	.8287	.8293	.8299	.8306	.8312	.8319
6.8	.8325	.8331	.8338	.8344	.8351	.8357	.8363	.8370	.8376	.8382
6.9	.8388	.8395	.8401	.8407	.8414	.8420	.8426	.8432	.8439	.8445
7.0	.8451	.8457	.8463	.8470	.8476	.8482	.8488	.8494	.8500	.8506
7.1	.8513	.8519	.8525	.8531	.8537	.8543	.8549	.8555	.8561	.8567
7.2	.8573	.8579	.8585	.8591	.8597	.8603	.8609	.8615	.8621	.8627
7.3	.8633	.8639	.8645	.8651	.8657	.8663	.8669	.8675	.8681	.8686
7.4	.8692	.8698	.8704	.8710	.8716	.8722	.8727	.8733	.8739	.8745
7.5	.8751	.8756	.8762	.8768	.8774	.8779	.8785	.8791	.8797	.8802
7.6	.8808	.8814	.8820	.8825	.8831	.8837	.8842	.8848	.8854	.8859
7.7	.8865	.8871	.8876	.8882	.8887	.8893	.8899	.8904	.8910	.8915
7.8	.8921	.8927	.8932	.8938	.8943	.8949	.8954	.8960	.8965	.8971
7.9	.8976	.8982	.8987	.8993	.8998	.9004	.9009	.9015	.9020	.9025
8.0	.9031	.9036	.9042	.9047	.9053	.9058	.9063	.9069	.9074	.9079
8.1	.9085	.9090	.9096	.9101	.9106	.9112	.9117	.9122	.9128	.9133
8.2	.9138	.9143	.9149	.9154	.9159	.9165	.9170	.9175	.9180	.9186
8.3	.9191	.9196	.9201	.9206	.9212	.9217	.9222	.9227	.9232	.9238
8.4	.9243	.9248	.9253	.9258	.9263	.9269	.9274	.9279	.9284	.9289
8.5	.9294	.9299	.9304	.9309	.9315	.9320	.9325	.9330	.9335	.9340
8.6	.9345	.9350	.9355	.9360	.9365	.9370	.9375	.9380	.9385	.9390
8.7	.9395	.9400	.9405	.9410	.9415	.9420	.9425	.9430	.9435	.9440
8.8	.9445	.9450	.9455	.9460	.9465	.9469	.9474	.9479	.9484	.9489
8.9	.9494	.9499	.9504	.9509	.9513	.9518	.9523	.9528	.9533	.9538
9.0	.9542	.9547	.9552	.9557	.9562	.9566	.9571	.9576	.9581	.9586
9.1	.9590	.9595	.9600	.9605	.9609	.9614	.9619	.9624	.9628	.9633
9.2	.9638	.9643	.9647	.9652	.9657	.9661	.9666	.9671	.9675	.9680
9.3	.9685	.9689	.9694	.9699	.9703	.9708	.9713	.9717	.9722	.9727
9.4	.9731	.9736	.9741	.9745	.9750	.9754	.9759	.9763	.9768	.9773
9.5	.9777	.9782	.9786	.9791	.9795	.9800	.9805	.9809	.9814	.9818
9.6	.9823	.9827	.9832	.9836	.9841	.9845	.9850	.9854	.9859	.9863
9.7	.9868	.9872	.9877	.9881	.9886	.9890	.9894	.9899	.9903	.9908
9.8	.9912	.9917	.9921	.9926	.9930	.9934	.9939	.9943	.9948	.9952
9.9	.9956	.9961	.9965	.9969	.9974	.9978	.9983	.9987	.9991	.9996

TABLE 3 Natural Logarithms

n	0.0	0.1	0.2	0.3	0.4	0.5	0.6	0.7	0.8	0.9
0*		7.697	8.391	8.796	9.084	9.307	9.489	9.643	9.777	9.895
1	0.000	0.095	0.182	0.262	0.336	0.405	0.470	0.531	0.588	0.642
2	0.693	0.742	0.788	0.833	0.875	0.916	0.956	0.993	1.030	1.065
3	1.099	1.131	1.163	1.194	1.224	1.253	1.281	1.308	1.335	1.361
4	1.386	1.411	1.435	1.459	1.482	1.504	1.526	1.548	1.569	1.589
5	1.609	1.629	1.649	1.668	1.686	1.705	1.723	1.740	1.758	1.775
6	1.792	1.808	1.825	1.841	1.856	1.872	1.887	1.902	1.917	1.932
7	1.946	1.960	1.974	1.988	2.001	2.015	2.028	2.041	2.054	2.067
8	2.079	2.092	2.104	2.116	2.128	2.140	2.152	2.163	2.175	2.186
9	2.197	2.208	2.219	2.230	2.241	2.251	2.262	2.272	2.282	2.293
10	2.303	2.313	2.322	2.332	2.342	2.351	2.361	2.370	2.380	2.389

* Subtract 10 if $n < 1$; for example, $\ln 0.3 \approx 8.796 - 10 = -1.204$.

TABLE 2 Natural Exponential Function

x	e^x	e^{-x}	x	e^x	e^{-x}
0.00	1.0000	1.0000	2.50	12.182	0.0821
0.05	1.0513	0.9512	2.60	13.464	0.0743
0.10	1.1052	0.9048	2.70	14.880	0.0672
0.15	1.1618	0.8607	2.80	16.445	0.0608
0.20	1.2214	0.8187	2.90	18.174	0.0550
0.25	1.2840	0.7788	3.00	20.086	0.0498
0.30	1.3499	0.7408	3.10	22.198	0.0450
0.35	1.4191	0.7047	3.20	24.533	0.0408
0.40	1.4918	0.6703	3.30	27.113	0.0369
0.45	1.5683	0.6376	3.40	29.964	0.0334
0.50	1.6487	0.6065	3.50	33.115	0.0302
0.55	1.7333	0.5769	3.60	36.598	0.0273
0.60	1.8221	0.5488	3.70	40.447	0.0247
0.65	1.9155	0.5220	3.80	44.701	0.0224
0.70	2.0138	0.4966	3.90	49.402	0.0202
0.75	2.1170	0.4724	4.00	54.598	0.0183
0.80	2.2255	0.4493	4.10	60.340	0.0166
0.85	2.3396	0.4274	4.20	66.686	0.0150
0.90	2.4596	0.4066	4.30	73.700	0.0136
0.95	2.5857	0.3867	4.40	81.451	0.0123
1.00	2.7183	0.3679	4.50	90.017	0.0111
1.10	3.0042	0.3329	4.60	99.484	0.0101
1.20	3.3201	0.3012	4.70	109.95	0.0091
1.30	3.6693	0.2725	4.80	121.51	0.0082
1.40	4.0552	0.2466	4.90	134.29	0.0074
1.50	4.4817	0.2231	5.00	148.41	0.0067
1.60	4.9530	0.2019	6.00	403.43	0.0025
1.70	5.4739	0.1827	7.00	1096.6	0.0009
1.80	6.0496	0.1653	8.00	2981.0	0.0003
1.90	6.6859	0.1496	9.00	8103.1	0.0001
2.00	7.3891	0.1353	10.00	22026.0	0.00005
2.10	8.1662	0.1225			
2.20	9.0250	0.1108			
2.30	9.9742	0.1003			
2.40	11.0232	0.0907			

25 $(1 - \sqrt{t})/t$ **27** $2x/y^2$ **29** $(1 - 2\sqrt{x} + x)/(1 - x)$
31 (a) 9.37×10^{10} (b) 4.02×10^{-6} **33** 2.75×10^{13}
35 $x^4 + x^3 - x^2 + x - 2$ **37** $-x^2 + 18x + 7$
39 $3y^5 - 2y^4 - 8y^3 + 10y^2 - 3y - 12$ **41** $a^4 - b^4$
43 $6a^2 + 11ab - 35b^2$ **45** $169a^4 - 16b^2$
47 $8a^3 + 12a^2b + 6ab^2 + b^3$
49 $81x^4 - 72x^2y^2 + 16y^4$ **51** $10w(6x + 7)$
53 $(14x + 9)(2x - 1)$ **55** $(2w + 3x)(y - 4z)$
57 $8(x + 2y)(x^2 - 2xy + 4y^2)$
59 $(p^4 + q^4)(p^2 + q^2)(p + q)(p - q)$
61 $(w^2 + 1)(w^4 - w^2 + 1)$
63 $x^{12} - 18x^{10}y + 135x^8y^2 - 540x^6y^3 + 1215x^4y^4$
  $- 1458x^2y^5 + 729y^6$
65 $-\frac{63}{16}b^{12}c^{10}$ **67** $(3x - 5)/(2x + 1)$
69 $(3x + 2)/x(x - 2)$ **71** $(5x^2 - 6x - 20)/x(x + 2)^2$
73 $-(2x^2 + x + 3)/x(x + 1)(x + 3)$
75 $(x^2 + 1)^{1/2}(x + 5)^3(7x^2 + 15x + 4)$

CHAPTER 2

Exercises 2.1, page 62

1 $-\frac{7}{4}$ **3** $5\sqrt{2}/2$ **5** $\frac{5}{2}$ **7** $\frac{8}{13}$
9 $-\frac{12}{19}$ **11** -6 **13** $\frac{34}{25}$ **15** $\frac{10}{9}$
17 1 **19** $\frac{31}{79}$ **21** $\frac{9}{38}$ **23** $-\frac{20}{39}$ **25** 3
27 $\frac{89}{48}$ **29** No solution **31** No solution **33** $\frac{5}{9}$
35 $-\frac{2}{3}$ **37** No solution **45** $\frac{5}{7}$
47 Choose any a and b such that $b = -5a/3$. For example, take $a = 3$, $b = -5$, etc.
49 (a) $x + 1 = x + 2$ (b) 2

Exercises 2.2, page 73

1 $P = I/rt$ **3** $h = 2A/b$ **5** $w = (P - 2l)/2$
7 $z = (d - ax - by)/c$ **9** $I = E/R$
11 $h = 3V/\pi r^2$ **13** $r = (S - a)/S$
15 $R_2 = RR_1R_3/(R_1R_3 - RR_3 - RR_1)$
17 $P = S/(1 + rt)$ **19** $r = (3V + \pi h^3)/3\pi h^2$
21 $r = (a - Sl)/(l - S)$ **23** $b_1 = (2A - hb_2)/h$
25 88 **27** -40 **29** Approximately 23 weeks
31 \$12,000 at 13%; \$18,000 at $15\frac{1}{2}$%
33 64 sec; 96 m and 128 m, respectively
35 After an additional 50 games **37** 60.3 g

39 $\frac{14}{3}$ oz. of the 30% glucose solution and $\frac{7}{3}$ oz. of water.
41 194.6 g of British sterling silver and 5.4 g of copper
43 55 ft **45** 6 mph **47** 36 min **49** 400 mi
51 (a) 4050 ft^2 (b) 2592 ft^2 (c) 3600 ft^2
53 36 min **55** (a) $\frac{5}{9}$ mph (b) $2\frac{2}{9}$ mi
57 (a) 125 (b) 21 **59** 180 months (or 15 yr)

Exercises 2.3, page 82

1 $\frac{2}{3}, -\frac{5}{2}$ **3** $\frac{7}{5}, \frac{1}{4}$ **5** $-\frac{5}{4}, -6$
7 $\frac{3}{4}$ **9** $\frac{5}{6}, -\frac{2}{3}$ **11** $\frac{3}{2}, -1$
13 $-1 \pm \sqrt{7}$ **15** $(2 \pm \sqrt{14})/2$ **17** $\frac{5}{2}$
19 $(-9 \pm \sqrt{21})/10$ **21** $(3 \pm \sqrt{13})/2$ **23** $0, -\frac{4}{9}$
25 (a) $x = (y \pm \sqrt{2y^2 - 1})/2$ (b) $y = -2x \pm \sqrt{8x^2 + 1}$
27 $r = \sqrt{3\pi hV}/\pi h$ **29** $d = \sqrt{gm_1m_2F}/F$
31 $t = (-v_0 + \sqrt{v_0^2 + 2gs})/g$ **33** $y = \pm (b/a)\sqrt{x^2 - a^2}$
35 2 cm decrease **37** 2 ft
39 (a) $d = \sqrt{(400t)^2 + (200t + 100)^2} = 100\sqrt{20t^2 + 4t + 1}$
  (b) 3:30 P.M.
41 24 in., 76 in. **43** 1 ft
45 Until 9:24 A.M. **47** 7 mph
49 1.5 in. at sides and top, 3 in. at bottom
51 (a) After 1 sec and after 3 sec (b) after 4 sec
  (c) 64 ft
53 (a) 206.25 ft (b) 40 mph
55 (a) 4320 m (b) 96.86 °C **57** 8 teams

Calculator Exercises 2.3, page 85

1 $-0.166, -3.728$ **3** $0.748, -3.584$

Exercises 2.4, page 91

1 $-2 + 6i$ **3** $-2 + 4i$
5 $7 - 5i$ **7** $4 + 7i$
9 $-6 + 3i$ **11** $-10 + 5i$
13 $20 - 10i$ **15** $-5 + 12i$
17 $-72 - 36i$ **19** 100
21 -1 **23** $\frac{3}{13} - \frac{2}{13}i$
25 $\frac{35}{61} + \frac{42}{61}i$ **27** $-\frac{1}{5} - \frac{11}{10}i$
29 $-\frac{7}{13} + \frac{17}{13}i$ **31** $-7 - 21i$
33 $x = 10, y = 3$ **35** $(3 \pm \sqrt{31}i)/2$
37 $-1 \pm 2i$ **39** $(-1 \pm \sqrt{47}i)/8$
41 $5, (-5 \pm 5\sqrt{3}i)/2$ **43** $2, -2, -1 \pm \sqrt{3}i, 1 \pm \sqrt{3}i$
45 $\pm 2i, \pm\frac{3}{2}i$ **47** $0, (-3 \pm \sqrt{7}i)/2$

Exercises 2.5, page 96

1 $-3, -\frac{3}{2}, \frac{3}{2}$ **3** $0, -\frac{5}{3}, \pm\sqrt{2}/2$ **5** $0, 16$

7 1 **9** $-\frac{57}{5}$ **11** $\frac{9}{5}$

13 $\pm\sqrt{62}/2$ **15** 6 **17** -1

19 $-\frac{5}{4}$ **21** 3 **23** $0, 4$

25 $\frac{1}{2}, -\frac{1}{2}, 3, -3$

27 $\pm\frac{1}{6}\sqrt{30 + 6\sqrt{13}}, \pm\frac{1}{6}\sqrt{30 - 6\sqrt{13}}$

29 $\frac{1}{27}, -27$ **31** $\frac{16}{9}, 25$

33 $3, \frac{1}{2}$ **35** $-\frac{5}{6}, -\frac{3}{2}$

37 $h = \sqrt{S^2 - \pi^2 r^4}/\pi r$ **39** $x = \pm(a/b)\sqrt{b^2 - y^2}$

41 $y = (a^{2/3} - x^{2/3})^{3/2}$ **43** $2\sqrt[3]{432/\pi} \approx 10.3$ cm

45 40 ft by 30 ft

Calculator Exercises 2.5, page 97

1 $0, 106.17$ **3** $3.13, -0.02$

5 There are two possible routes corresponding to $x \approx 0.6743$ and $x \approx 2.2887$ mi.

Exercises 2.6, page 103

1 (a) $-1 < 1$ (b) $-11 < -9$ (c) $-3 < -2$

(d) $3 > 2$

3 $(2, 5)$ **5** $(-1, 3]$ **7** $[1, 4]$

9 $(-1, \infty)$ **11** $(-\infty, 2]$

13 $-1 < x < 7$ **15** $8 < x \le 9$

17 $5 < x$ **19** $(\frac{17}{5}, \infty)$

21 $[-2, \infty)$ **23** $(\frac{5}{2}, \infty)$

25 $(-\infty, \frac{44}{3}]$ **27** $(-3, 1)$

29 $[1, 5]$ **31** $[2, \frac{8}{3})$

33 $(\frac{6}{38}, \infty)$ **35** $(-\frac{2}{3}, \infty)$

37 $(-\frac{7}{3}, \infty)$ **39** $(\frac{1}{4}, \infty)$

41 $(-\infty, 1)$ or $(1, \infty)$ **43** $\frac{140}{9} \le C \le \frac{80}{3}$

45 $\frac{20}{9} \le x \le 4$ **47** $0 < p < \frac{60}{7}$

49 The result is false if $a < 0$ (consider, for example, $a = -2$ and $b = 2$).

Exercises 2.7, page 111

1 $(-2, 2)$ **3** $(6, \infty) \cup (-\infty, -6)$

5 $[-10, 10]$ **7** $(9.95, 10.05)$

9 $[-10, 2]$ **11** $(-\frac{2}{3}, 4)$

13 $(-\infty, -6) \cup (2, \infty)$ **15** $(-\infty, 1] \cup [4, \infty)$

17 $(-2, 3)$ **19** $(-\infty, -\frac{5}{2}) \cup (1, \infty)$

21 $[0, 10]$ **23** $(-\infty, -\frac{1}{2}) \cup (\frac{10}{3}, \infty)$

25 $(-3, 3)$ **27** $(4, \infty) \cup (-4, 0]$

29 $[-5, -\frac{3}{2}) \cup (\frac{3}{2}, 5]$ **31** $(-\infty, \frac{1}{2}) \cup (\frac{7}{3}, \infty)$

33 $(-\infty, -1) \cup (0, 1)$ **35** $[-2, 1] \cup [2, \infty)$

37 $(-\infty, -1] \cup (1, 2] \cup (3, \infty)$ **39** $[\frac{1}{2}, 4]$

41 $0 \le v < 30$ **43** $0 < S < 4000$

45 $1 < t < 4$

Exercises 2.8, page 113

1 $-\frac{5}{6}$ **3** -32

5 All $x > 0$ **7** $(-2 \pm \sqrt{19})/3$

9 $(1 \pm \sqrt{21})/2$ **11** $125, -27$

13 $(1 \pm i\sqrt{14})/5$ **15** $\pm(\sqrt{14}/2)i, \pm(2\sqrt{3}/3)i$

17 $\frac{1}{4}, \frac{1}{9}$ **19** 2

21 5 **23** The interval $(-\frac{11}{4}, \frac{9}{4})$

25 $(-\infty, -\frac{3}{10})$ **27** $(-\infty, \frac{11}{3}] \cup [7, \infty)$

29 $(-\infty, -\frac{3}{2}) \cup (\frac{2}{5}, \infty)$ **31** $(-\infty, -\frac{3}{2}) \cup (2, 9)$

33 $(1, \infty)$ **35** $R = (S/\pi s) - r$

37 $R = \sqrt[4]{8nVL/\pi P}$

39 $R_2 = (n - 1)R_1 R/(R_1 - R(n - 1))$

41 $-1 + 8i$ **43** $-55 + 48i$ **45** $-\frac{9}{53} - \frac{48}{53}i$

47 $36 - 25\sqrt{2}$ cm ≈ 0.645 cm **49** 64 mph

51 315.8 g of ethyl alcohol and 84.2 g of water

53 $R_2 = \frac{10}{3}$ ohms **55** 75 mi **57** 9 in. by 11 in.

59 $d = \sqrt{4 - 200t + 2900t^2}$;

$t = (5 \pm 2\sqrt{19603})/145 \approx 1.97$ hr after 10:00 A.M. or approximately 11:58 A.M.

61 (a) $2\sqrt{2}$ ft (b) 2 ft **63** $7\frac{2}{3}$ yr **65** $4 \le p \le 8$

CHAPTER 3

Exercises 3.1, page 120

1

3 The line bisecting quadrants I and III.

5 (a) The line parallel to the y-axis that intersects the
x-axis at $(3, 0)$
(b) The line parallel to the x-axis that intersects the
y-axis at $(0, -1)$
(c) All points to the right of, and on, the y-axis
(d) All points in quadrants I and III
(e) All points under the x-axis

7 (a) $\sqrt{29}$ (b) $(5, -\frac{1}{2})$ **9** (a) $\sqrt{13}$ (b) $(-\frac{7}{2}, -1)$
11 (a) 4 (b) $(5, -3)$ **13** Area $= 28$
17 $(13, -28)$ **19** $d(A, P) = d(B, P)$
21 $\sqrt{x^2 + y^2} = 5$. A circle of radius 5 with center at the
origin
23 $(0, 3 + \sqrt{11}), (0, 3 - \sqrt{11})$ **25** $(-2, -1)$
27 $a > 4$ or $a < \frac{2}{5}$

Exercises 3.2, page 126

1   **3**   **5**

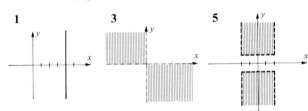

7   **9**   **11**

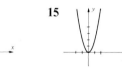

13   **15**   **17**

19   **21**   **23**

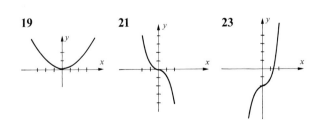

25

27

29

31

33 Circle of radius 4, center at the origin
35 Circle of radius $\frac{1}{3}$, center at the origin
37 Circle of radius 2, center at $(2, -1)$
39 Circle of radius 3, center at $(0, 3)$
41 $(x - 3)^2 + (y + 2)^2 = 16$
43 $(x - \frac{1}{2})^2 + (y + \frac{3}{2})^2 = 4$ **45** $x^2 + y^2 = 34$
47 $(x + 4)^2 + (y - 2)^2 = 4$
49 $(x - 1)^2 + (y - 2)^2 = 34$ **51** $(-1, 5); 4$
53 $(0, 3); 2$ **55** $(5, -4); \sqrt{41}$ **57** $(-1, 1); \frac{1}{2}$
59 $(\frac{1}{2}, -\frac{1}{3}); \frac{1}{6}$

Calculator Exercises 3.2, page 127

1   **3**

5   **7**

Exercises 3.3, page 134

1 $3, 9, 4, 6$ **3** $2, \sqrt{2} + 6, 12, 23$
5 (a) $5a - 2$ (b) $-5a - 2$ (c) $-5a + 2$
   (d) $5a + 5h - 2$ (e) $5a + 5h - 4$ (f) 5
7 (a) $2a^2 - a + 3$ (b) $2a^2 + a + 3$
   (c) $-2a^2 + a - 3$ (d) $2a^2 + 4ah + 2h^2 - a - h + 3$
   (e) $2a^2 - a + 2h^2 - h + 6$ (f) $4a + 2h - 1$

9 (a) $3/a^2$ (b) $1/(3a^2)$ (c) $3a^4$
   (d) $9a^4$ (e) $3a$ (f) $\sqrt{3a^2}$
11 (a) $2a/(a^2 + 1)$ (b) $(a^2 + 1)/2a$
   (c) $2a^2/(a^4 + 1)$ (d) $4a^2/(a^4 + 2a^2 + 1)$
   (e) $2\sqrt{a}/(a + 1)$ (f) $\sqrt{2a^3 + 2a}/(a^2 + 1)$
13 $[\frac{5}{3}, \infty)$ 15 $[-2, 2]$
17 All real numbers except 0, 3, and -3
19 All nonnegative real numbers except 4 and $\frac{3}{2}$
21 $\frac{9}{7}; (a + 5)/7; \mathbb{R}$
23 19; $a^2 + 3$; all nonnegative real numbers
25 $\sqrt[3]{4}; \sqrt[3]{a}; \mathbb{R}$ 27 Odd 29 Even
31 Even 33 Neither 35 Neither
37 $V = 4x^3 - 100x^2 + 600x$
39 (a) $y = 500/x$ (b) $C = 100[3x + (1000/x) - 3]$
41 $d = 2\sqrt{t^2 + 2500}$
43 (a) $y = \sqrt{2rh + h^2}$ (b) 1280.6 mi
45 $S = 4\pi r(5 + r)$ 47 $d = \sqrt{90,400 + x^2}$
49 (a) $h = 3(4 - r)$ (b) $V = 3\pi r^2(4 - r)$

Exercises 3.4, page 144

In Exercises 1–19, D denotes the domain of f and E denotes the range.

1 $D = \mathbb{R}, E = \mathbb{R}$;
   increasing on $\mathbb{R}$

3 $D = \mathbb{R}, E = \mathbb{R}$;
   decreasing on $\mathbb{R}$

5 $D = \mathbb{R}, E = (-\infty, 3]$;
   increasing on $(-\infty, 0]$,
   decreasing on $[0, \infty)$

7 $D = \mathbb{R}, E = [-4, \infty)$;
   decreasing on $(-\infty, 0]$,
   increasing on $[0, \infty)$

9 $D = [-4, \infty), E = [0, \infty)$; increasing on $[-4, \infty)$
11 $D = [0, \infty), E = [2, \infty)$; increasing on $[0, \infty)$

9   11

13 $D = (-\infty, 0) \cup (0, \infty) = E$; decreasing on $(-\infty, 0)$
    and on $(0, \infty)$
15 $D = \mathbb{R}, E = [0, \infty)$; decreasing on $(-\infty, 2]$, increasing
    on $[2, \infty)$

13   15

17 $D = \mathbb{R}, E = [-2, \infty)$;
    decreasing on $(-\infty, 0]$,
    increasing on $[0, \infty)$

19 $D = (-\infty, 0) \cup (0, \infty)$,
    $E = \{-1, 1\}$;
    Neither increasing nor
    decreasing

21   23   25

27   29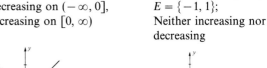

31 (a) (b) (c) (d)

(e) (f) (g) (h)

    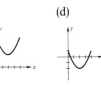

33

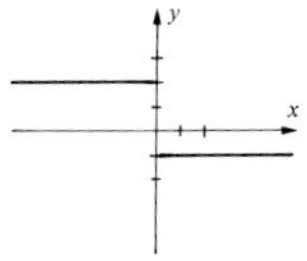

35

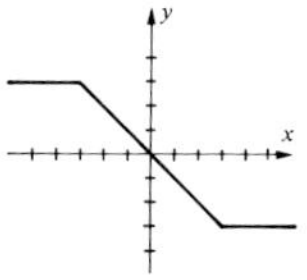

37

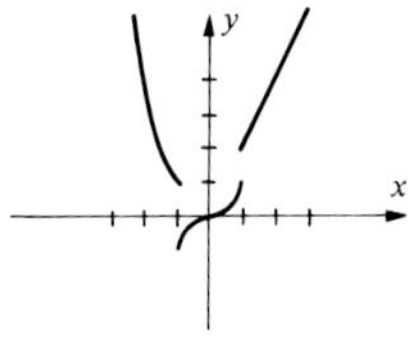

39

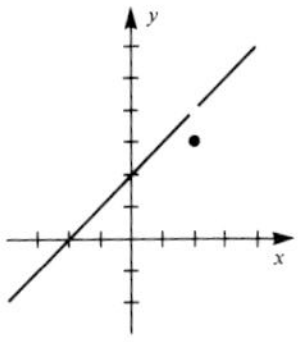

41 (a)

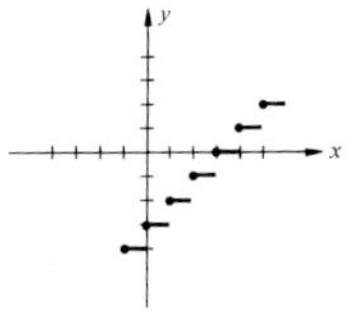

(b)

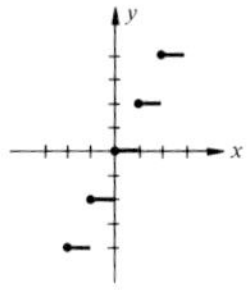

(c)

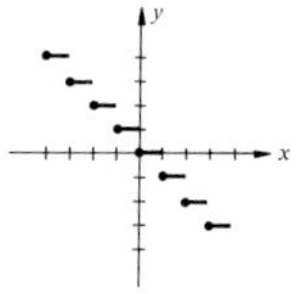

Exercises 3.5, page 156

1 4 **3** The slope does not exist.
5 The slopes of opposite sides are equal.
7 *Hint:* Show that opposite sides are parallel and that two adjacent sides are perpendicular.
9 $(-12, 0)$ **11** $x - 2y - 14 = 0$
13 $3x - 8y - 41 = 0$ **15** $x - 8y - 24 = 0$
17 (a) $x = 10$ (b) $y = -6$ **19** $5x + 2y - 29 = 0$
21 $5x - 7y + 15 = 0$
23 $x + 6y - 9 = 0$; $3x - 5y + 5 = 0$; $4x + y - 4 = 0$; $(\frac{15}{23}, \frac{32}{23})$
25 $m = \frac{3}{4}, b = 2$

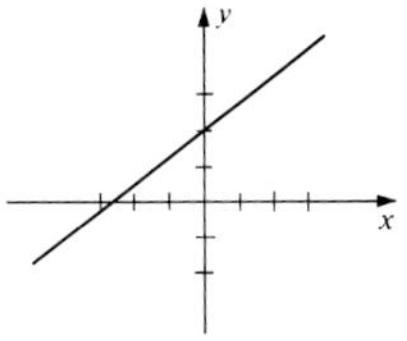

27 $m = -\frac{1}{2}, b = 0$

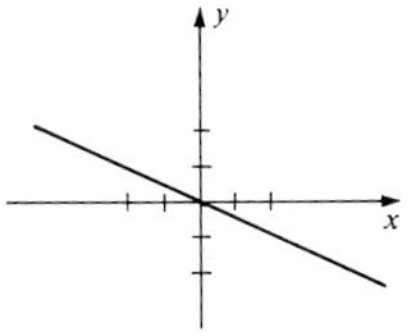

29 $m = 0, b = 4$

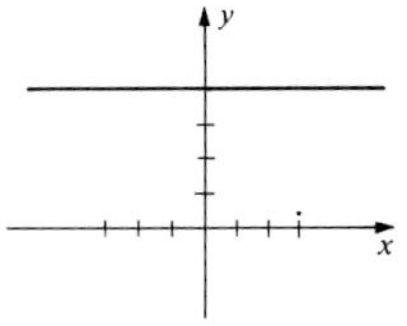

31 $m = -\frac{5}{4}, b = 5$

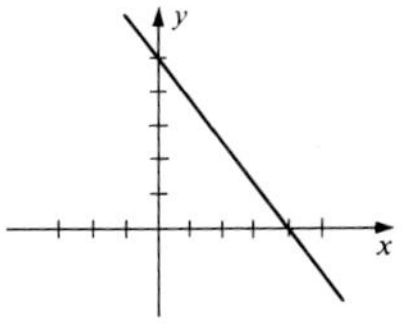

33 $m = \frac{1}{3}, b = -\frac{7}{3}$

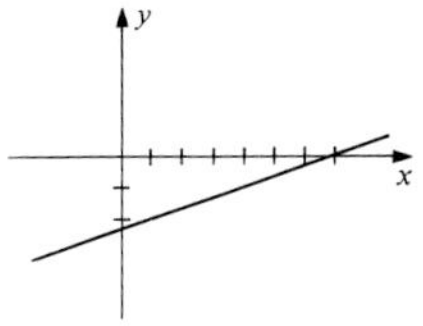

35 $k = -3$
37 $x/\frac{3}{2} + y/(-3) = 1$ **39** $r < 1$ or $r > 2$
41 $y = 59{,}000 + 6000x$, where y is the value of the house and x is the number of years after the purchase date; 2 years and 4 months after the purchase date.
43 (a) R_0 is the resistance when $T = 0\,°C$ (b) $\frac{1}{273}$ (c) $273\,°C$
45 (a) $y = -4000x + 60{,}000$ (b) \$4000
47 (a) 8.2 tons (b) as large as 3.4 tons
49 (a) $M = 0.45R - 900$ (b) \$2000 per month
51 If the player shoots when the plane is at P, the creature at $x = 3$ will be hit. All targets will be missed if the player shoots when the plane is at Q.

Exercises 3.6, page 163

1 $3x^2 + 1/(2x - 3)$; $3x^2 - 1/(2x - 3)$; $3x^2/(2x - 3)$; $3x^2(2x - 3)$
3 $2x$; $2/x$, $x^2 - (1/x^2)$, $(x^2 + 1)/(x^2 - 1)$
5 $2x^3 + x^2 + 7$; $2x^3 - x^2 - 2x + 3$; $2x^5 + 2x^4 + 3x^3 + 4x^2 + 3x + 10$; $(2x^3 - x + 5)/(x^2 + x + 2)$
7 $6x - 1, 6x + 3$ **9** $36x^2 - 5, 12x^2 - 15$
11 $12x^2 - 8x + 1, 6x^2 + 4x - 1$
13 $x^3 - 1, x^3 - 3x^2 + 3x - 1$
15 $x + 9 + 9\sqrt{x + 9}, \sqrt{x^2 + 9x + 9}$
17 $x^2/(2 - 5x^2), 4x^2 - 20x + 25$ **19** $5, -5$
21 $1/x^4, 1/x^4$ **23** x, x **25** $A = 36\pi t^2$
27 $h = 5\sqrt{t^2 + 8t}$
29 $h = (14/\sqrt{821})t + 2 \approx 0.4886t + 2$

Exercises 3.7, page 169

1 Yes **3** No **5** Yes
7 No **9** No **11** Yes
13 and 15 Show that $f(g(x)) = x = g(f(x))$.

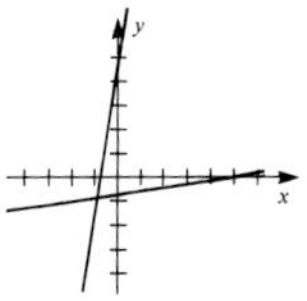

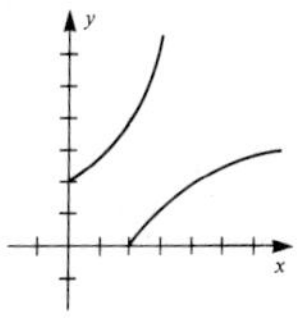

17 $f^{-1}(x) = (x + 3)/4$ **19** $f^{-1}(x) = (1 - 5x)/2x, x > 0$

21 $f^{-1}(x) = \sqrt{9 - x}, x \le 9$ **23** $f^{-1}(x) = \sqrt[3]{(x + 2)/5}$

25 $f^{-1}(x) = (x^2 + 5)/3, x \ge 0$ **27** $f^{-1}(x) = (x - 8)^3$

29 $f^{-1}(x) = x$

31 (a) $f^{-1}(x) = (x - b)/a$ (b) No (not one-to-one)

33 If g and h are both inverse functions of f, then
$f(g(x)) = x = f(h(x))$ for all x. Since f is one-to-one,
this implies that $g(x) = h(x)$ for all x, that is, $g = h$.

35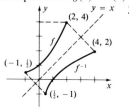

37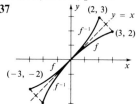

25 $D = \mathbb{R}, E = \mathbb{R}$;
decreasing on $\mathbb{R}$

27 $D = [-1, \infty), E = (-\infty, 1]$
decreasing on $[-1, \infty)$

29 $D = \mathbb{R}, E = \{1000\}$

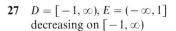

31 (a) (b) (c)

(d) (e) (f)

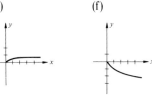

Exercises 3.8, page 172

1 $a = kv, k = \frac{2}{5}$ **3** $r = ks/t, k = -14$

5 $y = kx^2/z^3, k = 27$ **7** $c = ka^2b^3, k = -\frac{2}{49}$

9 295 lb/ft^2 **11** $\frac{50}{9}$ ohms **13** $3\sqrt{3}/2$ sec

15 223 days **17** 60.6 mph **19** 154 lb

21 2.05 times as hard

33 $18x^2 + 9x - 1; 6x^2 - 15x + 5$

35 $f^{-1}(x) = (10 - x)/15$ **37** $V = C^3/8\pi^2$

Exercises 3.9, page 174

1 Area $= 10$ **3** The points in quadrants II and IV

7 $(x + 5)^2 + (y + 1)^2 = 81$

9 (a) $18x + 6y - 7 = 0$ (b) $2x - 6y - 3 = 0$

11   **13**   **15**

17 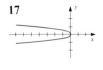 **19**   **21**

23 (a) $\frac{1}{2}$ (b) $-\sqrt{2}/2$ (c) 0 (d) $-x/\sqrt{3 - x}$
(e) $-x/\sqrt{x + 3}$ (f) $x^2/\sqrt{x^2 + 3}$ (g) $x^2/(x + 3)$

39 (a) $y = 20 - \frac{4}{5}x$ (b) $V = 4x(20 - \frac{4}{5}x)$

41 (a) $C = \frac{1}{16}x$ (b) $C = \frac{5}{88}x + 50$ (c) 8800 mi

43 (a) $V = 200h^2$ for $0 \le h \le 6$; $V = 7200 + 3200(h - 6)$
for $6 \le h \le 9$
(b) $V = 10t; h = \sqrt{t/20}$ for $0 \le t \le 720$;
$h = 6 + (t - 720)/320$ for $720 \le t \le 1680$

45 10,125 watts

CHAPTER 4

Exercises 4.1, page 183

1 (a)   (b)

(c)   (d)

3 Vertex: $(0, -9)$ **5** Vertex: $(0, 9)$

7 $3, -\frac{1}{4}$ **9** $\frac{2}{3}$

11 $2(x - 4)^2 - 9$; min: $f(4) = -9$

13 $-5(x + 1)^2 + 8$; max: $f(-1) = 8$

15 min: $f(-\frac{5}{2}) = -\frac{9}{4}$ **17** max: $f(4) = 4$

19 min: $f(-\frac{1}{2}) = \frac{11}{4}$

21 min: $f(2) = 4$

23 max: $f(-1) = 1$

   19   **21**   **23**

 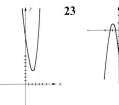

25 $y = \frac{4}{27}x(9 - x)$

27 (a) $x^2 = 500(y - 10)$ (b) 282 ft

29 $y = 2x^2 - 3x + 1$ **31** 424 ft, 100 ft **33** 10.5 lb

35 (a) $y = 250 - \frac{3}{4}x$ (b) $A = x(250 - \frac{3}{4}x)$

    (c) $166\frac{2}{3}$ ft by 125 ft

37 (a) $y = 12 - x$ (b) $A = x(12 - x)$

39 (a) $y = -\frac{7}{160}x^2 + x$ (b) 17.5 ft

Exercises 4.2, page 191

1 (a)   (b)

(c)   (d)

3 $f(x) > 0$ if $x > 2$ **5** $f(x) > 0$ for all x

   $f(x) < 0$ if $x < 2$

7 $f(x) > 0$ if $-3 < x < 0$ or $x > 3$

   $f(x) < 0$ if $x < -3$ or $0 < x < 3$

9 $f(x) > 0$ if $x < -2$ or $0 < x < 1$

   $f(x) < 0$ if $-2 < x < 0$ or $x > 1$

   7   **9**

11 $f(x) > 0$ if $-4 < x < 1$ or $x > 5$
$f(x) < 0$ if $x < -4$ or $1 < x < 5$
13 $f(x) > 0$ if $x < -2$ or $x > 2$
$f(x) < 0$ if $-2 < x < 2$

11    **13**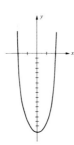

15 $f(x) > 0$ if $-1 < x < 1$
$f(x) < 0$ if $x < -1$ or $x > 1$
17 $f(x) > 0$ if $x < -3$, $-1 < x < 0$, or $x > 2$
$f(x) < 0$ if $-3 < x < -1$ or $0 < x < 2$

15   **17**

21 $k = -\frac{4}{3}$
29 $P(x) > 0$ if $-\sqrt{15}/5 < x < 0$ or $x > \sqrt{15}/5$;
$P(x) < 0$ if $x < -\sqrt{15}/5$ or $0 < x < \sqrt{15}/5$

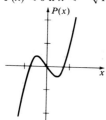

31 $V(x) > 0$ if $0 < x < 10$ or if $x > 15$
(which is not possible)

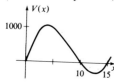

33 (a) $T > 0$ for $0 < t < 12$; $T < 0$ for $12 < t < 24$

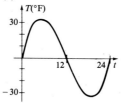

(b) $T(6) = 32.4$ and $T(7) = 29.75$
35 (a) $N > 0$ for $0 < t < 5$. The population becomes extinct after 5 years.
(b)

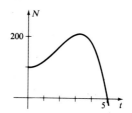

Calculator Exercises 4.2, page 193

In Exercises 3–9 there is a zero between the indicated numbers.

3 2.51 and 2.52   **5** -1.10 and -1.09
7 1.331 and 1.332   **9** -1.733 and -1.732

Exercises 4.3, page 198

1 $x^2 + x + 2$, $-2x + 13$ **3** $\frac{5}{2}x$, $-\frac{9}{2}x$
5 0, $7x^3 - 5x + 2$ **7** $2x^2 + x + 6$, 7
9 $x^2 - 3x + 1$, -8
11 $3x^4 - 6x^3 + 12x^2 - 18x + 36$, -65
13 $4x^3 + 2x^2 - 4x - 2$, 0
15 $x^{n-1} + x^{n-2} + \cdots + x + 1$, 0
17 95   **19** -23 **21** 0
23 -23 **25** 3277   **27** $8 + 7\sqrt{3}$
33 $k = \frac{17}{12}$ **35** $f(2) = 0$ **37** $f(c) > 0$
39 If $f(x) = x^n - y^n$, then $f(y) = 0$. If n is even, then $f(-y) = 0$.

Exercises 4.4, page 206

1 $-\frac{1}{15}x^3 + \frac{19}{15}x + 2$ **3** $-2x^3 + 12x^2 - 22x + 12$
5 $\frac{3}{10}x^3 - \frac{33}{10}x + 6$ **7** $x^4 - 2x^3 - 11x^2 + 12x + 36$

9 $\frac{2}{9}x^8 - \frac{4}{3}x^7 + \frac{8}{3}x^6 - \frac{16}{9}x^5$

11 -4 (multiplicity 3); $\frac{4}{3}$ (multiplicity 1)

13 0 (multiplicity 3); $5, -1$ (each of multiplicity 1)

15 $\pm\frac{5}{3}$ (each of multiplicity 4); $\pm 4i$ (each of multiplicity 1)

17 -2 (multiplicity 3); 1 (multiplicity 2); 2 (multiplicity 1)

19 $(x + 3)^2(x + 2)(x - 1)$ **21** $(x - 1)^5(x + 1)$

In Exercises 23–29 the types of possible solutions are listed in the following order: positive, negative, nonreal complex.

23 Either 3, 0, 0, or 1, 0, 2 **25** 0, 1, 2

27 Either 2, 0, 2; 2, 2, 0; 0, 2, 2; or 0, 0, 4

29 Either 2, 3, 0; 2, 1, 2; 0, 3, 2; or 0, 1, 4

31 Upper 5, lower -2 **33** Upper 2, lower -2

35 Upper 3, lower -3

7 **9** **11**
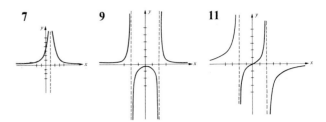

13 **15** **17**
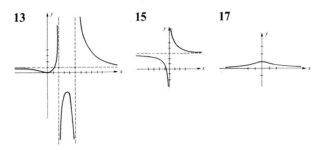

Exercises 4.5, page 214

1 $x^2 - 8x + 17$ **3** $x^3 - 10x^2 + 37x - 52$

5 $x^4 - 6x^3 + 30x^2 - 78x + 85$

7 $x^5 - 2x^4 + 27x^3 - 50x^2 + 50x$

9 No. If i is a root, then $-i$ is also a root. Hence, the polynomial would have factors $x - 1$, $x + 1$, $x - i$, $x + i$ and, therefore, would be of degree greater than 3.

11 $-1, -2, 4$ **13** $2, -3, \frac{5}{2}$

15 $4, -7, \pm\sqrt{2}$ **17** $4, -2, \frac{3}{2}$

19 $\frac{1}{2}, -\frac{2}{3}, -3$ **21** $-\frac{3}{4}, (-3 \pm 3\sqrt{7}i)/4$

23 $4, -3, \pm\sqrt{3}i$ **25** $1, -2, -\frac{4}{3}, \frac{2}{3}$

27 $3, -4, -2, \frac{1}{2}, -\frac{1}{2}$

37 Complex zeros occur in conjugate pairs.

41 The two boxes correspond to $x = 5$ and $x = 5(2 - \sqrt{2})$. The box corresponding to $x = 5$ has the smaller surface area.

43 $t = 4$ (10 A.M.) and 6.2020 (12:12 P.M.)

45 (c) The triangle has legs of length 5 and 12 ft.

47 (b) $x = 4$ ft

19
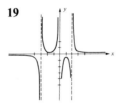

21 Vertical: $x = -1$ oblique: $y = x - 2$ **23** Vertical: $x = 0$ oblique: $y = -\frac{1}{2}x$

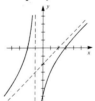 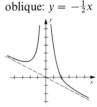

25 (a) $h = \dfrac{16}{(r + 0.5)^2} - 1$

(c) $r < 0$ and $r > 3.5$ must be excluded.

27 (a) $V = 50 + 5t$, $A = 0.5t$

(c) As $t \to \infty$, $c(t) \to 0.1$ lb of salt per gal.

29 (b)
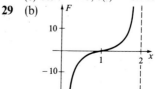

Exercises 4.6, page 226

1 **3** **5**
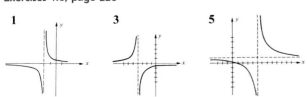

A24

Exercises 4.7, page 236

1 **3** **5**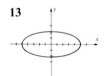

7 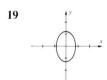 **9** **11**

13 **15** **17**

19 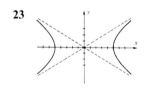 **21**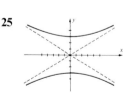

23 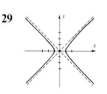 **25**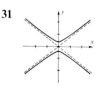

27 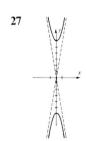 **29** **31**

33 Ellipse; center $(-5, 4)$, vertices $(-9, 4)$ and $(-1, 4)$, endpoints of minor axis $(-5, 7)$ and $(-5, 1)$

35 Hyperbola; center $(2, -7)$, vertices $(-4, -7)$ and $(8, -7)$, endpoints of conjugate axis $(2, 0)$ and $(2, -14)$

33 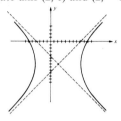 **35**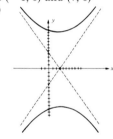

37 Ellipse; center $(-2, 2)$, vertices $(-2, 3)$ and $(-2, 1)$, endpoints of minor axis $(-\frac{5}{2}, 3)$ and $(-\frac{3}{2}, 3)$

39 Hyperbola; center $(3, 0)$, vertices $(3, \pm 10)$, endpoints of conjugate axis $(-1, 0)$ and $(7, 0)$

37 **39**

41 Ellipse; center $(-3, 5)$, endpoints of major axis $(-3, 9)$ and $(-3, 1)$, endpoints of minor axis $(-5, 5)$ and $(-1, 5)$

43 Ellipse; center $(6, 2)$, vertices $(6, 5)$ and $(6, -1)$, endpoints of minor axis $(5, 2)$ and $(7, 2)$

41 **43**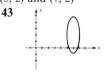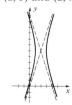

45 Hyperbola; center $(-5, -3)$, vertices $(-5, 1)$ and $(-5, -7)$, endpoints of conjugate axis $(-3, -3)$ and $(-7, -3)$

47 Hyperbola; center $(2, 6)$, vertices $(1, 6)$ and $(3, 6)$, endpoints of conjugate axis $(2, 3)$ and $(2, 9)$

45 **47**

49 $x^2 + 4y^2 = 100$ **51** $2\sqrt{21}$ ft
53 $A = 4a^2b^2/(a^2 + b^2)$
55 The graphs have the   **59** $\dfrac{(x-2)^2}{1} - \dfrac{y^2}{3} = 1$
same asymptotes.

61 $x^2 - 4y^2 = 9$

Exercises 4.8, page 239

1 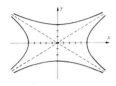   **3**   **5**

7    **9** 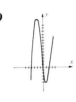   **11**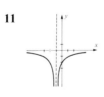

13 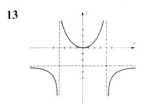   **15** Vertical: $x = -3$
oblique: $y = x - 1$

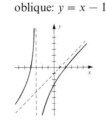

17 $f(x) = 5(x + 3)^2 + 4$; min: $f(-3) = 4$
19 Radius of semicircle is $1/(8\pi)$ mi; length of rectangle is $\frac{1}{8}$ mi.
21 $3x^2 + 2, -21x^2 + 5x - 9$ **23** $\frac{9}{2}, \frac{53}{2}$ **25** -132
27 $6x^4 - 12x^3 + 24x^2 - 52x + 104, -200$
29 $\frac{2}{41}x^3 + \frac{14}{41}x^2 + \frac{80}{41}x + \frac{68}{41}$ **31** $x^7 + 6x^6 + 9x^5$
33 1 (multiplicity 5); -3 (multiplicity 1)

35 (a) Either 3 positive and 1 negative or 1 positive, 1 negative, and 2 nonreal complex
(b) Upper bound 3; lower bound -1
39 $-\frac{1}{2}, \frac{1}{4}, \frac{3}{2}$

41   **43**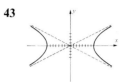

45 Ellipse; center $(-3, 2)$, vertices $(-6, 2)$ and $(0, 2)$, endpoints of minor axis $(-3, 0)$ and $(-3, 4)$
47 Hyperbola; center $(2, -3)$, vertices $(2, -1)$ and $(2, -5)$, endpoints of conjugate axis $(2 \pm \sqrt{2}, -3)$

45   **47**

49 $3x^2 + 7y^2 = 75$
51 The shelter is 12.5 ft tall and 10 ft long.
53 (a) 1 sec   (b) 4 ft   (c) On the moon, 6 sec and 24 ft.
55 $\dfrac{x^2}{10,000} + \dfrac{y^2}{960} = 1$; $8\sqrt{15} \approx 30.98$ ft

CHAPTER 5

Exercises 5.1, page 250

1   **3**   **5**

7   **9**   **11**

13 **15** **17**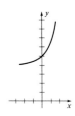

3 (a) 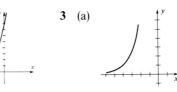 (b)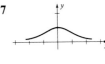

19 **21** **23**

5 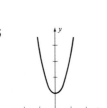 **7**

25 (a) 90 (b) 59 (c) 35

27 (a) 1039; 3118; 5400   **29** (a) 50 g, 25 g,

(b)

$(100)2^{-5/2} \approx 17.7$ g

(b)

9 -1 **11** $0, -\frac{3}{4}$ **13** $4/(e^x + e^{-x})^2$

Calculator Exercises 5.2, page 257

1 27.43 g   **3** 261.1 million   **5** 13.5%
7 (a) 8788 (b) 14,061   **9** 7.44 in.

31 $-1/1600$

33 (a) $1010.00 (b) $1020.10 (c) $1061.52
(d) $1126.83

35 (a) $7800 (b) $4790 (c) $2942

37 a^x is not always real if $a < 0$.

39 Reflection through the x-axis

Exercises 5.3, page 263

1 $\log_4 64 = 3$ **3** $\log_2 128 = 7$
5 $\log_{10}(0.001) = -3$ **7** $\log_t s = r$
9 $10^3 = 1000$ **11** $3^{-5} = \frac{1}{243}$ **13** $7^0 = 1$
15 $t^p = r$ **17** -2 **19** 2   **21** 5
23 13   **25** 27   **27** $\frac{1}{5}, -\frac{1}{5}$ **29** $\frac{7}{2}$
31 No solution   **33** 1
35 $2\log_a x + \log_a y - 3\log_a z$
37 $\frac{1}{2}\log_a x + 2\log_a z - 4\log_a y$
39 $\frac{2}{3}\log_a x - \frac{1}{3}\log_a y - \frac{5}{3}\log_a z$
41 $\frac{1}{2}\log_a x + \frac{1}{4}\log_a y + \frac{3}{4}\log_a z$
43 $\log_a (x^2 \sqrt[3]{x-2}/(2x+3)^5)$ **45** $\log_a (x^4/y^{5/3})$
47 $t = -1600 \log_2 (q/q_0)$ **49** $t = \log_3 (N/10^4)$

Calculator Exercises 5.1, page 251

1

3 (a) $1061.36 (b) $1126.49 (c) $1346.86
(d) $1814.02

5 (a) 5 yr (b) 8 yr (c) 12.5 yr   **7** $530.90

Exercises 5.2, page 256

1 (a) 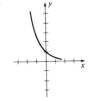   (b)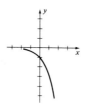

Exercises 5.4, page 266

1   **3**   **5**

7

9

11

13

15

17

19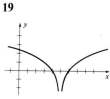

Exercises 5.5, page 271

1 4240 **3** 8.85 **5** 0.0237 **7** 9.97
9 1.05 **11** 0.202 **13** (a) 2 (b) 4 (c) 5
15 $A = 10^{(R + 7.5)/2.3} - 34,000$
17 (a) 10 (b) 30 (c) 40 **19** (a) 2.2 (b) 5 (c) 8.3
21 Acidic if pH $<$ 7, basic if pH $>$ 7
23 $t = -(L/R) \ln (I/20)$
25 $h = (\ln 29 - \ln P)/0.000034$ **27** $W = 2.4e^{1.84h}$
29 (a) $n = 10^{7.7 - 0.9R}$ (b) 12,589; 1585; 200
31 110 days **33** 139 months **35** In the year 2079

Exercises 5.6, page 277

1 (a) log 7 or ln 7/ln 10 (b) 0.85
3 (a) log 3/log 4 or ln 3/ln 4 (b) 0.79
5 (a) $4 - (\ln 5/\ln 3)$ (b) 2.54
7 (a) $(\ln 2 - \ln 81)/\ln 24$ (b) -1.16
9 (a) -3 (b) -3 **11** (a) 5 (b) 5
13 (a) $\frac{2}{3}\sqrt{\frac{1001}{111}}$ (b) 2.00 **15** 100, 1
17 10^{100} **19** 10,000
21 $x = \log (y \pm \sqrt{y^2 - 1})$
23 $x = \frac{1}{2} \log [(1 + y)/(1 - y)]$
25 $x = \ln (y + \sqrt{y^2 + 1})$
27 $x = \frac{1}{2} \ln [(1 + y)/(1 - y)]$
29 $t = -(L/R) \ln (1 - RI/E)$
31 86.4 m **33** (a) 7.21 hr (b) 3.11 hr

Exercises 5.7, page 278

1 -4 **3** 4 **5** 6 **7** $\frac{1}{2}$ **9** 5

11

13

15

17

19

21

23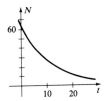

25 9 **27** 1 **29** $\log \frac{16}{3}/\log 2$ **31** $\log \frac{3}{8}/\log \frac{32}{9}$
33 $0, 1, -1$ **35** $4 \log x + \frac{2}{3} \log y - \frac{1}{3} \log z$
37 $x = \frac{1}{2} \log [(y + 1)/(y - 1)]$
39 (a) and (b) 78.3
41 (a) and (b) 6.05
43 (a) and (b) 1.887
45 2000; $2000 \sqrt[6]{3} \approx 2401$; $2000 \sqrt{3} \approx 3464$; 6000
47

49 (a) 11.39 yr (b) 6.3 yr **51** (a) $I = I_0 10^{x/10}$
53 $t = -\dfrac{1}{k} \ln \dfrac{a - y}{ab}$ **55** 1.814 yr

CHAPTER 6

Exercises 6.1, page 287

1 $(3, 5), (-1, -3)$ **3** $(1, 0), (-3, 2)$
5 $(0, 0), (\frac{1}{8}, \frac{1}{128})$ **7** $(3, -2)$ **9** No solutions
11 $(-4, 3), (5, 0)$ **13** $(-2, 2)$
15 $((-6 - \sqrt{86})/10, (2 - 3\sqrt{86})/10),$
     $((-6 + \sqrt{86})/10, (2 + 3\sqrt{86})/10)$
17 $(-4, 0), (\frac{12}{5}, \frac{16}{5})$ **19** $(0, 1), (4, -3)$
21 $(\pm 2, 5), (\pm\sqrt{5}, 4)$ **23** $(\sqrt{2}, \pm 2\sqrt{3}), (-\sqrt{2}, \pm 2\sqrt{3})$
25 $(2\sqrt{2}, \pm 2), (-2\sqrt{2}, \pm 2)$ **27** $x = 3, y = -1, z = 2$
29 $(1, -1, 2), (-1, 3, -2)$ **31** 8 in., 12 in.
33 There is a single solution that occurs at $x \approx 0.6412$.

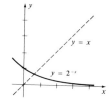

35 $r = 2$ in., $h = 50/\pi$ in.
37 (a) $a = 120{,}000$ and $b = 40{,}000$ (b) 77,143
39 $(0, 0), (0, 100)$, and $(50, 0)$. The fourth solution
    $(-100, 150)$ is not meaningful.

Exercises 6.2, page 294

1 $(4, -2)$ **3** $(8, 0)$ **5** $(-1, \frac{3}{2})$
7 $(\frac{76}{53}, \frac{28}{53})$ **9** $(\frac{51}{13}, \frac{96}{13})$ **11** $(\frac{8}{7}, -3\sqrt{6}/7)$
13 No solution
15 All ordered pairs (m, n) such that $3m - 4n = 2$
17 $(0, 0)$ **19** $(-\frac{22}{7}, -\frac{11}{5})$
21 313 students and 137 nonstudents
23 $x = (30/\pi) - 4 \approx 5.55$ and $y = 12 - (30/\pi) \approx 2.45$
25 $x = 20/\pi$ ft and $y = 10$ ft
27 \$6,650 at 6.5% and \$3,325 at 8%
29 40 g of 35% alloy, 60 g of 60% alloy
31 540 mph, 60 mph   **33** $v_0 = 10, a = 3$
35 20 sofas and 30 recliners
37 (a) $(c, \frac{4}{5}c)$ for an arbitrary $c > 0$ (b) \$16 per hour

Exercises 6.3, page 307

1 $(2, 3, -1)$ **3** $(-2, 4, 5)$
5 No solution   **7** $(\frac{2}{3}, \frac{31}{21}, \frac{1}{21})$

There are other forms for the answers in Exercises 9–15.
9 $(2c, -c, c)$ where c is any real number
11 $(0, -c, c)$ where c is any real number
13 $((12 - 9c)/7, (8c - 13)/14, c)$, where c is any real number
15 $((7c + 5)/10, (19c - 15)/10, c)$ where c is any real
    number
17 $(1, 3, -1, 2)$ **19** $x = 2, y = -1, z = 3, s = 4, t = 1$
21 $(\frac{1}{11}, \frac{31}{11}, \frac{3}{11})$ **23** $(-2, -3)$ **25** No solution
27 17 liters of 10%, 11 liters of 30%, 22 liters of 50%
29 4 hr for A, 2 hr for B, 5 hr for C
31 380 lb of G_1, 60 lb of G_2, 160 lb of G_3
33 (a) $I_1 = 0, I_2 = 2, I_3 = 2$ (b) $I_1 = \frac{3}{4}, I_2 = 3, I_3 = \frac{9}{4}$
35 (b) $\frac{5}{8}$ lb Columbian, $\frac{1}{8}$ lb Brazilian, $\frac{1}{4}$ lb Kenyan
37 $f(x) = -\frac{1}{2}x^2 + x + \frac{5}{2}$
39 $x^2 + y^2 - x + 3y - 6 = 0$

Exercises 6.4, page 314

1 $3/(x - 2) + 5/(x + 3)$ **3** $5/(x - 6) - 4/(x + 2)$
5 $2/(x - 1) + 3/(x + 2) - 1/(x - 3)$
7 $(3/x) + 2/(x - 5) - 1/(x + 1)$
9 $2/(x - 1) + 5/(x - 1)^2$
11 $(-7/x) + (5/x^2) + 40/(3x - 5)$
13 $-\frac{23}{25}/(2x - 1) + \frac{24}{25}/(x + 2) + \frac{2}{5}/(x + 2)^2$
15 $(5/x) - 2/(x + 1) + 3/(x + 1)^3$
17 $-2/(x - 1) + (3x + 4)/(x^2 + 1)$
19 $(4/x) + (5x - 3)/(x^2 + 2)$
21 $(4x - 1)/(x^2 + 1) + 3/(x^2 + 1)^2$
23 $2x + 3x/(x^2 + 1) + 1/(x - 1)$
25 $(2x + 3) + 2/(x - 1) - 3/(2x + 1)$

Exercises 6.5, page 323

1 $\begin{bmatrix} 9 & -1 \\ -2 & 5 \end{bmatrix}$, $\begin{bmatrix} 1 & -3 \\ 4 & 1 \end{bmatrix}$, $\begin{bmatrix} 10 & -4 \\ 2 & 6 \end{bmatrix}$, $\begin{bmatrix} -12 & -3 \\ 9 & -6 \end{bmatrix}$

3 $\begin{bmatrix} 9 & 0 \\ 1 & 5 \\ 3 & 4 \end{bmatrix}$, $\begin{bmatrix} 3 & -2 \\ 3 & -5 \\ -9 & 4 \end{bmatrix}$, $\begin{bmatrix} 12 & -2 \\ 4 & 0 \\ -6 & 8 \end{bmatrix}$, $\begin{bmatrix} -9 & -3 \\ 3 & -15 \\ -18 & 0 \end{bmatrix}$

5 $[11 \ \ -3 \ \ -3], [-3 \ \ -3 \ \ 7], [8 \ \ -6 \ \ 4],$
    $[-21 \ \ 0 \ \ 15]$

7 $\begin{bmatrix} -3 & 4 & 1 & 6 \\ 3 & 2 & 7 & -7 \end{bmatrix}$, $\begin{bmatrix} 3 & 4 & -1 & 0 \\ -1 & 2 & -7 & -3 \end{bmatrix}$,

    $\begin{bmatrix} 0 & 8 & 0 & 6 \\ 2 & 4 & 0 & -10 \end{bmatrix}$, $\begin{bmatrix} 9 & 0 & -3 & -9 \\ -6 & 0 & -21 & 6 \end{bmatrix}$

9 $\begin{bmatrix} 16 & 38 \\ 11 & -34 \end{bmatrix}, \begin{bmatrix} 4 & 38 \\ 23 & -22 \end{bmatrix}$

11 $\begin{bmatrix} .3 & -14 & -3 \\ 16 & 2 & -2 \\ -7 & -29 & 9 \end{bmatrix}, \begin{bmatrix} 3 & -20 & -11 \\ 2 & 10 & -4 \\ 15 & -13 & 1 \end{bmatrix}$

13 $\begin{bmatrix} 4 & 8 \\ -18 & 11 \end{bmatrix}, \begin{bmatrix} 3 & -4 & 4 \\ -5 & 2 & 2 \\ -51 & 26 & 10 \end{bmatrix}$

15 $\begin{bmatrix} 1 & 2 & 3 \\ 4 & 5 & 6 \\ 7 & 8 & 9 \end{bmatrix}, \begin{bmatrix} 1 & 2 & 3 \\ 4 & 5 & 6 \\ 7 & 8 & 9 \end{bmatrix}$

17 $[15], \begin{bmatrix} -3 & 7 & 2 \\ -12 & 28 & 8 \\ 15 & -35 & -10 \end{bmatrix}$ **19** $\begin{bmatrix} 4 \\ 12 \\ -1 \end{bmatrix}$

21 $\begin{bmatrix} 18 & 0 & -2 \\ -40 & 10 & -10 \end{bmatrix}$ **31** $\frac{1}{10}\begin{bmatrix} 3 & 4 \\ -1 & 2 \end{bmatrix}$

33 Does not exist. **35** $\frac{1}{8}\begin{bmatrix} 2 & 1 & 0 \\ -2 & 3 & 0 \\ 0 & 0 & 2 \end{bmatrix}$

37 $\frac{1}{3}\begin{bmatrix} -4 & -5 & 3 \\ -4 & -8 & 3 \\ 1 & 2 & 0 \end{bmatrix}$ **39** $\begin{bmatrix} \frac{1}{2} & 0 & 0 \\ 0 & \frac{1}{4} & 0 \\ 0 & 0 & \frac{1}{6} \end{bmatrix}$

41 $\frac{1}{6}\begin{bmatrix} -8 & -7 & 4 & -9 \\ -8 & -4 & 4 & -6 \\ -4 & -5 & 2 & -3 \\ 6 & 3 & 0 & 3 \end{bmatrix}$ **43** $ab \neq 0; \begin{bmatrix} 1/a & 0 \\ 0 & 1/b \end{bmatrix}$

47 $\left(\frac{13}{10}, -\frac{1}{10}\right)$ **49** $\left(-\frac{25}{3}, -\frac{34}{3}, \frac{7}{3}\right)$

Exercises 6.6, page 330

1 $M_{11} = -14, M_{21} = 7, M_{31} = 11, M_{12} = 10,$
$M_{22} = -5, M_{32} = 4, M_{13} = 15, M_{23} = 34, M_{33} = 6;$
$A_{11} = -14, A_{21} = -7, A_{31} = 11, A_{12} = -10,$
$A_{22} = -5, A_{32} = -4, A_{13} = 15, A_{23} = -34, A_{33} = 6$

3 $M_{11} = 0, M_{12} = 5, M_{21} = -1, M_{22} = 7; A_{11} = 0,$
$A_{12} = -5, A_{21} = 1, A_{22} = 7$

5 -83 **7** $.5$ **9** 2 **11** 0

13 -125 **15** 48 **17** -216 **19** $abcd$

31 (a) $x^2 - 3x - 4$ (b) $-1, 4$

33 (a) $x^2 + x - 2$ (b) $1, -2$

35 (a) $-x^3 - 2x^2 + x + 2$ (b) $1, -1, -2$

37 (a) $-x^3 + 4x^2 + 4x - 16$ (b) $4, 2, -2$

39 $-31i - 20j + 7k$ **41** $-6i - 8j + 18k$

Exercises 6.7, page 335

1 R_{23} **3** $(-1)R_1 + R_3$

5 2 is a common factor of rows 1 and 3.

7 Two rows are identical. **9** $(-1)R_2$

11 Every number in column 2 is 0. **13** $2C_1 + C_3$

15 -10 **17** -142 **19** -183

21 44 **23** 359

Exercises 6.9, page 345

1

3

5
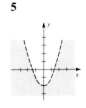

7
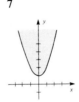

9
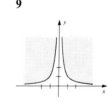

11

13

15

17
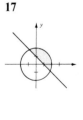

19

21

23
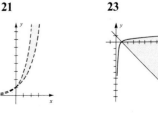

25 If x and y denote the number of brand A and brand B, respectively, then $x \geq 20$, $y \geq 10$, $x \geq 2y$, and $x + y \leq 100$. The graph is the region bounded by the triangle with vertices $(20, 10)$, $(90, 10)$, $\left(\frac{200}{3}, \frac{100}{3}\right)$.

27 If x and y denote the amounts in the first and second accounts, respectively, then $x \geq 2000$, $y \geq 2000$, $y \geq 3x$, and $x + y \leq 15000$. The graph is the region bounded by the triangle with vertices $(2000, 6000)$, $(2000, 13000)$, and $(3750, 11250)$.

29 $x + y \leq 9$, $y \geq x$, $x \geq 1$, where $x =$ height of the cone and $y =$ height of the cylinder.

31 If the plant is located at (x, y), then
$3600 \leq x^2 + y^2 \leq 10{,}000$ and
$3600 \leq (x - 100)^2 + y^2 \leq 10{,}000$ and $y \geq 0$
The graph is the region in the first quadrant that lies between the two concentric circles with center $(0, 0)$ and radii 60 and 100, and also between the two concentric circles with center $(100, 0)$ and radii 60 and 100.

Exercises 6.10, page 351

1 50 Double Fault and 30 Set Point
3 3.51 lb of S and 1 lb of T
5 Send 25 from W_1 to A and 0 from W_1 to B. Send 10 from W_2 to A and 60 from W_2 to B.
7 0 acres of alfalfa and 80 acres of corn
9 Minimum cost: 16 oz X, 4 oz Y, 0 oz Z; maximum cost: 0 oz X, 8 oz Y, 12 oz Z
11 2 vans and 4 small buses
13 Two answers lead to a maximum of 12,000 lb: Either stock 4000 bass or stock 3000 trout and 2000 bass.
15 60 small units and 20 large units

Exercises 6.11, page 353

1 $\left(\frac{19}{23}, -\frac{18}{23}\right)$ **3** $(-3, 5)$, $(1, -3)$
5 $(2\sqrt{3}, \pm\sqrt{2})$, $(-2\sqrt{3}, \pm\sqrt{2})$ **7** $\left(\frac{14}{17}, \frac{14}{27}\right)$
9 $\left(\frac{6}{11}, -\frac{7}{11}, 1\right)$
11 $(-2c, -3c, c)$ where c is any real number.
13 $(5c - 1, (-19c + 5)/2, c)$ where c is any real number

15 $\left(-1, \frac{1}{2}, \frac{1}{3}\right)$

17 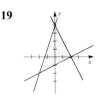   **19**

21 -6 **23** 48 **25** -84 **27** 120 **29** 0

31 $a_{11}a_{22}a_{33} \cdots a_{nn}$ **33** $\left(-\dfrac{1}{2}\right)\begin{bmatrix} 2 & 4 \\ 3 & 5 \end{bmatrix}$

35 $\dfrac{1}{7}\begin{bmatrix} 2 & 1 & 0 & 0 \\ -1 & 3 & 0 & 0 \\ 0 & 0 & 3 & 2 \\ 0 & 0 & -5 & -1 \end{bmatrix}$ **37** $\begin{bmatrix} 4 & -5 & 6 \\ 4 & -11 & 5 \end{bmatrix}$

39 $\begin{bmatrix} 0 & 4 & -6 \\ 16 & 22 & 1 \\ 12 & 11 & 9 \end{bmatrix}$ **41** $\begin{bmatrix} -12 & 4 & -11 \\ 6 & -11 & 5 \end{bmatrix}$

43 $\begin{bmatrix} a & 3a \\ 2b & 4b \end{bmatrix}$ **45** $\begin{bmatrix} 5 & 9 \\ 13 & 19 \end{bmatrix}$ **49** $-1 \pm 2\sqrt{3}$

53 $8/(x - 1) - 3/(x + 5) - 1/(x + 3)$
55 The field is $40\sqrt{5}$ ft by $20\sqrt{5}$ ft.
57 Tax $= \$18{,}750$; Bonus $= \$3125$
59 Pipe A: 30 ft^3 per hr; pipe B: 20 ft^3 per hr; pipe C: 50 ft^3 per hr
61 80 mowers and 30 edgers

CHAPTER 7

Exercises 7.2, page 372

1 $9, 6, 3, 0, -3; -12$ **3** $\frac{1}{2}, \frac{4}{5}, \frac{7}{10}, \frac{10}{17}, \frac{13}{26}; \frac{22}{65}$
5 $9, 9, 9, 9, 9; 9$
7 $1.9, 2.01, 1.999, 2.0001, 1.99999; 2.00000001$
9 $4, -\frac{9}{4}, \frac{5}{3}, -\frac{11}{8}, \frac{6}{5}; -\frac{15}{16}$ **11** $2, 0, 2, 0, 2; 0$
13 $\frac{2}{3}, \frac{2}{3}, \frac{8}{11}, \frac{8}{9}, \frac{32}{27}; \frac{128}{33}$ **15** $1, 2, 3, 4, 5; 8$
17 $2, 1, -2, -11, -38$ **19** $-3, 3^2, 3^4, 3^8, 3^{16}$
21 $5, 5, 10, 30, 120$ **23** $2, 2, 4, 4^3, 4^{12}$
25 $a_n = 2n + \frac{1}{24}(n - 1)(n - 2)(n - 3)(n - 4)(a - 10)$
(There are many other answers.)
27 -5 **29** 10 **31** 25
33 $-\frac{17}{15}$ **35** 61 **37** $10{,}000$
39 $(n^3 + 6n^2 + 20n)/3$ **41** $(4n^3 - 12n^2 + 11n)/3$

43 $\sum_{k=1}^{5}(4k-3)$ **45** $\sum_{k=1}^{4}k/(3k-1)$

47 $1+\sum_{k=1}^{n}(-1)^{k}x^{2k}/(2k)$ **49** $\sum_{k=1}^{7}(-1)^{k-1}/k$

51 $\sum_{n=1}^{99}1/n(n+1)$

Calculator Exercises 7.2, page 373

1 The terms of the sequence approach 1.

3 0.4, 0.7, 1, 1.6, 2.8

5 (a) $y_n = K$ for all n

(b) 400, 560, 425.6, 552.3, 436.8, 547.2, 443.9, 543.5, 448.9, 540.7; The terms appear to be oscillating about 500.

Exercises 7.3, page 377

1 18, 38, $4n-2$ **3** 1.8, 0.3, $3.3-0.3n$

5 5.4, 20.9, $(3.1)n-(10.1)$ **7** $\ln 3^5$, $\ln 3^{10}$, $\ln 3^n$

9 -8.5 **11** -9.8 **13** -105 **15** 30

17 530   **19** $\frac{423}{2}$ **21** 60; 12,780   **23** 24

25 $\frac{10}{3}, \frac{14}{3}, 6, \frac{22}{3}, \frac{26}{3}$ **27** 255

29 23.25, 22.5, 21.75, 21, 20.25, 19.5, 18.75

31 \$1200   **33** Four sides

Exercises 7.4, page 384

1 $\frac{1}{2}, \frac{1}{16}$; $8(\frac{1}{2})^{n-1}=2^{4-n}$

3 0.03, -0.00003; $300(-0.1)^{n-1}$ **5** 3125, 5^8; 5^n

7 $\frac{81}{4}, -3^7/2^5$; $4(-\frac{3}{2})^{n-1}$ **9** $x^8, -x^{14}$; $(-1)^{n-1}x^{2n-2}$

11 $2^{4x+1}, 2^{7x+1}$; 2^{nx-x+1} **13** $\frac{243}{8}$

15 $a_1 = \frac{1}{81}$, $S_5 = \frac{211}{1296}$ **17** $-\frac{3}{2}(1-3^{10}) = 88,572$

19 $-\frac{1}{3}(1-2^{-10})$ **21** $\frac{25}{256}$%

23 $10,000(\frac{6}{5})^t$, $10,000(\frac{6}{5})^{10}$ **25** $\frac{2}{3}$ **27** $\frac{50}{33}$

29 Sum does not exist $(|r|=\sqrt{2}>1)$.   **31** $\frac{23}{99}$

33 $\frac{2393}{990}$ **35** $\frac{5141}{999}$ **37** $\frac{16123}{9999}$ **39** 30 m

41 \$3,000,000   **43** (b) 375 mg

45 (a) $a_{k+1} = \frac{1}{4}\sqrt{10}a_k$

(b) $a_n = (\frac{1}{4}\sqrt{10})^{n-1}a_1$, $A_n = (\frac{5}{8})^{n-1}A_1$, $P_n = (\frac{1}{4}\sqrt{10})^{n-1}P_1$

(c) $[16/(4-\sqrt{10})]a_1$

Exercises 7.5, page 391

1 (a) 60 (b) 125   **3** 64   **5** 336   **7** 24

9 (a) $(2.34)10^6$ (b) $(2.16)10^6$ **11** 151,200; 5,760

13 1024   **15** 210   **17** 60,480

19 120   **21** 6   **23** 40,320

25 120   **27** (a) 60 (b) 125   **29** 9,000,000

Exercises 7.6, page 396

1 35   **3** 9   **5** n **7** 1

9 166,320   **11** 151,200   **13** 252   **15** 28, 56

17 8!5!4!3!   **19** 4,082,400

Exercises 7.7, page 402

1 (a) $\frac{1}{13}$ (b) $\frac{2}{13}$ (c) $\frac{3}{13}$ **3** (a) $\frac{1}{6}$ (b) $\frac{1}{6}$ (c) $\frac{1}{3}$

5 (a) $\frac{1}{18}$ (b) $\frac{5}{36}$ (c) $\frac{7}{36}$ **7** $\frac{1}{36}$ **9** $\frac{3}{8}$

11 $624/_{52}C_5 \approx 0.00024$

13 $(13)_{13}C_4/_{52}C_5 \approx 0.0036$ **15** $(4)_{13}C_5/_{52}C_5 \approx 0.002$

17 (a) $\frac{1}{256}$ (b) $_8C_7/256 = \frac{1}{32}$ (c) $_8C_6/256 = \frac{7}{64}$ (d) $\frac{37}{256}$

19 (a) Consider $P(E \cup E')$ (b) $1-(_{48}C_5/_{52}C_5) \approx 0.34$

21 (a) $\frac{2}{9}$ (b) $\frac{1}{9}$ **23** (a) $\frac{1}{32}$ (b) $\frac{31}{32}$

25 (a) $\frac{1}{24}$ (b) $\frac{1}{4}$ **27** (a) 0 (b) $\frac{1}{9}$

29 (a) $\frac{1}{16}$ (b) $\frac{3}{8}$

Exercises 7.8, page 404

7 $5, -2, -1, -\frac{20}{29}, -\frac{7}{19}$ **9** $2, \frac{1}{2}, \frac{5}{4}, \frac{7}{8}, \frac{65}{64}$

11 $10, \frac{11}{10}, \frac{21}{11}, \frac{32}{21}, \frac{53}{32}$ **13** $9, 3, \sqrt{3}, \sqrt[4]{3}, \sqrt[8]{3}$

15 75   **17** 1000   **19** $\sum_{k=1}^{5}3k$

21 $\sum_{k=0}^{4}(-1)^k 5(20-k)$ **23** $-5-8\sqrt{3}, -5-35\sqrt{3}$

25 $-31, 50$ **27** 64   **29** $4\sqrt{2}$

31 570   **33** 2041   **35** $\frac{5}{7}$

37 (b) The other four pieces have lengths $1\frac{1}{4}$ ft, 2 ft, $2\frac{3}{4}$ ft, and $3\frac{1}{2}$ ft.

39 $\frac{1}{2}; \frac{1}{4}$ **41** $_{52}P_{13}$; $(_{13}P_5)(_{13}P_3)(_{13}P_3)(_{13}P_2)$

43 495, 84   **45** (a) $\frac{1}{1000}$ (b) $\frac{1}{100}$ (c) $\frac{1}{20}$

47 (a) $(_6C_6 + _6C_5 + _6C_4)/2^6 = \frac{11}{32}$ (b) $\frac{21}{32}$

Exercises A.1, page A8

1 $2.5403, 7.5403-10, 0.5403$

3 $9.7324-10, 2.7324, 5.7324$

5 $1.7796, 5.7796-10, 2.7796$

7 $3.3044, 0.8261, 6.6956-10$

9 $9.9235-10, 0.0765, 9.9915-10$ **11** 4.0428

13 2.0878   **15** 8.1462   **17** 4240   **19** 8.85

21 162,000   **23** 0.543   **25** 0.0677   **27** 189

29 0.0237   **31** 1.4062   **33** 3.7294

35 $7.1000-10$ **37** 5.0913   **39** $9.8913-10$

41 2.5851   **43** $9.9760-10$ **45** 4.8234

47 $8.6356-10$ **49** 0.4776   **51** 27.78

53 49,330   **55** 0.1467   **57** 0.006536

59 234.7   **61** 1.367   **63** 0.09445

65 109,900   **67** 0.001345   **69** 0.2442

Index